Biology and Molecular Biology of Plant-Pathogen Interactions

NATO ASI Series

Advanced Science Institutes Series

A series presenting the results of activities sponsored by the NATO Science Committee, which aims at the dissemination of advanced scientific and technological knowledge, with a view to strengthening links between scientific communities.

The Series is published by an international board of publishers in conjunction with the NATO Scientific Affairs Division

A Life Sciences **B Physics**	Plenum Publishing Corporation London and New York
C Mathematical and Physical Sciences	D. Reidel Publishing Company Dordrecht, Boston and Lancaster
D Behavioural and Social Sciences **E Applied Sciences**	Martinus Nijhoff Publishers Boston, The Hague, Dordecht and Lancaster
F Computer and Systems Sciences **G Ecological Sciences** **H Cell Biology**	Springer-Verlag Berlin Heidelberg New York London Paris Tokyo

Series H: Cell Biology Vol. 1

Biology and Molecular Biology of Plant-Pathogen Interactions

Edited by

John A. Bailey

Long Ashton Research Station
Long Ashton, Bristol, United Kingdom

Springer-Verlag
Berlin Heidelberg New York London Paris Tokyo
Published in cooperation with NATO Scientific Affairs Divison

Proceedings of the NATO Advanced Research Workshop on Biology and Molecular Biology of Plant-Pathogen Interactions held at Dillington College, Ilminster, U.K., September 1–6, 1985.

ISBN-13: 978-3-642-82851-5 e-ISBN-13: 978-3-642-82849-2
DOI: 10.1007/978-3-642-82849-2

Softcover reprint of the hardcover 1st edition 1986

2131/3140-543210

PREFACE

This book is a collection of papers presented at a NATO Advanced Research Workshop on "Biology and Molecular Biology of Plant-Pathogen Interactions" which was held at Dillington College, Ilminster, UK, 1-6 September 1985. It had been preceded by Advanced Study Institutes at Porte Conte, Sardinia in 1975 and at Cape Sounion, Greece in 1981.

In recent years, methods for the manipulation and transfer of genes have revolutionized our understanding of gene structure and function. It was thus opportune to bring together scientists from distinct disciplines, e.g. plant pathology, cytology, biochemistry and molecular biology to discuss our present understanding of cellular interactions between plants. We also explored how the potential offered by the newer molecular technologies could best be realized. It soon became evident at the Workshop, and is a repeated theme of this publication, that future research will need concentrated multi-disciplinary programmes. Many of the new approaches will be valuable. For example, immunocytochemistry does, for the first time, allow molecules to be located precisely within infected tissues. Equally, the methods of DNA isolation and gene transformation will facilitate the isolation and characterization of genes associated with pathogenesis and specificity. The description at the Workshop of immunocytochemical protocols and of transformation systems for pathogenic fungi have already stimulated an upsurge in research on plant-pathogen relationships.

The papers discuss many interactions between plants and fungal and bacterial pathogens, but also provide a comparison with mycorrhizal and symbiotic relationships, and those involving mycoparasites. It is interesting to consider how often it is necessary for cells to survive penetration if microorganisms are to infect and colonize plant tissues. The mechanisms used by plants to resist invasion and by microorganisms to achieve infection are also described. Already biochemical and molecular studies indicate that both fungi and plants utilize gene transcription to effect their responses.

The success of any venture depends greatly on the help and advice of many colleagues and friends. A few deserve special mention. Firstly, Professors J. Friend, N.T. Keen and M.C. Heath who laid many of the foundations on which the Workshop was based. I am indebted to many people at Long Ashton Research Station: Professor K.J. Treharne for allowing me to use various facilities of the Research Station; Dr J.A. Hargreaves and Mr J.P.R. Keon, for help during the Workshop; and especially Dr R.K. Atkin whose experience and energy was invaluable to me.

I also thank several people for their help in preparing this book: my wife, Doreen, for ensuring accurate and consistent typescripts; Miss Beverley Jones (The Secretary Machine, Bristol) for preparing the camera-ready copies and Mr E.G.R. Chenoweth for help with the preparation of the diagrams.

Lastly and on behalf of all who participated in and profited from the Workshop, I thank all the staff at Dillington College, particularly Peter Epps, Bruce Dick and Rachel Pyke, for providing a style of living that is now all too rare. Similar thanks go to the Special Programme Panel on Cell to Cell Signals in Plants and Animals of the Scientific Affairs Division of NATO for their generous financial support.

J.A. Bailey

June 1986

CONTENTS

Introductory Comments On Host-Parasite Interactions 1
R.K.S. Wood

BIOLOGY OF PLANT-FUNGUS INTERACTIONS

Fundamental Questions Related to Plant-Fungal Interactions: Can Recombinant DNA Technology Provide the Answers? 15
M.C. Heath

Cellular Modifications during Host-Fungus Interactions in Endomycorrhizae 29
V. Gianinazzi-Pearson

Cellular Interactions between *Phaseolus vulgaris* and the Hemibiotrophic Fungus *Colletotrichum lindemuthianum* 39
R.J. O'Connell and J.A. Bailey

Ultrastructural Localization of Carbohydrates. Recent Developments in Cytochemistry and Affinity Methods 49
B. Vian

Attachment of a Mycoparasite with Host but not with Nonhost *Mortierella* Species 59
M.S. Manocha, R. Balasubramanian and S. Enskat

BIOLOGY OF PLANT BACTERIAL INTERACTIONS

The Biology of Interactions between Plants and Bacteria 71
J.W. Mansfield and I.R. Brown

Surface Interactions between *Rhizobium* and Infected Pea Root Nodule Cells, as Revealed by using Monoclonal Antibodies 99
N.J. Brewin, E.A. Wood, D.J. Bradley, G.W. Butcher and G. Galfre

MECHANISMS OF RESISTANCE

Biochemical Mechanisms Involved in Resistance of Plants to Fungi . . . 107
E.W.B. Ward

Cell Wall Modifications Associated with the Resistance of Cereals to Fungal Pathogens 133
J.A. Hargreaves and J.P.R. Keon

Mechanism of Resistance of Wheat against Stem Rust in the Sr5/P5 Interaction 141
H.J. Reisener, R. Tiburzy, K.H. Kogel, B. Moerschbacher and B. Heck

ELICITATION OF RESISTANCE

Elicitation of Active Resistance Mechanisms 149
P.J.G.M. de Wit

Evidence for both Induced Susceptibility and Induced Resistance in the Cladosporium fulvum - Tomato Interaction 171
V.J. Higgins, V. Miao and M. Arlt

Do Galacturonic Acid Oligosaccharides have a Role in the Resistance Mechanism of Cucumber towards Cladosporium cucumerinum? 177
B. Robertsen

Carbohydrate and Lipid-Containing Elicitors from Phytophthora infestans. Do they have a Common Mechanism of Action? 185
P.J. Keenan, J.S. Ellis, W.G. Rathmell and J. Friend

Differential Elicitation Activities of Fractions from Phytophthora spp. on Several Host-Plants 191
P. Ricci, P. Bonnet, P. Abad, P.M. Molot, P. Mas, M. Bruneteau, I. Fabre, O. Lhomme and G. Michel

Regulation of Steroid Glycoalkaloid and Sesquiterpenoid Stress Metabolite Accumulation in Potato Tubers by Inhibitors of Steroid Synthesis and Phytohormones in Combination with Arachidonic Acid 197
E.C. Tjamos, E. Nuckles and J. Kuč

Elicitation and Suppression of Necrosis in Potato Leaves by Culture Filtrate Components of Phytophthora infestans 205
L.C. Davidse, M. Boekeloo and A.J.M. van Eggermond

MOLECULAR BIOLOGY OF PLANT RESPONSES

Organization, Structure and Activation of Plant Defence Genes 207
T.B. Ryder, J.N. Bell, C.L. Cramer, S.L. Dildine, C. Grand, S.A. Hedrick, M.A. Lawton and C.J. Lamb

Molecular Targets for Elicitor Modulation in Bean (Phaseolus vulgaris) Cells 221
R.A. Dixon, G.P. Bolwell, M.P. Robbins and M.A.M.S. Hamdan

Accumulation of Hydroxyproline-Rich Glycoprotein mRNAs in Biologically Stressed Cell Cultures and Hypocotyls 235
A.M. Showalter, J.N. Bell, C.L. Cramer, J.A. Bailey, E. Varner and C.J. Lamb

Biochemical Study of Hydroxyproline-Rich Glycoproteins in Plant-Pathogen Interactions 245
D. Mazau, D. Rumeau and M.T. Esquerré-Tugayé

Biochemical Responses of Non-Host Plant Cells to Fungi and Fungal Elicitors 253
E. Kombrink, J. Bollmann, K.D. Hauffe, W. Knogge, D. Scheel, E. Schmelzer, I. Somssich and K. Hahlbrock

Pea Genes Associated with the Non-Host Resistance to Fusarium solani are also Induced by Chitosan and in Race-Specific by Pseudomonas syringae 263
L.A. Hadwiger, C. Daniels, B.W. Fristensky, D.F. Kendra and W. Wagoner

Race Variable Macromolecules Identified in Puccinia graminis f.sp. tritici; Correlations with Avirulence/Virulence to Host Genes for Resistance 271
N.K. Howes and W.K. Kim

MOLECULAR BIOLOGY OF BACTERIAL PATHOGENESIS

The Molecular Biology of Pectic Enzyme Production and Bacterial Soft Rot Pathogenesis 277
A. Collmer

Molecular Genetics of Pathogenicity of Xanthomonas campestris 291
M.J. Daniels

Genetic Control of Hypersensitivity and Pathogenicity in Pseudomonas syringae pv. phaseolicola 297
P.B. Lindgren and N.J. Panopoulos

Regulation of Bacterial Genes Involved in Bacterium-Plant Interactions by Plant Signal Molecules 307
R.J.H. Okker, H. Spaink, C. Wijffelman, A.A.N. van Brussel and B. Lugtenberg

Indoleacetic Acid and Cytokinins in the Olive Knot Disease. An Overview of their Role and their Genetic Determinants 315
G. Surico

MOLECULAR BIOLOGY OF FUNGAL PATHOGENESIS

Gene Expression during Infection Structure Development by Germlings of the Rust Fungi 331
R.C. Staples, O.C. Yoder, H.C. Hoch, L. Epstein and S. Bhairi

Isolation of Stage- And Cell-Specific Genes from Fungi 343
W.E. Timberlake

An Antibiotic Resistance Marker for Fungal Transformation 359
G. Turner and M. Ward

Analysis of Transcription - Control Signals in Aspergillus 365
C.A.M.J.J. van den Hondel, P.J. Punt, B.L.M. Jacobs-Meijsing, W. van Hartingsveldt, R.F.M. van Gorcom and P.H. Pouwels

Technology for Molecular Cloning of Fungal Virulence Genes 371
O.C. Yoder, K. Weltring, B.G. Turgeon, R.C. Garber and H.D. VanEtten

Molecular Evolution of Fungal Polygalacturonase 385
F. Cervone, G. de Lorenzo, G. Salvi and L. Camardella

The D^2-Factor in Ophiostoma ulmi : Expression and Latency 393
H.J. Rogers, K.W. Buck and C.M. Brasier

The Effects of Passaging on the Host Specificity of _Septoria nodorum_: A Cautionary Tale 401
A.E. Osbourn, C.E. Caten and P.R. Scott

Participants 409

Index 413

INTRODUCTORY COMMENTS ON HOST-PARASITE INTERACTIONS

R.K.S. Wood FRS

Department of Pure and Applied Biology
Imperial College of Science and Technology
London, UK

Although this Workshop is about the biology as well as the molecular biology of host/parasite relations I think that my comments would be spread too thinly were they on both. Therefore, and because of what I expect to be the preoccupation of the Workshop, I shall confine them very largely to interactions at cellular and sub-cellular levels. Even thus the scope and the substance are so huge that the survey I shall attempt must be somewhat superficial and is likely to achieve not much more than remind you of what is already known. I am afraid too that I ask all too many questions for which no-one has the answer. But this reflects the complexity of and our ignorance about the phenomena to be discussed. I shall start with the interactions about which most is known, between races and cultivars and continue but much more briefly with interactions at other levels.

HYPOTHESIS BASED ON INDUCTION OF RESISTANCE: RACE/CULTIVAR INTERACTIONS

Almost all research in depth on the physiology and biochemistry of host/parasite interactions has used a few diseases of crop plants in which susceptibility or resistance of cultivars to races of a pathogen conforms, or is assumed to conform to a pattern that would be expected if for each gene for resistance in the cultivars there were a matching gene for virulence in one or more races of the pathogen as shown by Flor [1] for flax and *Melampsora lini*. Here resistance is dominant to susceptibility, as it is in most other diseases, and low virulence (later, avirulence) is dominant to high virulence (later, virulence) as known for a few other diseases and assumed for many more on indirect evidence. It is not surprising that most of this research, certainly at the biochemical level, has used pathogens readily grown in culture and usually well on simple or not very complex media. When media contain suitable substrates such as substances from plant cell walls then pathogens usually produce enzymes that will degrade most of the polymers that comprise plant cells whether these are of susceptible or resistant cultivars. Therefore it is likely that a race of a pathogen will be able to get from any cultivar that which it needs for growth particularly if in early growth on reserves it can damage protoplasts by increasing their permeability or, and better, kill them. It is unlikely therefore that nutrition as such is important in race/cultivar interactions. For obligate parasites the story may be different because of the probable complexity of their nutrition and the almost invariable occurrence of haustoria. But even

NATO ASI Series, Vol. H1
Biology and Molecular Biology of Plant-Pathogen Interactions
Edited by J. Bailey

here perhaps one should not ignore the possible significance of substances outside host-protoplasts and cells particularly in the early growth of the pathogen and in its interactions with host cells.

Apart from requirements of essential nutrients, parasites may be affected by substances in host plants which disrupt their metabolism particularly when the effects are adverse so that the substances are toxins. In spite of much searching there still is little good evidence that toxins present in plants before infection are important in interactions although, prima facie, there seems no reason why they should not be. There are toxins in many plants but either they occur in cultivars in similar concentrations, have similar toxicity to races, or both. In contrast, there is abundant evidence that substances, as phytoalexins, accumulate much more rapidly around avirulent than virulent races during early stages of infection; the phytoalexins have similar toxicity to the races but differences in rates of accumulation are enough to be a reasonable explanation of the decreased growth of the avirulent but not of the virulent race, at least in a few diseases, e.g. Bailey et al., [2]. Therefore the specificity of interactions depends not on the phytoalexins but on events that lead to their accumulation. Because they do not occur in significant amounts, if at all, in uninfected plants, accumulation may be taken as synonymous with new synthesis well in excess of any degradation that may occur. Thus we come to the hypothesis that has been the basis of much of the research on interactions - that resistance depends on a reaction between the product of a (dominant) avirulence gene (say AV1) and that of a (dominant) resistance gene (say R1) which results in synthesis and then accumulation of a phytoalexin sufficient to decrease growth of the pathogen to the observed levels. It then is inferred or assumed that AV2 x R2 and so on does the same. In the absence of matching AV x R, synthesis of phytoalexins does not occur and the pathogen is able to grow.

Products of AVn were christened elicitors some years ago presumably because they are a cause which makes the plant do something that normally it does not do. The essence of the hypothesis is that the elicitors are specific as are their matching products in plants, usually called receptors. Therefore, it probably would have been better if from the beginning elicitor had been qualified to make this clear such as race/cultivar (r/c) elicitors or something similar. This is because the word soon was adopted for the many other substances, and even for an agent, such as UV radiation, which cause accumulation of phytoalexins, and which have been known almost from the state of research on them. These elicitors are non-specific, certainly when applied to cultivars.

Another difficulty arose when it was found, not unexpectedly in light of the above, that cells damaged or killed other than by a pathogen, released elicitors called constitutive or endogenous because they are wholly products of the plant [3]. It then was reasonable to suppose that r/c elicitors, by damaging plant cells, released an endogenous elicitor and that it was this that triggered synthesis of phytoalexins (elaborated by Keen [4]). If so then it is likely that the r/c and the endogenous elicitor affect host cells quite differently so it is, perhaps, unfortunate to have the same word for both unless each is qualified. I shall not suggest ways out of this and related difficulties because elicitor is now too well established to be abandoned. But I do suggest when elicitor is used that there should be no doubt about what is being elicited.

The two stage hypothesis, the first stage specific, the second (or more) non-specific, is simple and attractive and certainly a good basis for research with non-obligate parasites. But even for these it poses questions and faces difficulties when its implications are considered in detail. Here, briefly, are some of them and in relation to non-obligates unless stated otherwise.

1. Receptors as Rn gene products usually are assumed to be in plasma membranes of host cells partly no doubt from comparisons with recognition phenomena in animal cells. But may they as well not be in cells walls which contain many substances of types that could be receptors, and particularly when the pathogens are bacteria which cannot penetrate cell walls as do fungi? There is also the possibility that Rn genes are not expressed except in response to infection by an avirulent race so that uninfected cells do not carry Rn products.

2. r/c elicitors with the required specificity have been remarkably difficult to isolate as extracellular products in culture so are they firmly bound to cells of pathogens at least when these are fungi (also, see later).

3. Although each AVn x Rn reaction is specific, are they similar in type, with similar effects on host cells and with the same endogenous elicitor and similar triggering of synthesis of phytoalexins? For non-obligates such evidence as there is implies that they are. But certain limited evidence for obligates suggests that AVn x Rn reactions may have different effects.

4. What corresponds to Rn products as receptors for elicitors other than r/c elicitors and do such elicitors have similar effects, similar endogenous elicitors and so on, particularly when one considers that they range in type from high molecular weight carbohydrates and other polymers, through simple sugars and salts to UV radiation?

5. Usually it is thought that an endogenous elicitor moves from killed or damaged cells to adjacent cells where they trigger synthesis of phytoalexins which then move back into the cells in which growth of the pathogen is prevented. The fact that phytoalexins in quite low concentrations also may be toxic to plant cells raises other possibilities. Thus, damage to cells short of killing may release an endogenous elicitor which triggers synthesis of phytoalexins in the same cells and the accumulation to levels that kill the cells and also prevent growth of the pathogen. If the endogenous elicitor passed to adjacent living cells and phytoalexins again killed cells and so on, then the result could be a lesion comprising a necrotic area larger than that containing the pathogen as in many diseases in which the necrosis is attributed to a toxin from the pathogen but with no explanation as to why the pathogen does not continue to grow in the tissue that has been killed. It would be intriguing to find in such diseases that the toxin is a phytoalexin. But I can think of arguments against this idea.

6. Returning to more usual interpretations of events, ideally the hypothesis would assume that when other than AVn, Rn genes are matched, then there is no or not the right type of damage, the endogenous elicitor is not released and there is no synthesis of phytoalexins. However, particularly for non-obligates, susceptibility means that virulent races do damage and kill many more cells than do avirulent races, albeit somewhat later, at least in the diseases that have been studied in this context. And we know for a few diseases that phytoalexins do accumulate in susceptible responses although, of course, not at rates sufficient to decrease growth of the pathogen and killing of more cells (but note that lesions in susceptible plants rarely occupy all the tissue that would seem to be available). This implies that living cells at edges of lesions in susceptible responses are killed before they can respond to endogenous elicitor from cells killed earlier. This is plausible enough an explanation and for one or two diseases there may be evidence to support it. But one should like to see more for other diseases because of the critical importance for the hypothesis.

7. Another point about susceptible responses is that pathogens in their early growth in cultures containing host cell walls or substances from them readily produce enzymes that degrade pectic polysaccharides and rapidly kill protoplasts. They may also do so constitutively although to a very limited extent which, nevertheless, may be significant at the cellular level. No doubt pathogens also produce other toxic substances. Why, apparently, do they not produce such substances early in infection in those diseases, probably many, in which they grow in susceptible plants without killing cells so that the endogenous elicitor is not produced? So far as I know there are few data in the literature on this apparent anomaly which has become the more striking with the discovery that enzymes that degrade pectic polysaccharides yield products that are very active elicitors [5,6,7].

8. The assumption that AVn x Rn is determinative in race/cultivar interactions has led to many attempts to obtain from races substances that cause in cultivars effects that parallel those of the races themselves in causing necrosis or, more commonly, accumulation of phytoalexins. There certainly has been little difficulty in obtaining substances from pathogens that in realistic concentrations cause such effects in cultivars but they do so quite non-specifically or with not enough specificity to meet the hypothesis. Such non-specific elicitors are interesting on their own account particularly when active at very low concentrations, and for what their activity may tell about effects of damage other than those that trigger synthesis of phytoalexins. But they hardly are relevant for the hypothesis.

In contrast, the many attempts to isolate r/c elicitors with the required specificity have not met with much success except for the recent work with Fulvia fulva and tomato [8]. Here success depended on using cell free fluids from leaves inoculated with virulent races which grow biotrophically and for unusually long periods for a pathogen that readily can be grown in culture. Preliminary work suggests that one of the substances is a peptide of MW c. 5500 [9].

It is somewhat ironic that success with Fulvia fulva depended on growing it in a leaf because the attraction of research with such pathogens was that they can be grown so readily in culture away from the complexities of host tissues particularly for bacteria which, presumably, cause specific effects while growing in intercellular spaces and separated from protoplasts by cell walls which they do not penetrate. We, and no doubt, many others, failed to get specifically active substances from leaves infiltrated with avirulent bacteria. Clearly we also should have used virulent bacteria. Also I should mention that we have failed to get active extracts in experiments similar to those of De Wit but with hypocotyls of Phaseolus vulgaris and conidia of Colletotrichum lindemuthianum. I hope that other similar work will be more successful.

9. De Wit's work [8,9] implies that host cells may condition production of AVn products. Along somewhat different but related lines is the research which showed that a β-1,3-endoglucanase from soybean cells releases an elicitor from cell walls of Phytophthora megasperma var. glycinea, presumably non-specifically [10,11]. I believe that specificity of the elicitor at significant levels still has to be demonstrated.

10. Little has been reported about endogenous elicitors. This is disappointing in view of their importance for the hypothesis. Presumably we shall be looking in the first place for substances likely to be released by damaged host cells, with regulatory functions, and of types common to many plants. If so, then we may have something to learn from research on the responses of cells and tissues to wounding which at last seems to be attracting research and with some unexpected and striking results as, for

example, the new synthesis of proteinase inhibitors some distance from wounds in leaves [12], and the evidence that it may be a rhamnogalacturonan [13]. Low molecular weight pectic polysaccharides also are active elicitors [5,7,14]. There is, too, the fact that chitinases may be synthesized in response to wounding and to infection [15], and that chitosans of types occurring in cell walls of fungi are non-specific elicitors [16]. Is there a case for considering pectic polysaccharides or chitinases (by releasing chitosans) as candidate endogenous elicitors, that is, as the hypothetical "second messengers", AVn products presumably being the first? There is room here for much argument and speculation.

So much for points and questions about the two stage hypothesis based on specific induction of the synthesis of phytoalexins, or of other mechanisms which decrease growth in the resistance responses considered to be determinative. Here are some contrary views.

ARGUMENTS AGAINST SPECIFIC INDUCTION OF RESISTANCE

In Vanderplank's recent book called "Disease Resistance in Plants" [17], "phytoalexin" has one entry in the index, to a short paragraph on page 7 which reads "An erroneous belief in specific resistance, as distinct from specific resistance genes, has led to much "barking up the wrong tree". Among the examples of misdirected "barking" are the phytoalexin and hypersensitivity theories of host-pathogen specificity". Not much argument there or elsewhere in the book but there is in earlier books by Vanderplank. Others also have commented about difficulties and inconsistencies that are not easily reconciled with the hypothesis and results of research on phytoalexins. Thus Ellingboe [19] argues that the genetics of race/cultivar interactions demand that the mechanism of resistance be the first reaction between AVn and Rn gene products and that they do not allow both for a determinative event and others, involving more proteins, which act to produce the mechanism of resistance such as accumulation of phytoalexins.

A role for phytoalexins was also discounted in research on resistant responses of potato to *Phytophthora infestans* which led to the proposal that an early reaction between host cell and pathogen killed the pathogen and that the host cell was killed later as a consequence; any phytoalexins that accumulated had little to do with resistance to the pathogen [20]. This research apart, there is good evidence for other diseases that cells are damaged or killed before the pathogen stops growing. For this and other reasons I shall not press what I think to be persuasive arguments against this proposal, at least for non-obligate parasites. But there remains the case based on the genetics. If it cannot be met, then the hypothesis is invalidated and the research on phytoalexins may have been not much more than interesting diversions at least in the context of race/cultivar interactions.

Apart from such specific points there is the general one that if all that is known to happen after the primary event is irrelevant to inhibition of growth of the pathogen that causes it, then it may seem a little odd that a plant should be so profligate as regularly to synthesize, in large amounts on a cell basis, substances which could benefit the plant. Further, the plant produces these potentially beneficial secondary metabolites only when it would be sensible to do so. Is it not a little hard to accept that phytoalexins do not function in resistance to non-obligate parasites while agreeing that there are important points about them and induction of their synthesis which still are to be resolved?

For the obligate parasite, the story may be different because it does not continue to grow on dead cells. Therefore death of cells caused directly by the first and specific reaction in itself could explain why the pathogen does not grow. Secondary events are not needed. There are few data and little speculation on how obligate parasites do or could kill cells in this manner. Also it would not be difficult to devise a scheme in which the two stage hypothesis and phytoalexins, or another resistance mechanism, would explain resistance to obligate parasites. Indeed the excellent research of Mayama [21] has implicated avenalumins I, II, and III as phytoalexins in the resistance of oats to Puccinia coronata. But in this connexion it must be remembered that whether or not phytoalexins function in resistance, it would be surprising if they were not synthesized as one of the responses to the death of cells so common in high resistance responses to obligate parasites.

INDUCTION OF SUSCEPTIBILITY

The essence of the hypothesis discussed above is that induction of resistance is determinative and that a plant is susceptible when it does not occur. The opposite view is that resistance will be induced non-specifically whenever a potential parasite starts to grow in a plant unless either induction is prevented or its consequences are nullified. This is called, perhaps a little oddly, induction of susceptibility rather than suppression of resistance. Suppression of induction could occur somewhat as follows [22,23,24]. A pathogen produces an elicitor which has the potential to bind to a receptor in the host. If it does bind there is a defence response such as accumulation of phytoalexins that decreases growth of the pathogen. The pathogen also produces a suppressor which by binding to the host receptor prevents binding of the elicitor and, therefore, also the defence response. This allows the pathogen to grow. The suppressor has a number of sites, Pn, which do not at first affect its fit to the receptor. The receptor then is altered by resistance gene R1 which by matching P1 prevents binding with the suppressor so that the receptor again can fit the elicitor and thus cause a defence response. P1 would correspond to avirulence gene AV1. Then P1 may change to p1 so that is it no longer affected by R1, suppressor fits receptor so that there is no fit with elicitor, and no defence response; p1 corresponds to a gene for virulence. Therefore specificity is based on mutable receptors in the host and mutable suppressors from the pathogen that affect binding of an elicitor from the pathogen. This is one step less simple at the specificity-determining stage than is the other hypothesis. For Phytophthora infestans and potato there is preliminary evidence for glucan suppressors that nullify specifically responses caused non-specifically by other glucans in ways that could be related to resistance [25]. Also, a cultivar of genotype expected to confer resistance to a particular race can be made susceptible by inoculating it earlier with a virulent race [26,27,28]. This is the reverse of induction of resistance to a virulent pathogen by inoculating earlier with an avirulent race. Effective pre-inoculation periods are considerably longer for induced susceptibility than for induced resistance; this may be relevant in considering the merits of the two hypotheses.

The hypothesis based on induced susceptibility faces most of the difficulties which apply to the first hypothesis including those posed by Ellingboe [19], and also a few more.

Vanderplank [18] also has advocated that susceptibility determines specificity but along quite different lines, at least for biotrophs, which relate specificity to nutrition. Thus, if a specific protein of a pathogen can co-polymerize with a matching protein in a host, the infected cell remains alive in a susceptible response in which more of the host protein is synthesized and becomes available for nutrition of the parasite. If the protein of the pathogen cannot co-polymerize then it is free to damage the host cell which then cannot sustain growth of the pathogen. Note that whereas Ellingboe proposes that union of the two proteins is the cause of resistance, Vanderplank claims that it is the basis of susceptibility.

Even for biotrophs there are difficulties in Vanderplank's hypothesis which is still more difficult to accept for non-biotrophs with nutrition such that they would be expected to grow on cells killed in resistance responses as, of course, they do in susceptible responses.

It is perhaps unfortunate that induction of susceptibility has attracted so much less research than has induction of resistance.

HOST-SPECIFIC TOXINS

I shall refer only briefly to these. So far and with the odd exception they are known only in diseases caused by species of *Helminthosporium* and *Alternaria*. I think it unlikely that this is particularly significant at least as regards similar diseases caused by species of other genera. But even if such diseases are rare and oddities, this does not detract from their interest particularly when they are compared with the many more diseases in which the toxin characteristic for each disease and probably the cause of most of the symptoms is as damaging to non-host as to host plants. In such diseases specificity is probably determined by one of the mechanisms based on induction of resistance. Do such mechanisms operate in resistant cultivars in diseases in which the toxin is host-specific? If they do, then the toxin may not be involved in the resistant response although infected cells are killed. When resistance is not induced then the pathogen can grow but now it only kills cells susceptible to the toxin. What then is the advantage of specificity at this stage? There are a few other difficulties which need to be resolved before we can accept the simple explanation of specificity based wholly on resistance or susceptibility of host cells to the specific toxin.

GENES IN POPULATIONS

There has been plenty of speculation but little firm evidence about the nature of the products of race/cultivar genes that determine specificity. May we learn something about them from their behaviour in populations? Thus take the introduction into a cultivar of a crop grown on a large scale in a particular region of a new gene R1, quite possibly from another but related species growing remotely from the crop plant. The new gene can be effective in the crop only if the prevalent race/s of the pathogen already carry the matching gene for avirulence (AV1) which quite probably has persisted for many generations of the pathogen on cultivars not carrying the new gene. The gene AV1 could hardly have been hanging around and waiting for plant breeders to provide the matching gene for resistance (R1) that would justify its existence. Therefore the avirulence gene AV1 almost certainly codes for something the pathogen finds useful in ways other than in resistance. But what is it, bearing in mind that all growth of biotrophs and most growth of non-biotrophs occurs in susceptible plants and not saprophytically. In the presence of the R1 gene the pathogen can grow only if AV1 changes to av1 with no function in resistance. But what about the other functions which probably were beneficial. Are they also lost or are they changed in relation to these other functions only in ways that do not impair seriously their usefulness? There is some evidence that the change from AV1 to av1 indeed may carry penalties for fitness because av genes tend to decrease, at least in some circumstances, in successive generations in populations of cultivars not carrying the matching genes for resistance.

There may be the corresponding anomaly for resistance/susceptibility genes because for some diseases resistance which was present originally may decrease when plants are selected over many generations in the absence of pathogens. This at least hints at the possibility that resistance genes have functions which, other than in resistance, are less beneficial to the plant than are the corresponding genes for susceptibility. Note, however, the availability and therefore the persistence of certain resistance genes in populations of wild plants from which they may be transferred to crop plants. Unfortunately we know all too little about the behaviour of resistance genes in wild populations with a few notable exceptions in which it appears that there are plenty of them and of the matching genes for avirulence as in race/cultivar interactions, so that in this respect populations of wild and crop plants may not be all that different.

RACE NON-SPECIFIC RESISTANCE

I have dwelled for so long on race specific resistance because it is the resistance upon which almost all research has been based and about which at least we know a good deal of the genetics. In contrast is the resistance of cultivars to races which usually is called race non-specific because the pattern of interactions is not differential (or, better perhaps, it not yet known to be) so that it functions against all races for which it has been tested, although usually to different degrees. It is thought to be controlled by a number of genes, probably many. This possibly is why it has been studied comparatively little at physiological and biochemical levels although it is so important in the breeding and selection of resistant cultivars. In speculating in the absence of much experimental evidence I must confess that I find this type of resistance difficult to conceive other than in the context of gene-for-gene relations because these relate to how most plant pathologists think about disease - as a conflict in which one barrier in a potential host which could decrease growth of a potential pathogen has to be made ineffective by a suitable change if the pathogen is to become virulent, and so on for other barriers. If the barrier and the change, or things that depend on them, are controlled by single genes then we have the essentials of a gene-for-gene relation. In race-specific resistance the effect of each gene in resistance is recognized as large so that the genes are sometimes called major. It tends to be assumed that the effect essentially is similar for the different major gene pairs but for certain biotrophs it may be different although, of course, with the same final effect in decreasing growth of the pathogen. In both, however, the effects can hardly be additive when host and pathogen contain more than one pair of matching genes because a single pair confers high resistance as a near maximum effect. If race non-specific resistance is controlled by many minor gene pairs each with a small effect then how do the effects of each compare with those of major gene pairs in race specific resistance such as those which lead to the accumulation of phytoalexins as the cause of decreased growth of the avirulent pathogen? And how do the effects of the many minor gene pairs in race non-specific resistance add up to whatever it is that decreases growth of all races? Unfortunately the mechanisms of resistance have been little studied; research on them has been largely descriptive and in terms of rates of growth, amount of sporulation and the like.

As to the genes controlling race non-specific resistance it has been suggested for cultivars of crop plants that they were genes for high resistance to avirulent races which retain residual minor effects in relation to the alleles of corresponding virulent races. Polygenic, race non-specific resistance then becomes the sum of these minor "failed" resistances. But there is, in fact, very little evidence for this and there are some persuasive arguments against it.

Briefly here are some other points about polygenic resistance. If based on pairs of matching genes, and if avirulence genes are mutable, as they are and very characteristically in race specific resistance, then presumably resistance would be eroded as virulence genes accumulated. There is evidence that this can occur. Next, we know at least for a few species that wild plants carry genes that function in race specific resistance on transfer to crop plants. One must assume that they function against a background of polygenic resistance in wild plants. What does this imply particularly in relation to the resistance characteristic of the land races of less developed agricultures? To continue would labour what I need not emphasize - the extent of our ignorance about race non-specific resistance. Unfortunately because of the difficulties and uncertainties about its genetic control it is unlikely that it will attract much more research in the future than it has in the past in spite of its importance. So I shall turn now and finally to a third type of resistance which is the response of plants to microorganisms that are not its pathogens. This usually is called non-host resistance.

NON-HOST RESISTANCE

The complexity of this type of resistance is well illustrated by surveying the parasitism of higher plants by the many species of rust fungi, a very homogeneous group of pathogens, especially when the hosts of monokaryotic and dikaryotic cells of the same parasite are so widely separated taxonomically as often they are. Such complexities and the technical difficulties of working on non-host resistance particularly to biotrophs may explain why with a few exceptions, so few of us speculate on let alone attempt to study it. And we are not encouraged to do so when geneticists suggest that there may not be much point in such research because almost always a host and a non-host of a pathogen cannot be crossed. Nevertheless I believe that there are points about non-host resistance worth considering and investigating by research.

A striking feature is the frequency with which a pathogen infects a non-host plant with effects very similar to those caused by an avirulent race of a pathogen (of the plant) in a highly resistant cultivar and where the response is controlled by a single pair of matching genes. The similarity may extend even to accumulation of the same phytoalexins. Indeed some of the earliest experiments which led to the recognition of phytoalexins were done with non-host plants. In the race/cultivar, high resistance response, sceptics notwithstanding, there is reasonable evidence that synthesis of phytoalexins follows, at one or more removes, a reaction between the product of an avirulence and of a resistance gene. What then precedes synthesis of the same phytoalexin when the plant is infected by a pathogen for which it is not a host? If there be another gene pair, then the avirulence gene must be different in type if only because it never changes to a virulence gene. Being thus different, is its receptor in the non host plant also different in type from the product of the cultivar gene for high resistance? Extending this to other non-pathogens and non-hosts would mean non-host resistance depending on multitudes of matching gene pairs and an idea which most of us would consider too complicated to pursue [29]. But I am not sure that schemes which seek to explain non-host resistance in other ways are much less complex [22,23,24]. In such schemes I find it difficult to conceive how what may be called a propathogen counters non-host resistance and becomes a pathogen and especially a biotroph i.e. how it establishes the "basic compatibility" proposed by Ellingboe [30]. Unfortunately it is very unlikely that we ever shall learn anything significant about the evolution of parasitism from fossil records so it may not be profitable to speculate at length on how a propathogen became a pathogen. But we may ask whether it is possible to learn about non-host resistance by searching for parasitism now evolving in wild or in crop plants, as presumably it is, by attempting to change a saprophyte into a parasite, or by extending the host range of a parasite. Here are a few points in this connexion.

1. In its life cycle a pathogen such as *Venturia inaequalis* lives first as a biotroph or almost so, then as a typical foliar pathogen causing necrosis, and finally as a saprophyte on dead leaves. What controls these very different life styles and modes of nutrition?

2. A saprophyte is not a pathogen of any plant species. The same is true for a highly specialized parasite except for the very few species for which it is a parasite. Are the factors operating between saprophytes and living plants the same as those between the specialized parasite and almost all of these plants? Or are there particular attributes common to all parasites but which saprophytes lack? Also, a saprophyte may grow on moribund tissues of plants. What happens in senescent cells that allows growth of saprophytes?

3. A related point is that parasites, including and possibly especially highly specialized ones, often lose most of their pathogenicity in culture although they may continue to grow well enough. Effectively they now become saprophytes that cannot parasitize any plant. Are the properties lost in culture anything to do with those that make it a parasite? Probably not but perhaps they are worth studying in this context especially if one considers the claims that the original pathogenicity may be restored by passing the very low virulence form through a suitable host plant.

4. What are the differences between saprophytic and parasitic forms of the same species particularly of those such as *Fusarium oxysporum* with many special forms, each highly specialized parasites? What are the chances of altering a saprophytic *F. oxysporum* into a parasite, or one special form into another particularly when one considers that one special form may colonize extensively a non-host plant but escape attention because it does not cause symptoms visible to the naked eye. How common is such cryptic parasitism?

Of related interest is parasitism by species of *Helminthosporium* and *Alternaria* with some species highly specialized and virulent but with others much less so with most of their growth in plants as saprophytes.

5. A taxonomic group comprising species that are similar in many respects, morphologically and otherwise may, indeed often does, include a minority of species that are much less specialized parasites than are the rest. Would comparisons of their parasitism be likely to reveal anything significant about non-host resistance?

Tempting as is speculation about the nature of non-host resistance and specialization of parasitism if only because they are the most challenging problems in plant pathology, it may be somewhat indulgent if it cannot be related to research. Also it seems to me that the prospects of getting support for such research are not bright and perhaps understandably so because it is not easy to see how such research could be exploited for practical ends in the production of resistant crop plants unless recombinant DNA technology with all its achievements and potential clears the way for research along lines quite different from those we have followed so far.

RECOMBINANT DNA TECHNOLOGY AND HOST-PARASITE INTERACTIONS

I think we must admit that progress towards an understanding of these interactions has been disappointingly slow even for those for which the genetics are known and simple as in race/cultivar interactions controlled by pairs of matching genes. We are still largely in the dark about the nature of the products controlling such interactions and on how they determine their final outcome as resistance or susceptibility. In contrast there is the remarkable success of plant breeders in moving genes for resistance between plants and in developing multitudes of useful cultivars by repeatedly and judiciously crossing particular plants and selecting from segregating populations. But even if we had been more successful in elucidating host-parasite interactions success probably would not have helped very much the plant breeder in his use of major genes for resistance, satisfying as it would have been in other ways. Lack of success, despite the amount of good work and the skill of those who have done it, means that the methods used in the research have not been able to cope very productively with even the simplest of host/parasite interactions. They have, of course, been based essentially on the purification of substances likely to be present only in very low concentrations and among many other substances of similar type in infected tissues. Now, however, the prospects for success may be transformed by the use of recombinant DNA technology comprising the range of new, powerful and versatile techniques which already are being applied to a number of plant diseases. So far and not surprisingly, the emphasis has been on bacterial pathogens because of the immediate applicability to them of methods developed for bacteria in other connexions. But the technology is being adapted and applied to fungal pathogens and the diseases they cause. No doubt a major attraction will be the major genes controlling race/cultivar interactions with the tempting prospects of cloning the genes, using them to obtain and then characterize their products with all that this may lead to, mutating them specifically to see how this alters their expression and function and thus to the much better understanding of these interactions which so far has eluded us. Also there are now, as there could hardly have been from earlier research, the potentially immense practical benefits of having single specific genes for resistance in amounts sufficient for transfer to susceptible but otherwise desirable plants either with vectors or possibly by injecting directly into protoplasts. Such genes have the advantage of being types that would be expected to occur and function naturally so that expression and function are not likely to be a major problem. There are other problems but when these are resolved then the transfer of genes between plants may be made much easier than it is at present by the use of methods which although very succesful still are somewhat cumbersome and often take a long time to achieve their objectives. The technology may also allow the transfer of resistance genes which cannot now be used such as genes in plants that cannot be parents. Note, however, that the main benefits of cloning are likely to be in the greater rapidity and versatility with which the resistance genes can be deployed. The problems of using such genes against pathogens such as *Phytophthora infestans* will remain although some, no doubt, will be mitigated.

For major gene resistance in race/cultivar interactions the way ahead in the use of recombinant DNA technology seems reasonably clear and hopeful. It is less so for race non-specific resistance so far as this depends on many genes each with a minor effect which would be difficult to recognize. There are, however, strong economic motives for directing the technology at this type of resistance because of its importance and because it is difficult to use in breeding programmes by conventional techniques. The economic reasons for research with the new technology on non-host resistance are less compelling, at least in the not too distant future, but the longer term prospects are enticing because of the invariability of the resistance. This, however, is the quality that makes it so intractable a problem to investigate. With all that recombinant DNA technology now makes and will make possible, it still will take a great deal of ingenuity to deploy it so as to reveal the identity and functioning of the determinants in non-host resistance and in the complementary parasite/host specificity.

REFERENCES

1. Flor, H.H., 1942, Inheritance of pathogenicity of Melampsora lini, Phytopathology, 32: 653-669.
2. Bailey, J.A., Rowell, P.M. and Arnold, G.M., 1980, The temporal relationship between host cell death, phytoalexin accumulation and fungal inhibition during hypersensitive reactions of Phaseolus vulgaris to Colletotrichum lindemuthianum, Physiol. Plant Pathol., 17: 329-339.
3. Hargreaves, J.A. and Bailey, J.A., 1978, Phytoalexin production in hypocotyls of Phaseolus vulgaris in response to constitutive metabolites released by damaged cells, Physiol. Plant Pathol., 13: 89-100.
4. Keen, N.T., 1982, Specific recognition in gene-for-gene host-parasite systems, in: "Advances in Plant Pathology", Vol. 1, D.S. Ingram and P.H. Williams, eds., pp. 35-81, Academic Press, New York, London.
5. Darvill, A.G. and Albersheim, P., 1984, Phytoalexins and their elicitors - A defense against microbial infection in plants, Annu. Rev. Plant Physiol., 35: 243-275.
6. Nothnagel, E.A., McNeil, M., Albersheim, P. and Dell, A. 1983, Host-pathogen interactions. XXII. A galacturonic acid polysaccharide from plant cell walls elicits phytoalexins, Plant Physiol., 71: 916-926.
7. West, C.A., Bruce, R.J. and Jin, D.F., 1984, Pectic fragments of plant cell walls as mediators of stress responses, in: "Structure, Function and Biosynthesis of Plant Cell Walls", W.M. Dugger and S. Bartnicki-Garcia, eds., pp. 359-379, American Society of Plant Physiologists, Guithersberg.
8. De Wit, P.J.G.M. and Spikman, G., 1982, Evidence for the occurrence of race and cultivar-specific elicitors of necrosis in intercellular fluids of compatible interactions of Cladosporium fulvum and tomato, Physiol. Plant Pathol., 21: 1-11.
9. De Wit, P.J.G.M., Hofman J.E., Velthuis, G.C.M. and Kuć, J., 1985, Isolation and characterization of an elicitor isolated from intercellular fluids of compatible interactions of Cladosporium fulvum (syn. Fulvia fulva) and tomato, Plant Physiol., 77: 642-647.
10. Keen, N.J. and Yoshikawa, M., 1983, β-1,3-Endoglucanase from soybean releases elicitor-active carbohydrates from fungus cell walls, Plant Physiol., 71: 460-465.
11. Yoshikawa, M., Matama, M. and Masago, H., 1981, Release of a soluble phytoalexin elicitor from cell walls of Phytophthora megasperma var. sojae by soybean tissues, Plant Physiol., 67: 1032-1035.
12. Green, T.R. and Ryan, C.A., 1973, Wound-induced proteinase inhibitor in plant leaves, Plant Physiol., 51: 19-21.
13. Ryan, C.A., Bishop, P., Pearce, G., Darvill, A.G., McNeil, M. and Albersheim, P., 1981, A sycamore cell wall polysaccharide and a chemically related tomato leaf polysaccharide possess similar proteinase inhibitor-inducing activities, Plant Physiol., 68: 616-618.
14. Bruce, R.J. and West, C.A., 1982, Elicitation of casbene synthetase activity in castor bean. The role of pectic fragments of the plant cell in elicitation by a fungal endopolygalacturonase, Plant Physiol., 69: 1181-1188.
15. Pegg, G.F. and Young, D.H., 1982, Purification and characterization of chitinase enzymes from healthy and Verticillium albo-atrum- infected tomato plants and from V. albo-atrum, Physiol. Plant Pathol., 21: 389-409.

16. Walker-Simmons, M., Hadwiger, L. and Ryan, C.A., 1983, Chitosans and pectic polysaccharides both induce accumulation of the antifungal phytoalexin pisatin in pea pods and antinutrient proteinase inhibitors in tomato leaves, Biochem. Biophys. Res. Commun., 110: 194-199.
17. Vanderplank, J.E., 1984, "Disease Resistance in Plants", Academic Press, London, New York, 194 pp.
18. Vanderplank, J.E., 1982, "Host-pathogen Interactions in Plant Disease", Academic Press, London, New York, 207 pp.
19. Ellingboe, A.H., 1982, Genetical aspects of active defense, in: "Active Defense Mechanisms in Plants", R.K.S. Wood, ed., pp. 179-192, Plenum Press, London, New York.
20. Király, Z., Barna, B. and Ersek, T., 1972, Hypersensitivity as a consequence not the cause, of plant resistance to infection, Nature 239: 456-457.
21. Mayama, S., 1983, The role of avenalumin in the resistance of oats to crown rust, Memoirs of Faculty of Agriculture Kagawa University, 42: 1-62.
22. Bushnell, W.R. and Rowell, J.B., 1981, Suppressors of defense reactions: a model for roles in specificity, Phytopathology, 71: 1012-1014.
23. Heath, M.C., 1981, A generalized concept of host-parasite specificity, Phytopathology, 71: 1121-1123.
24. Heath, M.C., 1982, The absence of active defense mechanisms in compatible host-pathogen interactions, in: "Active Defense Mechanisms in Plants", R.K.S. Wood, ed., pp. 327-338, Plenum Press, London, New York.
25. Doke, N. and Tomiyama, K., 1980, Suppression of the hypersensitive response of potato tuber protoplasts to hyphal wall components by water soluble glucans isolated from Phytophthora infestans, Physiol. Plant Pathol., 16: 177-186.
26. Bahamish, H.S. and Wood, R.K.S., 1985, Induction of susceptibility and resistance of wheat leaves by preinoculation with high and low virulence races of Puccinia recondita f.sp. tritici, Phytopath. Z., 113: 97-112.
27. Heath, M.C., 1983, Relationship between development stage of the bean rust fungus and increased susceptibility of surrounding bean tissue to the cowpea rust fungus, Physiol. Plant Pathol., 22: 45-50.
28. Ouchi, S. and Oku, H., 1982, Physiological basis of susceptibility induced by pathogens, in: "Plant Infection : The Physiological and Biochemical Basis", Y. Asada et al., eds., pp. 117-136, Japan Scientific Society Press, Tokyo/Springer-Verlag, Berlin.
29. Wood, R.K.S., 1976, Specificity - an assessment, in: "Specificity in Plant Diseases", R.K.S. Wood and A. Graniti, eds., pp. 327-338, Plenum Press, London, New York.
30. Ellingboe, A.H., 1976, Genetics of host-parasite interactions, in: "Encylopedia of Plant Physiology", Vol. 4, R.H. Heitefuss and P.H. Williams, eds., pp. 761-778, Springer-Verlag, Berlin.

FUNDAMENTAL QUESTIONS RELATED TO PLANT-FUNGAL INTERACTIONS: CAN RECOMBINANT DNA TECHNOLOGY PROVIDE THE ANSWERS?

M.C. Heath

Botany Department
University of Toronto
Toronto, M5S 1A1, Canada

CHARACTERISTICS OF THE INTERACTIONS BETWEEN PLANTS AND FUNGAL PLANT PATHOGENS

As shown diagrammatically in Figure 1, fungal plant pathogens may enter susceptible plant tissues directly through epidermal cells (Fig. 1b, h, g), between the anticlinal walls of such cells (Fig. 1e), through stomata (Fig. 1c, d) or other natural openings, or via wounds (Fig. 1a) [1]. In most cases, such infections originate from spores on or near the plant surface, and the germ tube commonly differentiates to form a swollen appressorium from which the infection peg develops that enters the plant. Most pathogens show a preference in their mode of penetration, and for those that do not enter through wounds, cytological studies usually suggest that the fungus adheres tightly to the plant surface prior to entry regardless of whether an appressorium develops. For some rust fungi, such adhesion has been shown to be mandatory for the induction of appressoria [2].

Once inside susceptible tissue, some fungi continue their growth intercellularly, irrespective of whether they entered the tissue through the cytoplasm of epidermal cells. Others grow primarily within the cells, while many fungi produce undifferentiated hyphae that grow within and between the cells with equal preference (Fig. 1a). In other cases, the intracellular portions of the mycelium are differentiated from hyphae in the intercellular spaces. Such differentiation is particularly marked for many biotrophic fungi (which obtain their nutrients from living host cells) where the intracellular haustoria are morphologically and structurally distinct from the intercellular hyphae (Fig. 1c, e) [3,4].

Even in a susceptible plant, a given fungal pathogen is usually localized to certain parts or tissues, and may be able to colonize these only when they are of a certain age. An extreme form of localization is that shown by certain powdery mildew fungi which colonize only the surface of the leaf, obtaining their nutrients from the haustoria that penetrate epidermal cells (Fig. 1h). Vascular wilt fungi, as another example, preferentially grow within the xylem elements of the plant (Fig. 1f). The restriction of other types of fungi to certain areas within a plant part often is attributed to the presence of physical barriers to pathogen spread [5].

Fungal colonization may progress until all suitable tissue is infected, or may appear to be self-limiting. In general, a successful infection is one

NATO ASI Series, Vol. H1
Biology and Molecular Biology of Plant-Pathogen Interactions
Edited by J. Bailey

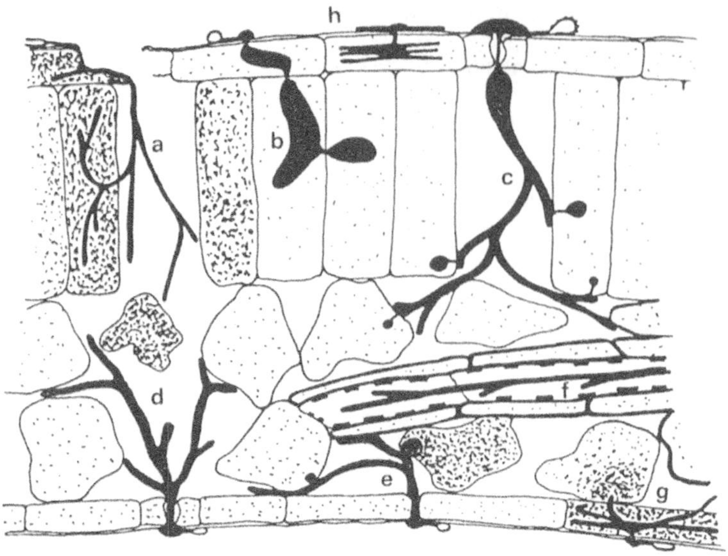

Figure 1. For explanation, see adjacent page.

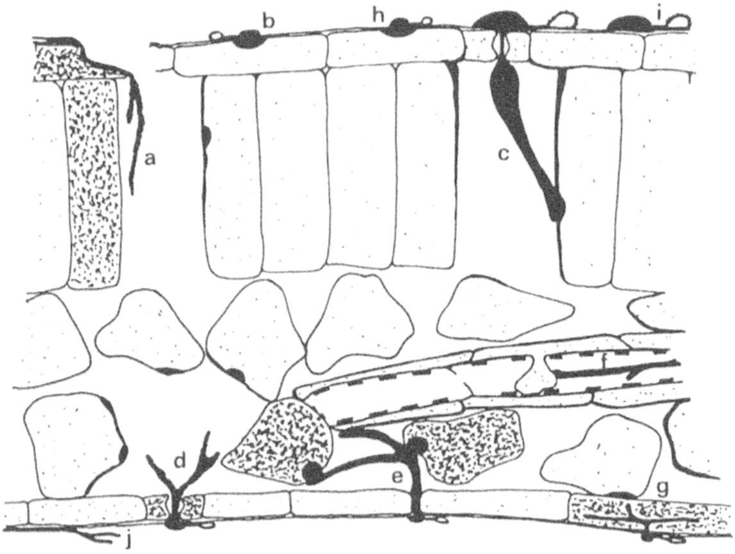

Figure 2. For explanation, see adjacent page.

Figure 1. A diagrammatic representation of some of the types of interaction that fungal pathogens may have with susceptible leaf tissue (shown in cross section). a) a wound-penetrating necrotroph that kills cells before invasion b) an intracellular biotrophic fungus; c) a stoma-penetrating, mainly intercellular biotrophic fungus that forms morphologically differentiated, intracellular, haustoria; d) an intercellular biotrophic fungus; e) a fungus that penetrates between anticlinal epidermal walls and which has a short biotrophic relationship with each haustorium-invaded cell; f) a vascular pathogen; g) an intracellular, nonbiotrophic fungus; h) a haustorium-producing biotroph that forms its extracellular mycelium on the leaf surface.

Figure 2. A diagrammatic representation of some of the types of interactions that fungal pathogens may have with resistant leaf tissue (cf. Figure 1). a) and d) inhibited fungal growth associated with necrosis, and deposits on and in the walls, of neighbouring plant cells; b) and h) lack of fungal penetration associated with localized deposits on and in epidermal walls; c) cessation of fungal growth associated with the formation of wall deposits and the lack of haustorium formation; e) reduced fungal growth associated with faster death of haustorium-invaded cell; f) limited fungal growth in xylem associated with the presence of tyloses; g) fungal growth restricted to the necrotic, first-invaded cell; i) fungus unable to penetrate due to appressorium being induced to form away from a stoma; j) fungus unable to penetrate because plant surface does not induce infection structures to form.

where the fungus manages to sporulate and such sporulation may occur at various positions within the tissue, depending on the fungus and type of spore. Wind or insect dispersed spores are formed, or become exposed, on the plant surface, while resting spores and other structures are often embedded in the tissue and released only after the tissue decays.

Biotrophic fungi usually cause little obvious damage to the plant cells around them, although changes in plant growth, suggestive of disturbances in hormone levels, may occur [6]. However, nonbiotrophs cause varying degrees of host cell death, with necrotrophs (or perthotrophs) by Luttrell's definition [7] causing such death in advance of fungal growth (Fig. 1a). Several fungi have a short biotrophic phase where they seem to exist in a compatible association with the plant cells before the latter die (Fig. 1e).

The limited host range of most fungal pathogens means that all plants are resistant to most potential pathogens (even when the fungus tries to infect its "preferred" plant part). This resistance can be divided into two types [8]. The first is that shown by specific genotypes of the host species and will be called host, or cultivar, resistance. The second is that shown by species not considered to be hosts for the fungus in question, and will be referred to as nonhost resistance. It has been suggested that cultivar resistance may be superimposed on a "basic compatibility" [9] between a fungus and its host species that is achieved by the ability of the pathogen to negate the nonspecific defences that govern nonhost resistance [8].

These two forms of resistance, when expressed by a given plant, may or may not resemble each other [10]. In some cases, nonhost resistance seems to be based on the presence of preformed features such as secondary metabolites that inhibit fungal growth [11] or surface phenomena that affect the formation of infection structures (Fig. 2i, j) [2,10]. However, there are many investigated examples of host and nonhost resistance that seem to be the result of induced changes in the plant. A comparison of these changes in different plant-fungus interactions shows that overall, all resistant tissues, whether acting as resistant hosts, nonhosts, or inappropriate plant parts, respond to attempted invasion in rather similar ways, and that these responses may be seen to a lesser extent in even susceptible tissue [12]. In infections of nonvascular tissue, such responses can be divided into those that are associated with the cell wall and those that are associated with the cell contents [12]. Examples of the former include the deposition on the cell wall of callose (rich in β-1,3 glucans) and the accumulation in the wall of such compounds as lignin, suberin, low molecular weight phenolic compounds, and silica. Examples of changes in cell contents are the accumulation of antimicrobial phytoalexins and other stress metabolites. However, not all responses involve controlled biosynthetic activity, and a variety of chemical changes may occur in dying cells caused by the intermingling of vacuolar contents with the cytoplasm after the breakdown of cellular compartmentalization [11]. For fungi that inhabit the vascular system, plant responses include the accumulation of phytoalexins and the plugging of the xylem vessels with gels or balloon-like extension (tyloses) (Fig. 2f) of the adjacent parenchyma cell walls [12]. In some situations, not only are there responses of pre-existing cells, but also cell proliferation is induced so that a tightly packed periderm layer of suberized or lignified cells develops around the infected area [13]. For responses that can be detected cytologically, it is common that many of these can be detected at a single infection site [14].

Many of the plant responses just listed have the potential to inhibit or impede fungal growth, often in several ways. For example, lignification of the cell wall, in theory, may prevent its degradation by fungal enzymes, impede fungal penetration into the cell, restrict efflux of nutrients from the cell, and release fungitoxic precursers to the fungus [15]. However, whether such lignification is going to be an effective barrier to the fungus in question depends greatly on its life style. If a fungus can degrade this material, or can grow faster than the cells can lignify, the inhibitory effect of lignification may be slight. Indeed, there is little evidence to suggest that the plant "tailor-makes" its responses with respect to bringing about the cessation of fungal growth. As discussed by Ward (this volume), many of these responses are the same as those induced by other forms of damage. It is possible, therefore, that most of the plant's responses to infection are a direct or indirect consequence of cellular damage rather than responses to the fungus itself. The fact that phytoalexin accumulation and plant cell necrosis can be elicited by components of plant walls likely to be released by the action of fungal, and possibly plant, enzymes [16] seems to support such a view. However, non-enzymic components of fungal walls may also trigger these responses [16] and the two extremes of many hypotheses are i) that the fungal compounds elicit responses indirectly via cell damage or ii) that they enter the plant nucleus and directly alter gene expression. Which of the variety of responses typically expressed during resistance acts as a defence reaction in a given situation presumably depends on the nature of the fungus and how it tries to interact with the plant. In theory it also is possible that the cessation of fungal growth is brought about in some or all situations by some feature of the plant that has yet to be discovered.

The identity of effective resistance mechanisms is only one of the many questions that arises from studies of plant-pathogen interactions. Yet despite considerable research over the last 50 years or so, many of the more fundamental questions concerning the way in which fungal plant parasites interact with their hosts or nonhosts remain unanswered. In the area of resistance to infection, the problem often lies with the multitude of potentially inhibitory features that accompany resistance. In other cases, our lack of progress stems from the difficulty of distinguishing cause from effect. For example, it has been hotly debated whether cell death is a cause of resistance towards biotrophic fungi, or whether it is a consequence of the death of the fungus caused by other features [17]. As Ellingboe [9,18] has pointed out, the use of mutations and other genetic manipulations are often the best, if not the only way, of investigating the role in the interaction of particular features of plant and pathogen, but on the whole, his advice has been little heeded. However, with the advent of recombinant DNA technology, it is now possible to ask questions, and to perform genetic manipulations that were previously impossible. The current surge of activity in applying these techniques to virtually all disciplines, including plant pathology, reflects the myriad of specific questions that recombinant DNA technology can help address. In addition, this upsurge of interest in genes, rather than their products, has re-directed thought towards more genetical approaches to many problems. The beginning of such a trend already can be seen in plant pathology and the purpose of the rest of this paper is to consider in greater detail the more fundamental questions associated with plant-fungal interactions, and to examine the role in answering them that molecular genetics may play. The questions chosen are those which seem, to me, to address problems common to most types of interaction. It should be realised, however, that each interaction may provide different answers, and that there are many more specific questions that could be asked for each particular plant-fungus combination.

FUNDAMENTAL QUESTIONS ASSOCIATED WITH THE INTERACTIONS BETWEEN PLANTS AND THEIR FUNGAL PATHOGENS

What Attributes are Required for Pathogenesis?

One of the more fundamental questions in plant pathology is what are the attributes that distinguish a pathogen from a saprophyte. Superficially these seem to be the ability to penetrate the plant tissue, the ability to colonize the tissue after penetration, and the ability to sporulate in the environment provided by the tissue. What these attributes represent in molecular terms, however, is not well understood, and must vary with the fungus and tissue invaded.

For plant pathogens that normally enter through an intact plant surface, a primary feature that seems to distinguish them from saprophytes is that they attempt to penetrate while saprophytes do not [19]. In some cases, it appears that the stimulus is merely contact with any solid surface [20], but for fungi that show a preference in the plant part or cellular position (e.g. at cell junctions or over stomata) at which penetration is attempted, more specific cues must be needed from the plant. Some specific questions that arise from these observations are i) what is the exact nature of the cues that the fungus responds to, ii) how does the fungus perceive them and how is this perception translated into a response, iii) is penetration of the plant wall purely mechanical, as used to be believed [20], or are exogenous activities (e.g. secretion of degradative enzymes) needed and iv) how many of these features or responses are lacking in obligate saprophytes? Currently, the most significant progress is being made in answering questions ii) and iii) as discussed by Staples in this volume, and Kolattakudy [21].

The ability to colonize the tissue is another attribute that seems to distinguish a pathogen from a true saprophyte, as the latter usually will not grow extensively even if introduced into a wound [22]. However, considerable growth of the saprophyte may occur if the metabolism of the plant is impaired [22], suggesting that it is not the lack of suitable substrates, but the presence of inhibitory features that prevent growth. Why pathogens are not similarly inhibited in susceptible tissues is a subject of much debate. Explanations include the release by the fungus of toxins and other factors such as wall degrading enzymes that kill or narcotize the cells before active defences can be induced, the secretion of specific suppressors of defence reactions, the degradation of inhibitory compounds, insensitivity to such compounds, and the fact that active defences are not elicited [23]. However, unequivocal proof for most of these explanations is lacking. It will be noticed that many of these hypotheses rely on some specific activity of the pathogen. It is also possible that certain activities of the pathogen are required for colonization for reasons other than the negation of defence reactions. For example, the degradation of host cell walls may be necessary to allow an intercellular fungus to grow between adpressed cells. For obligately biotrophic pathogens, the ability to colonize the tissue must also depend on the presence or availability of whatever it is that these fungi require from living cells. On the whole, there is little data concerning how, on a cellular level, a fungus obtains its nutrients from its host.

One question that may be important in exploiting our knowledge of host-parasite interactions for disease control, is what are the cues that a fungus needs to initiate sporulation? Conceivably sporulation may occur when the fungus reaches a certain physiological age or stage of development, but it also seems likely that external factors are involved, particularly when spore initials are located at specific positions in the tissue (e.g. under stomata [24] or on only one surface of the leaf). To my knowledge, this is an area of host-parasite relations that is rarely investigated.

Why Does a Resistant Plant Respond to Infection?

Another important question in fungal plant pathology is what are the attributes of the pathogen that trigger responses considered to be defence reactions in resistant host and nonhost plants? Recently, much emphasis has been placed on the ability of fungal glycans, chitosan, glycoproteins and lipids, and plant cell wall fragments, to elicit the synthesis and accumulation of phytoalexins [16]. Fungal [25] and plant [26] products have also been isolated that cause necrosis. However, with a few exceptions [25, 27 and see De Wit, this volume], there is little evidence that these elicitors are released during the plant-fungus interaction. Moreover, the nonspecificity of the majority of these elicitors leaves one with no clear explanation of why successful defence reactions are not elicited in susceptible plants. Important questions that remain are i) what are the natural elicitors of responses in resistant (and susceptible) plants, ii) are different responses elicited by different elicitors, as suggested by some data [16] but not others [28], iii) are the same elicitors involved in triggering responses in host and nonhost plants, and iv) how do elicitors elicit? Currently, the greatest research activity seems to be concerned with answering the last question [Ryder, this volume].

Where resistance of the host species is expressed only in cultivars possessing a given gene for resistance, and is elicited only if the pathogen possesses a specific gene for avirulence, a gene-for-gene interaction is said to exist. The nature of the products of these genes has been much debated. Opposing views exist as to whether the gene for resistance in the plant produces a product which by itself, or in combination with the product of the gene for avirulence, is inhibitory to the fungus [29], or whether it is a regulatory gene that controls many of the physiological responses of the plant that are associated with resistance [30]. Support for the latter hypothesis stems primarily from the belief that some of the observed plant responses must be responsible for resistance and from the observation that under suitable conditions, similar responses may be triggered in plants lacking a given gene for resistance [31]. Other questions that come to mind are i) how do the products of the genes for avirulence relate to the elicitors of putative defence reactions that can be extracted from the fungus, ii) how different are the fungal genes and gene products controlling virulence and avirulence, iii) how do the fungal genes and gene products differ between different genes for avirulence, iv) what role do these genes products play other than governing virulence and avirulence, v) how similar are the genes and gene products that control plant resistance towards the same or different pathogen species, and vi) do saprophytes have genes equivalent to the genes for avirulence in parasites? Also of interest is the situation, commonly found in fungi grown extensively in culture, where a strain of the pathogen loses its ability to infect successfully any cultivar of the host species. Has this strain acquired many genes for avirulence or has it lost some attribute necessary for successful colonization of the tissue [32]? None of these questions yet have any answers.

What are the Mechanisms of Host and Nonhost Resistance?

As mentioned earlier, in resistant host cultivars and nonhost plants, there is an enormous amount of data concerning the expression of resistance but few unequivocal conclusions concerning the mechanisms of resistance. The basic questions that need to be answered for each plant-fungus interaction are i) what feature or features are responsible for the cessation of fungal growth, and ii) how in gene-for-gene systems do these features relate to the gene for resistance?

THE APPLICATION OF RECOMBINANT DNA TECHNOLOGY TO FUNDAMENTAL QUESTIONS IN PLANT PATHOLOGY

One of the biggest problems with using recombinant DNA technology in the area of fungal plant pathology is the current inability to apply fully this technology to filamentous fungi or the majority of crop plants. For example, relatively few crop plants can, as yet, be routinely transformed as single cells and regenerated into entire plants. Moreover, few fungal pathogens fit the role of the ideal fungus that grows and sexually reproduces easily in culture without losing its pathogenicity or viability, and for which a reliable transformation system is available. However, given the progress that has been made in manipulating higher plants and in "domesticating" and manipulating fungi such as *Neurospora* and *Aspergillus* spp., it seems likely that the technical problems associated with manipulating each fungus and plant will be overcome with time. Nevertheless, significant progress in investigating plant-pathogen interactions using recombinant DNA technology in the short term, is going to depend on the type of plant and pathogen chosen. Many questions specific to certain types of interactions, for example those involving biotrophic fungi which cannot be grown away from the living plant, are likely to take longer to answer because of the extra technical difficulties in manipulating the fungus.

Before considering which of the questions discussed above are amenable to being answered using recombinant DNA techniques, it is worthwhile reviewing the types of information that these techniques can provide which may be of use to plant pathologists. If a gene product is available, such as mRNA or a polypeptide, then oligonucleotide probes can be synthesized to locate the gene in a cloned genomic library. A specific gene also can be identified by using a component of a genomic library to "rescue" a mutant deficient in the attribute in question, or by disrupting gene function by transposon mutagenesis. Once a gene has been isolated and cloned, it can be sequenced, compared with other genes, and flanking sequences examined for indications of how expression may be controlled. The nature and function of unknown products of a cloned gene may be investigated by using it to identify the corresponding mRNA, followed by the synthesis of complementary DNA (cDNA) which may be transcribed and translated *in vitro*. In theory, the function of the resulting polypeptide may be determined by comparing its sequence with that of known proteins and by using immunological techniques to locate its site in the cell. Finally, a cloned gene may be transferred from one organism to another.

Probably the technique most easily applied to fungal plant pathogens at the moment, is the identification of a gene following the isolation of its product [33]. In this manner, the gene coding for any fungal enzyme or polypeptide thought to be important in pathogenesis could be identified, cloned and sequenced. Its regulation may be investigated and sequence homology searched for in avirulent strains of the same fungus, in other pathogens, and in saprophytes. One outcome of such comparisons may be information useful in our understanding of the relationships between fungi and the evolution of parasitism. An example of a suitable fungal product would be the monooxygenase produced by *Nectria haematococca*, as genetic evidence suggests that the ability of the fungus to parasitize peas seems to depend partly on the ability of this enzyme to degrade the pea phytoalexin, pisatin [34]. Cutinase is another fungal enzyme for which the corresponding gene could be cloned as part of an investigation into its role in fungal penetration and into its distribution among pathogens and saprophytes. The progress that has been made on cloning such genes is described by Yoder (this volume).

Along similar lines, any plant enzyme or polypeptide thought to be important in resistance towards a given pathogen could be used to identify the gene involved, and homologies searched for in susceptible and resistant cultivars of the same plant species, or in nonhost species. How fruitful such an approach will be, given that the available evidence suggests that the genetic information for most types of responses is present in all plants, remains to be seen.

This type of approach assumes that a polypeptide important in pathogenesis has been identified. If this is not the case, another avenue that is being tried [35] is to use comparative changes in mRNA synthesis to identify those genes expressed during resistance or susceptibility. However, it is important to realise that given the myriad of responses that accompany resistance, none of the procedures described so far will answer any of the fundamental questions raised earlier. None will lead to unequivocal proof that any given fungal feature is important in pathogenesis, that any single plant response is the primary mechanism of resistance, or that a particular gene is the same as that identified by genetic studies as the gene for resistance. However, one area where the investigation of temporal changes in gene transcription may be particularly informative is in the study of the development of fungal infection structures in the manner that other forms of fungal morphogenesis have been studied [Timberlake, this volume]. That genetic variation may exist between fungal individuals in their mode of infection is illustrated by the observation that different races of *Phytophthora megasperma* f.sp. *glycinea* have different preferences in their cellular site of penetration [36]. Studies of such natural variants and laboratory-induced mutants, coupled with comparative studies of gene expression during the synchronous formation of infection structures on artificial substrates, may help elucidate the genes and products involved. Homologous gene sequences could then be searched for in related saprophytes.

While valuable information may be obtained by the procedures just described, answers to most of the questions listed earlier rely on the availability of a reliable transformation system for both fungus and plant. With such systems, the most obvious problems to be investigated are those that concern the nature and role of plant and fungal genes involved in gene-for-gene interactions. Only when a gene for avirulence or resistance has been isolated and identified by using it to transform individuals lacking such a gene can one begin to examine its function. Once the function of these genes are known, then perhaps a reconciliation can be made between the currently irreconcilable genetic and physiological data. Moreover, with available transformation systems, the importance of putative defence mechanisms in host and nonhost plants can be investigated by transforming plants with a mutated version of an appropriate gene that would modify a particular response. With a defence mechanism such as phytoalexin accumulation that involves the activation of a complex biosynthetic pathway catalyzed by many different enzymes, the chosen enzyme must be far enough along the biosynthetic pathway that other vital biosynthetic processes of the plant would not be seriously affected. As an alternative to transforming the plant with an altered gene, a natural, chemical- or transposon-induced mutant lacking the enzyme could be rescued by the normal gene in question, and the resistance of both mutant and "rescued" plant investigated.

Similar procedures would allow the investigation of the role of certain fungal products in pathogenesis. Not only could suitable avirulent mutants or strains be transformed with a given gene to see if virulence is restored, but one also could transform saprophytes to evaluate the importance of various fungal attributes in attaining pathogenicity.

From this brief summary, it can be seen that, in theory, recombinant DNA technology can provide answers to many of the questions raised earlier. That it also may do so in practice is suggested by the significant progress that already has been made with bacterial pathogens in terms of cloning genes of avirulence [Lindgren, this volume] and investigating the role of bacterial products in pathogenesis [Collmer, this volume]. The cloning of plant and fungal genes thought to be involved in resistance and pathogenesis, respectively, has begun in many laboratories [e.g. 35 and Yoder, this volume]. In addition, the widespread investigation of the biosynthetic response of cultured cells to fungal and plant wall components, coupled with recombinant DNA technology, is likely to lead to a better understanding of how these components regulate gene function [Ryder, this volume] even if the importance of such components in the plant-parasite interaction is not clear. Overall, therefore, I predict that the next 10 years will be exciting ones in terms of our increased understanding of how fungal pathogens interact with higher plants.

GENETIC ENGINEERING OF PLANTS FOR DISEASE PROTECTION

With the ability to transform fungus and plant comes the potential ability to manipulate the host-parasite interaction with the aim of producing disease control. On the whole, it seems easier to change the plant than to alter the natural fungal population, and it has become popular to suggest that plants may be transformed to possess novel and more durable forms of resistance [33,37]. Certainly, it seems possible that genes eventually will be transferred between plant species with an ease that cannot currently be achieved by the plant breeder. Often stressed is the potential for creating new forms of resistance by moving genes between unrelated species. Day et al. [37] have suggested that one could "shot-gun" DNA from a nonhost species into one that is host for the pathogen in question, and search for resistance. Such an empirical approach may well pay off, but there is no reason why any resulting resistance, merely because it is derived from a nonhost plant, should be any more durable or useful in the field that that produced by conventional means [38]. The beauty of recombinant DNA technology, as I see it, is the potential for "tailor-making" new forms of resistance based on a comprehensive understanding of the host-parasite interaction. For example, it may be possible to provide a plant with a novel regulatory gene that turns on several defence mechanisms in response to a basic feature of most fungal plant pathogens that the fungi would find lethal to lose or modify by mutation. Many other ideas can be conceived, but the important feature is to ensure durability by producing a form of resistance that the fungus would have difficulty in adapting to by the effects of random mutation, either because a product with a high degree of specificity has to be secreted to negate the resistance, or because the fungus cannot afford to lose the inducing attribute [38]. Whether we currently know enough about the genetic potential of any pathogen to predict how easily it could adapt to overcome a given form of resistance is debatable. However, as our knowledge of the more fundamental aspects of plant-fungal interactions increases, so should our ideas on how durable resistance may be engineered in a rational, controlled, and economically feasible manner.

THE NEED FOR AN INTERDISCIPLINARY APPROACH

While the role that recombinant DNA technology is going to play in the future of plant pathology cannot be underestimated, it is also obvious that the technique cannot stand alone. Questions that it can be used to answer can only be asked if there is a basic knowledge of the plant-parasite interaction. For example, one cannot investigate that role in resistance of changes in the plant cell wall if one does not know that such changes exist; nor can one identify genes coding for enzymes of certain biosynthetic pathways if these pathways have not been elucidated. Furthermore, if resistance relies on more than one mechanism, as is likely in nonhost resistance [38], altering one of them is not going to result in complete susceptibility; cytological and biochemical studies will have to accompany molecular genetic studies to ensure the detection of any change in the expression of resistance. A similarly comprehensive approach will have to be taken to dissect the steps involved in tissue penetration and colonization.

Not only are cytological and biochemical data going to be needed, but for many investigations, classical genetic studies and the availability of mutants are a prerequisite for further experimentation [33]. It is evident, therefore, that recombinant DNA technology is likely to make the biggest impact on those systems for which there already is a substantial body of genetical, cytological and biochemical data, or where a multidisciplinary approach to a particular problem is followed.

Finally, it must be realized that there also are phenomena involved in plant-interactions, such as the reactions that take place during the loss of cellular compartmentalization and induction of callose deposition [28], that may not require immediate changes in gene expression. Such phenomena, and others such as the nature of the interactions that take place across the haustorium-plant cell interface in infections of obligately biotrophic fungi still must be investigated at the level of the cell, rather than the genes.

REFERENCES

1. Dodman, R.C., 1979, How the defences are breached, in: "Plant Disease: an Advanced Treatise. Vol. IV. How Pathogens Induce Disease", J.G. Horsfall and E.B. Cowling, eds., pp. 135-153, Academic Press, New York.
2. Wynn, W K. and Staples, R.C., 1981, Tropisms of fungi in host recognition, in: "Plant Disease Control: Resistance and Susceptibility", R.C. Staples and G.H. Toenniessen, eds., pp. 45-69, John Wiley and Sons, New York.
3. Harder, D.E. and Chong, J., 1984, Structure and physiology of haustoria, in: "The Cereal Rusts", W.R. Bushnell and A.P. Roelfs, eds., pp. 431-476, Academic Press, Orlando.
4. Littlefield, L.J. and Heath, M.C., 1979, "Ultrastructure of Rust Fungi", Academic Press, New York.
5. Akai, S. and Fukutomi, M., 1980, Preformed internal physical defenses, in: "Plant Disease: an Advanced Treatise. Vol. V. How Plants Defend Themselves", J.G. Horsfall and E.B. Cowling, eds., pp. 139-159, Academic Press, New York.
6. Pegg, G.F., 1984, The role of growth regulators in plant disease, in: "Plant Diseases: Infection, Damage and Loss", R.K.S. Wood and G.J. Jellis, eds., pp. 29-48, Blackwell Scientific Publications, Oxford.
7. Luttrell, E.S., 1974, Parasitism of fungi in vascular plants, Mycologia, 66: 1-15.
8. Heath, M.C., 1981, A generalized concept of host-parasite specificity, Phytopathology, 71: 1121-1123.
9. Ellingboe, A.H., 1976, Genetics of host-parasite interactions, in: "Encyclopedia of Plant Physiology Vol. 4. Physiological Plant Pathology", R. Heitefuss and P.H. Williams, eds., pp. 761-778, Springer-Verlag, Berlin.
10. Heath, M.C., 1981, Nonhost resistance, in: "Plant Disease Control: Resistance and Susceptibility", R.C. Staples and G.H. Toenniessen, eds., pp. 201-217, John Wiley and Sons, New York.
11. Schönbeck, F. and Schlösser, E., 1976, Preformed sustances as potential protectants, in: "Encyclopedia of Plant Physiology Vol. 4. Physiological Plant Pathology", R. Heitefuss and P.H. Williams, eds., pp. 653-678, Springer-Verlag, Berlin.
12. Heath, M.C., 1980, Reactions of nonsuscepts to fungal pathogens, Annu. Rev. Phytopathology, 18: 211-236.
13. Beckman, C.H., 1980, Defences triggered by the invader: physical defences, in: "Plant Disease: An Advanced Treatise. Vol. V. How plants Defend Themselves", J.G. Horsfall and E.B. Cowling, eds., pp. 225-245, Academic Press, New York.

14. Fernandez, M.R. and Heath, M.C., 1986, Cytological responses induced by five phytopathogenic fungi in the nonhost plant Phaseolus vulgaris, Can. J. Bot., in press.
15. Ride, J.P., 1978, The role of cell wall alterations in resistance to fungi, Ann. App. Biol., 89: 302-306.
16. Darvill, A.G. and Albersheim, P., 1984, Phytoalexins and their elicitors - a defence against microbial infection in plants, Ann. Rev. Plant Physiol., 35: 243-275.
17. Heath, M.C., 1976, Hypersensitivity, the cause or the consequence of rust resistance?, Phytopathology, 66: 935-936.
18. Ellingboe, A.H. and Gabriel, D.W., 1977, Induced conditional mutants for studying host/pathogen interactions, in: "Induced Mutations against Plant Disease", pp. 35-44, International Atomic Energy Agency, Vienna.
19. Yoder, O.C. and Scheffer, R.P., 1969, Role of toxin in early interactions of Helminthosporium victoriae with susceptible and resistant oat tissue, Phytopathology, 59: 1954-1959.
20. Brown, W. and Harvey, C.C., 1927, Studies in the physiology of parasitism. X. On the entrance of parasitic fungi into the host plant, Ann. Bot., 41: 643-662.
21. Kolattakudy, P.E. and Soliday, C.L., 1985, Effect of stress on the defensive barriers of plants, in: "Cellular and Molecular Biology of Plant Stress", J.L. Key and T. Kosuge, eds., pp. 381-400, Alan R. Liss, New York.
22. Klarman, W.L., 1965, Heat-induced susceptibility of soybeans to nonpathogenic fungi, Phytopathology, 55: 505.
23. Heath, M.C. 1982, The absence of active defence mechanisms in compatible host-pathogen interactions, in: "Active Defence Mechanisms in Plants", R.K.S. Wood, ed., pp. 143-156, Plenum Press, New York and London.
24. McCain, J.W. and Hennen, J.F., 1984, Development of the uredial thallus and sorus in the orange coffee rust fungus, Hemileia vastatrix, Phytopathology, 74: 714-721.
25. De Wit, P.J.G.M., Hofman, A.E., Velthuis, G.C.M. and Kuć, J.A., 1985, Isolation and characterization of an elicitor of necrosis isolated from intercellular fluids of compatible interactions of Cladosporium fulvum (Syn. Fulvia fulva) and tomato, Plant Physiol., 77: 642-647.
26. Yamazaki, N., Fry, S.C., Darvill, A.G. and Albersheim, P., 1983, Host-pathogen interactions. XXIV. Fragments isolated from suspension-cultured sycamore cell walls inhibit the ability of the cells to incorporate ^{14}C leucine into proteins, Plant Physiol., 72: 864-869.
27. Keen, N.R. and Yoshikawa, M., 1983, β-1,3-endoglucanase from soybean releases elicitor-active carbohydrates from fungus cell walls, Plant Physiol., 71: 460-465.
28. Köhle, H., Jeblick, W., Poten, F., Blaschek, W. and Kauss, H., 1985, Chitosan-elicited callose synthesis in soybean cells as a Ca^{2+}-dependent process, Plant Physiol., 77: 544-551.
29. Ellingboe, A.H., 1982, Genetical aspects of active defence, in: "Active Defence Mechanisms in Plants", R.K.S. Wood, ed., pp. 179-192, Plenum Press, New York and London.
30. Callow, J.A., 1977, Recognition, resistance and the role of plant lectins in host-parasite interactions, in: "Advances in Botanical Research Vol. 4.", R.D. Preston and H.W. Woolhouse, eds., pp. 1-49, Academic Press, London.
31. Ward, E.W.B. and Buzzell, R.I., 1983, Influence of light, temperature and wounding on the expression of soybean genes for resistance to Phytophthora megasperma f. sp. glycinea, Physiol. Plant Pathol., 23: 401-409.
32. Rutherford, F.S., Ward, E.W.B. and Buzzell, R.I. , 1985, Variation in virulence in successive single-zoospore propagations of Phytophthora megasperma f. sp. glycinea, Phytopathology, 75: 371-374.

32. Gilchrist, D.G. and Yoder, O.C., 1984, Genetics of host-parasite systems: A prospectus for molecular biology, in: "Plant-Microbe Interactions. Molecular and Genetic Perspectives. Vol. 1.", T. Kosuge and E.W. Nester, eds., pp. 69-90, Macmillan Publishing Company, New York.
33. VanEtten, H.D., 1982, Phytoalexin detoxification by monooxygenases and its importance for pathogenicity, in: "Plant Infection: the Physiological and Biochemical Basis", Y. Asada, W.R. Bushnell, S. Ouchi and C.P. Vance, eds., pp. 315-327, Springer-Verlag, Berlin.
34. Fristensky, B., Riggleman, R.C., Wagoner, W. and Hadwiger, L.A., 1985, Gene expression in susceptible and disease resistant interactions of peas induced with Fusarium solani pathogens and chitosan, Physiol. Plant Pathol., in press.
35. Stössel, P., Lazarovits, G. and Ward, E.W.B., 1981, Differences in the mode of penetration of soybean hypocotyls by two races of Phytophthora megasperma var. sojae, Can. J. Bot., 59: 1117-1119.
36. Day, P.R., Barrett, J.A. and Wolfe, M.S., 1983, The evolution of host-parasite interaction, in: "Genetic Engineering of Plants. An Agricultural Perspective", T. Kosuge, C.P. Meredith and A. Hollaender, eds., pp. 419-430, Plenum Press, New York and London.
37. Heath, M.C., 1985, Implications of nonhost resistance for understanding host-parasite interactions. in: "Genetic Basis of Biochemical Mechanisms of Plant Disease" , J.V. Groth and W.R. Bushnell, eds., A.P.S. Symposium Book 4, American Phytopathological Society, St. Paul, in press.

CELLULAR MODIFICATIONS DURING HOST-FUNGUS INTERACTIONS IN ENDOMYCORRHIZAE

V. Gianinazzi-Pearson

Laboratoire de Phytoparasitologie
Station d'Amélioration des Plantes, INRA
BV 1540, 21034 Dijon Cedex
France

INTRODUCTION

Although evolution has produced a general state of resistance to 'non-self' in plants, more than 90% of plant taxa in fact form compatible root associations, mycorrhizae, with certain soil fungi; these are no doubt the most frequent examples of susceptibility in plants to fungi. In endomycorrhizae, which concern most plant species, host-fungus interactions are structurally complex as there is development of an important intracellular biotrophic phase where cellular relationships are functionally compatible. This usually leads to bidirectional nutrient exchange between the two partners and highly beneficial effects on plant growth [1,2].

The fungi forming endomycorrhizae have clear-cut host ranges but the nature and degree of specificity between associates are quite variable from one type of endomycorrhiza to another [3]. Ericoid endomycorrhizae, for example, can be considered as quite specific as they are formed by a very limited number of species in the Ericaceae and by fungi belonging to, or closely related to, one ascomycetous species. At the other extreme, vesicular-arbuscular endomycorrhizae lack host specificity and several genera of fungi from one zygomycetous family can infect a wide range of plant species from Thallophytes and Pteridophytes to Gymnosperms and Angiosperms. Endomycorrhizae only form in differentiated unsuberized tissues of host roots; fungi do not infect meristematic or vascular regions and intracellular haustoria are limited to the parenchymatous cells surrounding the central cylinder. The pattern of fungal development within host tissues varies between ericoid and vesicular-arbuscular(VA) endomycorrhizae; in the former the intracellular haustoria form as dense hyphal coils whilst in the latter they are complex and much-branched structures (arbuscules). In both cases, however, endomycorrhiza formation is accompanied by ultracytological modifications in both the fungal and plant cells. Recent studies of cellular relationships between endomycorrhizal associates and of events occurring at their surfaces during infection has considerably improved knowledge about the structural and physiological interactions between host and fungus in these symbiotic associations.

NATO ASI Series, Vol. H1
Biology and Molecular Biology of Plant-Pathogen Interactions
Edited by J. Bailey

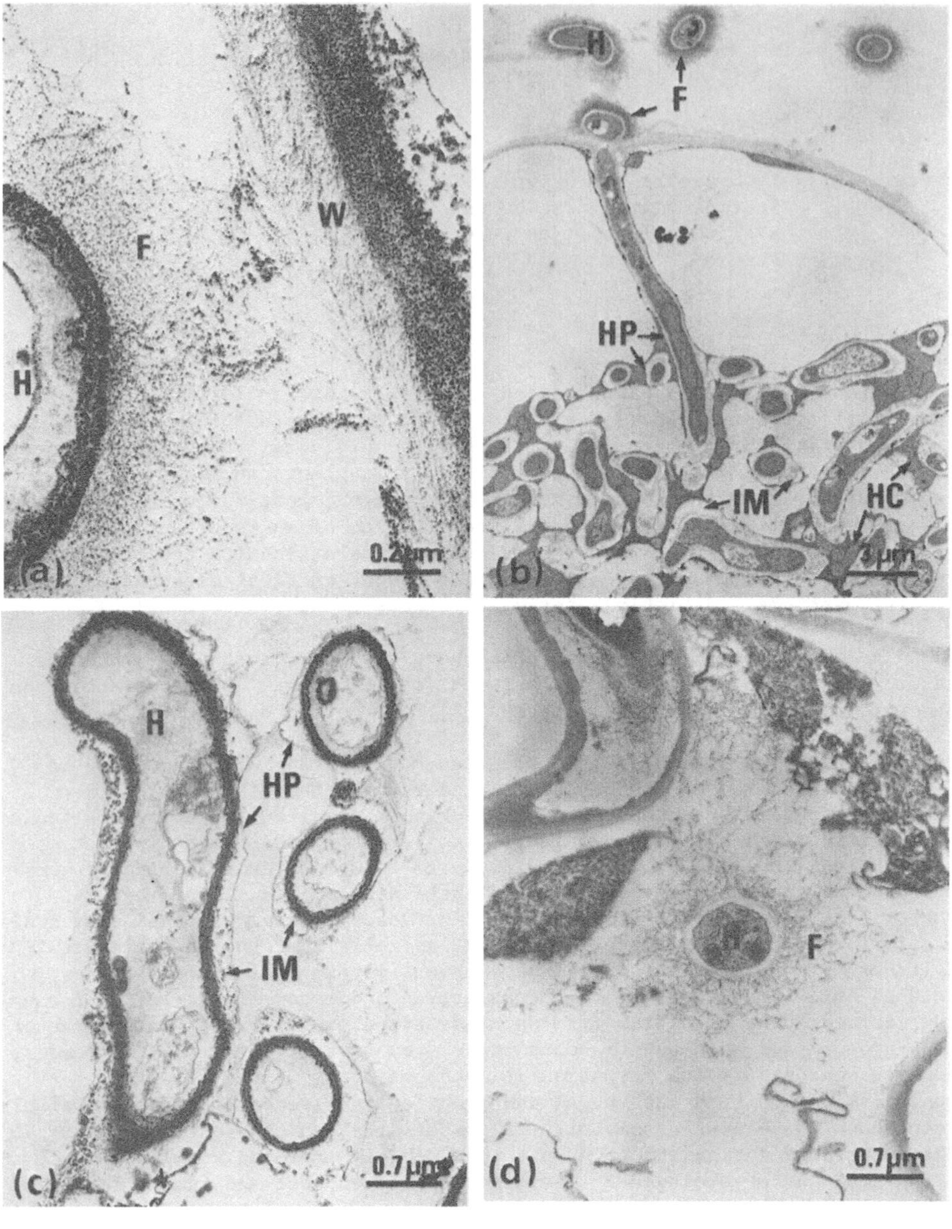

Figure 1. Electron micrographs showing interactions between an ericoid mycorrhizal fungus, Pezizella ericae, and cells of host (heather) and non-host (clover) plants. (a) Fibrils (F) of a hypha (H) adhere to the outer host wall (W) (PATAg reaction). (b) and (c) Adhesion is followed by fungal penetration and development in a host cell. Host plasmalemma (HP) and cytoplasm (HC) surround each hypha ; the interfacial matrix (IM) contains scattered fibrils and vesicles ((c)-PATAg reaction). (d) Fibrils around a hypha in a necrosing non-host cell.

CELL TO CELL CONTACT AND FUNGAL MORPHOGENESIS

Ericoid Endomycorrhizae

There is recent evidence that cell surface phenomena may be involved in the establishment of ericoid endomycorrhizal associations and that one of the first steps in the infection process is adhesion of the external hyphae to the root surface. Walls of ericoid mycorrhizal fungi in pure culture are probably composed of an inner chitin and an outer protein-polysaccharide layer [4]. Glucose, mannose and N-acetylglucosamine have been localized on the hyphal surface using lectin markers; the distribution of these carbohydrate moieties, however, differs between fungal isolates and this may be related to their infectiveness [5]. When the fungi grow around roots, they develop an abundant protein-polysaccharide fibrillar sheath which extends out from the hyphae wall and closely adheres to the outer surface of host cells as hyphae come into contact with the root (Figs. 1a and b) [4]. Those ericoid mycorrhizal fungi which produce little or no external fibrils on their hyphal walls have poor mycorrhiza-forming abilities [6]. This fibril production and wall adhesion are not phenomena specific to host plants as they also occur in presence of non-host roots [7]. Cells of host plants do, however, appear to exert a specific controlling influence on the production of the fibrillar sheath by hyphae once the fungus has entered a living host cell, since it cannot be detected around intracellular hyphal coils (Figs. 1b and c) but it is always evident in moribund host cells or in cells of non-hosts (Fig. 1d) [7].

Vesicular-arbuscular Endomycorrhizae

Such adhesion phenomena have not been observed in VA endomycorrhizal associations and cell-to-cell contact appears to be more fortuitous than in ericoid endomycorrhiza. Extraradical hyphae of VA fungi have thickened, stratified walls (Fig. 2a) which, according to cytochemical analyses, are composed of chitin, proteins and mainly alkali-insoluble polysaccharides [8, 9,10]. When VA fungi infect host root tissues, the hyphal walls thin out (Table 1) (Figs. 2b, c and d), but in hyphae forming coils in the outer cell layers of the root the ultrastructural organization of the fungal wall does not change greatly. Deeper in the inner parenchymal layers, where the intracellular haustoria (arbuscules) develop, walls undergo marked modifications which lead to a simplification of their structure and composition. The thin fungal walls become amorphous in structure, contain proteins, N-acetylglucosamine and mainly alkali-soluble polysaccharides but chitin chains cannot be detected in them [9,10,11].

Table 1. Fungal wall thickness (μm) during various stages of VA endomycorrhiza development.

	VA endomycorrhizal association:	
	Gigaspora margarita/ *Trifolium pratense* [12]	*Glomus fasciculatum*/ *Vitis vinifera* [9]
Extraradical hyphae	0.25 - 0.48	0.30 - 0.50
Coiled hyphae (outer cortex)	0.05 - 0.15	0.16 - 0.30
Intercellular hyphae	0.06 - 0.10	0.12
Arbuscular branches	0.01 - 0.03	0.05

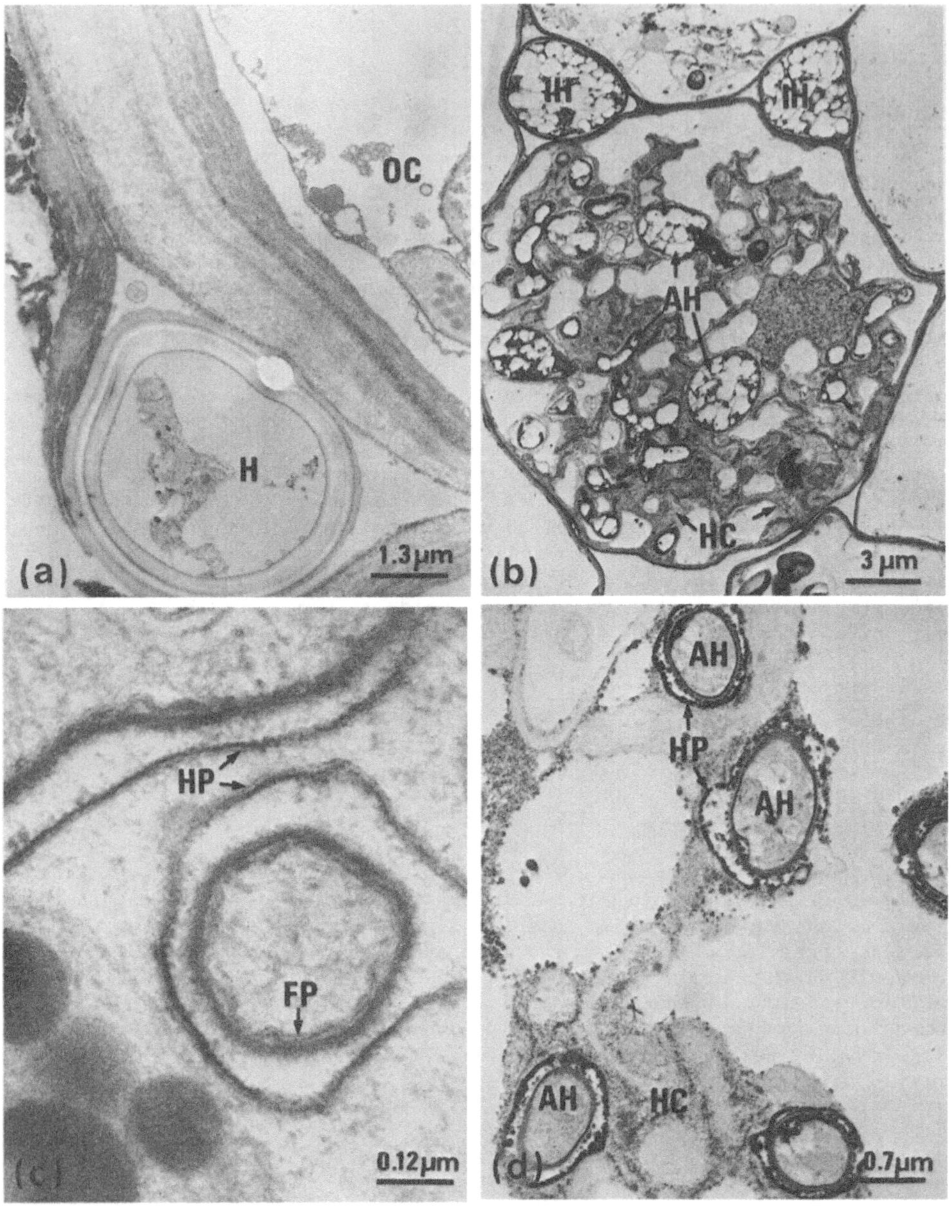

Figure 2. Electron micrographs of VA endomycorrhizal associations. (a) A thick-walled extraradical hypha (H) penetrates the outer cell (OC) layer of a host root (*Glomus intraradices*/gentian). (b) Thin-walled hyphae develop intercellularly (IH) and intracellularly (AH) ; host cytoplasm (HC) surrounds the hyphae (*Gigaspora margarita*/clover) (courtesy of F. Pons). (c) Host (HP) and fungal plasmalemmae (FP) have a three-layered structure (*G. mosseae*/soybean) (phosphotungstic acid staining ; courtesy of J. Dexheimer). (d) ATPase is localized (black precipitate) along the host plasmalemma around intracellular hyphae (*Glomus spp.*/onion).

Such alterations in the macromolecular organization of the fungal wall may lead to an increase in its plasticity, as has been suggested for apical walls of growing fungi where crystalline chitin is also absent [13]. These observations seem to indicate that the host cell somehow influences regulation of the biosynthesis of certain fungal wall components, as in ericoid endomycorrhiza, and the resulting wall modifications may in turn be responsible for the changes in fungal morphogenesis (arbuscule formation), as has been proposed for other fungi [14].

INTRACELLULAR INTERACTIONS AND HOST CELL RESPONSE

In endomycorrhizal associations, the infecting hypha arises from a more or less well-defined appressorium and either penetrates directly the outermost root cells or root hairs, or enters the host tissues via the intercellular spaces. In ericaceous roots, which have a very simple anatomy, the fungus forms hyphal coils in each living cell that it infects [4,7,15, 16], whereas in VA mycorrhizae intense intracellular fungal development leading to arbuscule formation only occurs in cells of the inner root cortex [9,17,18,19,20,21]. These haustorial structures have received considerable attention since they are generally assumed to be involved in nutrient transfer between fungus and host.

In spite of the taxonomical and morphological differences between ericoid and VA endomycorrhizae, the sequence of cytological events occurring in the host cells where fungal haustoria develop is quite similar. When hyphae enter host cells they penetrate the cell wall, probably by a combination of enzymic processes and physical pressure, and push back the host plasmalemma which elongates to surround closely each hypha or branch of the developing haustorium (Figs. 1b, c and 2b, c, d). This extension of the host plasmalemma cannot result from mere stretching of the existing membrane and de novo synthesis must occur to accommodate the important increase in its surface area [22,23]. The host cell responds to the fungal infection by laying down material in the interfacial matrix between its plasmalemma and the wall of the invading hypha. This material is similar to and continuous with the inner cell wall [4,24] and is thought to represent modified primary host wall. Such wall building activity by an extending plasmalemma is comparable to that found in elongating cells where wall formation is actively occurring. Furthermore, plasmalemma-bound neutral phosphatases, suggested to participate in cell wall building processes in plants [25,26,27,28] and found in differentiating cells showing active wall growth but absent once cells have differentiated, can be localized along the host membrane surrounding the haustorial hyphae [29,30,31]. As the intracellular hyphae develop, the quantity of wall material deposited in the host-fungus interface decreases, the interfacial matrix becomes electron transparent and often contains only scattered fibrils (Figs. 1c and 2c); the plasmalemma-bound neutral phosphatases, however, persist. Such observations seem to indicate that the host plasmalemma retains its ability to produce cell wall precursors around the actively growing hyphae but that the process of cell wall organization is somehow affected. This is directly related to the presence of living fungus since with hyphal senescence in VA endomycorrhizae, host material accumulates again around the fungal remains [19,24].

The increase in surface area of the host protoplast, due to extension of the plasmalemma around the fungal haustorium, is accompanied by a dramatic increase in the volume of host cytoplasm [22,23] which often forms a continuous mass in which the endomycorrhizal fungus is embedded (Figs. 1b and 2b). The host cytoplasm contains an elevated number of organelles and a well-developed endoplasmic reticulum whilst the cell vacuole is considerably reduced. These are all signs of a metabolically active condition recalling the situation in juvenile plant cells. During this intracellular biotrophic phase of endomycorrhizal associations, the hyphal contents of the fungi show typical cytological properties (Figs. 1b and 2b) with an active protoplasm rich in organelles and vacuoles which often contain dense granules, believed to contain polyphosphate [2,32,33].

Cytological studies of the fungal and host plasmalemmae indicate that both are active membranes. The fungal haustorium has a three-layered (Fig. 2c), PATAg-positive plasmalemma which in VA endomycorrhizae is also the site of both ATPase and glucose-6-phosphatase activities [30,34,35]. The newly synthesized host plasmalemma closely surrounding the haustorial hyphae frequently forms membranous configurations and vesicles (Fig. 1c) [34]; it has a typical tripartite structure and stains with phosphotungstic acid (Fig. 2c) and the PATAg test for polysaccharides (Fig. 1c) [24]. It possesses ATPase as well as neutral phosphatase activities (Fig. 2d) [29,30,35]; in VA endomycorrhizae this ATPase activity is specific to the plasmalemma around the fine hyphal branches of the arbuscule and little or no enzyme activity can be detected along the peripheral membrane of infected cells.

The establishment of an intracellular relationship which is compatible with the life of both the fungal and host cell must be of key importance for the physiology of endomycorrhizal associations. The formation, during the biotrophic phase, of a living interface where wall material of both host and fungus is reduced to a minimum and where energy generating enzyme systems exist for active transmembrane transport must greatly facilitate bidirectional nutrient exchange between the associates. Any factors affecting host-fungus relationships at this level will therefore have important repercussions on functional compatibility in endomycorrhizae.

The way in which this intracellular biotrophic phase degenerates differs between the two types of endomycorrhizae. In VA endomycorrhizal associations, fungal hyphae senesce when arbuscule development has reached its maximum in the cell but it is not yet clear whether this process is somehow provoked by the host plant. The fungal cytoplasm undergoes autolysis, hyphae collapse and their walls aggregate into clumps. Host membrane integrity is maintained during this process, neutral phosphatase activity persists but ATPase activity disappears, and with death of the fungus, the cytology of the host cell returns to that before infection. In ericoid endomycorrhizae, the intracellular hyphae usually outlive the host cytoplasm but it is not known whether the fungus enhances senescence of the host cell.

FUNGAL RELATIONSHIPS WITH NON-HOST PLANTS

Studies of interactions between endomycorrhizal fungi and roots of non-host plants have shown that cellular compatibility is host linked. VA fungi can penetrate non-host roots but infection is usually restricted to intercellular hyphae and no intracellular development into arbuscules occurs [36,37]. Furthermore, a recent ultrastructural study [38] indicates that the fungi only develop in tissues where cells are dead and that hyphae in contact with living non-host cells become vacuolate and senesce. Ericoid endomycorrhizal fungi can, on the contrary, penetrate the living cells of roots of non-host plants. The fungus extensively colonizes the root tissues, including the central cylinder, but does not form hyphal coils in the non-host cells. The plant plasmalemma does not extend around the invading hyphae and there is no increase in the cytoplasmic content of the cells in response to the fungal invasion. The fungus often seems to rupture the plant membranes and a necrotrophic relationship develops at the cellular level with rapid degeneration of the cell contents of the non-host plant (Fig. 1d) [7].

The mechanisms underlying this cellular incompatibility in non-host plants are as obscure as those which determine host-fungus compatibility in the endomycorrhizal associations. Glenn et al. [38] have suggested that the lack of penetration by VA fungi into living non-host cells may be due to the absence of necessary (nutritional) stimuli rather than to the production of an inhibitory factor by the cell, but this has yet to be proven. In ericoid endomycorrhizae, it is evident that the host cell exerts an influence on fungal activity and it has been proposed that the host plasmalemma plays a decisive role in the specificity and the establishment of a compatible relationship in this type of endomycorrhiza (7].

CONCLUSION

When appropriate endomycorrhizal associates come into close contact, they develop a structurally complex and functionally compatible relationship. This must depend on a series of recognition processes and interactions, some of which are no doubt related to detection of a nutrient source by the fungal partner whilst others probably involve molecular events occurring on the walls or membranes of the plant or fungus. Furthermore, although fungitoxic isoflavonoid phytoalexins can accumulate in endomycorrhizal roots [39], the amounts produced are always low which is also consistent with the existence of compatibility between the associates and with extensive development of the fungus within the host tissues. In conclusion, endomycorrhizal associations could represent an interesting model for studying the cellular and molecular basis of compatibility and recognition of non-self between fungi and plants.

REFERENCES

1. Gianinazzi-Pearson, V. and Gianinazzi, S., 1983, The physiology of vesicular-arbuscular mycorrhizal roots, Plant and Soil, 71: 197-209.
2. Read, D.J., 1983, The biology of mycorrhiza in the Ericales, Can. J. Bot., 61: 958-1004
3. Gianinazzi-Pearson, V., 1984, Host-fungus specificity, recognition and compatibility in mycorrhizae, in: "Plant Gene Research. Genes Involved in Microbe-Plant Interactions", D.P.S. Verma and T. Hohn, eds., pp. 225-253, Springer-Verlag, Wien and New York.
4. Bonfante-Fasolo, P. and Gianinazzi-Pearson, V., 1982, Ultrastructural aspects of endomycorrhiza in the Ericaceae. III. Morphology of the dissociated symbionts and modifications occurring during their reassociation in axenic culture, The New Phytologist, 91: 691-704.
5. Perrotto, S. and Bonfante-Fasolo, P., 1986, Lectin binding surfaces of ericoid mycorrhizal fungi, in: "Physiological and Genetical Aspects of Mycorrhizae", V. Gianinazzi-Pearson and S. Gianinazzi, eds., in press.
6. Gianinazzi-Pearson, V. and Bonfante-Fasolo, P., 1986, Variability in wall structure and mycorrhizal behaviour of ericoid fungal isolates, in: "Physiological and Genetical Aspects of Mycorrhizae", V. Gianinazzi-Pearson and S. Gianinazzi, eds., in press.
7. Bonfante-Fasolo, P., Gianinazzi-Pearson, V. and Martinengo, L., 1984, Ultrastructural aspects of endomycorrhiza in the Ericaceae. IV. Comparison of infection by *Pezizella ericae* in host and non-host plants, The New Phytologist, 98: 329-333.
8. Bonfante-Fasolo, P., 1984, I funghi micorrizici vesiculo-arbuscolari nel suolo : aspetti morfologici della fase extraradicale, Giornale Botanico Italiano, 118: 53-61.
9. Bonfante-Fasolo, P. and Grippiolo, R., 1982, Ultrastructural and cytochemical changes in the wall of a vesicular-arbuscular mycorrhizal fungus during symbiosis, Can. J. Bot., 60: 2303-2312.
10. Grippioli, R., Bonfante-Fasolo, P. and Testa, B., 1985, Localizzazione ultrastrutturale della chitina per mezzo di un complesso oro colloidale-chitinasi, Microscopia elettronica, 6: A28.

11. Bonfante-Fasolo, P., 1982, Cell wall architectures in a mycorrhizal association as revealed by cryoultramicrotomy, Protoplasma, 111: 113-120.
12. Pons, F., 1984, L'endomycorhization VA in vitro de quelques espèces herbacées et ligneuses : aspects ultrastructuraux de l'association, 3ème Cycle Thesis, Dijon University, France, 116 pp.
13. Vermeulen, L.A. and Wessels, J.G.H., 1984, Ultrastructural differences between wall apices of growing and non-growing hyphae of Schizophyllum commune, Protoplasma, 120: 123-131.
14. Farkas, V., 1979, Biosynthesis of cell walls of fungi, Microbiol. Rev. 43: 117-144.
15. Bonfante-Fasolo, P. and Gianinazzi-Pearson, V., 1979, Ultrastructural aspects of endomycorrhiza in the Ericaceae. I. Naturally infected hair roots of Calluna vulgaris L. Hull, The New Phytologist, 83: 739-744.
16. Duddridge, J.A. and Read, D.J., 1982, An ultrastructural analysis of the development of mycorrhizas in Rhododendron ponticum, Can. J. Bot., 60: 2345-2356.
17. Bonfante-Fasolo, P. and Fontana, A., 1985, VAM fungi in Gingko biloba roots:their interactions at the cellular level, Symbiosis 1: 53-67.
18. Carling, D.E. and Brown, M.F., 1982, Anatomy and physiology of vesicular-arbuscular and nonmycorrhizal roots, Phytopathology, 72: 1108-1114.
19. Gianinazzi-Pearson, V., Morandi, D., Dexheimer, J. and Gianinazzi, S., 1981, Ultrastructural and ultracytochemical features of a Glomus tenius mycorrhiza, The New Phytologist, 88: 633-639.
20. Holley, J.D. and Peterson, R.L., 1979, Development of a vesicular-arbuscular mycorrhizae in bean root, Can. J. Bot., 57: 1960-1978.
21. Kinden, D.A. and Brown, M.F., 1975, Electron microscopy of vesicular-arbuscular mycorrhizae of yellow poplar. I.. Characterisation of endophytic structures by scanning electron microscopy, Can. J. Microbiol., 21: 989-993.
22. Cox, G. and Tinker, P.B. 1976, Translocation and transfer of nutrients in vesicular-arbuscular mycorrhizas. I. The arbuscule and phosphorus transfer : a quantitative ultrastructural study, The New Phytologist, 77: 371-378.
23. Toth, R. and Miller, R.M., 1984, Dynamics of arbuscule development and degeneration in a Zea mays mycorrhiza, Amer. J. Bot., 71: 449-460.
24. Bonfante-Fasolo, P., Dexheimer, J., Gianinazzi, S., Gianinazzi-Pearson, V. and Scannerini, S., 1981, Cytochemical modifications in the host-fungus interface during intracellular interactions in vesicular-arbuscular mycorrhizae, Plant Sci. Lett., 22: 13-21.
25. Goff, C.W., 1973, Localisation of nucleoside diphosphatase in the onion root tip, Protoplasma, 78: 397-416.
26. Joshi, P.A., Stewart, J. McD. and Graham, E.T., 1985, Localization of β-glycerophosphatase activity in cotton fiber during differentiation, Protoplasma, 125: 75-85.
27. Maruyama, K., 1974, Localization of polysaccharides and phosphatases in the Golgi apparatus of Tradescantia pollen, Cytologia, 39: 767-776.
28. Zerban, H. and Werz, G., 1975, Localization of nucleoside diphosphatases and thiamine pyrophosphatase at various stages of the life cycle of the green algae Acetabularia cliftonii and Acetabularia mediterranea, Cytologia, 12: 13-17.
29. Gianinazzi-Pearson, V., Dexheimer, J., Gianinazzi, S. and Marx, C., 1983, Role of the host-arbuscule interface in the VA mycorrhizal symbiosis : ultracytological studies of processes involved in phosphate and carbohydrate exchange, Plant and Soil, 71: 211-215.
30. Gianinazzi-Pearson, V., Dexheimer, J., Gianinazzi, S. and Jeanmaire, C., 1984, Plasmalemma structure and function in endomycorrhizal symbioses, Z. f. Pflanzenphysiologie, 114: 201-205.

31. Jeanmaire, C., Dexheimer, J., Marx, C., Gianinazzi, S. and Gianinazzi-Pearson, V., 1985, Effect of vesicular-arbuscular mycorrhiza infection on the distribution of neutral phosphatase activities in root cortical cells, J. Plant Physiol., 119: 285-293.
32. Cox, G., Moran, K.J., Sanders, F., Nockolds, C. and Tinker, P.B., 1980, Translocation and transfer of nutrients in vesicular-arbuscular mycorrhizas. III. Polyphosphate granules and phosphorus translocation, The New Phytologist, 84: 649-659.
33. Strullu, D.G., Gourret, J.P., Garrec, J.P. and Fourcy, A., 1981, Ultrastructure and electron-probe microanalysis of the metachromatic vacuolar granules occurring in Taxus mycorrhizas, The New Phytologist, 87: 537-545.
34. Dexheimer, J., Gianinazzi-Pearson, V. and Gianinazzi, S., 1982, Acquisitions récentes sur la physiologie des mycorhizes VA au niveau cellulaire, Les Colloques de l'INRA, 13: 61-73.
35. Marx, C., Dexheimer, J., Gianinazzi-Pearson, V. and Gianinazzi, S., 1982, Enzymatic studies on the metabolism of vesicular-arbuscular mycorrhizas. IV. Ultracytoenzymological evidence (ATPase) for active transfer processes in the host-arbuscule interface, The New Phytologist, 90: 37-43.
36. Hirrel, M.C., Mehravaran, H. and Gerdemann, J.W., 1978, Vesicular-arbuscular mycorrhizae in the Chenopodiaceae and Cruciferae : do they occur?, Can. J. Bot., 56: 2813-2817.
37. Ocampo, J.A., Martin, J. and Hayman, D.S., 1980, Influence of plant interactions on vesicular-arbuscular mycorrhizal infections. I. Host and non-host plants grown together, The New Phytologist, 84: 27-35.
38. Glenn, M.G., Chew, F.S. and Williams, P.H., 1985, Hyphal penetration of Brassica (Cruciferae) roots by a vesicular-arbuscular mycorrhizal fungus, The New Phytologist, 99: 463-472.
39. Morandi, D., Bailey, J.A. and Gianinazzi-Pearson, V., 1984, Isoflavonoid accumulation on soybean roots infected with vesicular-arbuscular mycorrhizal fungi, Physiol. Plant Pathol., 24: 357-364.

CELLULAR INTERACTIONS BETWEEN PHASEOLUS VULGARIS AND THE HEMIBIOTROPHIC FUNGUS COLLETOTRICHUM LINDEMUTHIANUM

R.J. O'Connell and J.A. Bailey

Department of Agricultural Sciences
University of Bristol
Long Ashton Research Station
Long Ashton
Bristol BS18 9AF, UK

INTRODUCTION

The modern techniques of molecular biology are expected to provide fresh insights into the mechanisms which regulate plant-pathogen interactions, especially those controlling fungal pathogenicity and specific race-cultivar resistance and susceptibility. However, for the proper interpretation of the resulting biochemical and genetic data, a detailed understanding of the cytology of such interactions will be essential. One plant-pathogen combination which has become increasingly appropriate for such research is that between Phaseolus vulgaris and Collectotrichum lindemuthianum, a hemibiotrophic fungus.

C. lindemuthianum was the first plant pathogen to be shown to exist as various physiological races exhibiting distinct virulence and avirulence on different cultivars of a host species [1]. This specificity can be obtained routinely and reproducibly by inoculating uninjured seedling hypocotyls with conidia. The present paper outlines the morphological processes which distinguish resistant and susceptible reactions, and indicates how cytochemistry can reveal further information about the nature of successful and failed infections. The advantages of using the well-documented interaction between P. vulgaris and C. lindemuthianum in studies of the expression of genes associated with synthesis of phytoalexins are discussed in this volume by Ryder [see also 2].

SPORE GERMINATION AND PENETRATION

Conidia of C. lindemuthianum germinate on the surface of bean hypocotyls to produce germ-tubes, which are generally very short and differentiate to form appressoria close to the spore [3, see also Staples in this volume]. Appressoria adhere firmly to the plant cuticle and effect direct penetration of epidermal cells by means of narrow infection pegs (Figs. 1 and 2). There is no change in the ultrastructure of the cuticle to suggest the involvement of either enzymic dissolution or mechanical pressure in cuticle penetration (Fig. 1), although other lines of evidence indicate that both may be involved [4,5]. Subsequent penetration of the epidermal cell wall appears to be mediated by enzymes, since a narrow zone of the cell wall adjacent to the

NATO ASI Series, Vol. H1
Biology and Molecular Biology of Plant-Pathogen Interactions
Edited by J. Bailey

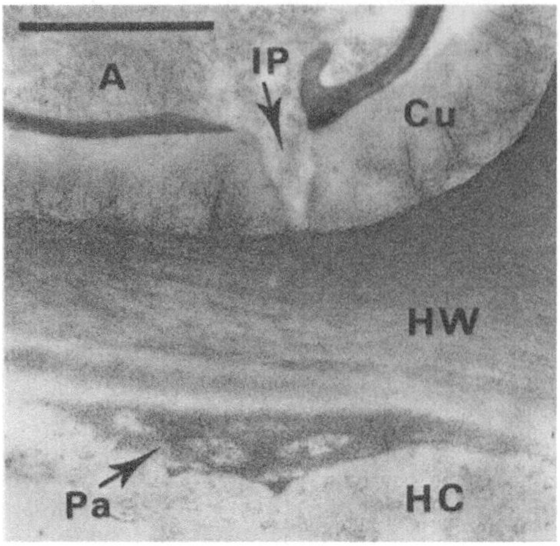

Figure 1. Penetration of cuticle (Cu) by infection peg (IP). Serial sectioning showed host wall (HW) not penetrated. Note papilla (Pa) within epidermal cell beneath site of penetration.
A, appressorium; HC, host cytoplasm. Bar = 1 µm.

infection peg shows enhanced reactivity with the periodic acid-chromic acid-phosphotungstic acid (PACP) [6] and periodic acid-silver methenamine (PA-AgMe) [7] staining procedures (Figs. 2a and b), indicating modification of the wall polysaccharides. As in infections of plants by obligate biotrophic fungi [8], degradation of the wall is highly localized around the infection peg.

The processes of germination, differentiation of appressoria and penetration are similar on susceptible and resistant bean cultivars. Race-specific responses of host cells leading to successful or failed pathogenesis appear to occur after penetration of epidermal cells.

SUSCEPTIBLE INTERACTIONS

Establishment of Infection and Colonization of Tissue

In susceptible tissue, penetrated epidermal cells remain alive initially, and infection pegs enlarge to form globose intracellular infection vesicles (Fig. 3). Many other fungi which penetrate epidermal cells directly, e.g. some Oomycete and rust fungi [9,10,11,12] also produce a swollen hypha or vesicle within the cell lumen. Although it is not clear whether the differentiation of vesicles is a consequence or a cause of host cell survival, infection vesicles may be required to establish a functional biotrophic relationship with host cells. As noted by Aist [13], the possible physiological significance of these specialized infection structures merits more attention.

Intracellular primary hyphae emerge from infection vesicles and colonize adjacent epidermal and cortical cells. Primary hyphae are greatly constricted where they penetrate from cell to cell, and as with infection pegs, penetration appears to involve localized enzymic degradation of the cell wall [3,14].

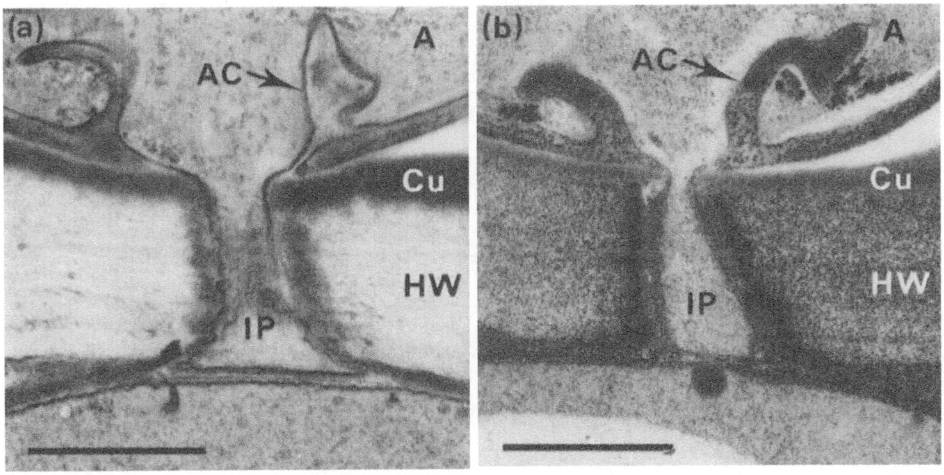

Figure 2. Host wall (HW) adjacent to infection peg (IP) shows enhanced staining with:
a) periodic acid-chromic acid-phosphotungstic acid
b) periodic acid-silver methenamine
A, appressorium; AC, appressorial cone; Cu, cuticle. Bar = 1 μm.

For 24-48 h after penetration by vesicles and primary hyphae, host cells retain normal ultrastructure (Fig. 3) and function, as indicated by their ability to plasmolyse. Later, infected cells show evidence of osmotic disturbance, such as dilation of the endoplasmic reticulum and an increase in volume of cytoplasm. Infected cells also lose the ability to plasmolyse normally in hypertonic media, and are unable to exclude the permeability tracer tannic acid [3]. These results suggest that the semi-permeable properties of the host plasmalemma are altered before the plasmalemma and other cellular membranes lose their structural integrity. Subsequently the host cytoplasm gradually degenerates, and about 60 h after penetration only membranous debris remains [3].

As each host cell is colonized by primary hyphae, the sequence of a transient biotrophic phase, followed by gradual senescence and death, is repeated. Thus, recently colonized cells at the advancing margin of the primary mycelium appear intact, while earlier infected cells near the centre are severely damaged or dead [3]. Although many host cells die during development of the primary mycelium they do not become brown and isoflavonoid phytoalexins do not accumulate [15]. Furthermore, the enzymes involved in phytoalexin biosynthesis (and the mRNAs that encode them) show no increase in activity at this stage [2].

The Plant-Pathogen Interface

During the intracellular development of C. lindemuthianum, invagination of living host protoplasts around infection vesicles and primary hyphae produces an interface of large surface area. This interface comprises the fungal cell wall, the host plasmalemma, and an intervening matrix. In previous studies, the composition and biological activity of extracts from C. lindemuthianum cell walls have received most attention [16]. Here we consider the matrix and host plasmalemma.

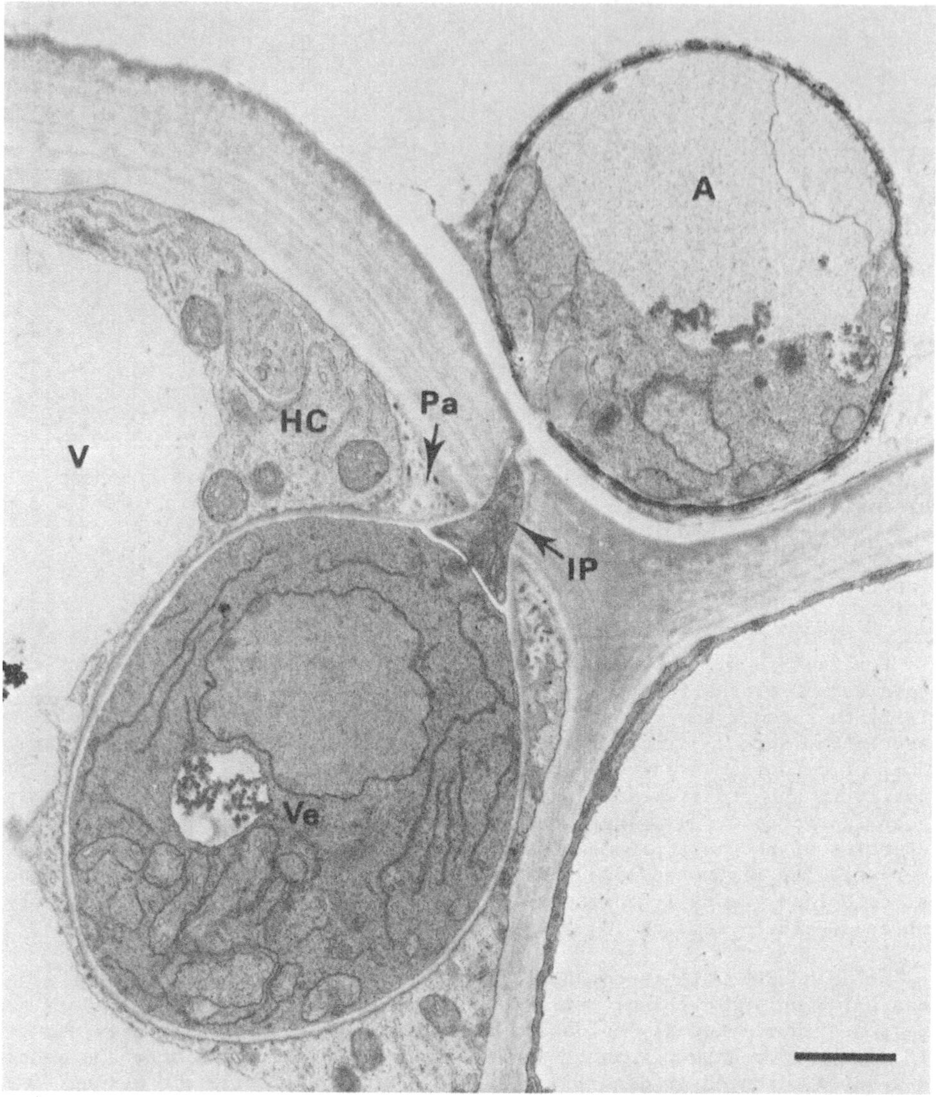

Figure 3. Infection vesicle (Ve) within epidermal cell. Host cytoplasm (HC) shows normal ultrastructural. Note papilla (Pa) surrounding neck of vesicle.

A, appressorium; IP, infection peg; V, vacuole. Bar = 1 μm.

(a) Interfacial Matrix. Infection vesicles and primary hyphae are surrounded by an amorphous layer of material of variable thickness [3]. This matrix reacts positively when stained with PA-AgMe and thus consists at least partly of polysaccharides or glycoconjugates containing vicinal hydroxyl groups. A polysaccharide matrix also occurs at the interface between plant cells and the intracellular hyphae of many other biotrophic fungi, e.g. haustoria of Oomycetes [17]; rusts [18]; powdery mildews [19] and endomycorrhizae [Gianinazzi-Pearson, this volume]. Micro-autoradiographic studies suggest that extrahaustorial matrices are solely of host origin [20,21]. However, both host and fungus contribute to the matrix surrounding hyphae of *C. lindemuthianum*. Products of host Golgi activity appear to be released into the matrix by means of exocytosis, while immunocytochemistry has provided direct evidence that fungal proteins occur in the matrix [22].

The function of the matrix can only be speculated upon. The presence of a similar structure around haustoria suggests a role in nutrient uptake. Alternatively, the matrix may be important to the survival of penetrated cells, perhaps by immobilizing toxic fungal metabolites [23], or by interfering with cell-cell recognition. Insight into the possible function of the matrix will require more data regarding its composition, which may come from the use of labelled antibody and lectin probes [see Vian, this volume].

(b) Invaginated host plasmalemma. The intracellular hyphae of *C. lindemuthianum* cause extensive invagination of the host plasmalemma. To accommodate this, new membrane material appears to be incorporated by fusion of host cytoplasmic vesicles with the plasmalemma.

The host plasmalemma invaginated around haustoria of some obligate biotrophic fungi has been shown to become highly modified. Freeze fracture analysis reveals changes in membrane ultrastructure [23]; the adenosine triphosphatase activity normally associated with the plant plasmalemma cannot be detected [17]; and the membrane is poorly stained by the PACP procedure [17,23]. In addition, the transition from the normal plasmalemma to the specialized extrahaustorial membrane is generally marked by a "neckband" structure, which binds the membrane to the wall of the haustorial neck [24]. In complete contrast, the host plasmalemma surrounding the vesicles and primary hyphae of *C. lindemuthianum* is, using the techniques described above, indistinguishable from that lining the host cell wall. Furthermore, there is no evidence that hyphae of *C. lindemuthianum* have neckbands [3].

Modification of the host plasmalemma and neckbands are believed to be important in determining the duration of biotrophy in infections by haustorial parasites [17]. The transient nature of the biotrophic relationship of *C. lindemuthianum* with host cells may thus be related to the lack of a specialized interface.

Lesion Formation

After a large primary mycelium is established, the morphology and mode of parasitism of *C. lindemuthianum* changes. Six to seven days after inoculation narrow secondary hyphae branch out from primary hyphae within dead host cells. Unlike the intracellular primary hyphae which cause minimal damage to host cell walls, secondary hyphae frequently develop intramurally [3]. The colonized cell walls swell, show reduced affinity for cytological stains, and the tissue becomes macerated. In addition, host protoplasts are completely disrupted [3]. These effects are all characteristic of pectolytic enzyme activity [25,26], and they extend several hundred microns ahead of the secondary hyphae. The product of this destructive necrotrophic phase of growth is a dark brown spreading lesion.

Isoflavonoid phytoalexins are produced by the host at this lesion stage, but they do not accumulate to high concentrations [15]. It seems likely that phytoalexin accumulation is prevented, and hence pathogen growth facilitated, because host cells are killed before inhibitory amounts of phytoalexin can form. The mechanisms contolling the abrupt change in mode of parasitism are not understood, but the switch coincides with increased secretion by the fungus of an endo-pectin lyase, an enzyme shown to degrade walls and to cause ion leakage when applied to bean tissue [27].

RESISTANT INTERACTIONS

Resistance of young French bean tissues to C. lindemuthianum typically involves localized death and browning of host cells, which is accompanied by accumulation of phytoalexins and restriction of pathogen development [28]. The spectrum of symptoms produced ranges from single dead cells containing minute infection hyphae to larger areas of necrosis (limited lesions) containing extensive primary mycelia. In this discussion we have distinguished between resistance involving the early death of epidermal cells without a detectable biotrophic phase, and resistance in which cell death follows an initial, albeit short, period of biotrophy.

Resistance Associated with Early Death of Epidermal Cells

In many resistant cultivars, single epidermal cells beneath appressoria die as early as 40-60 h after inoculation, as indicated by their granular cytoplasm and failure to plasmolyse. These cells quickly become darkly pigmented, and intracellular hyphae are often not visible by light microscopy [3,29]. However, electron microscopy indicates that all such cells are penetrated [3]. Most intracellular hyphae are dead and very small, suggesting that fungal growth is arrested soon after penetration. Five to ten per cent of intracellular hyphae are larger and have cytoplasm of normal ultrastructure, suggesting that some hyphae remain alive [see also 28]. The consistent feature of these responses is that the pathogen is always restricted within a single dead cell. There appears to be no period of biotrophy: infection vesicles are not produced and the small intracellular hyphae which do form have never been found in contact with living host protoplasts. This type of resistance, which seems dependent on the localized accumulation of phytoalexins, has been referred to as an example of hypersensitivity based on host cell incompatibility [30].

Resistance Associated with Delayed Death of Infected Cells

In other resistant cultivars the initial events resemble the susceptible interaction: epidermal cells survive penetration, and typical infection vesicles develop. However, after a biotrophic phase of varying duration, the infected cells die rapidly and become brown. Further fungal growth is prevented [31,32,33, O'Connell, unpublished data], The extent of symptom and pathogen development appears to depend on when the infected cells die, turn brown and accumulate phytoalexins. In some race-cultivar combinations, infection vesicles or vesicles with short primary hyphae are confined to single dead epidermal cells, producing a symptom that resembles the hypersensitive response described above. In other race-cultivar combinations, primary hyphae may colonize two, three or a small group of epidermal and cortical cells before resistance is expressed. In yet others, extensive primary mycelia are produced but secondary hyphae are generally absent, and the fungus is confined to a limited lesion. In all these resistant interactions, some surrounding uninfected cells also die and become brown.

Resistance can also be induced by modifying host physiology or by environmental factors. For example, the presence of cotyledons and/or leaves confers resistance to hypocotyls which without these organs are extremely susceptible. Again, resistance appears to involve premature death and browning of the infected cells [34].

It is generally assumed that race-specific resistance of *P. vulgaris* to *C. lindemuthianum* is based on a gene-for-gene interaction. However, this brief description of the variety of different resistant phenotypes should serve to illustrate the difficulty of relating symptoms to an underlying genetic mechanism. One possible explanation of the yes:no relationship required by a gene-for-gene interaction lies in a model based on the initial death or survival of infected epidermal cells [30].

PAPILLAE

In both resistant and susceptible cultivars, bean epidermal cells respond to initial infection by *C. lindemuthianum* by forming wall appositions (papillae) of varying size at the site of penetration [3,14,29,33]. Even before penetration of the epidermal cell wall is complete, a thin homogeneous layer of finely granular electron-opaque material is deposited between the host plasmalemma and the wall (Fig. 1). This suggests that papilla formation may be initiated by diffusible fungal metabolites or by products of cuticle and/or wall degradation. The necks of infection vesicles are, in addition, surrounded by one or more layers of electron-lucent material with irregular electron-opaque inclusions (Fig. 3). It is not clear if these collar-like structures are deposited before or after vesicle formation. Papillae do not normally prevent infection, but a small proportion (5 to 15%) of intracellular hyphae become completely encased by papillae, and fail to develop further (Fig. 4a). Encasement is not a race specific response. It occurs with similar frequencies in all interactions, whether resistant or susceptible [3,29]. It is interesting that in resistant race-cultivar combinations that normally give the rapid hypersensitive response, epidermal cells containing encased hyphae remain alive, as indicated by their ability to plasmolyse normally (Fig. 4b). This suggests that papillae may interfere with the mechanism by which the hyphae trigger cell death, perhaps by isolating the host plasmalemma from factors on the surface, or diffusing ahead of, the fungus.

SUMMARY

Cytological and biochemical studies of the *Phaseolus*:*Colletotrichum* interaction have resolved many of the temporal relationships between the development of the pathogen and responses of the host. It is clear that the expression of both resistance and susceptibility involves a series of precisely regulated processes. These include fungal differentiation on the plant surface; penetration of the cuticle and epidermal cell wall; formation of papillae; survival or death of the initially infected epidermal cell; establishment of biotrophy in each newly colonized cell; necrotrophic growth involving extensive cell wall dissolution; and the rapid accumulation of phytoalexins associated with cessation of pathogen development.

Several important questions arise from these studies, especially with regard to activities of the pathogen. What regulates differentiation of the pathogen on the plant surface? What are the mechanisms which permit cells to be invaded, especially when they survive infection? What triggers cell death? Why is biotrophy transient, and what mechanisms are responsible for the switch to necrotrophic growth? These questions arise from traditional

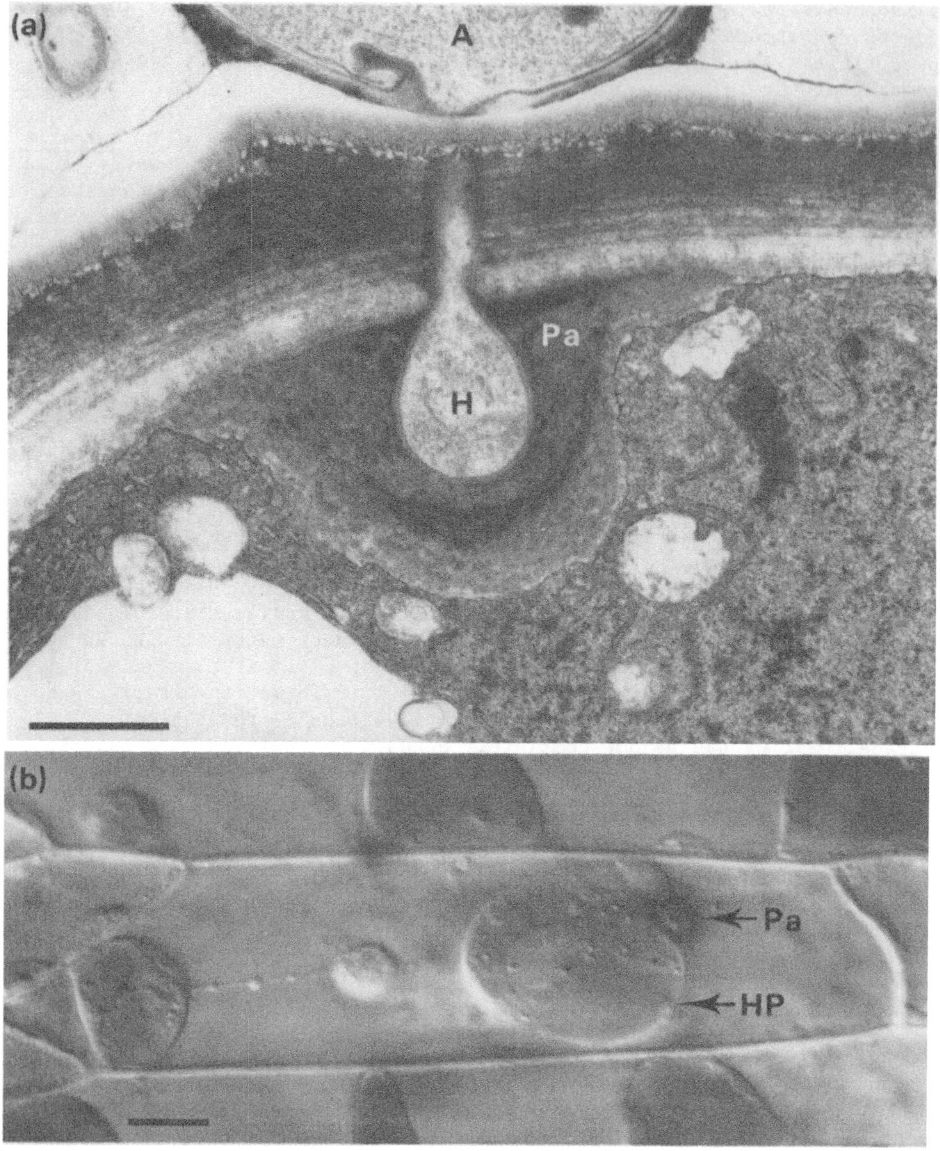

Figure 4.

a) Intracellular hypha (H) encased by multi-layered papilla (Pa) within epidermal cell. Tissue mordanted with tannic acid.

A, appressorium. Bar = 1 μm.

b) Epidermal cell containing papilla (Pa) plasmolysed normally in 0.8M KNO_3.

HP, host plasmalemma. Bar = 10 μm.

cytological and biochemical studies. One must anticipate that some of the answers will be provided by the advances in fungal genetics and cell biology described in this volume.

REFERENCES

1. Barrus, M.F., 1911, Variation in varieties of bean in their susceptibility to anthracnose, Phytopathology, 1: 190-195.
2. Lamb, C.J., Bell, J.N., Cramer, C.L., Dildine, S.L., Grand, C., Hedrick, S.A., Ryder, T.B. and Showalter, A.M., 1986, Molecular response of plants to infection, in: "Biotechnology for Solving Agricultural Problems", J.B. St. John, ed., Beltsville Agricultural Research Center, Symposium X, in press.
3. O'Connell, R.J., Bailey, J.A. and Richmond, D.V., 1985, Cytology and physiology of infection of Phaseolus vulgaris by Colletotrichum lindemuthianum, Physiol. Plant Pathol., 27: 75-98.
4. Dickman, M.B., Patil, S.B. and Kolattukudy, P.E., 1982, Purification, characterization and role in infection of an extracellular cutinolytic enzyme from Colletotrichum gloeosporioides Penz. on Carica papaya L., Physiol. Plant Pathol., 20: 333-347.
5. Wolkow, P.M., Sisler, H.D. and Vigil, E.L., 1983, Effect of inhibitors of melanin biosynthesis on structure and function of appressoria of Colletotrichum lindemuthianum, Physiol. Plant Pathol., 22: 55-71.
6. Roland, J.C., Lembi, C.A. and Morré, D.J., 1972, Phosphotungstic acid-chromic acid as a selective electron-dense stain for plasmamembranes of plant cells, Stain Technology, 47: 195-200.
7. Thiéry, J.P., 1967, Mise en évidence des polysaccharides sur coupes fines en microscopie électronique, Journale de Microscopie, 6: 987-1018.
8. Cooper, R.M., 1983, The mechanism and significance of enzymic degradation of host cell walls by parasites, in: "Biochemical Plant Pathology", J.A. Callow, ed., pp. 101-136, John Wiley and Sons, London.
9. Bonde, M.R., Bromfield, K.R. and Melching, J.S., 1982, Morphological development of Physopella zeae on corn, Phytopathology, 72: 1489-1491.
10. Coffey, M.D. and Wilson, U.E., 1983, Histology and cytology of infection and disease caused by Phytophthora, in: "Phytophthora: its Biology, Taxonomy, Ecology and Pathology", D.C. Erwin, S. Bartnicki-Garcia and P.H. Tsao, eds., pp. 289-301, The American Phytopathological Society, St. Paul, Minnesota.
11. Gold, R.E. and Mendgen, K., 1984, Cytology of basidiospore germination, penetration, and early colonization of Phaseolus vulgaris by Uromyces appendiculatus var. appendiculatus, Can. J. Bot., 62: 1989-2002.
12. Tommerup, I.C., 1981, Cytology and genetics of downy mildews, in: "The Downy Mildews", D.M. Spencer, ed., pp. 121-142, Academic Press, London.
13. Aist, J.R., 1981, Development of parasitic conidial fungi in plants, in: "Biology of Conidial Fungi", G.T. Cole and B. Kendrick, eds., pp. 75-110, Academic Press, New York.
14. Mercer, P.C., Wood, R.K.S. and Greenwood, A.D., 1975, Ultrastructure of the parasitism of Phaseolus vulgaris by Colletotrichum lindemuthianum, Physiol. Plant Pathol., 5: 203-214.
15. Bailey, J.A. and Deverall, B.J., 1971, Formation and activity of phaseollin in the interaction between bean hypocotyls (Phaseolus vulgaris) and physiological races of Colletotrichum lindemuthianum, Physiol. Plant Pathol., 1: 435-449.
16. Anderson, A.J., 1980, Studies on the structure and elicitor activity of fungal glucans, Can. J. Bot., 58: 2343-2348.
17. Woods, A.M. and Gay, J.L., 1983, Evidence for a neckband delimiting structural and physiological regions of the host plasmamembrane associated with haustoria of Albugo candida, Physiol. Plant Pathol., 23: 73-88.

18. Chong, J., Harder, D.E. and Rohringer, R., 1981, Ontogeny of mono- and dikaryotic rust haustoria : cytochemical and ultrastructural studies, Phytopathology, 71: 975-983.
19. Bushnell, W.R. and Gay, J.L., 1978, Accumulation of solutes in relation to the structure and function of haustoria in powdery mildews, in: "The Powdery Mildews", D.M. Spencer, ed., pp. 183-235, Academic Press, London.
20. Mendgen, K. and Heitefuss, R., 1975, Micro-autoradiographic studies on host-parasite interactions. I. The infection of Phaseolus vulgaris with tritium-labelled uredospores of Uromyces phaseoli, Arch. Microbiol., 105: 193-199.
21. Spencer-Phillips, P.T.N. and Gay, J.L., 1980, Electron microscope autoradiography of ^{14}C photosynthate distribution at the haustorium-host interface in powdery mildew of Pisum sativum, Protoplasma, 103: 131-154.
22. O'Connell, R.J., Bailey, J.A., Vose, I.R. and Lamb, C.J., 1986, Immunogold labelling of fungal antigens in cells of Phaseolus vulgaris infected by Colletotrichum lindemuthianum, Physiol. Mol. Plant Pathol., 28: 99-105.
23. Bracker, C.E. and Littlefield, L.J., 1973, Structural concepts of host-pathogen interfaces, in: "Fungal Pathogenicity and the Plant's Response", R.J.W. Byrde and C.V. Cutting, eds., pp. 159-313, Academic Press, London.
24. Coffey, M.D., 1983, Cytochemical specialization at the haustorial interface of a biotrophic fungal parasite, Albugo candida, Can. J. Bot., 61: 2004-2014.
25. Baker, C.J., Aist, J.R. and Bateman, D.F., 1980, Ultrastructural and biochemical effects of endopectate lyase on cell walls from cell suspension cultures of bean and rice, Can. J. Bot., 58: 867-880.
26. Bateman, D.F., 1976, Plant cell wall hydrolysis by pathogens, in: "Biochemical Aspects of Plant-Parasite Relationships", J. Friend and D.R Threlfall, eds., pp. 79-113, Academic Press, London.
27. Wijesundera, R.L.C., 1984, Cell wall degrading enzymes of Colletotrichum lindemuthianum : their role in the development of anthracnose of bean, Ph.D. Thesis, University of Bristol.
28. Bailey, J.A., 1982, Physiological and biochemical events associated with the expression of resistance to disease, in: "Active Defense Mechanisms in Plants", R.K.S. Wood, ed., pp. 39-65, Plenum Press, New York.
29. Skipp, R.A. and Deverall, B.J., 1972, Relationships between fungal growth and host changes visible by light microscopy during infection of bean hypocotyls (Phaseolus vulgaris) susceptible and resistant to physiological races of Colletotrichum lindemuthianum, Physiol. Plant Pathol., 2: 357-374.
30. Bailey, J.A., 1983, Biological perspectives of host-pathogen interactions, in: "The Dynamics of Host Defense", J.A. Bailey and B.J. Deverall, eds, pp. 1-32, Academic Press, London.
31. Elliston, J., Kuć, J. and Williams, E.B., 1976, A comparative study of the development of compatible, incompatible and induced incompatible interactions between Colletotrichum spp. and Phaseolus vulgaris, Phytopath. Z., 87: 289-303.
32. Mercer, P.C., Wood, R.K.S. and Greenwood, A.D., 1974, Resistance to anthracnose of French bean, Physiol. Plant Pathol., 4: 291-306.
33. Townsend, P.J., 1981, Aspects of host-parasite interactions, D.Phil. Thesis, University of Oxford.
34. Rowell, P.M. and Bailey, J.A., 1983, The influence of cotyledons, roots and leaves on the susceptibility of hypocotyls of bean (Phaseolus vulgaris) to compatible races of Colletotrichum lindemuthianum, Physiol. Plant Pathol., 23: 245-256.

ULTRASTRUCTURAL LOCALIZATION OF CARBOHYDRATES. RECENT DEVELOPMENTS IN CYTOCHEMISTRY AND AFFINITY METHODS

B. Vian

Laboratoire des Biomembranes et Surfaces Cellulaires
Végétales, Ecole Normale Supérieure
24 rue Lhomond
75231 Paris Cedex 05
France

INTRODUCTION

In plant cells carbohydrates represent one of the major components of the cell surface. Due to this peripheral situation carbohydrates are involved in many biological mechanisms, such as recognition phenomena, which are important in symbiosis and pathology. During the last 20 years, carbohydrate chemistry and biochemistry have expanded considerably. In particular, the determination of primary or covalent structure of many polysaccharides has been established and has revealed that their structural complexity is much more important than was generally thought [1].

This paper discusses the cytochemical tools capable of localizing and characterizing specific polysaccharide polymers both in situ and from isolated molecules. First, methods based on the detection of the chemical functions of polysaccharides will be outlined. Second, more recently developed methods based on affinity between macromolecules, e.g., lectins/sugar residues, enzyme/substrate, antigen/antibody, will be discussed.

CHEMICAL GROUPS OF CARBOHYDRATES CAN BE VISUALIZED BY MARKERS

It is known that a great variety of chemical functions occur along polysaccharide molecules (hydroxyl, carboxyl, carboxymethyl groups, etc...). Their visualization by means of cytochemical markers has been the object of many studies and now constitutes a series of more or less routine techniques [2].

Most functions occur on several classes of polysaccharides. For example, the vicinal-glycol functions are very abundant and easily detectable by the PATAg test (Fig. 1a). This method has allowed the cell wall components to be characterized from their sites of synthesis to their sites of delivery, i.e. in the wall itself. However the method does not distinguish between different classes of polysaccharides.

To enhance the specificity, a combination of the PATAg test and mild chemical or enzymic extractions is suitable. The advantage of such

NATO ASI Series, Vol. H1
Biology and Molecular Biology of Plant-Pathogen Interactions
Edited by J. Bailey

subtractive cytochemistry is twofold: the fine structure of the polysaccharide assemblies becomes evident (Fig. 1b) and at the same time the extracted components may be chemically analyzed with greater precision [3].

Other functions only present in one or several classes of carbohydrates can also be visualized. For example free carboxyl groups can be detected because of their capacity to bind with colloidal metal particles visible at the electron microscope level (Fig. 1c). When the markers are used at the level of the molecules themselves after their extraction and purification, the frequency and the type of distribution of the available sites along a single molecule can be assessed. For example, on Figure 2, one can compare the distribution of anionic groups on both pectin and hemicellulose molecules.

Generally speaking cytochemistry sensu stricto allows both a precise localization of the main structural polymers of cell walls and a reconstitution of their 3-dimensional organization. However limitations remain due to the fact that such methods do not distinguish between sequences which differ by only one or a few sugar building blocks, or by the occurrence of sugars in α or β configuration, or by the occurrence of sugars in pentose or hexose forms.

SUGAR RESIDUES OF CARBOHYDRATES CAN BE VISUALIZED BY LECTINS

In view of the increasing evidence of the structural complexity of carbohydrates and of the role of some oligosaccharide sequences in cell signalling, it appears necessary to find new probes for a more specific identification and characterization of these components. The lectins constitute one valuable tool. Lectins are proteins or glycoproteins of non-immune origin, in which specificity is directed to a single monosaccharide or to a sequence or series of monosaccharides.

Briefly, lectins can be used in two ways: a direct method, in which an electron dense label is conjugated to the lectin itself; an indirect method in which the lectin and its corresponding markers are used sequentially. For some time, lectins have been used mainly to detect glycoconjugates of cell surface membranes because of their poor penetration into cells and tissues. However many possibilities now exist to circumvent this problem, such as labelling cryosections [4,5], or using low temperature embedding resins such as Lowicryl K4M [6].

Although a lack of specificity is often claimed, lectins have been extensively used both at the light and the electron microscope level from various types of cell walls, as for example maize root slimes [7], and fungal cell walls [4,5,8,9]. For the particular study of higher plant cell walls, two facts may however limit the use of lectins as probes: on the one hand, lectins help to visualize only free sugars or terminal sugars of carbohydrates and do not provide a labelling of the backbone of complex polymers; on the other hand, no lectins are available which will bind to monosaccharides such as xylose or arabinose, both frequent components of wall polymers in higher plant cells.

COVALENT LINKAGES OF CARBOHYDRATES CAN BE VISUALIZED BY ENZYMES

Enzyme-gold cytochemistry is another way to investigate the cell wall macromolecules. This method is based on the specific affinity properties existing between an enzyme and its substrate. It was first proposed by Bendayan [10] for the localization of nucleic acid in animal cells, and has

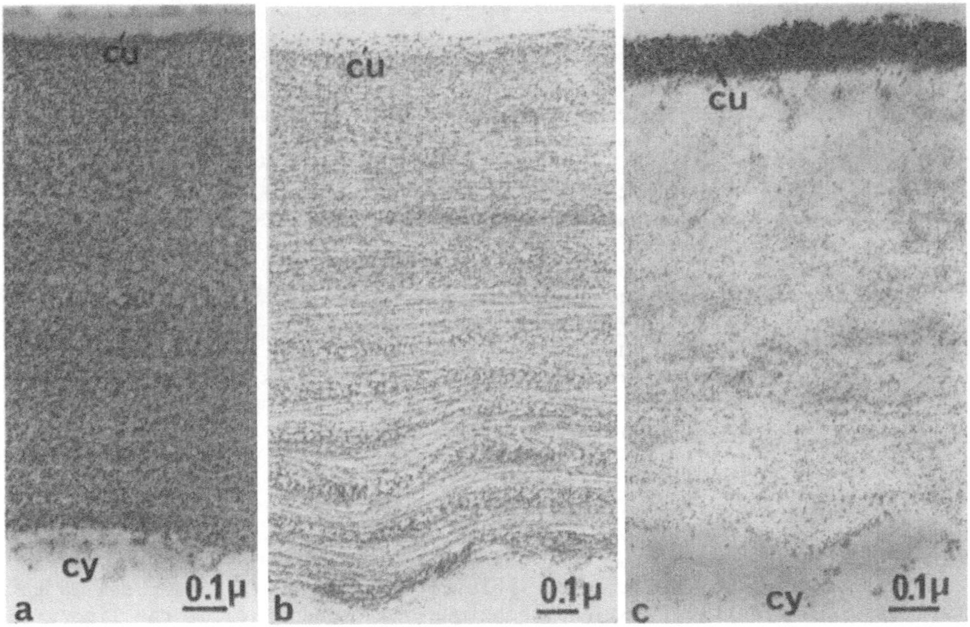

Figure 1. Detection of the chemical functions of polysaccharides. Epidermis of mung bean hypocotyl. a: PATAg test. Massive staining of the whole wall; b: PATAg test associated with methylamine extraction. Visualization of the polylamellate texture of the wall; c: detection of anionic groups of both cell wall and cuticle by cationized ferritin. cu: cuticle; cy: cytoplasm.

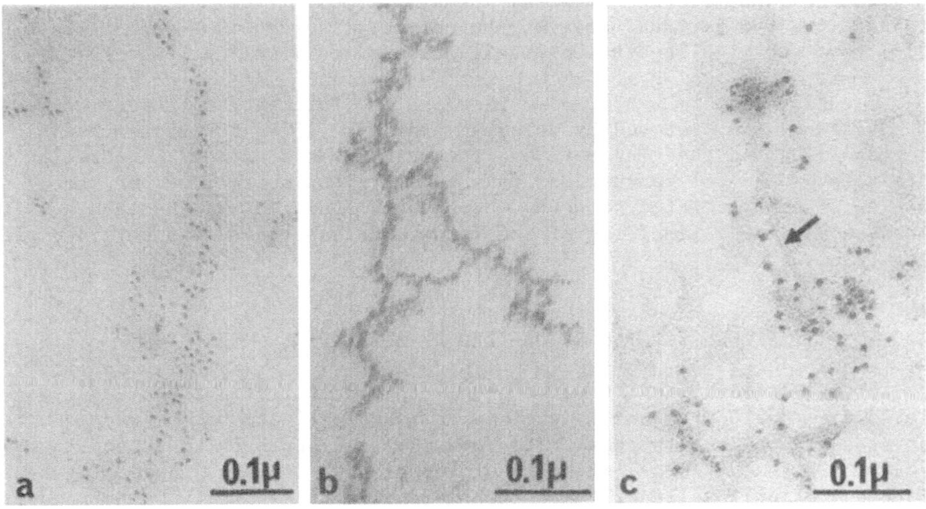

Figure 2. Labelling of carboxyl groups of isolated molecules. a and b: EDTA extracted pectin molecule labelled respectively with cationized ferritin and colloidal iron. The labelling is continuous along the molecule; c: KOH extracted glucuronoxylan molecule labelled with colloidal iron. The labelling is random and spaced. The arrow points out the molecule itself.

been adapted to the detection of cell wall polysaccharides of plant cells [11,12].

The enzyme-gold is a direct one-step postembedding technique. The principle is given in Figure 3. The enzyme is bound to colloidal gold as a dense marker, providing an enzyme-gold complex which retains a great part of its biological activity. Upon incubation with a tissue the gold particles interact with the substrate molecules exposed at the surface of the section. Details for the steps are described by Bendayan [13]. Figures 4 and 5 provide examples of the labelling of hemicellulose molecules by means of the enzyme-gold technique.

β-galactosidase was used as an enzyme specific for the β-galactoside linkage of side chains of xyloglucans. The size of the gold particles was 15nm. Labelling was performed on epon sections of tamarind seeds (4 days after sowing). A uniform labelling was found over the xyloglucan-rich part of the wall while no labelling was found over the primary wall and the cytoplasm. Controls (incubation of sections at 4 °C) were devoid of particles.

Xylanase was used as an enzyme specific for the β-1-4-xylopyranose linkages available along the backbone of xylan molecules [11]. Two sizes of gold particles were used: 15nm for the in situ labelling of xylans (Fig. 5b, c and d), 5nm for the labelling of isolated xylan molecules. Figures 5b to d illustrate the specificity of the technique. In wood fibre of the lime tree only the secondary wall is labelled while in the pith cell of rush only the primary wall is labelled with gold particles. This corresponds to the actual distribution of xylans in both dicotyledon and monocotyledon cell walls.

Many substrates have been localized with the enzyme-gold technique in animal cells [13], and in plants other examples have been provided which concern cell wall polymers. By means of a 1-4-β-D glucan cellobiohydrolase, Chanzy et al. [14] labelled cellulose microfibrils or microcrystals. Chitin was also specifically localized using chitinase-gold complex in cell walls of fungi [9, Bonfante-Fasolo, personal communication]. From the results obtained with the various enzymes, the enzyme-gold technique appears as a useful tool for high resolution localization of cell wall polysaccharides.

Recently another method, also based on the interaction between enzyme and substrate, was proposed by Joseleau and Ruel [15]. The sequence of reaction is called the ETAg method. The method makes use of a controlled hydrolysis of the polysaccharides by a specific enzyme followed by the staining of newly created reducing end-groups. Using a xylanase and a mannanase, a direct labelling of the respective substrates was obtained with great precision.

CARBOHYDRATE ANTIGENS CAN BE VISUALIZED BY ANTIBODIES

In order to distinguish between one oligosaccharide and another closely related one, antibodies and in particular monoclonal antibodies are theoretically the best probes. The immunological approach to the cell wall is used by a few authors working on higher plants. Details concerning the immunology of cell walls will not be provided in this paper and the reader is referred to the excellent review of Roberts et al. [16]. Most results using immunological probes have concerned the carragheenans of red algae [17], and the mannans of yeast cells [18]. Recently Roberts et al. [16] raised monoclonal antibodies against cell wall glycoproteins of Chlamydomonas and demonstrated that some of them recognize carbohydrates or carbohydrate-containing determinants. For higher plants, antibodies have

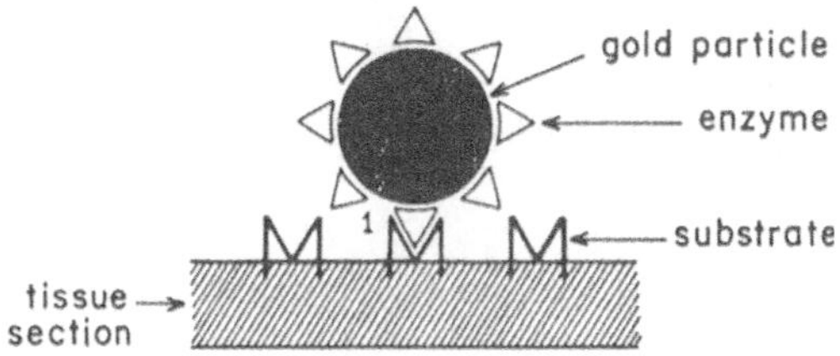

Figure 3. Diagram illustrating the principle of one step enzyme-gold technique, adapted from Bendayan [13].

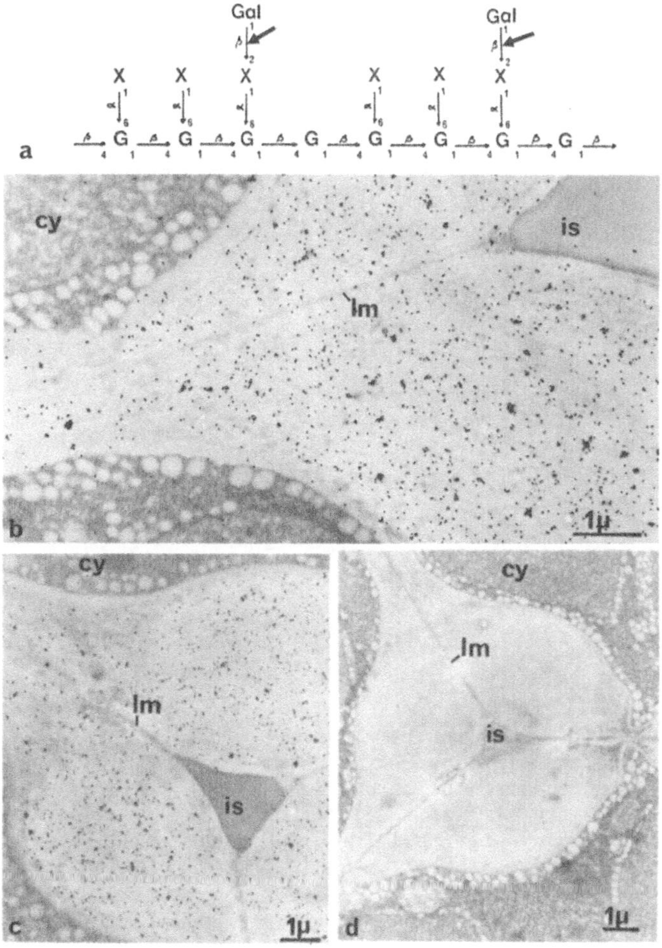

Figure 4. Labelling of xyloglucans of tamarind by β-galactosidase-gold complexes. a: structural features of xyloglucans; G: glucose; X: xylose; Gal: galactose. The arrows indicate the site of hydrolysis for the enzyme; b and c: labelling of ultrathin section. Gold particles are found over the wall thickenings; c: cytoplasm; is: intercellular space; d: control experiment (labelling at 4 °C).

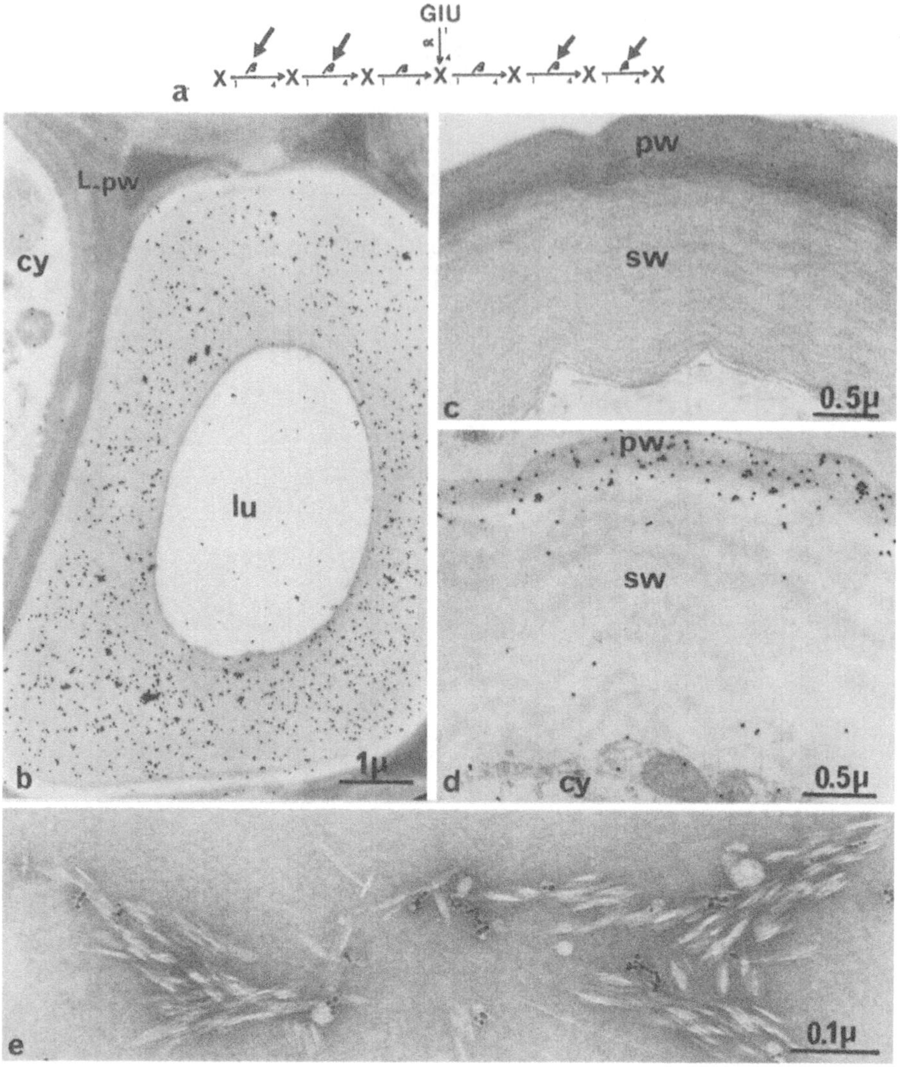

Figure 5. Labelling of xylans by xylanase-gold complexes.
a: structural feature of glucuronoxylans of hardwood. X: xylose; Me-glu: 4-O-methylglucuronic acid; the arrows indicate the site of hydrolysis for the enzyme. b: wood fibre of lime tree (*Tilia platyphyllos*). Gold particles are seen over the secondary wall, sw. L-pw: compound middle lamella-primary wall devoid of particles; lu: lumen; c and d: pith of rush. PATAg staining (c) and labelling with xylanase gold complexes (d). Gold particles are seen over the secondary wall; cy: cytoplasm: e: isolated molecules of glucuronoxylans of lime tree labelled with xylanase-gold complexes.

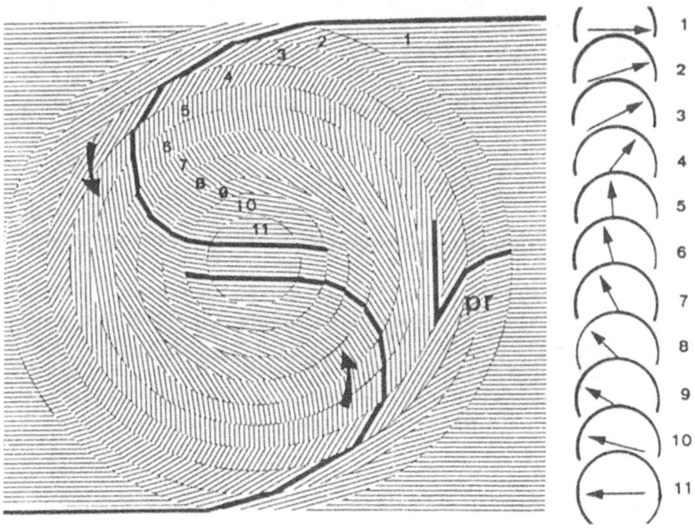

Figure 6. Helicoidal structure of cell walls. The diagram show that the arcs seen on sections correspond to the regular rotation of subunits (layers 1 to 11; pr; apparent reversion point adapted from Roland et al. [22]).

been raised against glycoproteins involved in cell recognition [19]. However the identification of the multiple and complex oligosaccharides of cell walls remains an enormous challenge.

CONCLUDING REMARKS

Carbohydrates are involved in a great number of ways in symbiotic interactions and in plant-pathogen interactions [20]. The presence of a great variety of carbohydrates in the cell wall, i.e. in the outer barrier of the cell, is one important aspect which has been emphasized in recent years. In particular, it is now clear that the walls of cells located in the external regions of organs (epidermis, cortical parenchyma, collenchyma ...) are defined constructions. They are generally helicoidal assemblies in which the component layers are regularly ordered in all directions of space (Fig. 6) [21,22,23,24]. The walls also appear flexible and susceptible to exogenous variations. In the perspective of surface interactions, if one supposes that targets for recognition factors may exist within the walls, it is important to localize them precisely in this ordered construction. In the near future it is likely that the development of new probes will greatly advance our knowledge of the spatial distribution of carbohydrate polymers and the significance of their biological functions.

REFERENCES

1. McNeil, M., Darvill, A.D., Fry, S.C. and Albersheim, P., 1984, Structure and function of the primary cell walls of plants, Annu. Rev. Biochem., 53: 625-663.
2. Roland, J.C., 1978, General preparation and staining of thin sections in: "Electron Microscopy and Cytochemistry of Plant Cells", J.L. Hall, ed., pp. 1-62, Elsevier, Amsterdam.

3. Reis, D., 1980-1981, Cytochimie ultrastructurale des parois en croissance par extractions ménagées. Effets comparés du diméthylsulfoxyde et de la méthylamine sur le démasquage de la texture, Ann. Sc. Nat. Bot., 2 and 3: 121-136.
4. Vian, B., 1981, The use of ultracryotomy for the study of the microfibrillar units of the cell wall, in: "Cell Walls 81", Proc. of 2nd Cell Wall Meeting, D.G. Robinson and H. Quader, eds., pp. 153-161, Wissenchaftliche Verlagasgesellschaft, Stuttgart.
5. Bonfante-Fasolo, P., 1982, Cell wall architectures in a mycorrhizal association as revealed by cryoultramicrotomy, Protoplasma, 111: 113-120.
6. Roth, J., 1983, Application of lectin-gold complexes for electron microscopic localization of glycoconjugates on thin sections, J. Histochem. Cytochem., 31: 987-999.
7. Rougier, M., 1984, Ultrastructural detection of fucosyl residues at the surface of axenically grown maize roots/sequential use of UeA lectin and fucosyl ferritin as the specific glycosylated marker, Eur. J. Cell Biol. 34: 45-51.
8. Tronchin, G., Poulain, D., Herbant, J. and Bignot, J., 1981, Localization of chitin in the cell wall of Candida albicans by means of wheat germ agglutinin. Fluorescence and ultrastructural studies, Eur. J. Cell Biol., 26: 121-128.
9. Chamberland, H., Charest, P.M., Ouellette, G.B. and Pauze, F.J., 1985, Chitinase-gold complex used to ultrastructurally localize chitin in tomato root cells infected by Fusarium oxysporum f.sp. radicis lycospersici, as compared with chitin specific gold conjugated lectins, Histochem. J., 17: 313-322.
10. Bendayan, M., 1981, Electron microscopical localization of nucleic acids by means of nuclease-gold complexes, Histochem. J., 13: 699-710.
11. Vian, B., Brillouet, J.M. and Satiat-Jeunemaitre, B., 1983, Ultrastructural visualization of xylans in cell walls of hardwood by means of xylanase-gold complex, Biol. Cell, 49: 179-182.
12. Ruel, K. and Joseleau, J.P., 1984, Use of enzyme-gold complexes for the ultrastructural localization of hemicelluloses in the plant cell wall, Histochemistry, 81: 573-580.
13. Bendayan, M., 1984, Enzyme-gold microscopic cytochemistry: a new affinity approach for the ultrastructural localization of macromolecules, J. Electr. Microsc. Techn., 1: 349-372.
14. Chanzy, H., Henrissat, B. and Vuong, R., 1984, Colloidal gold labelling of 1-4-β-D-glucan cellobiohydrolase adsorbed on cellulose substrates, Febs Lett., 172: 193-197.
15. Joseleau, J.P. and Ruel, K., 1985, A new cytochemical method for ultrastructural localization of polysaccharides, Biol. Cell, 53: 61-66.
16. Roberts, K., Phillips, J., Shaw, P., Grief, C. and Smith, E., 1985, An immunological approach to the plant cell wall, in: "Biochemistry of Plant Cell Walls", C.T. Brett and J.R. Hillman, eds., Cambridge University Press, Cambridge.
17. Gordon-Mills, E.M. and McCandless, E.L., 1975, Carragheenans in the cell walls of Chondrus crispus (Rhodophycae), Phycologia, 14: 275-281.
18. Ballou, C.E. and Raschke, W.C., 1974, Polymorphism of the somatic antigen of yeast, Science, 184: 127-133.
19. Harris, P.J., Anderson, M.A., Bacic, A. and Clarke, A.E., 1984, Cell-cell recognition in plants with special reference to the pollen stigma interactions, in: "Oxford Surveys of Plant Molecular and Cell Biology", Vol.1., B.J. Miflin, ed., pp. 161-203, Oxford University Press, Oxford.
20. Kosuge, T., 1981, Carbohydrates in plant pathogen interactions, in: "Plant Carbohydrates II", W. Tanner and K.A. Loewus, eds., pp. 584-623, Springer-Verlag, Berlin.

21. Roland, J.C. and Vian, B., 1979, The wall of the growing plant cell: its three dimensional organization, Int. Rev. Cytol., 61: 129-166.
22. Roland, J.C., Reis, D., Vian, B. and Satiat-Jeunemaitre, B., 1983, Traduction du temps en espace dans les cellules végétales, Ann. Sci. Nat. Bot. Biol. Vég., 5: 173-192.
23. Neville, A.C., 1985, Molecular and mechanical aspects of helicoid development in plant cell walls, Bio. Essays, 3: 4-8.
24. Sôulie, M.C., Vian, B. and Guillot-Salomon, T., 1985, Interactions hôte-parasite lors de l'infection par *Cercosporella herpotrichoides*, agent du piétin-verse: morphologie du parasite et ultrastructure des parois d'hôtes sensibles et résistants, Can. J. Bot., 63: 851-858.

ATTACHMENT OF A MYCOPARASITE WITH HOST BUT NOT WITH NONHOST *MORTIERELLA* SPECIES

M.S. Manocha, R. Balasubramanian and S. Enskat

Department of Biological Sciences
Brock University, St. Catharines
Ontario, Canada, L2S 3A1

INTRODUCTION

Piptocephalis virginiana, is a biotrophic, haustorial mycoparasite with a host range limited to the members of order Mucorales [1]. All mucoraceous fungi are not susceptible, a few are completely resistant; e.g. *Phascolomyces articulosus*, and some are nonhost to this mycoparasite. It seems that attachment of the parasite's germ tube to the host cell surface is a prerequisite for parasitism and its failure to attach to the cell surface results in a nonhost response [2]. Characterization of cell surfaces of three *Mortierella* species differing in their response to the mycoparasite showed no marked difference in the major constituents of their cell walls [3]. The cell wall composition of *Mortierella* species was typical of mucoraceous fungi with chitin and chitosan as major polysaccharides. Electron microscopy of host-parasite interactions revealed that the mycoparasite penetrated and formed haustoria in the hyphae of host species, *M. pusilla* and *M. isabellina*. However, the parasite germ tube failed to penetrate the hyphae of the nonhost, *M. candelabrum* [3].

The purpose of this investigation was to elucidate further the process of attachment in the mycoparasitic system and to determine whether the mechanism reported previously [2] is also operative in *Mortierella* species. Cell surfaces of *Mortierella* species were investigated with fluorescein-labelled lectins of different carbohydrate binding specificities and by SDS-gel electrophoresis of their proteins and glycoproteins.

MATERIALS AND METHODS

Cultures of *Mortierella candelabrum* V. Teigh and Le Monn, *M. hygrophila* Linnemann, *M. isabellina* Oudeman and *M. pusilla* Oudeman were grown routinely at 23 °C ± 1 °C) on a solid medium consisting of malt extract, 20 g; yeast extract, 2 g; agar, 20 g, in 1 litre of distilled water. An axenic population of spores of the biotrophic mycoparasite, *Piptocephalis virginiana* Leadbeater and Mercer, was obtained by growing the parasite on its susceptible host, *Choanephora cucurbitarum* (Berk. & Rav.) Thaxter, in complete, continuous darkness, which is known to inhibit sporulation of the host while the parasite sporulates normally [1]. The spores of the parasite were used to inoculate *Mortierella* species by mixing the former with spores

NATO ASI Series, Vol. H1
Biology and Molecular Biology of Plant-Pathogen Interactions
Edited by J. Bailey

Table 1. Per cent spore germination and penetration by *Piptocephalis virginiana* of *Mortierella* species at 20 h post inoculation

Host Species	Total spores counted*	P. virginiana spore germination, host and non-host interactions			
		Per cent germination	Per cent spores apparently contacted	Per cent spores developed appressoria	Per cent Penetrations
Mortierella pusilla	561	85.3	75.1	69.3	49.2
Mortierella isabellina	551	80.1	72.0	65.8	45.2
Mortierella candelabrum	663	86.3	78.2	0.0	0.0
Mortierella hygrophila	581	79.0	67.8	0.0	0.0

* Numbers are average of duplicate experiments with triplicate slide cultures in each experiment.

of the host and nonhost species and growing them at 23 °C ± 1 °C on the same medium as described above.

Light Microscopy

For examination of host-parasite interactions by light microscopy, a spore suspension of the parasite (1 x 10^7 spores ml^{-1}) and of a host or nonhost species (1 x 10^4 spores ml^{-1}) was inoculated on a glass slide which had been previously coated with malt-yeast-extract medium. The slide cultures were incubated in petri dishes with moist filter paper for 19-20 h at 23 °C ± 1 °C, stained with cotton blue and were examined using a Leitz Orthoplan microscope fitted with an orthomat camera attachment. Percentage germination of P. virginiana spores and the number of apparent contacts between the parasite germ tubes and host hyphae were recorded. Development of appressoria and penetrations of host cell surfaces by the parasite's germ tubes were also observed.

Attachment of germinating spores of the parasite to the isolated cell wall fragments of the host and nonhost hyphae was determined by the artificial inoculation and washing-off procedure. The details of this procedure and the method for preparation of isolated cell wall fragments are described in an earlier paper [2].

Fluorescent Microscopy

Lectin binding on surfaces of host and nonhost Mortierella species was investigated with fluorescein-labelled lectin (FITC conjugates) as specific markers for sugar residues. Details of the lectins used and the fluorescent microscopy methods employed have been reported elsewhere [2].

Protein Extraction and SDS-Polyacrylamide Gel Electrophoresis

Proteins, extracted from isolated cell walls into a solution of hot sodium dodecyl sulphate (SDS) and β-mercaptoethanol, were separated by electrophoresis on SDS-polyacrylamide gels according to the method of Orlowski [4]. Samples were heated for 5 min at 100 °C in the presence of 1% (w/v) SDS, 2.5% (v/v) β-mercaptoethanol, 25 mM Tris/HCl, pH 7.0 and 5% (w/v) sucrose.

Gels were prepared, as described [5], using 10% acrylamide for the separation gel and 5% for the stacking gel. Gels were fixed and stained for total proteins with Coomassie Brilliant Blue R-250 (Sigma Chemical Co., St. Louis, Mo., USA) and for glycoproteins with periodic acid-Schiff's reagent [6]. Gels were scanned with Auto Scanner Flur-Vis (Helena Laboratories, Beaumont, Texas, USA).

Protein in the isolated cell walls was quantitated by the method of Bradford [7] after extraction into 1N NaOH at 90 °C for 1 h [8].

Isolation of Agglutinin and Agglutination Assay

Cell wall fragments prepared according to the method of Manocha [2] were suspended in 0.05M phosphate buffer saline (PBS), pH 6.5, and homogenized in Sorvall Omni-Mixer (Ivan Sorvall Inc., Newton, CT, USA) for 1 min at 3400 rev. min^{-1}. The homogenate left overnight at 4 °C was centrifuged at 5,000 g for 20 min. To the supernatant solid ammonium sulphate was added to give 80% saturation. The precipitate obtained after centrifugation at 25,000 g for 30 min was dissolved in 2 ml of 0.05M PBS and dialyzed against PBS (4 x 1l) for 2 h. The agglutinating activity of the dialyzed solution was tested

Table 2. Fluorescence reactions of host and non-host *Mortierella* species

Fluorescent lectin tested	Sugar-binding specificity	Fluorescent reaction* with: *M. candelabrum*	*M. hygrophila*	*M. isabellina*	*M. pusilla*
Triticum vulgare (WGA)	N-acetylglucosamine oligomers	++	++	+++	++
Phytolacca americana (PaA)	N-acetylglucosamine oligomers	++	++	++	++
Ulex europeaus (UeA)	L-fucose	+	++	-	-
Lotus tetragonolobus (LtA)	L-fucose	+	+	+	+
Canavalia ensiformis (ConA)	D-mannose (D-glucose)	-	-	-	-
Pisum sativum (PsA)	D-mannose (D-glucose)	-	-	-	-
Glycine max (SBA)	N-acetyl-D-galactosamine	-	-	-	-
Ricinus communis I (RcA I)	N-acetyl-D-galactosamine	+	+	-	-
Ricinus communis II (RcA II)	D-galactose	+	+	-	-
Arachis hypogea (PNA)	D-galactose	-	-	-	-

* Fluorescence reactions recorded as: -, no fluorescence; + to +++, increasingly intense FITC green fluorescence.

against the cell wall fragments of the parasite as described by Young and Kauss [9].

Isolated cell walls of host and nonhost species were also treated with zymolyase (1 mg ml^{-1}) to liberate agglutinins. After centrifugation (5000 g) the supernatant was dialyzed against 0.05M phosphate buffer saline (pH 6.5). The agglutinating activity was tested as described above.

RESULTS

Parasite Spore Germination and Appressorium Formation

All light microscope observations were recorded from a series of events that occurred at 19-20 h post-seeding of the mixed spore suspension of the mycoparasite, *P. virginiana*, and of the host or nonhost species on malt-yeast-extract agar medium. The spores of the parasite germinated within 16-18 h and usually produced two germ tubes, occasionally one or three. Details of parasite spore germination have been described earlier [10]. The spores of the host species, *M. pusilla* and *M. isabellina* germinated within 10-12 h and those of nonhost species, *M. candelabrum* and *M. hygrophila*, within 4-5 h.

There was no marked difference in the percentage germination of *P. virginiana* spores in the presence of the host or nonhost *Mortierella* species (Table 1). The frequency of contacts between parasite germ tube and host or nonhost hypha was also the same. However, the parasite germ tube formed an appressorium at the point of contact on the host cell surface (Figs. 1-3), but not on the nonhost cell surface (Figs. 5-7). Occasionally the parasite germ tube crossed over or under the host hypha, probably in the region lacking receptor sites, turned around and formed an appressorium (Fig. 3). On the other hand, the parasite germ tube either crossed over or under, or sometimes continued to grow along the side of the nonhost hypha without any indication of appressorium formation (Figs. 5-7). To ascertain the remote possibility of delayed interaction between the parasite and the nonhost, cultures were also examined at 24 and 30 h after inoculation. Even after a long incubation period no appressoria were observed.

Attachment of Parasite to Isolated Cell Wall Fragments

Figure 4 shows the attachment of the parasite's germ tube to the cell wall fragments of the host, *M. isabellina*. The germ tube branch formed an appressorium at the point of contact on the cell wall fragment. Similar attachments on the cell wall fragments of nonhost species were not observed. Ungerminated spores or heat-killed germinated spores of the mycoparasite did not attach to the cell wall fragments of the host species.

Fluorescent Lectin Binding Assay

Table 2 summarizes the results of lectin binding assay recorded as absence (-) or presence (+, ++, +++ with increasing intensity) of fluorescent reaction with host and nonhost *Mortierella* species. All four species showed a positive reaction to wheat germ agglutinin (WGA) and *Phytollaca* lectin (PaA) both specific for N-acetylglucosamine oligomer. The lectin binding was inhibited by chitotriose and not by N-acetylglucosamine. Lotus lectin (LtA) specific for L-fucose showed binding with all four species but gorse lectin (UeA) was able to bind only with the nonhost species, *M. candelabrum* and *M. hygrophila*. Castor bean lectins (RcA I and RcA II) reacted positively with the nonhost species, whereas the soybean lectin (SBA) and peanut lectin

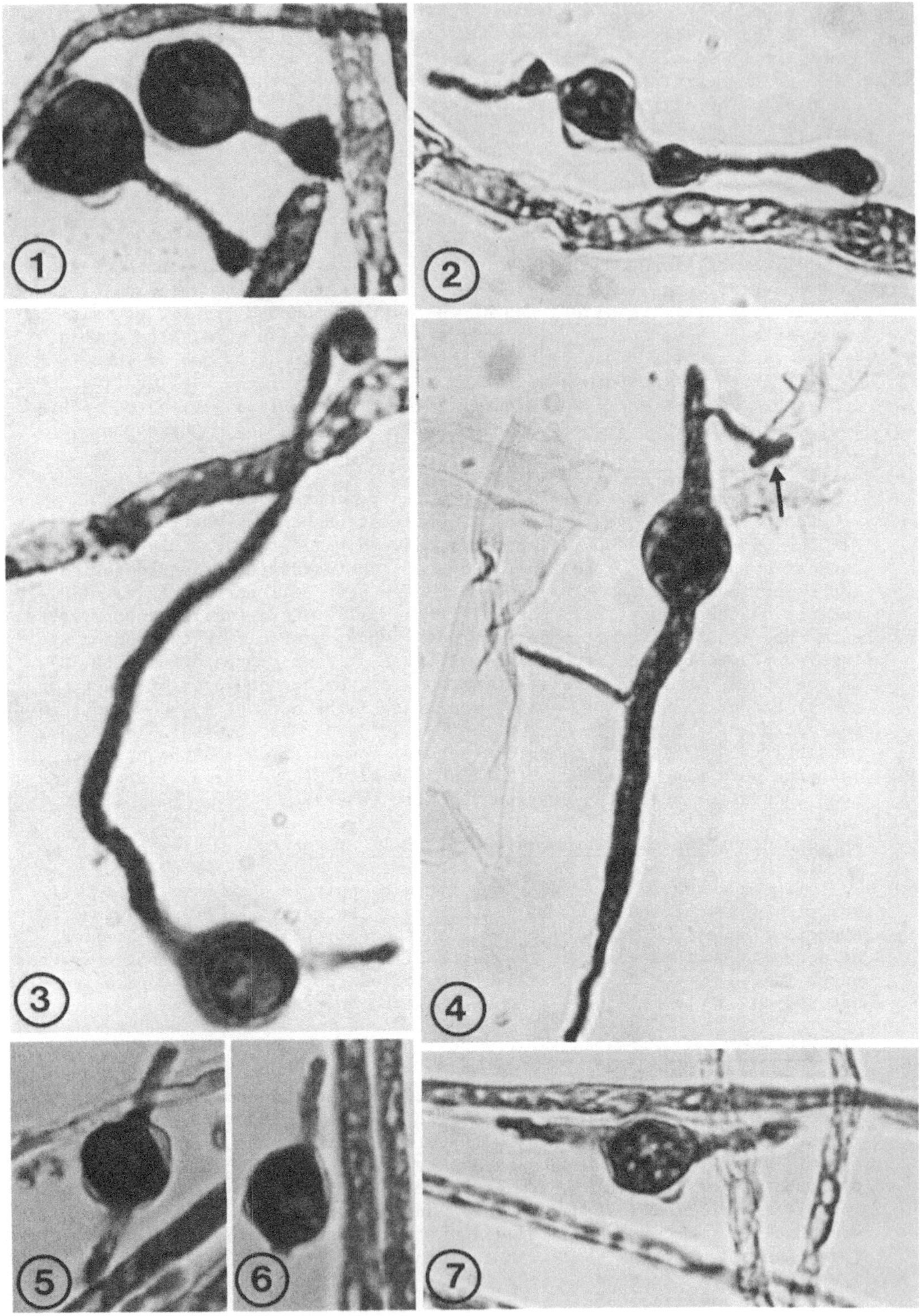

Figures 1-7. For explanation, see adjacent page.

Figures 1-4. Light microscopy of interaction between the parasite, *Piptocephalis virginiana*, and the host species. Note the parasite's germ tubes with prominent appressoria at the points of contact with host hyphae, *Mortierella pusilla* (Fig. 1) and *M. isabellina* (Fig. 2). Figure 3 shows the parasite's germ tube crossed over the host hypha, turned back and formed an appressorium at the point of contact in *M. isabellina*. Figure 4 shows the attachment of the parasite germ tube to the isolated cell wall fragments of the host, *M. isabellina*. Note the appressorium (arrow) at the tip of the parasite germ tube, x3500.

Figures 5-7. Light microscopy of the parasite, *P. virginiana* in close proximation to the hyphae of the nonhost species, *M. hygrophila* (Figs. 5 and 6) and *M. candelabrum* (Fig. 7). Note the absence of appressoria at the points of apparent contact between the germ tubes of the parasite and the nonhost hyphae, x3500.

(PNA), specific for N-acetylgalactosamine and D-galactose, respectively, failed to bind with any of the four species. No binding was observed with ConA and pea lectin (PsA), both specific for D-mannose.

There was no difference in the fluorescein-labelled lectin binding pattern between the young germ tubes (20-70 μm in length) and the 19-20 h old hyphae of the four Mortierella species tested here. Pre-treatment of hyphae with trypsin (1% for 20 min at 30 °C), known to expose receptor sites [11] did not change the pattern of lectin binding in any of the four species.

SDS-Polyacrylamide Gel Electrophoresis of Cell Wall Proteins and Glycoproteins

Figure 8 shows typical band patterns of proteins (A) and glycoproteins (B) obtained after the electrophoresis of cell wall extracts of the four Mortierella species. Comparison of band patterns with samples from host and nonhost species showed quantitative and qualitative differences. These differences are well marked in band patterns of glycoproteins (Fig. 8B) revealed after staining with periodic acid-Schiff's reagent. Two prominent peaks of slow electrophoretic mobility were observed in the cell wall extract of the host species, but not in the nonhost species.

Agglutinating activity

Crude extracts of agglutinin prepared from the cell walls of the host, M. isabellina, agglutinated cell wall fragments of the parasite, P. virginiana. Agglutination of the same degree was observed when the extract prepared from the nonhost, M. candelabrum, was used. Buffer control (without the extract) did not show any agglutination.

In preliminary studies agglutination tests were carried out using germinating spores of the parasite instead of the cell wall fragments. The results obtained were highly variable due to the tendency among the germ tubes to agglutinate in the buffer control. Extraction of agglutinin from the cell wall fragments of the host and the nonhost species by zymolyase treatment (1 mg ml^{-1}) did not show any difference between the two species. The possibility that the agglutination of cell wall fragments was caused by other glycoproteins was not examined in this study.

DISCUSSION

The spores of the mycoparasite, P. virginiana, germinated equally well in the presence of the four Mortierella species tested. The germ tubes apparently contacted the hyphae of the host, M. isabellina and M. pusilla, and the nonhost, M. candelabrum and M. hygrophila, species. Light microscopy of parasite interaction has shown that the parasite germ tube attached and formed a distinct appressorium at the point of contact with the host hyphal surface. Requirement of a living host by the parasite to form appressoria at the points of attachment was not absolute, since appressoria were also observed on isolated cell wall fragments. No attachment occurred or appressoria were formed when the parasite's germ tubes made apparent contact with the nonhost hyphae or their isolated cell wall fragments. These findings support an earlier work on the electron microscopy of parasite interaction with Mortierella species [3].

The lack of attachment of parasite's germ tubes could be due to the absence of binding sites on the hyphal surface of the nonhost species. Knox and Clarke [12] have suggested the use of lectins as a tool for probing the

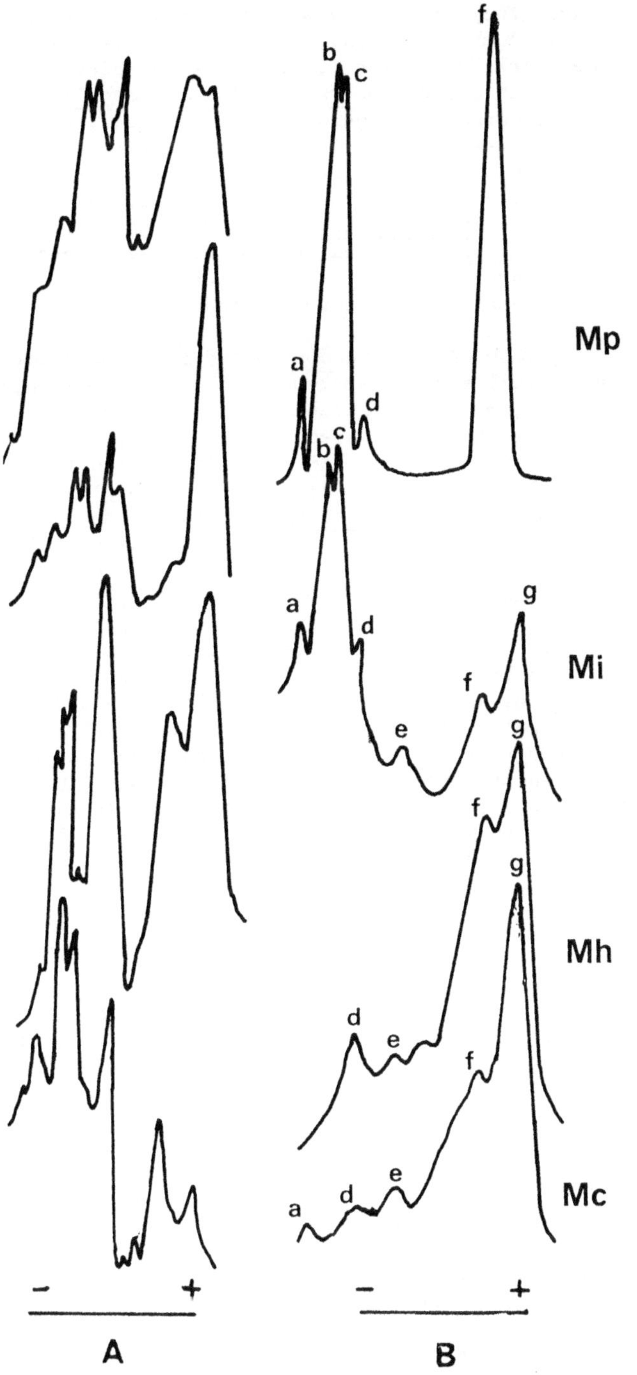

Figure 8. Comparison of band patterns of proteins (A) and glycoproteins (B) extracted from isolated cell walls of Mortierella pusilla (Mp), M. isabellina (Mi), M. hygrophila (Mh), and M. candelabrum (Mc). Note the two bands of glycoproteins marked b and c which are present in the host species but not in the nonhost species.

cell surface because they are highly specific reagents for carbohydrate cytochemistry. Lectin binding assay using ten different lectins specific for 5 sugars confirmed earlier results [2] obtained by using a different set of host and nonhost species, Choanephora cucurbitarum and Linderina pennispora, respectively. All mucoraceous species that have been tested showed a positive reaction of various intensity to fluorescein-labelled lectins specific for N-acetylglucosamine. Galactose and N-acetylgalactosamine residues were accessible at lectin binding sites only on the hyphal surfaces of the nonhost species. Mortierella species also showed specificity for fucose lectin which was not observed in case of C. cucurbitarum and L. pennispora. Whether such differences are of any significance in the taxonomy of mucoralean fungi remains to be investigated.

Cell surface glycoproteins are considered as prime candidates for cell-cell interaction in well-studied systems such as plant-pathogen [13] and plant-symbiont [14,15,16]. Specific binding between complementary surface macromolecules has also been suggested in mycoparasitic systems [2,17]. The present study with host and nonhost Mortierella species lends support to the previous work from this laboratory [2], where glycoproteins extracted from the cell walls of the host, C. cucurbitarum, and the nonhost, L. pennispora, showed markedly different patterns on SDS-polyacrylamide gel electrophoresis. Two bands of high molecular weight glycoproteins were observed in the extract of the host, but not in the extract of the nonhost species. However, crude extracts of the host and the nonhost agglutinins prepared by either $(NH_4)_2SO_4$ precipitation or zymolyase treatment did not show the specificity exhibited by the intact cell walls. These results are in agreement with those of Young and Kauss [9], who demonstrated non-specific agglutination of fungal cell wall fragments by plant extracts and by various proteins. It appears that extraction procedures result in loss of specificity exhibited by intact surfaces. Precise structure and location of agglutinins at the host cell surface may be responsible for host specific interaction. Further purification and characterization of the high molecular weight glycoproteins is necessary to evaluate their possible role in attachment.

ACKNOWLEDGEMENT

MSM thanks the Natural Sciences and Engineering Research Council of Canada for financial support in the form of an operating grant.

REFERENCES

1. Berry, C.R. and Barnett, H.L., 1957, Mode of parasitism and host range of Piptocephalis virginiana, Mycologia, 49: 374-387.
2. Manocha, M.S., 1985, Specificity of mycoparasite attachment to the host cell surface, Can. J. Bot., 63: 772-778.
3. Manocha, M.S., 1984, Cell surface characteristics of Mortierella species and their interaction with a mycoparasite, Can. J. Microbiol., 30: 290-298.
4. Orlowski, M., 1979, Changing pattern of cyclic AMP-binding proteins during germination of Mucor racemosus sporangiospores, Biochem. J., 182: 547-554.
5. Marriot, M.S., 1977, Mannan-protein location and biosynthesis in plasma membranes from the yeast form of Candida albicans, J. Gen. Microbiol., 103: 51-59.
6. Segrest, J.P. and Jackson, R.L., 1972, Molecular weight determination of glycoprotein by polyacrylamide gel electrophoresis in sodium dodecyl sulfate, Methods in Enzymology, 28B: 54-63.

7. Bradford, M., 1976, A rapid and sensitive method for the quantification of microgram quantities of protein utilizing the principle of protein dye-binding, Anal. Biochem., 72: 248-254.
8. Chapman, C.P., Nelson, R.K. and Orlowski, M., 1983, Chemical composition of spore case of the acellular slime mold Fuligo septica, Exp. Mycol., 7: 57-65.
9. Young, D.H. and Kauss, H., 1982, Agglutination of mycelial cell wall fragments and spores of Colletotrichum lindemuthianum by plant extracts, and by various proteins, Physiol. Plant Pathol., 20: 285-297.
10. Manocha, M.S. and Golesorkhi, R., 1981, Host-parasite relations in a mycoparasite. VII. Light and scanning electron microscopy of interactions of Piptocephalis virginiana with host and non-host species, Mycologia, 73: 976-987.
11. Kleinschuster, S.J. and Baker, R., 1974, Lectin-detectable differences in carbohydrate-containing surface moieties of macroconidia of Fusarium roseum 'Avenaceum' and Fusarium solani, Phytopathology, 64: 394-399.
12. Knox, R.B. and Clarke, A.E., 1978, Localization of proteins and glycoproteins by binding to labelled antibodies and lectins, in: "Electron Microscopy and Cytochemistry of Plant Cells", J.L. Hall, ed., pp. 149-185, Elsevier North Holland Inc., New York.
13. Whatley, M.H. and Sequeira, L., 1981, Bacterial attachment to plant cell walls, in: "The Phytochemistry of Cell Recognition and Cell Surface Interactions", F. Loewus and C. Ryan, eds., pp. 213-240, Plenum Press, New York and London.
14. Dazzo, F.B., 1980, Lectins and their saccharide receptors as determinants of specificity in the Rhizobium-legume symbiosis, in: "The Cell Surface: Mediator of Developmental Processes", S. Subtelny and N.K. Wessells, eds., pp. 277-304, Academic Press, London.
15. Graham, T.L., 1981, Recognition in Rhizobium-legume symbiosis, Int. Rev. Cytol. (Supplement), 13: 127-148.
16. Bauer, W.D., 1981, Infection of legumes by Rhizobia, Annu. Rev. Plant Physiol., 32: 407-449.
17. Barak, R., Elad, Y., Mirelman, D. and Chet, I., 1985, Lectins: A possible basis for specific recognition in the interaction of Trichoderma and Sclerotium rolfsii, Phytopathology, 75: 458-462.

THE BIOLOGY OF INTERACTIONS BETWEEN PLANTS AND BACTERIA

J.W. Mansfield and I.R. Brown

Department of Biological Sciences
Wye College
University of London
Ashford
Kent TN25 5AH, UK

INTRODUCTION

The Mollicutes (spiroplasmas and mycoplasma-like organisms) and fastidious vascular bacteria, few of which have been cultivated [1], will not be considered in this article. The plant pathogenic bacteria to be described in most detail here are members of the genera Agrobacterium, Erwinia, Pseudomonas and Xanthomonas. All are Gram negative bacteria which grow well in culture on simple media. Despite their common lack of exacting nutritional requirements the bacteria produce very different types of disease ranging from soft rots (Erwinia spp.) to tumorigenic hyperplasias (Agrobacterium spp.). The unifying feature of all phytopathogenic bacteria is their ability to multiply within the tissues of their host plants [2,3].

Debate on the evolution of microbial pathogenicity to plants has mainly been confined to consideration of the fungi [4,5]. With fungal pathogens, the distinction between obligate and facultative parasites is, in many cases, clearly reflected by the mode of nutrition of the parasite. Thus, the "fastidious" filamentous fungi such as the downy and powdery mildews are all biotrophic, obtaining nutrients from living cells by means of the production of intracellular haustoria [6]. By contrast, the facultative parasite Botrytis cinerea is necrotrophic, feeding on nutrients released from killed plant tissue [7,8]. Extension of the concept of nutritional mode to bacterial pathogenesis reveals that although the prokaryotes do not have exacting nutritional requirements there are examples of adaptations to biotrophy. For example, A. tumefaciens appears to grow biotrophically within crown gall tumours [9,10] and P. syringae pv. phaseolicola proliferates within the intercellular spaces of Phaseolus pod tissue causing watersoaking but not plant cell death (Fig. 1). Multiplication of A. tumefaciens and P. s. pv. phaseolicola within living host tissue contrasts with the behaviour of soft rotting erwinias which seem to be totally necrotrophic during successful invasion, killing cells in advance of colonization [11,12]. Such differences in modes of nutrient transfer clearly reflect differences in the physiological basis underlying successful parasitism of plants by bacteria.

Most phytopathogenic bacteria depend upon colonization of the host plant for survival and overwintering. The pathogens differ greatly in their

NATO ASI Series, Vol. H1
Biology and Molecular Biology of Plant-Pathogen Interactions
Edited by J. Bailey

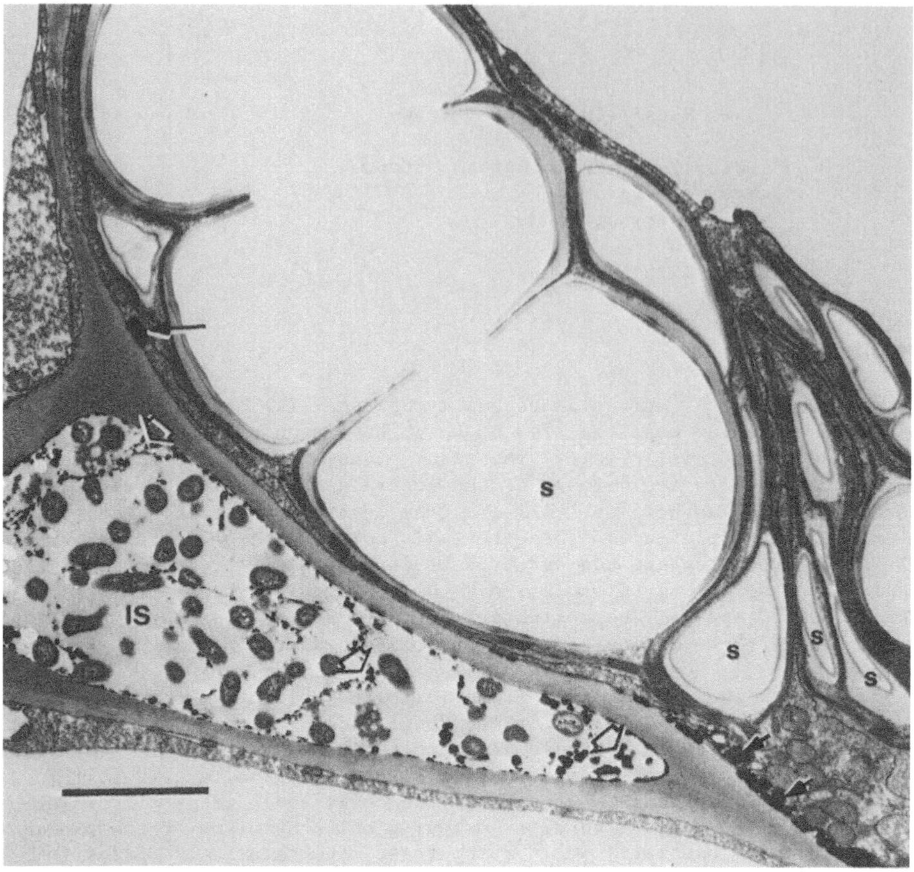

Figure 1. *P. s.* pv. *phaseolicola* (race 3) multiplying in pod tissue of *Phaseolus vulgaris* cv. Red Mexican 4 days after inoculation. Note that the bacterial colony fills the small intercellular space (IS) and that bacterial cells are surrounded by condensed and globular EPS (open arrows). Adjacent plant cells appear healthy apart from depletion of starch (S) in one chloroplast and the presence of dark vesicles (solid arrows) along the plasma membrane. The specimen was fixed in glutaraldehyde containing ruthenium red and post-fixed in 1% osmium tetroxide. (Bar = 1 μm).

success as saprophytes. Long term survival in the soil in the absence of host colonization is exceptional but has been established for A. tumefaciens, P. solanacearum and perhaps for certain Erwinia species e.g. E. carotovora subsp. carotovora [13,14,15]. Other pathogens may persist on the surfaces of symptomless plants as components of the rhizoplane and phylloplane microflora. P. s. pv. syringae has been reported to survive as an epiphyte on various non-host plants [13]. Experimental evidence is lacking on the degree of host-specificity expressed during the growth of pathogens on plant surfaces. Occurrence of the pathogen on its host (albeit on symptomless tissue) may simply reflect the close proximity of internal infections which have harboured rapidly multiplying populations of the bacteria.

Following the passive entry of bacteria into the plant through wounds or natural openings, disease development may be considered the end result of completion of the following stages of infection, i) avoidance of processes of resistance, ii) establishment of a nutritional relationship, iii) colonization of tissue and finally, iv) development of disease symptoms. The factors which cause disease symptoms need not be of direct importance to the establishment of a nutritional relationship between plant and bacterium. Similarly, although one factor may appear to have a major role in both nutrient transfer and symptom development it is never the sole determinant of pathogenicity in the organism [2,8].

MECHANISMS OF PATHOGENICITY

Mechanisms of pathogenicity are best illustrated by reference to specific examples. Major factors operating in four diseases are summarized in Table 1; each pathogen listed will be considered separately.

P. syringae pv. phaseolicola

Halo blight disease is usually recognized in the field by the presence of necrotic spots surrounded by chlorotic haloes in leaves. Chlorosis may sometimes become systemic, leaves becoming discoloured in the absence of bacterial invasion. In pods, lesions vary from necroses to patches of water soaking. P. s. pv. phaseolicola produces a low M.W. toxin, phaseolotoxin, which causes the chlorotic symptom [16] but toxin production is not necessary for the growth of the bacterium in the plant. Patil et al. [17] demonstrated that a toxin deficient ultra-violet-induced mutant multiplied as rapidly as the wild type strain in bean leaves. Thus, phaseolotoxin is required for disease development but not for successful parasitism. A similar conclusion has been reached for the role of tabtoxin in the wildfire disease of tobacco caused by the closely related P. s. pv. tabaci [18].

Following artificial inoculation, the bacterium multiplies rapidly in the intercellular spaces without killing plant cells. In leaves this period of biotrophy lasts for one or two days when tissue at the inoculation site becomes necrotic, turns brown and desiccates [19,20]. In pods, large water soaked lesions are produced which may not turn brown for at least seven days after inoculation. As is illustrated in Figure 1, intercellular spaces within infected pods become packed with bacteria and extracellular polysaccharides (EPS). Plant cells within the water soaked lesion become depleted of starch and undergo changes in chloroplast and cytoplasmic structure similar to those occurring during senescence but there are no signs of rapid cytoplasmic disorganization. EPS, produced in abundance by the invading bacteria, probably act as a hydrophilic matrix (sponge) maintaining a supply of water essential for bacterial growth [2,21]. Production of EPS in vivo may therefore be considered an essential requirement for pathogenicity.

Table 1. Mechanisms of pathogenicity.

Mechanism (1) or factor produced (2-5)	Requirement for bacteria to colonize host plant			
	Agrobacterium tumefaciens	*Erwinia carotovora*	*Erwinia amylovora*	*Pseudomonas phaseolicola*
1. Plasmid transfer	+	-	-	-
2. Pectolytic enzymes	-	+	-	-
3. Low M.W. toxins	-	-	-	-*
4. Substance causing increased permeability	-	-	+	?
5. EPS	-	-	+	+

* Phaseolotoxin required for production of chlorotic haloes

We know little about the transfer of nutrients from bean cells to *P. s.* pv. *phaseolicola* during the biotrophic phase. The levels of nutrients within the intercellular spaces of healthy tissues are probably insufficient to support the rapid growth of bacteria. Changes in membrane properties adjacent to bacteria are indicated by occasional localized convolution of the plant cell plasma membrane and the paramural deposition of callose [22]. It may, however, be unnecessary to postulate the production of some membrane leakage inducing factor to explain nutrient transfer from plant to bacteria. Given the high metabolic rate of bacterial colonies, their presence may simply create a physiological sink for nutrients [6,23]. Production of EPS may be an integral component of such an effect by diverting carbohydrates into insoluble forms and thereby rendering them unavailable to the plant.

Erwinia amylovora

Fireblight disease of pome fruits and related trees and shrubs is caused by *E. amylovora*. The first sign of infection is usually the appearance of droplets of bacterial ooze soon followed by necrosis and wilt of blossoms and young shoots which turn dark brown or black [24,25]. Microscopical studies indicate that there may be a brief period of initial invasion during which the bacterium multiplies in a similar manner to that described for *P. s.* pv. *phaseolicola* in leaves but tissues soon become necrotic and symptoms are generally consistent with the accumulation of a phytotoxic metabolite. Attempts to isolate a toxin which will reproduce the disease syndrome have proved unsuccessful [25]. Unlike other *Erwinia* spp., *E. amylovora* does not appear to produce plant cell wall degrading enzymes [26]. The production of EPS (and associated capsulation) has been implicated in virulence. However, Bennett and Billing [24,27,28] in studies using various avirulent mutants have demonstrated that EPS production is not the sole determinant of pathogenicity. Their findings are summarized in Table 2. Two classes of mutants were obtained, one (P) which produced EPS but failed to cause water soaking and the accumulation of water droplets on the surface of pear fruit segments and another (S, L, Q and E) which failed to produce EPS *in vitro*, caused leakage in the pear fruit assay but did not reproduce the symptoms of leakage and milky ooze characteristic of fully virulent strains. Mixtures of mutants P and S, L, Q or E were found to reproduce the symptoms caused by

Table 2. Production of EPS *in vitro* and the formation of symptoms in immature pear chunks and apple shoots by cultures of *E. amylovora* (adapted from [24,27,28]).

Culture	Selection method	EPS *in vitro*	Pear fruit: Leakage[a]/ necrosis	Pear fruit: Milky ooze	Apple shoot infection
T	Wild type	+	+	+	++
P	Colony variant	+	-	-	-
S,L[b]	Phage resistant	-	+	-	-
Q	Colony variant	-	+	-	-
E	Colony variant	-	+	-	-
P+S, L, Q or E			+	+	+

[a] Production of water soaking and accumulation of water droplets on the pear surface.

[b] Mutants S and L produce EPS in the presence of galactose.

virulent isolates on pear fruit. Billing has postulated that the P mutant may lack a virulence factor causing an increase in cell permeability whereas the EPS deficient mutants lack the ability to maintain a supply of free water for bacterial growth [28].

Erwinia carotovora

The *Erwinia carotovora* group of pathogens includes the subspecies *atroseptica* (*Eca*), *carotovora* (*Ecc*) and *chrysanthemi* (*Echr*). The subspecies are closely related biochemically and serologically but differ in temperature optima. The maximum temperatures at which most strains of *Eca*, *Ecc* and *Echr* will grow are 27, 29 and 39°C respectively. These temperature characteristics are reflected in host range and geographical distribution [11,15].

The *E. carotovora* subspp. cause soft rots characterized by maceration and killing of parenchymatous and chlorenchymatous tissue [11,29]. The symptoms of infection have been reproduced with highly purified preparations of pectate lyase (PL) from *Ecc* and *Echr* [11,30,31]. Secretion of PL appears essential for pathogenicity; the enzyme kills tissue thereby releasing nutrients from plant cells and also facilitates the catabolism of pectic polymers which themselves provide a rich source of carbohydrates for bacterial growth. Genetical evidence that production of PL is the principal determinant of virulence was obtained by Chatterjee and Starr [29,32] who demonstrated the conjugational transfer of a gene (Pat) that controls secretion of pectate lyase to recipient strains with reduced lyase activity. Only the Pat^+ recombinants caused maceration of celery, potato and carrot tissues. Pat^- recombinants were unable to cause maceration even though they possessed polygalacturonase activity comparable to Pat^+ strains. Beraha and Garber [33] found that the restoration of mutants to virulence was associated with recovery of production of high levels of various pectic enzymes, cellulase and phosphatidase. In both these studies it is not clear if mutations were in structural genes or regulatory loci. Mutants defective in structural genes for pectate lyase have proved difficult to isolate probably because of the production of several isoenzymes by E. carotovora [34,35]. The regulation of synthesis and secretion of pectate lyase has been reviewed in detail by Collmer et al. [11].

Many bacteria are as efficient producers of pectic enzymes but unlike the E. carotovora subspp. they are not pathogenic to plants [11,15,29]. It has been suggested that the ability of the erwinias to invade plants is due to the rate of PL production at infection sites, and also to adaptation of the enzymes to particular galacturonan linkages critical for cell wall and tissue integrity in host plants [11]. Further understanding of PL regulation may help to explain the phenomenon of latent infections particularly well described for Eca in potato. The surface of tubers is often colonized by Eca but no soft rot develops. The transition from latent infection to development of an aggressive soft rot is affected by a number of factors which may interact quantitatively and qualitatively. Pérombelon [15] identifies the occurrence of mechanical damage, the presence of sufficient numbers of bacteria in potential infection courts and oxygen concentration as three important factors. Damage may release wall fragments which are potent inducers of PL synthesis in E. carotovora and at low oxygen levels the bacteria can grow as facultative anaerobes but oxygen dependent processes of resistance, for example phytoalexin biosynthesis, may be suppressed within potato tuber tissue [36,37]. If conditions favour the bacterium, then tissues will be macerated and killed before they are able to initiate the processes of resistance and a spreading rot will develop [38].

Agrobacterium tumefaciens

A. tumefaciens is a soil borne pathogen which infects most dicotyledonous plants at wound sites, at the base or crown of the plant, and causes the formation of crown gall tumours. All tumorigenic strains of the bacterium harbour one of a diverse group of tumour inducing (Ti) plasmids. During infection, bacteria become attached to the plant cell wall and a portion of the Ti plasmid, the T-DNA, is transferred to the plant cell where it is stably integrated into plant nuclear DNA [39]. Integration of T-DNA disrupts the normal metabolism of growth hormones within the infected tissue causing hyperplasia and hypertrophy. The overproduction of auxin and cytokinins by transformed cells allows tumours to grow in the absence of exogenously supplied hormones when placed in axenic tissue culture [10,40]. Of more direct nutritional benefit to the bacterium is that tumours also synthesize novel molecules (the opines) which serve as a unique food supply for A. tumefaciens. The opines have been divided into two major classes: the amino acid derivatives octopine, nopaline and agropine, and the sugar derivatives, the agrocipines [41]. Opines can be used as carbon sources only by the infecting bacteria that carry the Ti plasmid, and also induce conjugative transfer of the Ti plasmid to plasmidless strains of A. tumefaciens. Thus, any non-pathogenic agrobacteria present at the infection site will rapidly acquire the Ti plasmid, become able to utilize the opines and also become pathogenic [9]. Symptom development in the crown gall disease therefore appears to lead to the provision of nutrients for the pathogen which multiplies within the microenvironment its plasmid has created.

A. tumefaciens has often been described as a natural genetic engineer. It surely represents the most highly evolved microbial pathogen of plants. As Hepburn [42] concluded in a recent article ... "Agrobacterium tumefaciens through its interactions with higher plants and indeed with related strains of Agrobacterium, not only demonstrates the complexities within the natural world but also that no matter how devious humans are in their scientific manipulation of living systems, mother nature got there first!"

All Ti plasmids contain three regions associated with disease development; the T-DNA region involved in tumour development and opine production, a region involved in the uptake and utilization of opines and thirdly the vir region which appears to provide all the bacterial functions necessary for the transfer of Ti plasmid DNA from the bacterium to the plant cell and also includes some of the genes which control host range [43,44]. The vir genes can operate in trans for the transfer of T-DNA from bacteria harbouring a vir region and T region on separate plasmids [45], or if the T-DNA has been translocated into the bacterial chromosome [46]. With the exception of one vir gene that encodes a function involved in the level of indole-3-acetic acid production, no precise role has been ascribed to the vir genes. Hagiya et al. [43] recently identified the protein products of an operon (designated the host dependent variation or hdv locus) in the vir region of pTiC58. Genetic and nucleotide sequence analyses showed that the hdv operon is comprised of six genes which are transcribed clockwise towards the T-DNA region and are under the control of a chromosomal gene, ros, in A. tumefaciens. The vir genes were fully expressed in the presence of a mutation in the ros gene [43,47]. Studies of the precise role of the proteins encoded by vir genes should help elucidate the mechanism of T-DNA transfer and also the physiological basis underlying host range.

In addition to the regulatory role of chromosomal genes proposed by Kado [47], chromosal genes code for other functions necessary for virulence, notably the attachment of bacteria to plant cells [10,48,49]. Although it is generally accepted that attachment is essential for tumour formation there is conflicting evidence concerning the role of attachment in controlling virulence and host range [50]. Differing opinions on the significance of attachment may in part result from differences in assays used. For example Lippincott and Lippincott [51,52] have demonstrated the importance of attachment using assays involving the inhibition of tumour formation by avirulent bacteria or cell components usually applied simultaneously to Phaseolus vulgaris leaves; whereas Douglas et al. [53] have measured the numbers of ^{14}C labelled agrobacteria which bind to isolated mesophyll cells. The most striking difference of opinion between the two groups concerns the failure of agrobacteria to cause galls in monocotyledonous plants [52,54]. Lippincott and Lippincott [52] found that cell walls from several species of grasses did not inhibit tumour formation in the bean leaf assay and therefore concluded that monocotyledons do not bind to A. tumefaciens. By contrast, Douglas et al. [53] using direct binding assays, showed that attachment to bamboo cells in suspension culture was indistinguishable from attachment to cells of dicotyledons. Bacterial mutants defective in attachment to dicotyledons showed a similar failure to attach to bamboo and Douglas et al. [53] concluded that the resistance of monocotyledons is likely to be due to a subsequent step in pathogenesis for example a block in the transfer of T-DNA to the plant cell. Changes in the cell surface of attachment defective mutants are indicated by their failure to adsorb phages to which the wild type strain is sensitive; whether or not the altered surface features are lipopolysaccharide has yet to be confirmed [48,55].

MECHANISMS OF RESISTANCE

Expression of the resistance of plants to colonization by bacteria is summarized in Table 3. The hypersensitive reaction (HR) is characterized by the rapid confluent necrosis and collapse of sites inoculated with concentrated suspensions of bacteria (usually 10^8 viable cells ml^{-1}). The HR is the most common response of plants to avirulent phytopathogenic bacteria and Kennedy and Lacy [3] point out that ability to cause the HR in young tobacco leaves is a very reliable diagnostic test for phytopathogenicity. It is an attractive hypothesis, persuasively argued by

Table 3. Expression of resistance in plant/bacterium interactions

Interaction	Bacterium/plant	Macroscopic symptoms in leaves	Bacterial multiplication[a]	Examples[b]	Ref.
1. Disease development	Pathogen/host	Disease symptoms	+++	P.p/French bean	[20]
2. Non-host resistance	Saprophyte/plants in general	None	-	P.put/tobacco	[18]
	Pathogen/non-host plant	None or slight chlorosis	-	E.c.a./soybean	[66]
			-	P.tab/oats	[60]
		HR	+	P.pisi/tobacco	[61]
			++	P.p/soybean	[66]
3. Varietal resistance	Pathogen/resistant variety				
a) race specific		HR	++	P.p/French bean	[116]
b) race non-specific		Delayed appearance of disease symptoms	++	P.cor./oats	[60]
		Reduced severity of disease symptoms	++	P.p/French bean	[124]

[a] Multiplication at the inoculation site

[b] Abbreviations; P. = Pseudomonas syringae pv.; P.put = Pseudomonas putida; p. = phaseolicola; tab. = tabaci cor. = coronafaciens, E.c.a. = Erwinia carotovora subsp. atroseptica

Klement [57] that the factor inducing the HR in non-hosts might contribute to pathogenicity in the host plant. Such a factor may cause controlled leakage of nutrients from host plant without causing rapid necrosis. Experiments with avirulent mutants of E. amylovora summarized in Table 2 lend support to this proposal: mutant P which lacks the ability to cause leakage, also fails to induce the HR in tobacco. Bauer and Beer [58] recently claimed to have cloned a gene from E. amylovora involved in induction of both hypersensitivity and pathogenicity. Transposon (Tn5) induced mutants with reduced virulence in pear failed to cause the HR in tobacco leaves. From one mutant (Ea 322 T 101) a 21 Kb Tn5 containing EcoRI endonuclease digest fragment was cloned in plasmid pBR322. Homologies to the Tn5 containing fragment were identified by colony hybridization with genomic libraries. Three recombinant plasmids were found to restore pathogenicity and the ability to induce the HR when introduced into Ea 322 T 101. The inserts of DNA that complemented the mutation were 17, 18 and 29 Kb in length. Bauer and Beer [58] concluded that genes required for induction of the HR may also be essential for pathogenicity. Similar loss of virulence and ability to induce the HR in tobacco has also been reported for Tn5 derived mutants of P. s. pv. phaseolicola [59]. It should not be overlooked, however, that certain groups of plant pathogens do not induce the HR in tobacco, e.g., tumour inducing agrobacteria, Erwina spp. causing soft rots and P. marginalis which causes leaf rot in lettuce and soft rotting in tissues of other plants [57]. Bacteria causing the HR in tobacco and other dicotyledons may not cause the reaction in the temperate cereals. For example, inoculation of barley, oats and wheat with P. s. phaseolicola or P. s. tabaci causes, at most, faint chlorosis [60].

The failure of bacteria to cause the development of disease symptoms in resistant plants is due to their failure to multiply and spread as rapidly as in susceptible tissues. The multiplication of bacteria noted in Table 3 refers to growth at the site of inoculation. Although bacteria may multiply before the HR is expressed they do not spread from the responding tissue and even at inoculation sites, usually fail to reach the same population densities as those found in susceptible plants [20,57,60,61].

Ultrastructural studies

In many studies on the processes of varietal resistance and the resistance of non-host plants to fungi, quantitative microscopy has revealed the timing of the cessation of fungal development and plant responses such as papilla formation and plant cell death. This type of information provides an essential background for the biochemical analysis of determinants of host specificity. The value of microscopy to studies on plant/fungus interactions has in part been due to the existence of clearly defined stages in fungal development characteristic of successful parasitism, e.g. formation of penetration pegs and haustoria. In contrast to the success achieved with fungi, attempts to use microscopy to develop an understanding of interactions between plants and bacteria have produced less conclusive results. A major problem when studying bacterial development in planta is that bacterial products, notably extracellular polysaccharides (EPS), may be lost during fixation or not revealed by the stains used for conventional transmission (TEM) or scanning electron microscopy (SEM). Colonies may also be dislodged from cell walls and dispersed during fixation by vacuum infiltration [62].

We have investigated fixation methods for SEM of Phaseolus vulgaris inoculated with races of P. s. pv. phaseolicola and P. s. pv. coronafaciens. Conventional glutaraldehyde fixation and critical point drying failed to preserve EPS. A fibrillar matrix of EPS surrounding bacteria was preserved following treatment with ruthenium red (a cytochemical stain for acidic polysaccharides) and post fixation with osmium and thiocarbohydrazide [63]

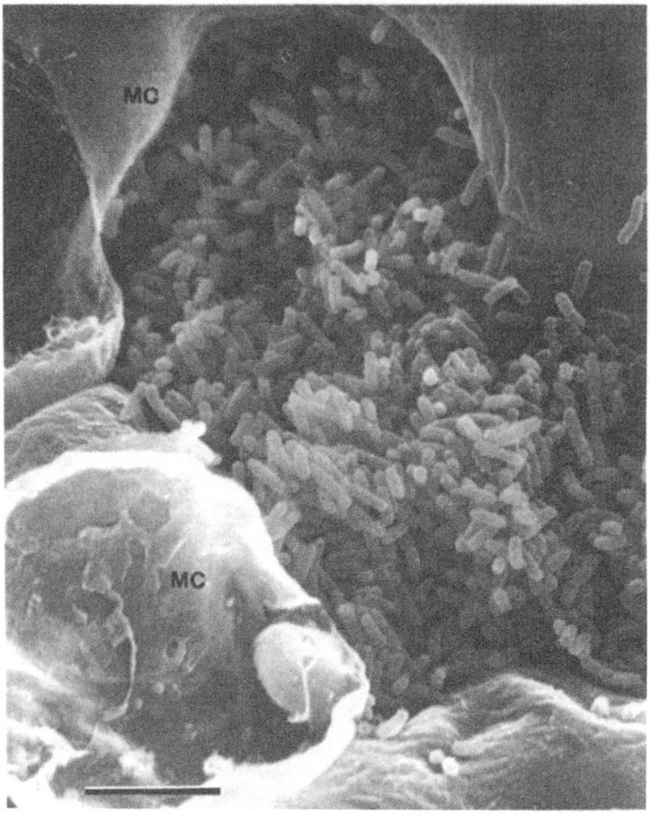

Figure 2. A large colony of P. s. pv. phaseolicola (race 2) in a leaf of bean cv. Prince 3 days after inoculation. Note that the intercellular space between mesophyll cells (M) is filled with bacteria. The specimen was fixed in glutaraldehyde without ruthenium red and prepared for SEM by critical point drying. (Bar = 5 μm).

(Figs. 2 and 3). A new and more accurate image of bacterial colonies was obtained by cryo preservation and use of a Hexland low temperature SEM system which overcomes the requirement for dehydration and critical point drying [64]. Colonies appeared as jelly-like drops embedded with bacteria (Fig. 4).

Ruthenium red staining to reveal EPS was also used for TEM studies of interactions between pseudomonads and leaves of oats and Phaseolus in which tissues were fixed without vacuum infiltration. Some results from a quantitative time course examination of the proportions of bacteria attached in various ways to plant cell walls and associated with fibrillar EPS are summarized in Table 4.

In oats, some cells of the saprophyte P. fluorescens and non-pathogens P. s. pv. tabaci and P. s. pv. coronafaciens var. atropurpurea were apparently embedded in thickened regions of the plant cell wall (Fig. 5). The embedded bacteria correspond to those described as encapsulated or immobilized in other systems for example P. s. pv. pisi in tobacco [45] and E. c. subsp. atroseptica in soybean [66].

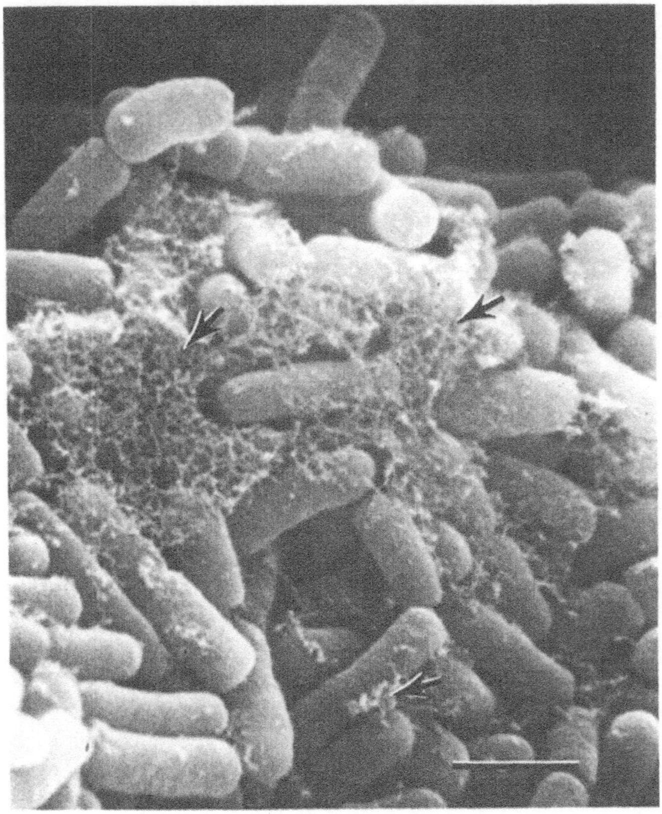

Figure 3. Part of a colony of P. s. pv. phaseolicola (race 2) in a leaf of bean cv. Prince 1 day after inoculation. (Bar = 1 μm). Note that a fibrillar matrix (arrowed) has been preserved by sequential treatment with ruthenium red, osmium tetroxide, thiocarbohydrazide and osmium tetroxide before critical point drying.

Results such as those obtained for oats (Table 4a and [67]) have been used to support proposals that the agglutination and encapsulation of bacteria on walls may restrict their multiplication [65,68]. It has also been suggested that attachment (as observed with P. s. pv. coronafaciens var. atropurpurea) may be essential for the transfer of substances triggering the HR [65, 69]. The virulence of P. s. pv. coronafaciens and other bacteria has been attributed to their ability to produce EPS in the host plant and thereby to prevent close attachment and transfer of a putative elicitor of the HR [21,65,69].

In Phaseolus cv. Red Mexican both P. s. pv. coronafaciens and the avirulent race 1 of P. s. pv. phaseolicola caused a hypersensitive response but neither bacterium was encapsulated on plant cell walls (Table 4). Time course studies showed that the morphology of colonies of avirulent and virulent bacteria, as revealed by ruthenium red/lead citrate and uranyl acetate staining, was very similar in P. vulgaris both in terms of interactions with the plant cell wall and accumulation of EPS. The absence of ultrastructural features of bacterium/wall interactions that distinguish between avirulent and virulent bacteria has also been reported in soybean [66,70].

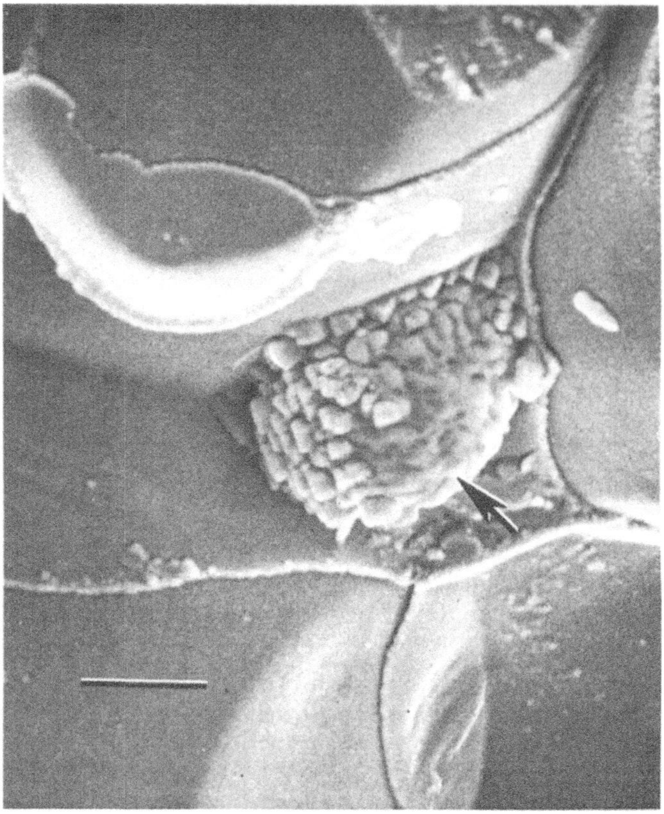

Figure 4. Colony of P. s. pv. phaseolicola (race 2) in pod tissue of bean cv. Red Mexican 36 h after inoculation. Bacteria are embedded in a hydrophilic matrix (allowed). The specimen was examined using the Hexland cryo. SEM system and is in the frozen hydrated state. (Bar = 5 μm).

It could be argued that the importance of attachment and encapsulation varies in different plants. It is, however, perhaps significant that encapsulation is observed only when there is little or no multiplication of the inoculated bacteria. Changes in plant cell wall structure presumed to immobilize bacteria may follow the operation of some unknown process of resistance which inhibits bacterial growth. Presumably some factor other than immobilization restricts the growth of P. fluorescens in oats because although a high proportion of bacterial cells were closely attached to the plant cell wall many bacteria were found free in the intercellular spaces. It seems likely that (in the absence of supportive EPS) the free bacteria had been dislodged from the plant cell wall during fixation even without the use of vacuum infiltration.

Table 4. Examples of quantitative analyses of bacteria/plant cell wall interactions (from [67] and unpublished data).

Bacterium	Plant reaction[a]	Bacterial multiplication	% bacteria in categories				% bacteria surrounded by EPS
			Close to wall			Free in intercellular space[b]	
			Embedded	Close contact	Adjacent		
a) Oats cv. Milford, 10 h after inoculation							
P. s. pv. coronafaciens	S	++++	0	14	33	53	98
P. s. pv. coronafaciens var. atropurpurea	HR	+	24	30	7	39	39
P. s. pv. tabaci	NS	-	35	31	13	21	0
P. fluorescens	NS	-	29	41	16	14	0
b) French bean cv. Red Mexican 12 h after inoculation							
P. s. pv. phaseolicola race 2	S	++++	0	17	35	48	94
P. s. pv. phaseolicola race 1	HR	+++	0	16	33	51	80
P. s. pv. coronafaciens	HR	+++	0	15	35	50	96

[a] S = susceptible plant, disease symptoms develop; [b] Bacteria at least 0.5 μm away from plant cell wall

HR = hypersensitive reaction; NS = no symptoms

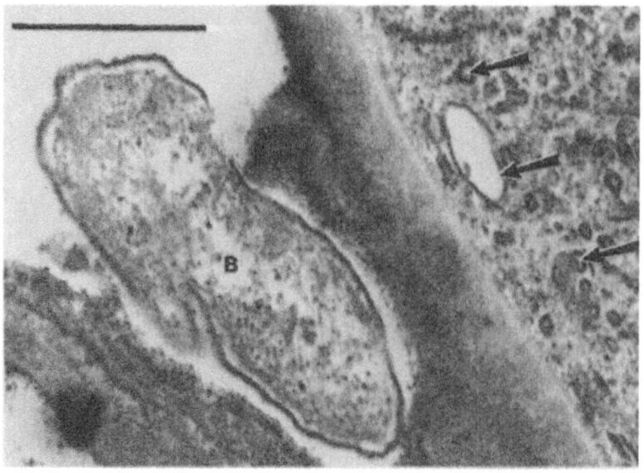

Figure 5. Attachment of P. s. pv. tabaci to oat cell walls 24 h after inoculation. Note the accumulation of paramural vesicles (arrowed) in the mesophyll cell adjacent to the embedded bacterium (B). The specimen was prepared without ruthenium red staining. (Bar = 0.5 μm).

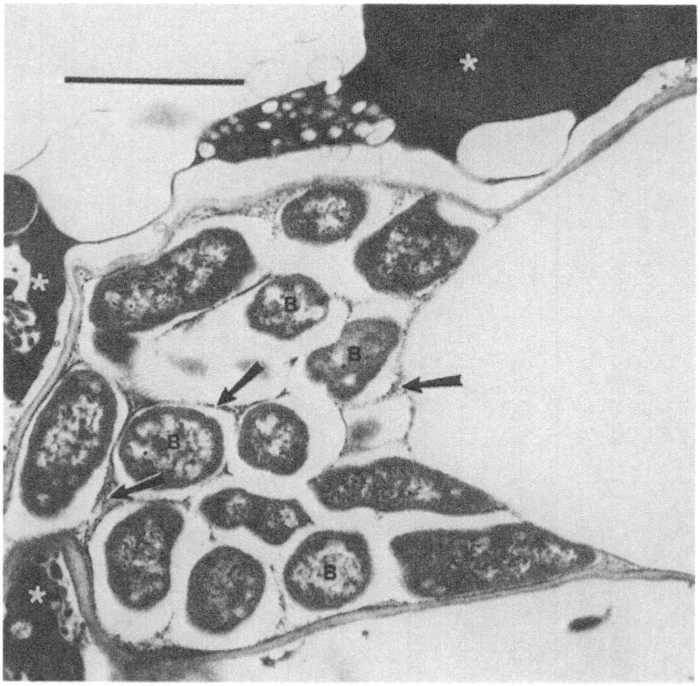

Figure 6. Colony of P. s. pv. coronafaciens in a leaf of bean cv. Prince 12 h after inoculation. Note the filbrillar matrix of EPS (arrowed) surrounding bacteria (B) and the dark, degenerating cytoplasm of adjacent plant cells (asterisks). The specimen was fixed in glutaraldehyde containing ruthenium red and post-fixed in osmium tetroxide. (Bar = 1 μm).

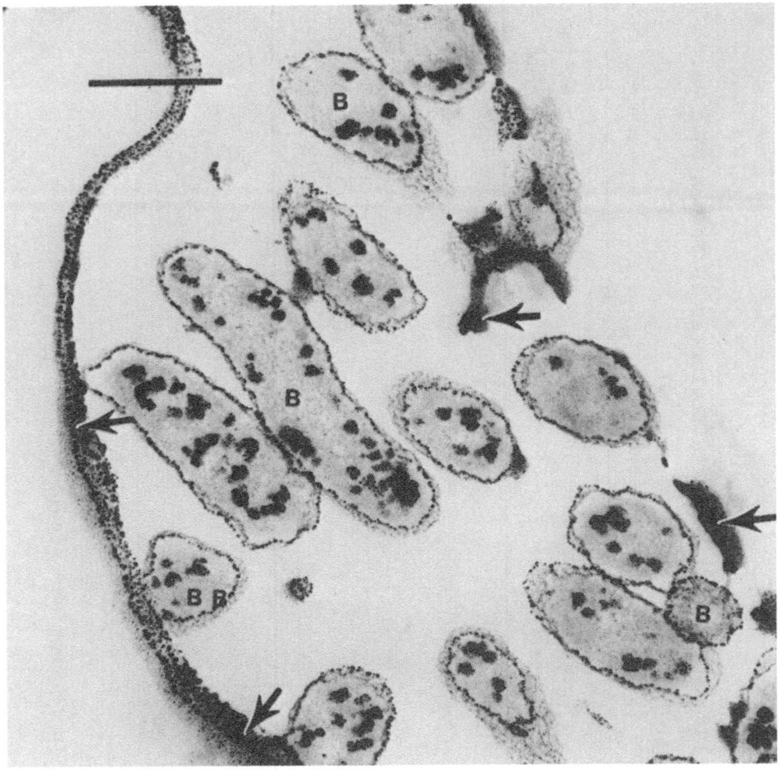

Figure 7. As in Figure 6, except that the specimen was fixed in formaldehyde and stained by the PATAg procedure [71]. Note the dark staining deposits (arrowed) in the plant cell wall and between bacteria. (Bar = 0.5 μm).

Differences in the morphology of colonies of avirulent and virulent bacteria in Phaseolus are revealed if tissues are stained for carbohydrates by the periodic acid, thiocarbohydrazide, silver proteinate (PATAg) procedure which is specific for free vicinal glycols as occur in the unbranched 1-4 linked polysaccharides in plant cell walls [71]. As EPS are predominantly composed of highly branched polymers they stain poorly with PATAg. Colonies of avirulent but not virulent bacteria were found to contain considerable amounts of PATAg positive material, and many sites of intense staining were observed in adjacent cell walls per se during the incompatible interaction as illustrated in Figures 6 and 7. Quantitative analyses of the occurrence of PATAg staining material in colonies, adjacent walls and as a component of paramural papillae in adjacent plant cell cytoplasm are given in Table 5, interestingly, more papillae were observed in the compatible interactions.

Detailed time course studies are needed on the accumulation of the PATAg positive materials before their possible significance can be determined but use of the routine histochemical staining method has allowed yet another view of the structure and composition of the bacterial colony. Further application of histochemical methods may clarify the nature of polysaccharides and other materials at the bacteria/plant cell wall interface and provide valuable clues to the biochemical basis underlying control of the HR.

Table 5. Occurrence of silver proteinate staining material in bacterial colonies in French bean cv. Prince, 12 h after inoculation.

Bacterium	Plant reaction[a]	% colonies with material[b] 0	+	++	+++	% with papillae in adjacent plant cells
P. s. pv. phaseolicola race 1	S	62	38	0	0	7
P. s. pv. phaseolicola race 2	S	69	31	0	0	7
P. s. pv. coronafaciens	HR	8	21	62	10	3

[a] S = susceptible variety, disease symptoms develop

HR = hypersensitive reaction

[b] 0 = no material detected, + ⟶ +++ = increasing amounts observed

Antibacterial compounds

The close attachment of bacteria to plant cell walls observed by TEM indicates the presence of constitutive agglutinating factors and also the accumulation of agglutinins around bacteria following infection. Extracts from a variety of uninfected plants have been shown to possess agglutinating activity. For example, Jasalavich and Anderson [72] found that an agglutinin with the properties of a glycoprotein was present in the water extract from leaves, roots and stems of Phaseolus bean and in leaves of mung bean, soybean and lima bean. Both crude and purified preparations showed strong agglutinating activity against the saprophytes P. aeruginosa, P. fluorescens and P. putida but did not affect plant pathogenic bacteria or non-pseudomonad saprophytes. Slusarenko et al. [73] found that acidic pectic polysaccharides from leaves of Phaseolus vulgaris agglutinated cells of several plant pathogens and saprophytes including pseudomonads, erwinias and agrobacteria. Similar cell wall fractions from several varieties of French bean, mung bean, broad bean, tulip and a Citrus sp. all agglutinated cells of race 1 and 2 of P. s. pv. phaseolicola. Cells of race 1 were agglutinated more actively than were those of race 2 but there was no correlation between resistance to the two races and agglutination [74].

There is, therefore, good biochemical evidence for the presence of constitutive agglutinating factors but there is disagreement about the specificity of agglutination [75,76]. One system in which agglutination does appear specific for avirulent isolates is the interaction between P. solanacearum and potato. Avirulent (EPS deficient) strains bind rapidly to an hydroxyproline-rich glycoprotein from potato tubers [77].

It is not clear if agglutination per se can prevent bacterial multiplication in planta. Al-Issa and Sigee [78] have demonstrated that cells of P. s. pv. pisi bound to tobacco cell walls remain metabolically active. However, agglutination may have a subtle suppressive effect on bacterial activity. The agglutinins isolated from uninfected plants may not be the most active materials. It would be interesting to examine the antibacterial activities of the hydroxyproline-rich glycoproteins now known to be synthesized in plants following infection [79]. It should prove possible to locate the sites of accumulation of the glycoproteins using immunocytochemical techniques.

During plant/bacterium interactions in which there is no HR and little bacterial multiplication, a large proportion of bacterial cells is often bound to the plant cell wall (Tables 3 and 4) but there remain the unbound bacteria which do not multiply. The cause of inhibition of the unbound bacteria is unknown. The well known phytoalexins and constitutive inhibitors appear not to be involved because their synthesis or release in whole plants is closely associated with plant cell death, in the former case as part of the HR [37,80, 81,82,83]. Rathmell and Sequeira [84] demonstrated the presence of a low MW bacteriostatic compound in the intercellular fluids of tobacco leaves inoculated with heat-killed cells of P. solanacearum. They suggested that accumulation of the unidentified inhibitor might explain the failure of virulent or avirulent bacteria to multiply rapidly and cause symptoms of either disease or the HR respectively in tissues previously inoculated with heat-killed cells [85]. It is possible that the compound isolated by Rathmell and Sequeira may be an example of a class of antibiotic which may be of widespread occurrence in plants. The biological evidence certainly suggests that there is a requirement for production of antibacterial compounds in the absence of the HR.

There have been a number of reports of the accumulation of antibacterial compounds in tissues undergoing the HR but not during susceptible reactions.

One of the earliest and most convincing studies was by Stall and Cook [86] who examined the response of pepper leaves to X. campestris pv. vesicatoria. Inhibitors were recovered from intercellular spaces, the sites of bacterial multiplication, using the technique devised by Klement [87]. Inhibitors accumulated more rapidly and reached higher concentrations in extracts from resistant leaves undergoing the HR than in fluids from susceptible infected tissues. The antibacterial factor in the intercellular fluids from pepper has not been identified.

Phytoalexins, recognized by their antifungal activity, can accumulate to high concentrations during hypersensitive reactions. Compounds such as the isoflavonoids glyceollin, phaseollin and kievitone have been strongly implicated in the reduction in multiplication rates of bacteria observed following the appearance of the HR [19,66,88,89,90,91,92]. There are, however, conflicting reports on the activity of phytoalexins against Gram negative bacteria [91,93]. Antibacterial activity is not expressed under certain assay conditions and this has raised doubts about the efficacy of the compounds in vivo. The isoflavonoid phytoalexins (notably phaseollin) can occur in high concentrations in aqueous diffusates from infected tissues [94] but a major problem in bioassays with the purified compounds is their very poor solubility in aqueous media. Some solubilizing factor seems to be present in vivo and it is therefore possible that the apparent lack of activity in vitro is anomalous.

Rudolph [21] has argued that the desiccation of tissues undergoing the HR in response to high concentrations of bacteria creates a physical environment unsuited for bacterial multiplication. Where low concentrations of bacteria are used and isolated cells respond hypersensitively within largely intact tissue [61,92,95,96] desiccation is unlikely to be a major cause of restricted growth. Essenberg and colleagues have examined the phenomenon of localized bacteriostasis in cotton leaves resistant to X. campestris pv. malvacearum and, using a combination of fluorescence microscopy and methanol extraction, have demonstrated the accumulation of the phytoalexins lacinilene C and lacinilene C 7-methyl ether in cells adjacent to bacterial colonies [97]. Very sensitive radioimmunoassay and laser microprobe mass analysis techniques for glyceollin measurement have been developed by Moesta et al. [98,99]. It would be valuable to use these techniques to examine the localization of the soybean phytoalexin in tissues infected with bacteria, for example the well studied races of P. s. pv. glycinea [96,100,101].

The protein synthesis inhibitor blasticidin S (BcS, 2 $\mu g.l^{-1}$) applied at the time of inoculation or up to 9 h later was found to prevent the accumulation of glyceollin and related compounds in soybean leaves inoculated with avirulent races of P. s. pv. glycinea or the pea pathogen P. s. pv. pisi [96, 102]. In the absence of phytoalexin accumulation the avirulent bacteria multiplied rapidly, at rates similar to virulent races of P. s. pv. glycinea. Results obtained were consistent with the hypothesis that the phytoalexin caused the normal restriction of multiplication of avirulent bacteria but BcS treatment also allowed multiplication of the saprophyte P. fluorescens which normally failed to grow but did not cause accumulation of glyceollin. The operation of additional protein synthesis dependent processes of resistance other than phytoalexin biosynthesis could not therefore be ruled out.

A most interesting feature of the careful studies with BcS by Keen and colleagues [102] is that both plant cell death and the accumulation of phytoalexins were inhibited. Lyon and Wood [103] working with Phaseolus bean and races 1 and 2 of P. s. pv. phaseolicola found that cycloheximide (like BcS an inhibitor of mRNA translation) prevented the leakage of electrolytes and browning of tissue associated with the HR to the avirulent race 1 in cv.

Red Mexican. It has been argued that inhibitors are not to be trusted because of non-target effects, indeed, both BcS and cycloheximide are phytotoxic at high concentrations, but this makes the suppression of cell death and membrane damage by the inhibitors even more significant. Yoshikawa [104] points out that although there are numerous reports that transcription and translation are required for phytoalexin biosynthesis, whether or not protein synthesis is required for the induction of hypersensitive cell necrosis is controversial. In several plant/fungus infections studied inhibitors do not prevent plant cell death during incompatible reactions although they may allow enhanced fungal development [105,106]. Hazen and Bushnell [107] however, found that heat shock and cytochalasin B treatments prevented plant cell death in barley coleoptiles inoculated with avirulent races of the powdery mildew fungus *Erysiphe graminis*.

In soft rots caused by *Erwinia* spp. successful infection is associated with the widespread and rapid killing of tissues in advance of colonization so that plant cells may die before they are able to activate induced antibacterial systems [15,36]. However, successful parasitism by the leaf spotting pseudomonads and xanthomonads (and perhaps also *E. amylovora*) requires colonization of their hosts without killing plant cells, at least during the early stages of infection. Susceptible tissues do not accumulate phytoalexins or activate other processes of resistance during the period of biotrophy. Resistance to these bacteria is usually expressed by the HR.

In any article on induced resistance (and there have been very many, e.g. 80,88,90,104,108,109,110,111), discussion inevitably leads to the elicitation of antimicrobial processes, in particular phytoalexin biosynthesis, and the HR. In this context it is important to redefine what is meant by the HR. The term "hypersensitive response" was first used by Stakman [112] to describe the rapid death of cells around penetration sites in varieties of wheat and other cereals resistant to *Puccinia graminis*. More recently HR has been used in a much wider sense to describe not only plant cell death but also the discoloration of cells and accumulation of phytoalexins during the expression of resistance [80,88,90,111]. Given our increasing understanding of the biochemical events occurring during incompatible interactions, including the activation of genes, it seems best to continue to use the term hypersensitive reaction in a broad sense. Thus, localized plant cell death and restriction of microbial growth are essential features of all hypersensitive reactions but there is scope for variation in other biochemical events associated with the HR *sensu lato*; for example, in some plants phytoalexins may accumulate whereas in others different antimicrobial systems may be activated.

Bailey [80] has argued that rapid decompartmentalization during necrosis may release endogenous elicitors which trigger phytoalexin biosynthesis and other processes of resistance in surrounding live cells. Although endogenous elicitors of phytoalexin accumulation have been isolated from plants [113,114] their role during hypersensitive reactions is far from clear. The endogenous elicitor theory and two alternative schemes for the control of events occurring during the HR are outlined in Figure 8. In each scheme there is a requirement for a phase in which the plant recognizes the avirulent bacterium. In i) recognition leads directly to plant cell death, and in ii) cell death is one of a cascade of responses concurrently induced following a single recognition process. It seems unlikely that scheme iii), in which induced responses are controlled by separate recognition events, operates during the HR as an expression of race specific resistance because this is usually conferred by a single gene in the host plant [109]. Nevertheless, some form of recognition must occur when resistance is expressed without plant cell death and this is accommodated most simply by scheme iii).

Figure 8. Relationship between recognition, plant cell death (rapid decompartmentalization) and the induction of antimicrobial systems including phytoalexin (PA) biosynthesis during the hypersensitive reaction.

i) Plant cell death as the cause of induced responses

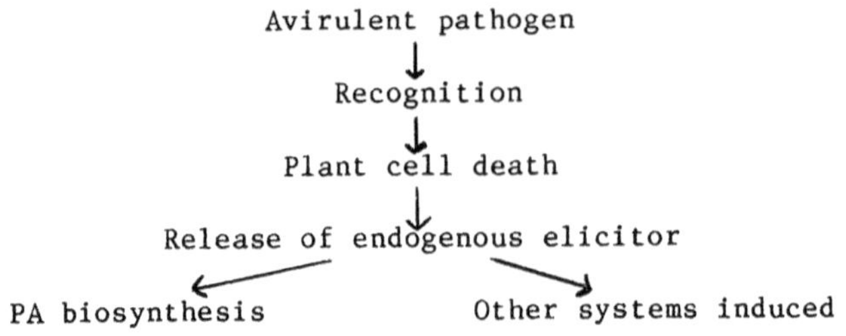

ii) Plant cell death as part of an induced response

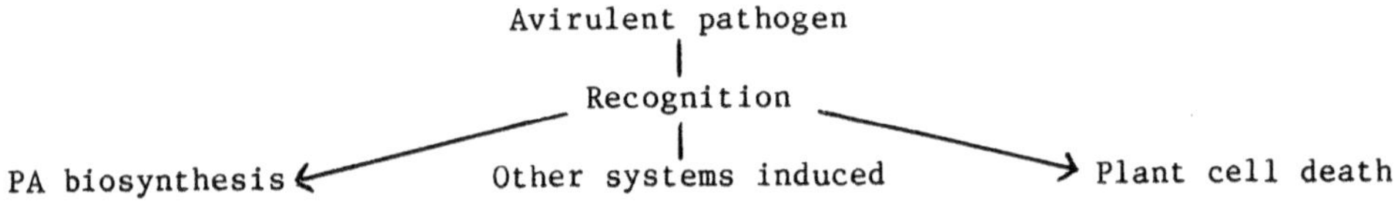

iii) Induced responses controlled by separate recognition events

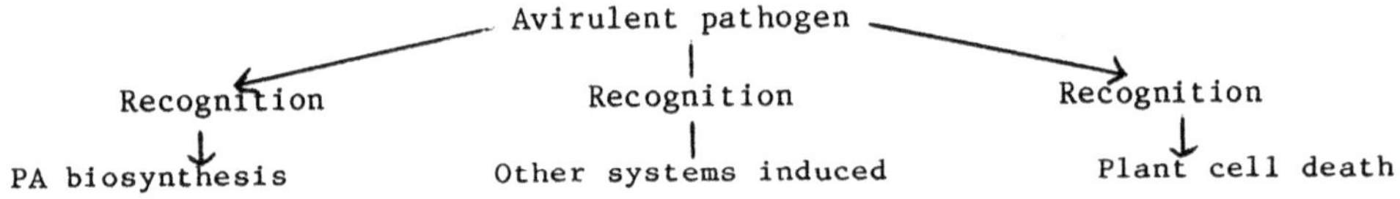

Elicitors and control of the HR

Speculation abounds on the molecules involved in recognition. Production of elicitors of the HR by avirulent bacteria rather than suppression of the reaction by virulent strains [67,106,115] is indicated by the results of experiments with mixed inocula. Long et al. [100] found that 1:1 mixtures of avirulent and virulent races of *P. s.* pv. *glycinea* gave a visible HR and glyceollin accumulation similar to the avirulent race alone. On *Phaseolus* pods the "dominance" of the avirulent bacterium is even more marked. Thus on cv. Tendergreen, HR was induced when virulent and avirulent races of *P. s.* pv. *phaseolicola* were mixed in proportions up to 10:1 (virulent : avirulent). A similar result was obtained when the avirulent race was replaced by the oat halo blight bacterium *P. s.* pv. *coronafaciens* (unpublished observations).

Attempts to isolate from bacteria, race and variety specific elicitors or indeed non-specific elicitors which will consistently induce the HR and/or phytoalexin accumulation have met with mixed success [57,90]. For example, in a very thorough study, Lyon and Wood [116] searched for elicitors of the HR produced by an avirulent race of *P. s.* pv. *phaseolicola* and by *P. s.* pv. *mors-prunorum* both of which caused a rapid HR when injected as suspensions of live bacteria into bean leaves. They found that the HR was not induced by bacteria killed by antibiotics, heat, formalization, ultrasonication or mickle tissue disintegration or by the liquid in which they were killed. Extracts of, or intercellular fluids from, leaves in which the HR was developing; isolated cell walls from bacteria or cell-free filtrates from cultures grown in various media including intercellular fluids, also failed to induce the HR following their injection into leaves [116]. Failure to detect elicitors in intercellular fluids recovered from beans infected with bacteria contrasts with the success achieved using this technique with the fungal pathogen Cladosporium fulvum. De Wit et al. [117,118] have isolated a peptide from infected tomato leaves which is a specific elicitor of necrosis in varieties of tomato with the Cf9 gene for resistance to leaf mould. Isolation of the peptide from intercellular fluids was achieved using cv. Sonatine which proved particularly sensitive to the elicitor. It is possible that elicitors of the HR in beans were not detected by Lyon and Wood [116] because their assay was not sufficiently sensitive.

Elicitors of the HR may be produced in very low concentrations and/or be unstable; in consequence their activity may be highly localized within the infected plant. The requirement for close contact between the bacterium and the plant cell wall (although not necessarily binding) for elicitation of the HR in pepper has been demonstrated experimentally by Stall and Cook [119]. It is possible that production of the elicitors may be induced by the presence of certain plant products. It has also been suggested that the molecules causing plant cell death and phytoalexin accumulation may be released from the plant following infection [88]. Davis et al. [120] have demonstrated that an endopolygalacturonan lyase from E. c. subsp. carotovora releases plant cell wall fragments which elicit phytoalexin accumulation in soybean. It is tempting to suggest that the PATAg positive material detected in colonies of avirulent bacteria in bean leaves (Fig. 7, Table 5) may have some role in triggering the HR. Although release of an active factor from the plant cell wall may explain the reaction of non-host plants, it is difficult to accommodate the specific release of plant cell wall fragments into the many bacteria/plant interactions involving single genes for race-specific resistance in the host [109].

Elicitors of glyceollin accumulation and browning in cut soybean cotyledons have been isolated from P. s. pv. glycinea and P. s. pv. pisi [121]. Following ultrasonication, factors with race non-specific activity were recovered from the cytosol fraction and specific elicitor activity was found in the cellular envelopes. Results obtained were rather variable and subsequent attempts to purify the specific elicitor have proved unsuccessful [56]. This work has encouraged the belief that bacterial wall components may play an important role in controlling the HR, but it is important to remember that elicitation of the HR during bacterial infection does not involve contact between the bacterial cell wall and the plant plasmalemma. There is a requirement for any elicitor to be released from the bacterium and cross the plant cell wall before reaching the plant protoplast. Bacterial products other than wall components should not be overlooked as possible elicitors. It is perhaps significant that C. fulvum grows entirely intercellularly, like a bacterial pathogen, during infection of tomato and the elicitor associated with specificity in tomato leaf mould has proved to be a small peptide [117].

It is evident that the somewhat traditional direct approaches to isolate determinants of specificity have met with limited success. Recent developments in molecular biology have, however, allowed the cloning of race specific avirulence genes associated with induction of the HR. Gabriel [122] has cloned several different avirulence genes from Xanthomonas campestris pv. malvacearum, a pathogen of cotton and Staskawicz et al. [101] have isolated three avirulence genes from the soybean pathogen P. s. pv. glycinea. The avirulence gene from race 6 of P. s. pv. glycinea appears to code for a single large protein. If the avirulence gene products are themselves elicitors which are recognized by plant resistance gene products, labelling of the bacterial molecules should allow isolation of the complementary plant molecules by affinity techniques [109,110].

CONCLUDING REMARKS

As Kennedy and Lacy [3] have pointed out, the fundamental weakness of the biochemical approach is that it requires a preconceived idea about what is important in the many interactions between bacteria and plants. Genetic dissection and molecular characterization of genes controlling pathogenicity and virulence require no such prejudice. Once a locus is identified it can, in theory, be isolated and characterized for its gene product. Perhaps because of the plasmid borne nature of most of the genes controlling pathogenicity in A. tumefaciens, the crown gall disease has proved particularly amenable to analysis by molecular genetics. Exciting advances have now also been made in our understanding of the genetical control of pathogenicity in X. campestris pv. campestris [123] and of race specific avirulence in P. s. pv. glycinea and X. c. pv. malvacearum. Given the development of a sound genetical background it seems probable that the biochemical mechanisms controlling the outcome of plant/bacterium interactions will be determined in the near future. We should soon be able to sort out fact from fiction in the many speculative schemes such as those presented in Figure 8, and turn our attention to applying our clear understanding of pathogenesis to the development of new approaches to durable disease control.

We wish to thank authors for supplying manuscripts before publication and to acknowledge support from the UK Agricultural and Food Research Council.

REFERENCES

1. McCoy, R.E., 1982, Chronic and insidious disease: the fastidious vascular pathogens, in: "Phytopathogenic Prokaryotes", Vol. 1., M.S. Mount and G.H. Lacy, eds., pp. 475-489, Academic Press, New York.
2. Billing, E., 1982, Entry and establishment of pathogenic bacteria in plant tissues, in: "Bacteria and Plants", M.E. Rhodes-Roberts and F.A. Skinner, eds., pp. 51-70, Academic Press, London.
3. Kennedy, R.W. and Lacy, G.H., 1982, Phytopathogenic prokaryotes: an over- view, in: "Phytopathogenic Prokaryotes", Vol. 1., M.S. Mount and G.H. Lacy, pp. 1-18, Academic Press, New York.
4. Cooke, R.C. and Whipps, J.M., 1980, The evolution of modes of nutrition in fungi parasitic on terrestrial plants, Biol. Rev., 55: 341-362.
5. Lewis, D.H., 1973, Concepts in fungal nutrition and the origin of biotrophy, Biol. Rev., 48: 261-278.
6. Manners, J.M. and Gay, J.L., 1985, The host-parasite interface and nutrient transfer in biotrophic parasitism, in: "Biochemical Plant Pathology", J.A. Callow, ed., pp. 163-195, John Wiley and Sons, Chichester.
7. Mansfield, J.W., 1984, Plant cell death during infection by fungi, in: "Cell Ageing and Cell Death", I. Davies and D.C. Sigee, eds., pp. 323-345, Cambridge University Press, Cambridge.
8. Wood, R.K.S., 1967, "Physiological Plant Pathology", Blackwell, Oxford.
9. Drummond, M., 1983, Crown gall disease: a case study, in: "Biochemical Plant Pathology", J.A. Callow, ed., pp. 65-76, John Wiley and Sons, Chichester.
10. Nester, E.W., Gordon, M.P., Amasino, R.M. and Yanofsky, M.F., 1984, Crown gall: a molecular and physiological analysis, Annu. Rev. Plant Physiol., 35: 387-413.
11. Collmer, A., Berman, P. and Mount, M.S., 1982, Pectate-lyase regulation and bacterial soft-rot pathogenesis, in: "Phytopathogenic Prokaryotes", Vol. I., M.S. Mount and G.H. Lacy, eds., pp. 395-422, Academic Press, New York.

12. Fox, R.T.V., Manners, J.G. and Myers, A., 1971, Ultrastructure of entry and spread of Erwinia caratovora var. atroseptica into potato tubers, Pot. Res., 14: 61-73.
13. Blakeman, J.P., 1982, Phylloplane interactions, in: "Phytopathogenic Prokaryotes", Vol. 1, M.S. Mount and G.H. Lacy, eds., pp. 308-333, Academic Press, New York.
14. De Boer, S.H., 1982, Survival of phytopathogenic bacteria in soil, in: "Phytopathogenic Prokaryotes", Vol. 1., M.S. Mount and G.H. Lacy, eds., pp. 285-306, Academic Press, New York.
15. Pérombelon, M.C.M., 1982, The impaired host and soft rot bacteria, in: "Phytopathogenic Prokaryotes", Vol. 2., M.S. Mount and G.H. Lacy, eds., pp. 55-71, Academic Press, New York.
16. Mitchell, R.E., 1976, Isolation and structure of a chlorosis-inducing toxin of Pseudomonas phaseolicola, Phytochemistry, 15: 1941-1947.
17. Patil, S.S., Hayward, A.C. and Emmons, R., 1974, An ultraviolet-induced nontoxigenic mutant of Pseudomonas phaseolicola of altered pathogenicity, Phytopathology, 64: 590-595.
18. Turner, J.G. and Taha, R., 1984, Contribution of tabtoxin to the pathogenicity of Pseudomonas syringae pv. tabaci, Physiol. Plant Pathol., 25: 55-69.
19. Lyon, F.M. and Wood, R.K.S., 1975, Production of phaseollin, coumestrol and related compounds in bean leaves inoculated with Pseudomonas spp., Physiol. Plant Pathol., 6: 117-124.
20. Omer, M.E.H. and Wood, R.K.S., 1969, Growth of P. phaseolicola in susceptible and resistant bean plants, Ann. App. Biol., 63: 103-106.
21. Rudolph, K., 1976, Models of interaction between higher plants and bacteria, in: "Specificity in Plant Disease", R.K.S. Wood and A. Graniti, eds., pp. 109-126, Plenum Press, New York and London.
22. Kohle, M., Jeblick, W., Poten, F., Blaschek, W. and Kauss, H., 1985, Chitosan-elicited callose synthesis in soybean cells, A Ca^{2+}-dependent process, Plant Physiol., 77: 544-559.
23. Ho, L.C. and Baker, D.A., 1982, Regulation of loading and unloading in long distance transport systems, Physiologia Plantarum, 56: 225-230.
24. Bennett, R.A., 1980, Evidence for two virulence determinants in the fireblight pathogen Erwinia amylovora, J. Gen. Microbiol., 116: 351-356.
25. Goodman, R.N., 1983, Fireblight - a case study, in: "Biochemical Plant Pathology", J.A. Callow, ed., pp. 45-63, John Wiley and Sons, Chichester.
26. Cooper, R.M., personal communication.
27. Bennett, R.A. and Billing, E., 1980, Origin of the polysaccharide component of ooze from plants infected with Erwinia amylovora, J. Gen. Microbiol., 116: 341-349.
28. Billing, E., 1984, Studies on avirulent strains of Erwinia amylovora, Acta Horticulturae, 151: 249-253.
29. Chatterjee, A.K. and Starr, M.P., 1980, Genetics of Erwinia species, Annu. Rev. Microbiol., 34: 654-676.
30. Bateman, D.F., 1976, Plant cell wall hydrolysis by pathogens, in: "Biochemical Aspects of Plant-Parasite Relationships", J. Friend and D.R. Threlfall, eds., pp. 70-103, Academic Press, London.
31. Stephens, G.J. and Wood, R.K.S., 1975, Killing of protoplasts by soft-rot bacteria, Physiol. Plant Pathol., 5: 165-181.
32. Chatterjee, A.K. and Starr, M.P., 1977, Donor strains of the soft rot bacterium Erwinia chrysanthemi and conjugational transfer of the pectolytic capacity, J. Bacteriol., 132: 862-869.
33. Beraha, L. and Garber, E.D., 1971, Avirulence and extracellular enzymes of Erwinia carotovora, Phytopath. Z., 70: 335-344.
34. Andro, T., Chambost, J-P., Kotoujansky, A., Cattaneo, J., Bertheau, Y., Barras, F., Van Gijsegem, F. and Coleno, A., 1984, Mutants of Erwinia Chrysanthemi defective in secretion of pectinase and cellulase, J. Bacteriol., 160: 1199-1203.

35. Kotoujansky, A., Diolez, A., Baccara, M., Bertheau, Y., Andro, T. and Coleno, A., 1985, Molecular cloning of Erwinia chrysanthemi pectinase and cellulase structural genes, EMBO J., 4: 781-785.
36. Lyon, G.D., Lund, B.M., Bayliss, C.E. and Wyatt, G.M., 1975, Resistance of potato tubers to Erwinia carotovora and formation of rishitin and phytuberin in infected tissue, Physiol. Plant Pathol., 6: 43-50.
37. Mansfield, J.W., 1982, Role of phytoalexins in disease resistance, in: "Phytoalexins", J.A. Bailey and J.W. Mansfield, eds., pp. 253-282, Blackie and Sons, Glasgow.
38. Lund, B.M. and Nicholls, J.C., 1970, Factors influencing the soft rotting of potato tubers by bacteria, Pot. Res., 13: 210-215.
39. Chilton, M-D., Saiki, R.K., Yadav, N., Gordon, M.P. and Quetier, F., 1980, T-DNA from Agrobacterium Ti-plasmids is in the nuclear DNA fraction of crown gall tumour cells, Proc. Nat. Acad. Sci. USA, 77: 4060-4064.
40. Buchmann, I., Marner, F-J., Schroder, G., Waffenschmidt, S. and Schroeder, J., 1985, Tumour genes in plants: T-DNA encoded cytokinin biosynthesis, EMBO J., 4: 853-859.
41. Guyon, P., Chilton, M.D., Petit, A. and Tempe, J., 1980, Agropine in "null type" crown gall tumors: evidence for generality of the opine concept, Proc. Natl. Acad. Sci. USA, 77: 2693-2697.
42. Hepburn, A., 1983, Mother nature got there first, New Scientist, 29 September 1983: 923-925.
43. Hagiya, M., Close, T.J., Tait, R.C. and Kado, C.I., 1985, Identification of pTiC58 plasmid-encoded proteins for virulence in Agrobacterium tumefaciens, Proc. Natl. Acad. Sci. USA, 82: 2669-2673.
44. Thomashow, M.F., Panagopoulos, C.G., Gordon, M.P. and Nester, E.W., 1980, Host range of Agrobacterium tumefaciens is determined by the Ti plasmid, Nature, 283: 794-796.
45. Hoekema, A., Hirsch, P.R., Hooykaas, P.J.J. and Schilperoort, R.A., 1983, A binary plant vector strategy based on separation of vir and T-region of the Agrobacterium tumefaciens Ti-plasmid, Nature, 303: 179-180.
46. Hoekema, A., Roelvink, P.W., Hooykaas, P.J.J. and Schilperoort, R.A., 1984, Delivery of T-DNA from the Agrobacterium tumefaciens chromosome into plant cells, EMBO J., 3: 2485-2490.
47. Kado, C.I., 1985, Molecular analysis of the virulence region of A. tumefaciens plasmid pTiC58, Abstract, EMBO workshop on The Molecular Biology of Bacterial Plant Pathogens, University of East Anglia, Norwich 1985.
48. Douglas, C.J., Halperin, W. and Nester, E.W., 1982, Agrobacterium tumefaciens mutants affected in attachment to plant cells, J. Bacteriol., 152: 1265-1275.
49. Garfinkel, D.J. and Nester, E.W., 1980, Agrobacterium tumefaciens mutants affected in crown gall tumorigenesis and octopine catabolism, J. Bacteriol., 144: 732-743.
50. Pueppke, S.G., 1984, Adsorption of bacteria to plant surfaces, in: "Plant-Microbe Interactions, Molecular and Genetic Perspectives", Vol. 1., T. Kosuge and E.W. Nester, eds., pp. 215-261.
51. Lippincott, B.B. and Lippincott, J.A., 1969, Bacterial attachment to a specific wound site as an essential stage in tumour initiation by Agrobacterium tumefaciens, J. Bacteriol., 97: 620-628.
52. Lippincott, J.A. and Lippincott, B.B., 1978, Cell walls of crown gall tumours and embryonic plant tissues lack Agrobacterium adherence sites, Science, 199: 1075-1077.
53. Douglas, C., Halperin, W., Gordon, M. and Nester, E., 1985, Specific attachment of Agrobacterium tumefaciens to bamboo cells in suspension cultures, J. Bacteriol., 161: 764-766.
54. De Cleene, M. and De Ley, J., 1976, The host range of crown gall, Bot. Rev., 42: 389-466.

55. Whatley, M.H., Margot, J.B., Schell, J., Lippincott, B.B. and Lippincott, J.A., 1978, Plasmid and chromosomal determination of Agrobacterium adherence specificity, J. Gen. Microbiol., 107: 395-398.
56. Barton-Willis, P.A., Wang, M.C., Holliday, J.J., Long, M.R. and Keen, N.T., 1984, Purification and composition of lipopolysaccharide from Pseudomonas syringae pv. glycinea, Physiol. Plant Pathol., 25: 387-398.
57. Klement, Z., 1982, Hypersensitivity, in: "Phytopathogenic Prokaryotes, Vol. 2., M.S. Mount and G.H. Lacy, eds., pp. 150-178, Academic Press, New York.
58. Bauer, D.W. and Beer, S.V., 1985, Cloning of a gene from Erwinia amylovora involved in induction of hypersensitivity and pathogenicity, Abstracts VIth International Conference on Plant Pathogenic Bacteria, p. 5., University of Maryland, Maryland, USA.
59. Anderson, D.M. and Mills, D., 1985, The use of transposon mutagenesis in the isolation of nutritional and virulence mutants in two pathovars of Pseudomonas syringae, Phytopathology, 75: 104-108.
60. Smith, J.J. and Mansfield, J.W., 1981, Interactions between pseudomonads and leaves of oats, wheat and barley, Physiol. Plant Pathol., 18: 345-356.
61. Sigee, D.C., 1984, Induction of leaf cell death by phytopathogenic bacteria, in: "Cell Ageing and Cell Death", I. Davies and D.C. Sigee, eds., pp. 295-322, Cambridge University Press, Cambridge.
62. Hildebrand, D.C., Alosi, M.C. and Schroth, M.N., 1980, Physical entrapment of pseudomonads in bean leaves by films formed at air-water interfaces, Phytopathology, 70: 98-109.
63. Platt-Aloia, K.A. and Thomson, W.W., 1980, Aspects of the three dimensional intracellular organization of mesocarp cells as revealed by scanning electron microscopy, Protoplasma, 104: 157-165.
64. Beckett, A. and Porter, R., 1982, Uromyces viciae-fabae on Vicia faba scanning electron microscopy of frozen-hydrated material, Protoplasma, 111: 28-37.
65. Goodman, R.N., Huang, P-Y. and White, J.A., 1976, Ultrastructural evidence for immobilization of an incompatible bacterium Pseudomonas pisi in tobacco leaf tissue, Phytopathology, 66: 754-764.
66. Fett, W.F. and Jones, S.B., 1984, Stress metabolite accumulation, bacterial growth and bacterial immobilization during host and nonhost responses of soybean to bacteria, Physiol. Plant Pathol., 25: 259-266.
67. Smith, J.J. and Mansfield, J.W., 1982, Ultrastructure of interactions between pseudomonads and oat leaves, Physiol. Plant Pathol., 21: 259-266.
68. Mazzucchi, U., 1983, Recognition of bacteria by plants, in: "Biochemical Plant Pathology", J.A. Callow, ed., pp. 299-324, John Wiley and Sons, Chichester.
69. Sequeira, L., Gaard, G. and De Zoeten, G.A., 1977, Interaction of bacteria and host cell walls: its relation to mechanisms of induced resistance, Physiol. Plant Pathol., 10: 43-50.
70. Jones, S.B. and Fett, W.F., 1985, Fate of Xanthomonas campestris infiltrated into soybean leaves: an ultrastructural study, Phytopathology, 65: 733-741.
71. Roland, J-C., 1978, General preparation and staining of thin sections, in: "Electron Microscopy and Cytochemistry of Plant Cells", J.L. Hall, ed., pp. 1-62, Elsevier/North Holland Inc., Amsterdam.
72. Jasalavich, C.A. and Anderson, A.J., 1981, Isolation from legume tissues of an agglutinin of saprophytic pseudomonads, Can. J. Bot., 59: 264-271.
73. Slusarenko, A.J., Epperlein, M. and Wood, R.K.S., 1983, Agglutination of plant pathogenic and certain other bacteria by pectic polysaccharides from various plant species, Phytopath. Z., 106: 337-343.
74. Slusarenko, A.J. and Wood, R.K.S., 1983, Agglutination of Pseudomonas phaseolicola by pectic polysaccharide from leaves of Phaseolus vulgaris, Physiol. Plant Pathol., 23: 217-227.

75. Fett, W.F. and Sequeira, L., 1980, A new bacterial agglutinin from soybean. I. Isolation partial purification and characterization, Plant Physiol., 66: 847-852.
76. Fett, W.F. and Sequeira, L., 1980, A new bacterial agglutinin from soybean. II. Evidence against a role in determining pathogen specificity, Plant Physiol., 66: 853-858.
77. Duvick, J.P. and Sequeira, L., 1984, Interaction of Pseudomonas solanacearum lipopolysaccharide and extracellular polysaccharide with agglutinin from potato tubers, App. Envir. Microbiol., 48: 192-198.
78. Al-Issa, A.N. and Sigee, D.C., 1982, The hypersensitive reaction in tobacco leaf tissue infiltrated with Pseudomonas pisi. 3. Changes in the synthesis of DNA in bacteria and mesophyll cells, Phytopath. Z., 105: 198-213.
79. Bolwell, G.P., Robbins, M.P. and Dixon, R.A., 1985, Metabolic changes in elicitor-treated bean cells. Enzymic responses associated with rapid changes in cell wall components, Eur. J. Biochem., 148: 571-578.
80. Bailey, J.A., 1983, Biological perspectives of host-pathogen interactions, in: "The Dynamics of Host Defence", J.A. Bailey and B.J. Deverall, eds., pp. 1-32, Academic Press, Sydney.
81. Mansfield, J.W., 1983, Antimicrobial compounds, in: "Biochemical Plant Pathology", J.A. Callow, ed., pp. 237-265, John Wiley and Sons, Chichester.
82. Rust, L.A., Fry, W.E. and Beer, S.V., 1980, Hydrogen cyanide sensitivity in bacterial pathogens of cyanogenic and non-cyanogenic plants, Phytopathology, 70: 1005-1008.
83. Schönbeck, F. and Schlösser, E., 1976, Preformed substances as potential protectants, in: "Encyclopedia of Plant Physiology", Vol. 4., R. Heitefuss and P.H. Williams, eds., pp. 653-678, Springer-Verlag, Berlin.
84. Rathmell, W.G. and Sequeira, L., 1975, Induced resistance in tobacco leaves: the role of inhibitors of bacterial growth in the intercellular fluid, Physiol. Plant Pathol., 5: 65-73.
85. Sequeira, L. and Hill, L.M., 1974, Induced resistance in tobacco leaves: The growth of Pseudomonas solanacearum in protected tissues, Physiol. Plant Pathol., 4: 447-455.
86. Stall, R.E. and Cook, A.A., 1968, Inhibition of Xanthomonas vesicatoria in extracts from hypersensitive and susceptible pepper leaves, Phytopathology, 58: 1584-1587.
87. Klement, Z., 1965, Method of obtaining fluid from the intercellular spaces of foliage and the fluid's merit as substrate for phytobacterial pathogens, Phytopathology, 55: 1033-1034.
88. Darvill, A.G. and Albersheim, P., 1984, Phytoalexins and their elicitors - a defence against microbial infection in plants, Annu. Rev. of Plant Physiol., 35: 243-275.
89. Gnanamanickam, S.S. and Patil, S.S., 1977, Accumulation of antibacterial isoflavonoids in hypersensitively responding bean leaf tissues inoculated with Pseudomonas phaseolicola, Physiol. Plant Pathol., 10: 159-168.
90. Keen, N.T. and Holliday, M.J., 1982, Recognition of bacterial pathogens by plants, in: "Phytopathogenic Prokaryotes", Vol. 2., M.S. Mount and G.H. Lacy, eds., pp. 179-220, Academic Press, New York.
91. Webster, D.M. and Sequeira, L., 1977, Expression of resistance in bean pods to an incompatible isolate of Pseudomonas syringae, Can. J. Bot., 55: 2043-2052.
92. Wyman, J.G. and Van Etten, H.D., 1982, Isoflavonoid phytoalexins and non- hypersensitive resistance of beans to Xanthomonas campestris pv. phaseoli, Phytopathology, 72: 1419-1424.
93. Smith, D.A., 1982, Toxicity of phytoalexins, in: "Phytoalexins", J.A. Bailey and J.W. Mansfield, eds., pp. 218-251, Blackie and Sons, Glasgow.

94. Cruickshank, I.A.M. and Perrin, D.R., 1971, Studies on phytoalexins. XI. The induction, antimicrobial spectrum and chemical assay of phaseollin, Phytopath. Z., 70: 209-229.
95. Essenberg, M., Hamilton, B., Cason, E.T., Jr., Brinkerhoff, L.A., Gholson, R.K. and Richardson, P.E., 1979, Localized bacteriostasis indicated by water dispersal of colonies of Xanthomonas malvacearum within immune cotton leaves, Physiol. Plant Pathol., 15: 69-78.
96. Holliday, M.J., Keen, N.T. and Long, M., 1981, Cell death patterns and accumulation of fluorescent material in the hypersensitive response of soybean leaves to Pseudomonas syringae pv. glycinea, Physiol. Plant Pathol., 18: 279-287.
97. Pierce, M., Cover, E.C. and Essenberg, M., 1983, Comparison of phytoalexin concentrations near sites of bacterial infection in leaves of resistant and immune cotton, Abstract, Fourth International Congress of Plant Pathology, Melbourne, Australia.
98. Moesta, P., Hahn, M.G. and Grisebach, M., 1983, Development of a radioimmunoassay for the soybean phytoalexin glyceollin I, Plant Physiol., 73: 233-237.
99. Moesta, P., Seydel, U., Lindner, B. and Grisebach, M., 1982, Detection of glyceollin on the cellular level in infected soybean by laser microprobe mass analysis, Z. Naturforsch., 37c: 748-751.
100. Long, M., Barton-Willis, P., Staskawicz, B.J., Dahlbeck, D. and Keen, N.T., 1985, Further studies on the relationship between glyceollin accumulation and the resistance of soybean leaves to Pseudomonas syringae pv. glycinea, Phytopathology, 75: 235-239.
101. Staskawicz, B.J., Dahlbeck, D. and Keen, N.T., 1984, Cloned avirulence gene of Pseudomonas syringae pv. glycinea determines race-specific incompatibility on Glycine max (L.) Merr., Proc. Natl. Acad. Sci. USA., 81: 6024-6028.
102. Keen, N.T., Ersek, T., Long, M., Bruegger, B. and Holliday, M., 1981, Inhibition of the hypersensitive reaction of soybean leaves to incompatible Pseudomonas spp. by blasticidin S, streptomycin or elevated temperature, Physiol. Plant Pathol., 18: 325-337.
103. Lyon, F.M. and Wood, R.K.S., 1977, Alteration of response of bean leaves to compatible and incompatible bacteria, Ann. Bot., 41: 359-367.
104. Yoshikawa, M., 1983, Macromolecules, recognition and the triggering of resistance, in: "Biochemical Plant Pathology", J.A. Callow, ed., pp. 267-298, John Wiley and Sons, Chichester.
105. Doke, N. and Tomiyama, K., 1975, Effect of blasticidin S on hypersensitive death of potato leaf petiole cells caused by infection with an incompatible race of Phytophthora infestans, Physiol. Plant Pathol., 6: 169-175.
106. Heath, M.C., 1982, The absence of active defense mechanisms in compatible host-pathogen interactions, in: "Active Defense Mechanisms in Plants", R.K.S. Wood, ed., pp. 143-156, Plenum Press, New York and London.
107. Hazen, B.E. and Bushnell, W.R., 1983, Inhibition of the hypersensitive reaction in barley to powdery mildew by heat shock and cytochalasin B., Physiol. Plant Pathol., 23: 421-438.
108. Callow, J.A., 1977, Recognition, resistance and role of plant lectins in host-parasite interactions, Adv. Bot. Res., 4: 1-49.
109. Ellingboe, A.H., 1982, Genetical aspects of active defence, in: "Active Defense Mechanisms in Plants", R.K.S. Wood, ed., pp. 179-192, Plenum Press, New York.
110. Keen, N.T., 1985, Novel approaches to improving disease resistance in plants, in: "Beltsville Symposium X, Biotechnology for Solving Agricultural Problems", in press.
111. Kuč, J. and Rush, J.S., 1985, Phytoalexins, Arch. Biochem. Biophys., 236: 455-472.

112. Stakman, E.C., 1915, Relations between Puccinia graminis and plants highly resistant to its attack, J. Agric. Res., 4: 193-200.
113. Hahn, M.G., Darvill, A.G. and Albersheim, P., 1981, Host-pathogen interactions. XIX. The endogenous elicitor, a fragment of a plant cell wall polysaccharide that elicits phytoalexin accumulation in soybeans, Plant Physiol., 68: 1161-1169.
114. Hargreaves, J.A. and Bailey, J.A., 1978, Phytoalexin production by hypocotyls of Phaseolus vulgaris in response to constitutive metabolites released by damaged cells, Physiol. Plant Pathol., 13: 89-100.
115. Bushnell, W.R. and Rowell, J.A., 1981, Suppressors of defense reactions: A model for roles in specificity, Phytopathology, 71: 1012-1014.
116. Lyon, F.M. and Wood, R.K.S., 1976, The hypersensitive reaction and other responses of bean leaves to bacteria, Ann. Bot., 40: 479-491.
117. De Wit, P.J.G.M., Hofman, A.E., Velthuis, G.C.M. and Kuć, J.A., 1985, Isolation and characterization of an elicitor of necrosis isolated from intercellular fluids of compatible interactions of Cladosporium fulvum (Syn. Fulvia fulva) and tomato, Plant Physiol., 77: 642-647.
118. De Wit, P.J.G.M. and Spikman, G., 1982, Evidence for the occurrence of race and cultivar-specific elicitors of necrosis in intercellular fluids of compatible interactions of Cladosporium fulvum and tomato, Physiol. Plant Pathol., 16: 391-408.
119. Stall, R.E. and Cook, A.A., 1979, Evidence that bacterial contact with the plant is necessary for the hypersensitive reaction but not the susceptible reaction, Physiol. Plant Pathol., 14: 77-84.
120. Davis, K.R., Lyon, G.D., Darvill, A.G. and Albersheim, P., 1984, Host-pathogen interactions XXV. Endopolygalacturonic acid lyase from Erwinia caratovora elicits phytoalexin accumulation by releasing plant cell wall fragments, Plant Physiol., 74: 52-60.
121. Bruegger, R.B. and Keen, N.T., 1979, Specific elicitors of glyceollin accumulation in the Pseudomonas glycinea-soybean host-parasite system, Physiol. Plant Pathol., 15: 43-51.
122. Gabriel, D.W., 1985, Molecular cloning of specific avirulence genes from Xanthomonas malvacearum, in: "Advances in Molecular Genetics of the Bacteria-Plant Interaction", A.A. Szalay and R.P. Legock, eds., pp. 202-204, Cornell University, Ithaca, USA.
123. Turner, P., Barber, C. and Daniels, M., 1985, Evidence for clustered pathogenicity genes in Xanthomonas campestris pv. campestris, Mol. Gen. Gen., 199: 338-343.
124. Rudolph, K. and Mendgen, K., 1985, Multiplication of Pseudomonas syringae pv. phaseolicola "in planta". II. Characterization of susceptible and resistant reactions by light and electron microscopy compared with bacterial countings, Phytopath. Z., 113: 200-212.

SURFACE INTERACTIONS BETWEEN *RHIZOBIUM* AND INFECTED PEA ROOT NODULE CELLS, AS REVEALED BY USING MONOCLONAL ANTIBODIES

N.J. Brewin[a], E.A. Wood[a], D.J. Bradley[b], G.W. Butcher[b] and G. Galfre[b]

[a]John Innes Institute
Colney Lane
Norwich
NR4 7UH, UK

[b]Monoclonal Antibody Centre
AFRC Institute of Animal Physiology
Babraham
Cambridge, UK

INTRODUCTION

At the centre of the legume root nodule are plant cells that each contain about 10,000 nitrogen-fixing *Rhizobium* bacteroids (Fig. 1). Both the bacteroids and the host cells which harbour them are specialized structures created in part by the operation of sets of genes that are only expressed within the nodule [1]. Besides the infected cells, the nodule also contains other specialized plant cells, for example the uninfected interstitial cells, which assimilate the fixed nitrogen derived from bacteroids [2], and the nodule meristem (formed in response to *Rhizobium* infection) which generates the other tissues of the root nodule [3]. This article will be concerned exclusively with the infected plant cells, and how they interact with the bacteria that they contain.

Some idea of the "pathology" of the *Rhizobium*-legume symbiosis can be gained by considering the origin and fate of these infected plant cells. Their ultimate fate is senescence and death, together with all the bacteria that they contain. The mechanism for this programmed senescence is unknown, but clearly it serves to limit the progress of *Rhizobium* infection within the plant. Moreover, senescence is accelerated when the plant is provided with an adequate supply of fixed nitrogen in the form of nitrate [4], or if the plant has been infected with a mutant form of *Rhizobium* which fails to fix nitrogen [5]. During the process of infection, bacteria are released into the plant cell by endocytosis from an infection thread, an inwardly growing tube originally derived by involution of the plant cell plasma membrane [3], and within which the growing and dividing bacteria are arranged in single file. Once released into the plant cytoplasm, the bacteria differentiate into the pleiomorphic bacteroid form, which alone is capable of nitrogen fixation, sustained by the respiration of dicarboxylic acid substrates derived from the plant. The bacteroids are enclosed individually by a peribacteroid membrane, which is ontogenetically related to the plasma

NATO ASI Series, Vol. H1
Biology and Molecular Biology of Plant-Pathogen Interactions
Edited by J. Bailey

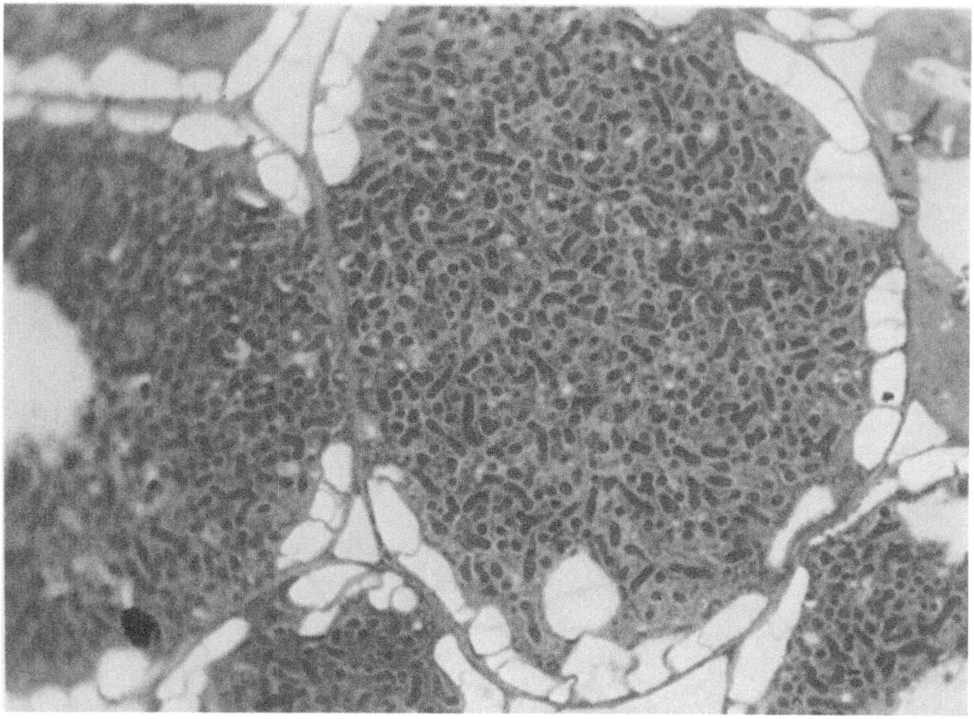

Figure 1. Light micrograph of a nodule thick section stained with toluidine blue, showing an infected plant cell with peripheral starch granules and large numbers of nitrogen-fixing Rhizobium bacteroids within the plant cytoplasm (x500).

membrane and the bounding membrane of the infection thread [6]. Thus, in a sense, even the endosymbiotic bacteroid form of Rhizobium is topologically 'outside' the infected plant cell.

From this brief outline of the interactions between Rhizobium and its host, it is clear that there are many parallels with plant pathogenesis in relation to infection mechanism, biotrophic interactions and the host's ability to limit the infection by programmed senescence. Moreover, the techniques available for the study of bacteria-plant interactions are the same for symbionts such as Rhizobium as for pathogens such as Pseudomonas. From the point of view of plant pathology, the most intriguing stage in the Rhizobium-legume symbiosis is perhaps that represented by the infected nodule cell, because here the interactions between the eukaryote and some 10,000 enclosed bacteroids are held, at least temporarily, in balance. As an approach to the study of this stage of the Rhizobium-plant interaction, we have embarked upon an analysis of the two interacting surfaces, the bacteroid cell wall, and the plant-derived peribacteroid membrane. Because of the difficulties of obtaining large quantities of these materials from nodules, the best strategy seemed to be to isolate monoclonal antibodies that reacted with components from either of these surfaces. Once obtained, these monoclonal antibodies could be used as sensitive and specific molecular reagents for the study of the cytology and biochemistry of the particular epitopes to which each monoclonal antibody (McAb) reacted.

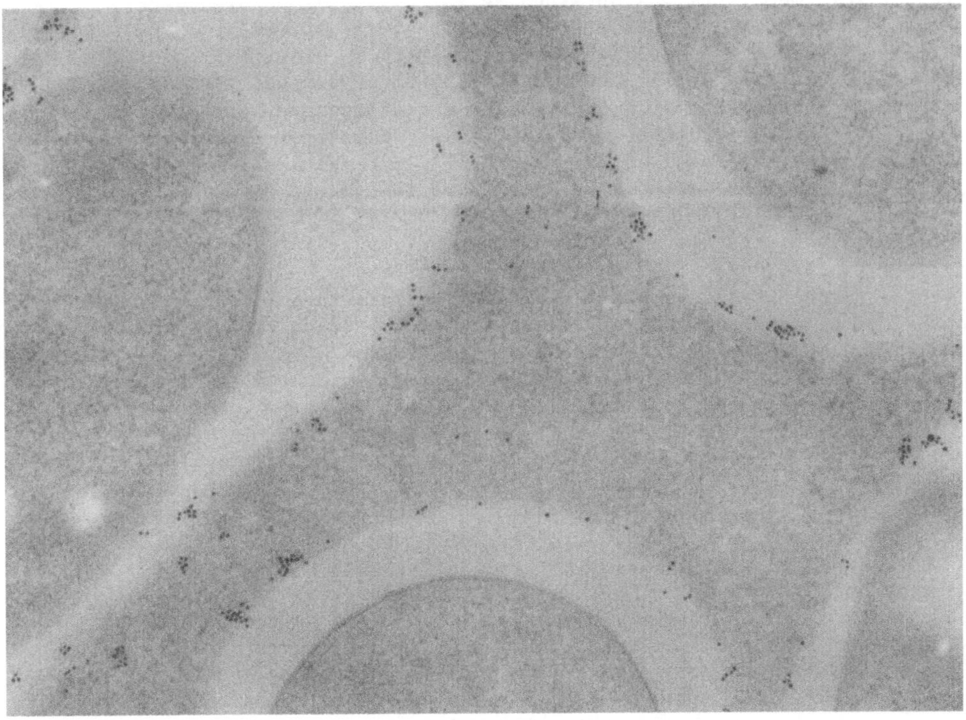

Figure 2. Electron micrograph of a nodule thin-section embedded in Lowicryl K4M resin and treated with MAC 64 antibody followed by goat anti-rat IgG conjugated to 10nm colloidal gold. The antigen labelled by MAC 64 and immunogold staining is distributed along the peribacteroid membrane adjacent to the bacteroids in the four corners of the picture (x35,000).

A major advantage of using monoclonal antibodies was that it was possible to immunize the rats with small quantities of peribacteroid membrane or bacteroid material in a form which was unpurified and biochemically uncharacterized [7]. By appropriate screening of the monoclonal antibody producing cell lines arising after a myeloma fusion experiment, it was possible to identify the McAbs that would be most useful for further study. These were, firstly, antibodies to the most abundant antigens from the peribacteroid membrane and bacteroid surfaces and, secondly, to the antigens (of either plant or bacterial origin) which were unique to the nodule state. The present account will be restricted to a discussion of two representatives of the first class, i.e. the most abundant antigen from the peribacteroid membrane and from the bacteroid surface. It will be shown how the corresponding monoclonal antibodies can be used to build up a picture of the components of these surfaces and how they might interact with each other.

PLANT MEMBRANE GLYCOPROTEIN

Figure 2 is an electron micrograph of a thin section of a pea nodule in which the peribacteroid membrane (PBM) has been highlighted by immunogold staining following treatment with the monoclonal antibody AFRC MAC 64 [7].

In order to obtain this and similar McAbs, a preparation enriched for peribacteroid membranes was obtained as follows. Pea nodules were homogenized in 50mM Tric HCl pH7.4 containing 0.5M sucrose and fractionated by differential sucrose gradient centrifugation in order to isolate bacteroids that were still enveloped by the plant-derived peribacteroid membrane. A subsequent mild osmotic shock treatment ruptured the PBM and, once the bacteroids had been removed by centrifugation, the supernatant fraction formed the basis of the material used as antigen to raise and to screen for monoclonal antibodies against peribacteroid membrane.

From several separate fusion experiments, six different McAbs were isolated that reacted with the peribacteroid fraction by dot immunoassay and with the peribacteroid membrane by immunogold electron microscopy. Subsequent biochemical analysis revealed that, in each case, the McAb reacted with the carbohydrate moiety of the same glycoprotein species of polydisperse molecular weight [7]. Presumably therefore, these McAbs had all been raised to the antigen from peribacteroid membrane that was immunodominant under the immunization and screening procedures used.

So far, the monoclonal antibody MAC 64 has been useful in two ways. First, it was demonstrated by immunogold electron microscopy that the carbohydrate epitope is found not only in the peribacteroid membrane but also in the plasma membrane of all plant cells (infected and uninfected) and in the membranes of the Golgi apparatus: these observations reinforce the previous cytological evidence concerning the relatedness of these membrane systems, and offers the opportunity to study the patterns of membrane flow within the plant cell. Moreover, the immunofluorescence studies with isolated membrane-enclosed bacteroids indicate that the carbohydrate epitope may be oriented exclusively on the lumenal face of the peribacteroid membrane, i.e. the face adjacent to the bacteroid surface, and thus the monoclonal antibody may prove useful in studying the orientation of plant membranes.

A second application of MAC 64 has been as an immunochemical marker for the peribacteroid membrane in studies concerning possible interactions between the bacteroid surface and PBM. Immunofluorescence studies with washed isolated bacteroids did not reveal any long-term attachment of PBM (i.e. MAC 64 antigen) on the bacteroid surface (such as might have been expected if, for example, a plant lectin on the PBM interacted with a carbohydrate component of the bacteroid surface). However, a more surprising result has indicated that components of the bacteroid outer membrane may become incorporated (to a limited extent) into the PBM, which must therefore be thought of as a hybrid membrane rather than being composed exclusively of plant-derived material. The experiments which have led to this conclusion will be discussed below, after the section dealing with the monoclonal antibodies to bacteroid lipopolysaccharide.

BACTEROID LIPOPOLYSACCHARIDE

Figure 3 is an electron micrograph of an isolated bacteroid, showing the typically pleiomorphic shape of these cells. The bacteroid outer membrane of this intact negatively-stained bacterium is clearly visible by immunogold staining following treatment with AFRC MAC 57, a monoclonal antibody that reacts with lipopolysaccharide from this strain. The antigen used to raise and to screen for MAC 57 and similar monoclonal antibodies was simply a sonicated bacteroid preparation that had been centrifuged to remove cell debris. McAbs that reacted with lipopolysaccharide (LPS) were the predominant class to be isolated from this screen.

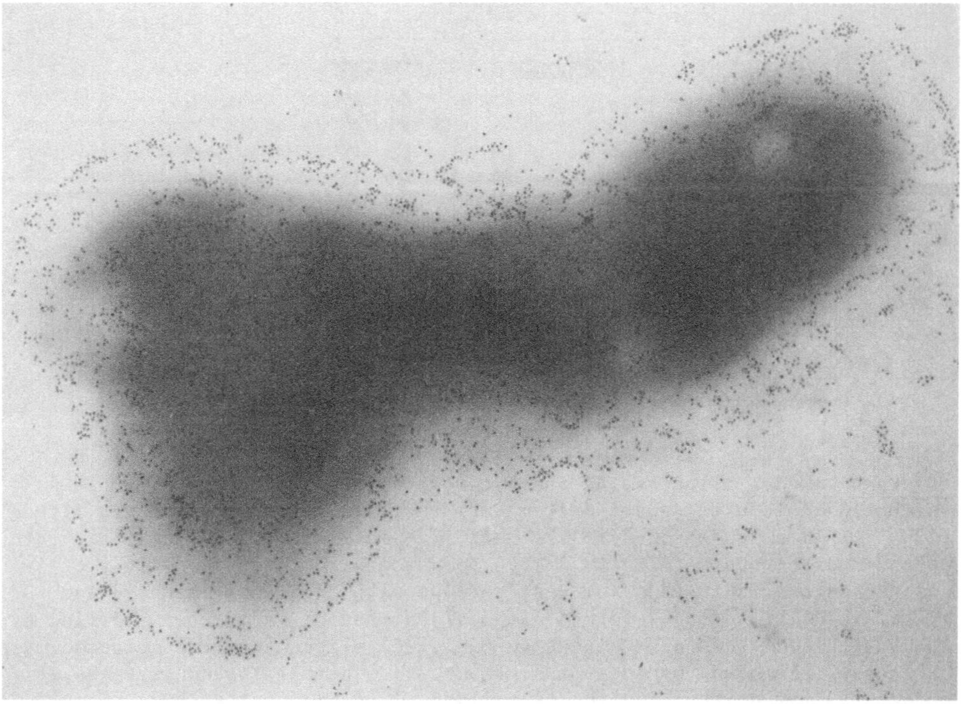

Figure 3. Electron micrograph of an isolated intact bacteroid following treatment with MAC 57 antibody, 10nm gold anti-rat lgG, and negative staining with ammonium molybdate. The lipopolysaccharide antigen highlighted by immunogold staining clearly defines the bacteroid outer membrane (x 17,000).

The size distribution of LPS molecules can be examined following cell lysis and SDS polyacrylamide gel electrophoresis [8]: a 'ladder' of LPS bands is normally interpreted as being due to the successive addition of O-antigen oligosaccharide units to a core lipopolysaccharide structure (see Fig. 4). By a modification of the gel electrophoresis technique, we were able to examine the size distribution of the LPS molecules reacting with MAC 57. SDS gels were transferred to nitrocellulose sheets by electroblotting, treated with MAC 57, and antibody binding was detected by the indirect immunoperoxidase method [7]. When preparations from free-living cultures of Rhizobium were used, the typical ladder of immunostaining bands was observed, but preparations from bacteroids showed only the fast-migrating (rough) form of LPS, which presumably lacked the O-antigen side chains (Fig. 4). This distinction between the form of LPS in bacteroids and in free-living cultures was also seen for the several other strains of R. leguminosarum examined.

What might be the significance of this alteration in the form of LPS during the infection process? Carlson [9] has shown that the O-antigen side-chains of R. leguminosarum are composed mainly of rhamnose, fucose, mannose and galactose, whereas the core polysaccharide consists principally of a polygalacturonic acid chain, linked to lipid A through the 7-carbon sugar keto-3-deoxyoctonate. Thus, the bacteroid outer membrane (in which LPS is embedded) would be expected to carry a high density of negative surface charge and would resemble in some respects the structure of pectin, the

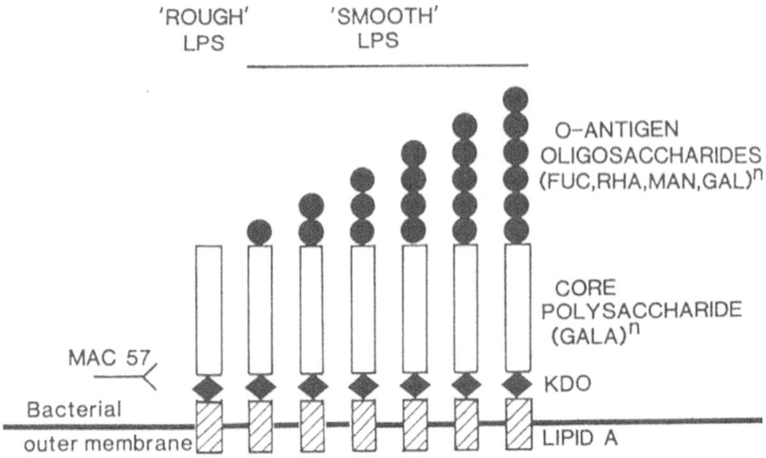

Figure 4. Model illustrating the heterogeneity of Rhizobium lipopolysaccharide molecules (adapted from Carlson [9]; Lugtenberg and van Alphen [12]. Lipid A is embedded in the bacterial outer membrane and linked through keto-3-deoxyoctonate (KDO) to a core polysaccharide composed mainly of galacturonic acid (GALA) residues, perhaps with additional mannose and galactose. This "rough" form of lipopolysaccharide is further modified by successive additions of a repeating O-antigen oligosaccharide composed mainly of fucose, rhamnose, mannose, and galactose. The heterogeneous forms of smooth LPS may be fractionated as a 'ladder' of bands after SDS polyacrylamide gel electrophoresis. In R. leguminosarum 3841 all forms of LPS react with MAC 57 antibody: there is a single unique antibody-binding site per LPS molecule and a possible position for this binding site is indicated in the diagram. In free-living cultures of Rhizobium, rough LPS and the heterogeneous smooth forms of LPS co-exist, but in bacteroids isolated from pea root nodules only the 'rough' form of LPS is found.

polygalacturonic acid containing component of the primary plant cell wall. Moreover, the absence of LPS O-antigen side-chains and the absence of capsular polysaccharide in bacteroids would serve to increase the level of interaction between the LPS core galacturonic residues of the bacteroid outer membrane and the adjacent peribacteroid membrane. In other words, this may be an example of 'host mimicry' in which the bacteroid avoids being detected, like a plant pathogen, by presenting a surface with physical-chemical properties similar to those of the primary plant cell wall itself.

INTERACTIONS BETWEEN BACTEROID AND PERIBACTEROID MEMBRANES

The McAb MAC 57 has also proved useful as a cytochemical marker with which to study the distribution of LPS within the infected nodule cell. Two lines of evidence indicate that about 1% of the LPS antigen (and presumably the associated bacteroid outer membrane) becomes incorporated into the plant-derived peribacteroid membrane within infected nodule cells. One line of evidence comes from a careful examination of electron micrographs of nodule thin-sections treated with MAC 57 and stained with colloidal gold-labelled anti-rat IgG which clearly indicates the presence of a small amount of LPS antigen in the peribacteroid membrane. The second line of evidence makes use of a sandwich ELISA assay (Fig. 5) in which purified MAC 57 antibody was used as a marker for LPS from the bacteroid outer membrane and purified MAC 64 was used as a marker for peribacteroid membrane.

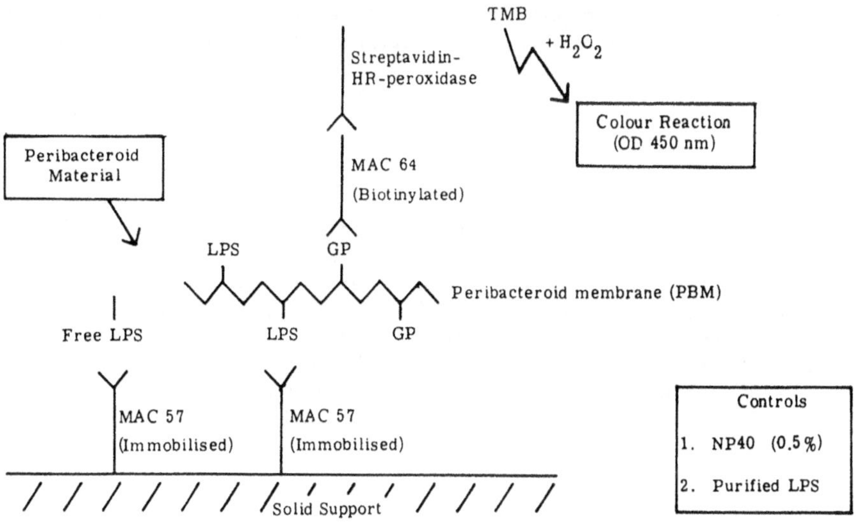

Figure 5. Sandwich immuno-assay for the detection of a physical association between lipopolysaccharide (LPS) derived from the bacteroid outer membrane and peribacteroid membrane isolated after sucrose gradient centrifugation of pea nodule homogenates. Purified monoclonal antibody MAC 57 (which reacts with bacteroid LPS) was immobilized on microtiter plates and then exposed to peribacteroid material. The plates were washed and then treated with biotinylated MAC 64 antibody, which reacts with a glycoprotein component (GP) of the peribacteroid membrane. After subsequent washing, the persistence of peribacteroid membrane in the microtiter plates was detected by an 'ELISA' reaction. In control experiments, the peribacteroid material was pre-treated for 1 h with either detergent (0.5% NP 40) or purified LPS: each of these treatments eliminated the persistence of peribacteroid membrane in the assay.

It was noted that, when membrane-enclosed bacteroids were isolated and the peribacteroid membrane ruptured by mild osmotic shock treatment, about 1% of the LPS antigen was released with the peribacteroid material. The status of this LPS antigen was examined by the sandwich ELISA experiment outlined in Figure 5. It was found that bacteroid LPS (i.e. MAC 57 antigen) was linked to peribacteroid membrane (i.e. MAC 64 antigen), through a linkage that could be dissociated by a detergent treatment (0.5% NP40), which did not inhibit the individual antigen-antibody reactions.

On the basis of these observations, it is suggested that the bacteroid membrane may bleb off [10,11,12], releasing vesicles into the peribacteroid space, and these vesicles subsequently fuse into the peribacteroid membrane [13]. This model has two interesting implications. Firstly, it provides a mechanism for the transport of proteins and other macromolecules from the periplasmic space of the bacteroid across the peribacteroid space and into the plant cytoplasm. Secondly, the model implies that the peribacteroid membrane is in fact a hybrid membrane containing minor components of bacterial origin: if the proportion of bacterial material increases with time, this could possibly be a factor contributing to the eventual senescence of infected nodule cells.

CONCLUSIONS

The Rhizobium-legume symbiosis provides an interesting example of a transiently stable bacterial-plant interaction. By making use of monoclonal antibodies to describe surface components of the bacteroid and the adjacent plant derived peribacteroid membrane, considerable progress has already been made towards an understanding of the molecular biology of this interaction within the infected plant cell. In the future, it may be possible to extend these studies to include, on the one hand, the mechanism of plant cell infection by Rhizobium and, on the other hand, the process of host cell senescence, which confines the Rhizobium infection to a small zone of tissue within the legume root nodule.

REFERENCES

1. Bisseling, T., Been, C., Klugkist, J., van Kammen, A. and Nadler, K., 1983, Nodule-specific host proteins in effective and ineffective root nodules of Pisum sativum, EMBO J., 2: 961-966.
2. Bergmann, H., Preddie, E. and Verma, D.P.S., 1983, Nodulin 35: a subunit of specific uricase (uricase II) induced and localised in the uninfected cells of soybean nodules, EMBO J., 2: 2333-2339.
3. Newcomb, W., Sippel, D. and Peterson, R.L., 1979, The early morphogenesis of Glycine max and Pisum sativum root nodules, Can. J. Bot., 57: 2603-2616.
4. Carroll, B.J., McNeil, D.L. and Gresshoff, P.M., 1985, Isolation and properties of soybean mutants which nodulate in the presence of high nitrate concentrations, Proc. Natl. Acad. Sci. USA., 82: 4162-4166.
5. Ma, Q-S., Johnston, A.W.B., Hombrecher, G. and Downie, J.A., 1982, Molecular genetics of mutants of R. leguminosarum which fail to fix nitrogen, Mol. Gen. Genet., 187: 166-171.
6. Robertson, J.G. and Lyttleton, P., 1984, Division of peribacteroid membranes in root nodules of white clover, J. Cell Sci., 69: 147-157.
7. Brewin, N.J., Robertson, J.G., Wood, E.A., Wells, B., Larkins, A.P., Galfre, G. and Butcher, G.W., 1985, Monoclonal antibodies to antigens in the peribacteroid membrane from Rhizobium-induced root nodules of pea cross-react with plasma membranes and Golgi bodies, EMBO J., 4: 605-611.
8. Hitchcock, P.J. and Brown, T.M., 1983, Morphological heterogeneity among Salmonella lipopolysaccharide chemotypes in silver-stained polyacrylamide gels, J. Bacteriol., 154: 269-277.
9. Carlson, R.W., 1984, Heterogeneity of Rhizobium lipopolysaccharide, J. Bacteriol., 158: 1012-1017.
10. Bal, A.K., Shantharam, S. and Verma, D.P.S., 1980, Changes in the outer cell wall of Rhizobium during development of the root nodule symbiosis in soybean, Can. J. Microbiol., 23: 573-582.
11. Wensink, J. and Witholt, B., 1981, Outer membrane vesicles released by normally growing E. coli contain very little lipoprotein, Eur. J. Biochem., 116: 331-335.
12. Lugtenburg, B. and van Alphen, L., 1983, Molecular architecture and functioning of the outer membrane of E. coli and other Gram-negative bacteria, Biochim. Biophys. Acta, 737: 51-115.

BIOCHEMICAL MECHANISMS INVOLVED IN RESISTANCE OF PLANTS TO FUNGI

E.W.B. Ward

Agriculture Canada, London Research Centre
University Sub Post Office, London
Ontario, N6A 5B7, Canada

INTRODUCTION

In order to survive, plants must be able to combat stresses placed upon them by a vast range of environmental agencies and overcome the effects of severe physical damage. The initial effect of wounding is to destroy the membranes of the ruptured cells. From this primary event processes are set in motion that lead to wound healing. In other words plants have some very basic mechanisms for dealing with the disorganization of cells and tissues that are initiated by the rupture of cell membranes. The possibility that the wound response is the basis of disease resistance has been raised by others [1,2,3]. This is not an unreasonable premise if wounding is considered to include membrane perturbation or damage or any disruption of normal cellular activities. It would be a reasonable economy if plants used common mechanisms to deal with stresses of different kinds, all of which ultimately disrupt tissue or cell function. Furthermore the wound is a common site of ingress for many foreign organisms that cannot infect intact tissue and again it would not be unreasonable that biochemical responses to wounding should have been selected for their usefulness in limiting invasion by microorganisms.

There is widespread evidence from a variety of host-pathogen interactions that the development of a resistant response is associated with damage to one or more host cells. This includes evidence from diseases caused by necrotrophs in which limited host-cell death permits the development of resistance [4], and from diseases caused by hemibiotrophs and biotrophs in which restricted hypersensitive host-cell death occurs in resistant hosts, and varying degrees of biotrophy with delayed host-cell death occur in susceptible hosts [5,6,7].

IMMEDIATE RESPONSES TO WOUNDING AND INFECTION

Membrane Depolarization

According to Kahl [8], the rupture of membranes in damaged cells results in depolarization of the membranes of surrounding cells; the message, presumably electrical in nature, being transmitted in the symplasm through the plasmodesmata. Koopowitz et al. [9] detected large changes in membrane

NATO ASI Series, Vol. H1
Biology and Molecular Biology of Plant-Pathogen Interactions
Edited by J. Bailey

potential in cells surrounding wounds 30 min following wounding of soybean cotyledons or Jerusalem artichoke tubers. Depolarization is accompanied by increases in ion movement through the membrane and presumably can provide the trigger for extensive physiological activity in the surrounding tissue. Cell-membrane damage following infection can be expected to initiate the same sequence of events.

The earlier literature concerning permeability changes in infected tissues was summarized by Wheeler and Hanchey [10]. Many of these studies were related to necrotrophs and the production and activity of toxins [11], and frequently changes in permeability were associated with susceptibility. However, Pellizari et al. [12] for apple leaves infected with Venturia inaqualis and Elnaghy and Heitefuss [13] for beans infected with Uromyces phaseoli reported permeability changes to be a feature of the hypersensitive response.

Recently Tomiyama and co-workers [14] investigated the effects of infection by Phytophthora infestans on the membrane potential of cells in potato shoots. Following penetration there was a gradual depolarization of the cell in the incompatible combination for about 1 h, then 30 min before cell death, depolarization became very rapid. No changes in membrane polarization were observed in the compatible combination during the same period of development. It would be most interesting to learn what changes occurred in the membrane potential of surrounding uninfected cells. One adaptation for compatibility may be to avoid producing changes in membrane potentials and consequent ion fluxes at least temporarily in hemibiotrophs, such as species of Phytophthora, and for long periods in the case of biotrophs such as the rusts.

Of related interest is the work of Kauss and his co-workers [15] on the stimulation of callose formation in soybean cell suspensions by treatment with chitosan. The authors concluded that callose formation occurs in this system only when chitosan and also poly-L-lysine and polymyxin B, in the presence of divalent calcium ions, cause membrane perturbations that result in electrolyte leakage. Callose synthesis starts within 20 min and is presumably too rapid a response to involve de novo enzyme synthesis. It is believed that 1,3-β-glucan synthase, a Ca^{2+} dependant enzyme, is activated by the influx of Ca^{2+} ions through the damaged membrane.

Thus evidence from wounding and mechanical injury indicates that rapid changes in membrane polarization occur and that these changes are transmitted to neighbouring cells. There is evidence that rapid depolarization occurs in a resistant reaction to Phytophthora infestans in potato and not in a susceptible reaction, and that one widespread response to infection, callose formation, is stimulated by membrane damage.

Membrane and Membrane Bound Components

An immediate consequence of wounding and associated membrane damage is the release of membrane components and the contents of membrane-bound organelles. Evidence from potato tubers (summarized by Galliard [16]) indicates that degradative enzymes released from vacuoles or lysosomes can digest membrane lipids. The free fatty acids that are the product of acyl hydrolase activity stimulate the process and can directly affect membranes. Fatty acid hydroperoxides formed by lipoxygenase activity may also be toxic to membranes. Thus the process becomes autocatalytic once started. Presumably similar activities occur in cells damaged by infection although this has received little attention in relation to disease resistance. Shephard and Pitt [17] observed that similar enzymes are produced by pathogens, and that both host and pathogen enzymes might contribute to tissue autolysis. Evidence for the activity of a phospholipase produced by Phytophthora infestans in infected leaves has also been presented recently [18]. Arachidonic and eicosapentaenoic acids have been shown to function as elicitors [19]. It would be of interest to determine whether fatty acids released in damaged cells play a role in the development of the hypersensitive response, possibly as "endogenous elicitors".

In addition to enzymes of lipid metabolism, enzymes of the phenolase complex are activated on injury. These enzymes react with substrates normally separated in the vacuole, resulting in rapid oxidation of phenolics to quinones and eventually to brown or black melanins. Similar effects occur widely in diseased plants, but it is not clear what role phenol oxidases may play in resistance. Activity primarily associated with the resistant response of soybeans to Phytophthora megasperma f. sp. glycinea appears to reflect more extensive host-cell damage in resistant responses [20]. However, the o-quinone products of the reaction are very strong oxidizing agents that can react with amino, amine and thiol groups of proteins and are inhibitory to microbial growth [21]. Their contribution to the restriction of pathogen growth at the early stages of the development of a resistant response warrants further consideration.

Ethylene Production

Wounding or injury causes an almost immediate rise in the release of ethylene in some tissues [22]. The significance of wound ethylene is not clear but a variety of effects have been reported or proposed. In sweet potato, Imazeki [23] reported that low levels of ethylene stimulated peroxidase, polyphenol oxidase and phenylalanine ammonia-lyase (PAL) activity. Ethylene applications greatly enhanced PAL activity in carrot tissue [24] and PAL and cinnamic acid-4-hydroxylase in pea tissues [25,26] and in lettuce leaves [27].

Evidence that ethylene-induced increases in enzyme activity in wounded tissue involve mRNA and protein synthesis has been provided for pea tissues [28]. Ethylene stimulates in vitro RNA synthesis in mungbean nucleoli [29] and rapid changes in translatable mRNA in intact soybeans within 2 h of treatment [30].

The production of ethylene by diseased tissue has been reported frequently and has been the subject of several reviews [eg. 31,32]. Amounts released at the critical early stages of infection may be very small and comparisons may not be achieved satisfactorily. Larger amounts may be produced later in disease development but may reflect continuing tissue damage with little significance with respect to mechanisms of resistance.

Enhanced ethylene production has been reported for sweet potato and carrot roots infected with Ceratocystis fimbriata [33], barley infected with powdery mildew [34], rust-infected wheat [35] and beans [36], apple infected with Monilinia fructicola [37], Cercospora-infected sugar beet [38], and melon infected with Colletotrichum lagenarium [39]. Daly et al. [35] demonstrated that much more ethylene was produced by rust-infected susceptible wheat leaves than similar resistant leaves. However, ethylene released at the earliest post-inoculation period was similar in both interactions and subsequent greater production in susceptible leaves may reflect primarily more extensive colonization. Similar conclusions can be drawn about differences in ethylene production by tomato plants infected with Verticillium albo-atrum [40] and by infected barley leaves susceptible and resistant to powdery mildew [34].

Montalbini and Elstner [36] demonstrated clearly that in rust-infected bean leaves there was a burst of ethylene production 13 h after inoculation coinciding with the development of substomatal vesicles in the stomatal cavities. The amount was greatest in a highly resistant cultivar. A second large release of ethylene occurred in a less resistant cultivar after 60 h, a time at which necrotic brown lesions first became visible. No comparable release of ethylene occurred in a susceptible cultivar. One interpretation of these results is that the differences in ethylene production reflect differences in the amount and timing of cell damage in the three interactions.

Attempts to determine the effect of treating infected tissues with ethylene on disease development and related activities have also yielded conflicting results. Stahmann et al. [41] reported that exogenously applied ethylene increased resistance of sweet potatoes to black rot, although this was not confirmed by Chalutz and DeVay [42]. Daly et al. [35] reported that ethylene treatment induced susceptibility to stem rust in wheat. However, the induction of resistance by ethylene treatment has been reported for tangerines to *Colletotrichum gloeosporioides* [43], for tomatoes to *Verticillium albo-atrum* [44] and for apples to *Monilinia fructicola* [37]. It is probable that, as with other hormones, responses are highly dependent on concentration and timing. Ethylene treatments have been reported to stimulate phytoalexin synthesis both in uninfected pea tissues [45] and in infected carrot [33] and potato tissues [46]. In potato tissues, ethylene enhanced phytoalexin production in incompatible interactions but not in compatible interactions. Ethylene has been reported also to stimulate biosynthesis of hydroxyproline-rich glycoproteins in cell walls of pea and melon seedlings, a response associated with resistance of the latter to *Colletotrichum lagenarium* [39,47].

Of related interest are the effects of elicitors on ethylene production. Interpretation of these findings must be made with caution because of the use of wounded tissues in which physiologically active levels of ethylene presumably have been released.

Glycoprotein elicitors from *C. lagenarium*, glucan elicitors from *Phytophthora* species and endogenous elicitors from cell walls of melon leaves stimulated ethylene production in melon petiole or hypocotyl segments [48]. A cell wall elicitor from *Phytophthora megasperma* f. sp. *glycinea* stimulated ethylene and glyceollin biosynthesis in wounded soybean cotyledons [49]. However, it was concluded that ethylene did not control glyceollin synthesis or PAL activity by demonstrating that inhibition of ethylene production by treatment with the inhibitor, aminoethoxyvinylglycine (AVG) did not affect these processes and also that stimulation of ethylene production by supplying the ethylene precursor, 1-aminocyclopropane-1-carboxylic acid (ACC) in the absence of elicitor did not promote glyceollin biosynthesis. On the other hand, Kimpel and Kosuge [50] reported that with wounded soybean hypocotyls AVG caused 50% inhibition of glyceollin production and this was reversed by ACC and concluded that ethylene does not control the initiation of the glyceollin response but does influence the flow of carbon into the biosynthetic pathway. However, in the melon system, inhibition of ethylene production by AVG was accompanied by a comparable reduction in the accumulation of hydroxyproline-rich glycoproteins that have been associated with resistance to *C. lagenarium* [48].

Further work is required to determine if ethylene has an unidentified primary function in stressed cells or if it influences a number of different processes independently. In view of the wide range of metabolic activities attributed to ethylene and its release both from wounded and infected tissues, it would be surprising if it did not play a role in the biochemical response of plants to infection.

Respiration

Wounding of tissue results in an immediate rise in respiration [51]. In potato tubers the substrate for this respiration is apparently lipid, derived from disorganized membranes. A second increase in respiration, based on carbohydrate substrates, occurs within the first 12 h following wounding. This coincides with the regeneration of membranes and is dependant on the transcription of mRNA and protein synthesis as indicated by inhibitor studies [51]. The main purpose of the induced respiration is to provide energy to drive the many biosynthetic processes involved in wound healing, protein synthesis and the manufacture of intermediate or secondary metabolites. However, there also are major shifts in respiratory mechanisms including increased activity of the pentose phosphate shunt with de novo synthesis of glucose-6-phosphate dehydrogenase and 6-phosphogluconate dehydrogenase [52]. The pentose shunt can be regarded as the starting point of the shikimic acid pathway leading to biosynthesis of aromatic amino acids and a range of secondary metabolites. There are also changes in electron transport systems resulting in development of cyanide resistant respiration. The role of this alternate system is not understood but it has been proposed that it may provide a source of superoxide and peroxide [53].

Daly [1] in reviewing reports of respiratory changes in plant diseases concluded that with the possible exception of systemic virus infections, stimulation of respiratory activity is characteristic of all plant diseases. As in the case of wounded tissues there are examples of changes in pathways of respiration and electron transport and possibly the involvement of alternate terminal oxidases. There is an obvious requirement for carbon skeletons and energy to support the many biosynthetic processes that accompany the development of a resistant response and to a degree these processes appear to be similar to those set in motion by tissue damage caused by wounding.

MECHANISMS OF WOUND HEALING AND DISEASE RESISTANCE

Following the stimulus of wounding a variety of metabolic activities are set in motion that eventually become organized and result in wound healing. These include a regeneration of membranes, enhanced respiration, protein synthesis, biosynthesis of phenyl-propanoids and changes in hormonal activity [8]. These activities lead to lignin and suberin formation, the deposition of hydroxyproline-rich glycoprotein (extensin) and the regeneration of cell walls, in some cases cell division, and finally the restoration of the tissue to its normal metabolic state. It is noteworthy that all these activities are set in motion by a simple physical stimulus that initiates in membrane disruption. With some variation in mechanisms due to the increased duration of the signal and contributions from metabolic activities of the pathogen, the response to infection can be presumed to have the same basic objectives: the sealing off of the diseased tissue and the establishment of normal functions in the surrounding tissue.

In wounded potato tuber tissue there is initially rapid protein synthesis based upon existing mRNA, and after 3-4 h protein synthesis is increasingly based upon newly synthesized messenger, ribosomal and transfer RNA [54]. Additional evidence for gene activation is provided by changes that occur in the quantity and quality of nonhistone chromosomal proteins and increases in activity of chromosomal enzymes [8].

Evidence for the involvement of protein synthesis in disease resistance is provided by experiments in which either application of inhibitors of protein synthesis [55,56,57,58,59] or the disruption of normal protein synthesis by heat shock [55,60,61] has rendered plants susceptible when they would otherwise be resistant, and by the demonstration of newly synthesized proteins in resistantly responding tissues that are not present in uninfected or susceptibly-responding tissues [62,63,64,65]. The demonstration of the necessity for, or the association of protein synthesis with resistance has been considered evidence that defence mechanisms are active or generated in response to infection. However, with the possible exception of the synthesis of enzymes of secondary metabolism, disease resistance mechanisms that are directly related to protein synthesis have not been identified, nor is there convincing evidence that the surge of protein synthesis following infection differs in principle from that following wounding. Most enzymes that have been shown to be synthesized following infection of resistant plants have also been demonstrated in wounded tissues.

Peroxidase

Peroxidase has been studied widely in wounded and in infected tissues. Generally following wounding there are increases in activity and the appearance of new isozymes [54]. The role that peroxidase plays in wounded tissue is by no means clear. It is believed to be involved in lignin synthesis [66], oxidation of indoleacetic acid [67], ethylene biosynthesis [68], flavonoid biosynthesis [69] and associated with the hydroxylation of proline incorporated into cell wall protein [70]; activities that also have been considered to have significance in disease resistance. Kahl [54] has summarized evidence that different isozymes of peroxidase may have different functions which may be related to their differential distribution within tissues.

A detailed study by Daly and co-workers [71,72] of peroxidase activity in wheat carrying the Sr6 or Sr11 alleles for resistance to stem rust demonstrated that increases in activity were associated primarily with incompatible interactions. However, temperature-induced changes in compatibility did not result in comparable changes in peroxidase levels and it was concluded that enhanced peroxidase activity was a result of non-specific injury associated with incompatibility. Recent studies of induced resistance associated with rapid lignification in cucumbers suggest that enhanced peroxidase activity plays a major role [73]. Enrichment of cell-walls with hydroxyproline has also been demonstrated in this system [74]. In wheat leaves inoculated with *Botrytis cinerea*, Thorpe and Hall [75] demonstrated increased peroxidase activity following inoculation and showed cytochemically that peroxidase was localized around the margins of infected wounds where lignification occurs. These authors also demonstrated enhanced peroxidase activity in wounded wheat leaves and concluded that inoculation augments the wound response by the production of a diffusible stimulus and not simply by increased physical damage. In soybeans, however, Partridge and Keen [76] found that wounding activated peroxidase but that this was not enhanced by inoculation of the wounds with compatible or incompatible races of *Phytophthora megasperma* f. sp. *glycinea*. Evidence for the generation of superoxide during hypersensitive responses of several plant species [77] indicates that following dismutation there maybe an abundant supply of H_2O_2 for enhanced peroxidase activity.

Peroxidases oxidize phenolics in the presence of H_2O_2, and involvement of the products in plant disease has been summarized [78]. Leatham et al. [21] observed that oxidized phenolics may be both highly reactive and frequently toxic to both plants and micro-organisms. They demonstrated that numerous phenols in the presence of peroxidase or polyphenol oxidase cross-linked several proteins examined. They suggested that this process could inactivate enzymes, eg. hydrolases, released by the pathogen, or that quinones in addition to being directly toxic could polymerize with proteins to produce barriers at infection sites.

The Phenylpropanoid Pathway

Wounding causes a rise in the numbers and concentration of phenolic compounds of low molecular weight derived from cinnamic acid. The function of these compounds in wounded tissue is not clear although it has been proposed that they provide a reserve for the synthesis of more complex derivatives and evidence that chlorogenic acid for example, can be mobilized to caffeic acid has been summarized by Rhodes and Wooltorton [79]. The possibility that phenols produced in infected tissues may serve a similar purpose has received little consideration [78] perhaps because plant pathologists have been preoccupied with the search for inhibitory compounds.

A key step in cinnamic acid biosynthesis is the elimination of NH_3 from shikimate-derived phenylalanine catalyzed by the enzyme phenylalanine ammonia-lyase (PAL). Stimulation of PAL activity has been demonstrated in several species of plants including potato and sweet potato in which activity increases rapidly after a lag of 3-6 h following wounding and reaches a maximum between 24 and 48 hours [54]. Evidence that the increase is due to de novo synthesis has been provided convincingly by studies with inhibitors of transcription and translation [54], and in sweet potato tissue, by immunochemical experiments [80].

Phenylalanine ammonia-lyase activity has been examined following infection in a limited number of diseases. In most cases greatest activity has been demonstrated in incompatible responses. These include soybeans inoculated with various fungi [81] including Phytophthora megasperma f. sp. glycinea [82], although Partridge and Keen [76] did not detect significant increases in PAL after inoculation of wounds with this pathogen; potato tuber tissue inoculated with Phytophthora infestans [83]; beans inoculated with Helminthosporium carbonum or races of Colletotrichum lindemuthianum [84]; wounded wheat leaves inoculated with Botrytis cinerea [75,85]; reed canary grass inoculated with Botrytis cinerea [86]; and unwounded oat leaves inoculated with an incompatible race of crown rust [58]. Occasionally maximum stimulation has been found in susceptible reactions as for example by Loschke et al. [87] for Fusarium-inoculated peas, but this may be related to the amount of tissue colonized by this necrotrophic pathogen.

Phenylalanine ammonia-lyase activity is stimulated also by UV light [88] and the application of a wide range of elicitors. These include chitosan [89], preparations from cell walls of Phytophthora megasperma f. sp. glycinea [50,90], culture filtrates from P. infestans [83], denatured ribonuclease [91], polypeptides [92], salts of heavy metals [87,93] and inhibitors of protein and RNA synthesis [87,93]. Although it has been proposed that these agents and infection directly derepress genes that code for PAL [94], their great variety and the facility with which activity is stimulated simply by wounding render this improbable. It appears that PAL activity is a highly sensitive indicator of disturbances in the normal state of the cell.

Lignin

The products of enhanced activity of PAL and enzymes involved in cinnamic acid metabolism differ in different plant species. In potato and sweet potato, caffeic and chlorogenic acids are the most prominent immediate products of this pathway that accumulate following either wounding or infection [79,95,96,97]. There is no general consensus about the role of these compounds in disease resistance [78], although in some examples production is correlated with resistance and they may serve as precursors for other more active compounds such as quinones, and for lignin. Lignin commonly accumulates following wounding [79] and there is considerable evidence that lignin or lignin-like compounds are associated with disease resistance. For example Friend and his co-workers [78,83,96] demonstrated that activation of PAL and deposition of lignin-like material occurred more rapidly in resistant than susceptible tubers inoculated with *Phytophthora infestans*. In wheat leaves, lignification was demonstrated in wounds inoculated with host non-pathogens [85,98] and the fungi were restricted to the lignified area in the wounds. Lignification occurred more slowly following inoculation with two wheat pathogens, *Septoria tritici* and *S. nodorum*, and the fungi spread from the wounds. The deposition of lignin in papillae produced in reed canary grass in response to infection by *Peronospora arundinacea* also appears to play a role in resistance [86]. The association of enhanced lignification with induced resistance in cucumbers has been referred to earlier [74]. The possible role of lignin in disease resistance has been reviewed by Vance et al. [99]. Proposed mechanisms include: resistance to mechanical penetration and enzymic degradation, the toxicity of lignin precursors and free radicals released during polymerization and the possibility that lignification of invading hyphal tips may inhibit growth by reducing plasticity. These possibilities and the widespread stimulation of the key enzymes PAL and peroxidase suggest that the potential of lignin biosynthesis as a disease resistance mechanism warrants wider study.

Phytoalexins

The isoflavonoid phytoalexins are also products of PAL activity and cinnamic acid metabolism. They appear to be characteristic of the Leguminosae and similar or related compounds occur not only as phytoalexins but as normal constituents. It is interesting that enhanced PAL activity in the Solanaceae that follows wounding or infection, does not lead to the accumulation of phenolics generally regarded as phytoalexins. Compounds considered to be phytoalexins in this family are principally terpenoids and are derived from mevalonic rather than shikimic acid.

Evidence that phytoalexins may be produced as a result of wounding is provided by the low background levels that have been observed in wounded control tissues in many studies but more particularly by phytoalexin accumulation in tissue damaged by localized freezing [61,100] or from tracer studies that demonstrate biosynthetic activity in wounded tissues [101,102]. Phytoalexin production in wounded tissue can be greatly enhanced by the application of a wide range of elicitors. Whether these act by aggravating the injury or have more specific modes of action is not known. That such diverse molecules can produce essentially the same result would argue for the former, however there are claims that some elicitors are active without causing cell damage [103,104].

As originally employed by Müller and Börger [105, see Deverall, 106, for discussion], the term phytoalexin was used for a postulated toxic principle produced or activated by the plant following infection that inhibits the development of the pathogen. A more recent definition [107] avoids the question of the function of these compounds, requiring only that they have some antimicrobial activity and, as discussed by Stoessl [108], falls short of the original concept. Nevertheless the rationale in most work on phytoalexins assumes or attempts to establish a role for these compounds in the defence of plants. The occurrence, isolation, activity, structure, biosynthesis and role of phytoalexins have been the subjects of extensive reviews and these should be consulted for details. General schemes for biosynthesis of the two main groups of phytoalexins are illustrated in Figure 1.

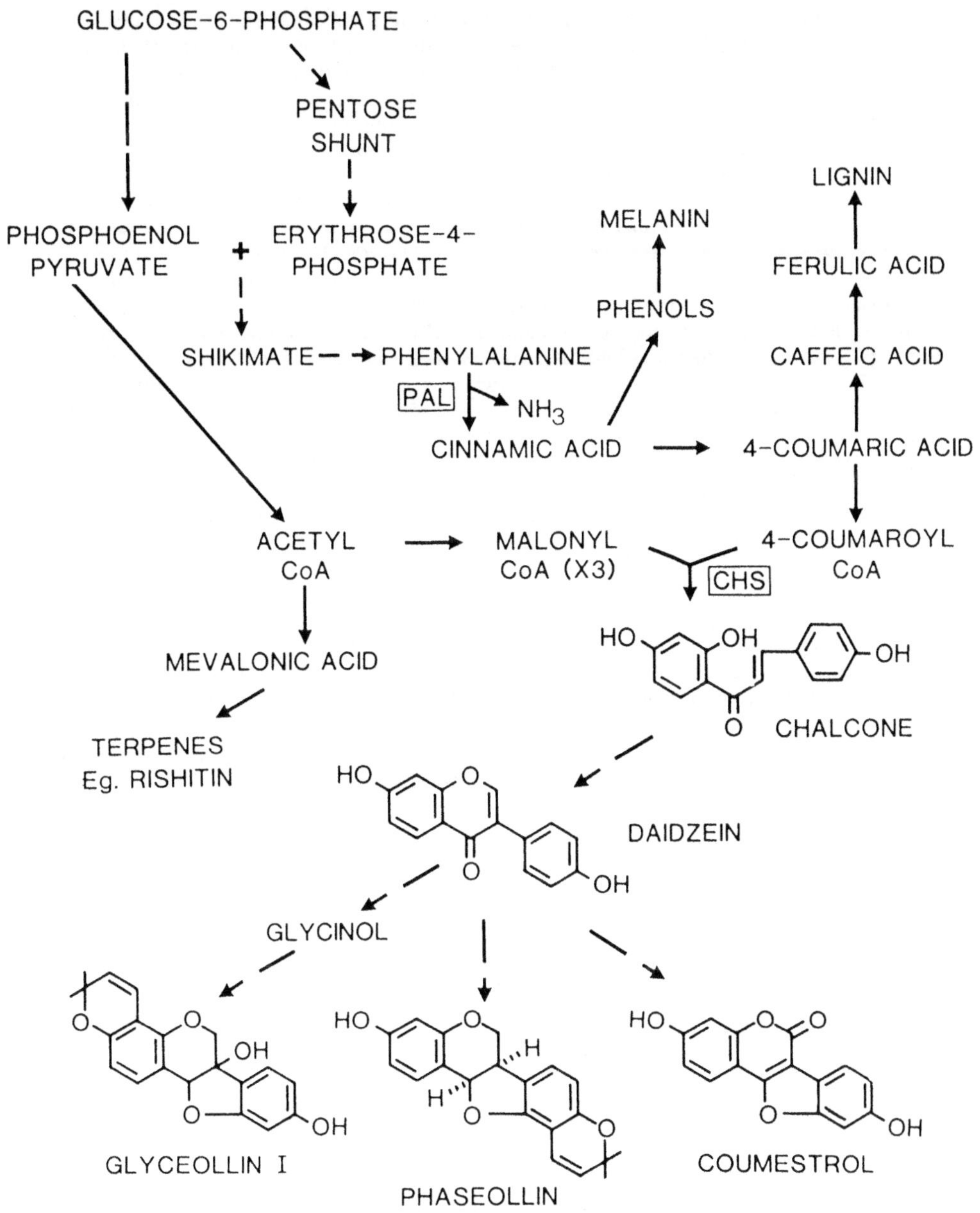

Figure 1. Biosynthetic scheme for sesquiterpene and isoflavonoid phytoalexins.

No attempt will be made here to cover this extensive literature. In general, close correlations have been demonstrated between restriction of pathogens in necrotic or hypersensitive lesions and the accumulation of fungitoxic levels of phytoalexins. Proof that in vivo phytoalexins are the agents that actually halt the spread of the fungus has been more difficult to obtain. Refined methods for detection of glyceollin in soybean or avenalumin in oats [109,110] confirm their localized accumulation but do not determine whether this is the cause or the result of limitation of hyphal growth, since glyceollin and other phytoalexins accumulate in interactions that would otherwise be compatible if growth of the pathogen is stopped by other means [111,112,113,114,115,116]. There are examples also of resistance or susceptibility that do not correlate with phytoalexin levels in infected tissues. Potato tubers retained race specific resistance even in fresh tubers in which phytoalexin accumulation was reduced greatly [117]. Compatible races of Phytophthora megasperma f. sp. glycinea appear to be able to survive in lesions in which high levels of glyceollin have accumulated and to spread from the lesions when favourable conditions are restored [115,118]. Aphanomyces euteiches is sensitive to pisatin in vitro but is apparently unaffected by the high concentrations that accumulate in diseased plants [119]. Thus the possibilities that toxicity of phytoalexins in vivo may not always reflect that in vitro, or that unidentified detoxification mechanisms may occur in vivo should not be overlooked.

Broadly speaking there are two systems which need not be alternatives, that could account for resistance and susceptibility based on phytoalexins. Phytoalexin production could occur in response to all invaders, the successful pathogens being those that either are insensitive to or can metabolize the phytoalexins to less toxic materials; or successful pathogens could avoid stimulating phytoalexin production to effective levels, while all unsuccessful invaders stimulate phytoalexin production. In critical work Van Etten and his colleagues have shown that inheritance of virulence in Nectria haematococca is closely linked to pisatin tolerance or to the ability to demethylate pisatin [120,121]. This work provides the most convincing evidence to date that phytoalexins can be essential factors in disease resistance. It may illustrate a system that is characteristic for necrotrophs. Examples of host-pathogen interactions in which the successful pathogen largely avoids stimulating phytoalexin production are Colletotrichum lindemuthianum on beans and Phytophthora species on potato and soybeans although there are conflicting views about the degradation of glyceollin in infected soybeans [102,122,123].

If phytoalexin biosynthesis is primarily a response to infection and not simply part of a general activation of phenolic biosynthesis, it would be anticipated that control of production would occur at branch points from common pathways. However, much research has been devoted to PAL, which can only signal phenol production in general, and to chalcone synthase (CHS), which directs p-coumaric acid into flavonoid biosynthesis, or other enzymes governing early steps in the pathway [82,91,124,125,126,127,128,129]. Enhanced PAL activity following infection correlates closely with the production of phytoalexins in soybeans according to Börner and Grisebach [82] but not according to Partridge and Keen [76]. The latter authors considered enhanced PAL activity a general response to wounding while the former did not observe enhanced activity following wounding. Both groups found differences between PAL activity in resistant and susceptible lesions to be either absent or to develop only at later stages of the interaction. Chalcone synthase activity was also the same in both types of lesions [82].

Bell et al. [124] have reported that in both incompatible and compatible interactions of beans with Colletotrichum lindemuthianum there are increases in the production of chalcone synthase mRNA and in the activity of chalcone synthase. The increases coincided with the onset of symptom development in both interactions and evidently are not exclusively expressions of resistance. They possibly reflect a common condition, such as cell disruption, that occurs early in the incompatible interactions and later and much more extensively in the compatible interaction. The correspondingly large increase in activity in the latter, however, did not lead to comparable increases in phaseollin production, indicating that other factors modulate phytoalexin production, at least in the compatible interaction.

Experiments with cell-suspensions, however, indicate that different stimuli can differentially stimulate enzyme synthesis and activity in different pathways. In parsley cells, UV light stimulated flavonoid biosynthesis [126], whereas a fungal elicitor stimulated the furanocoumarin pathway [130]. In bean cell-suspensions, fractions differing in monosaccharide composition obtained by gel filtration of a cell wall elicitor preparation from Colletotrichum lindemuthianum each differentially affected PAL induction [131]. Significantly, these differences did not correlate with the activity of the same fractions as elicitors of phytoalexins in bean hypocotyls. Further arguments that elicitor treatment in this system selectively enhances the activity of enzymes involved in phytoalexin production have been presented recently [104]. It is not known to what extent these findings can be extrapolated to the whole plant. In view of evidence that hormones may influence phenolic metabolism [132] there is the possibility that tissue culture conditions have been chosen that selectively favour phytoalexin production [133].

Rathmell [134] considered the stimulation of phenolic metabolism in infected beans to be directed primarily towards phaseollin production, and in general the levels of phytoalexins that accumulate indicate that plants commit extensive resources to phytoalexin biosynthesis. However, because of the emphasis on antimicrobial activity in the search for phytoalexins, many other changes induced by infection frequently are ignored. For example a large number of products were detected in extracts from Solanaceous plants challenged with Monilinia fructicola, only a few of which could be regarded as phytoalexins [135,136]. Biehn et al. [81] reported a major increase in phenols following inoculation of soybeans and beans with several non-pathogenic fungi. A major component appeared to have the UV absorption spectrum of the isoflavone, daidzein. The major incorporation of [^{14}C] phenylalanine in the ethanol soluble fraction from infected soybeans was in compounds other than glyceollin [102].

Daidzein is present in uninfected soybean hypocotyls [137], increases in concentration following infection [123,137] and is a precursor of the glyceollins [138]. It is also a common isoflavonoid intermediate in other plants. In Phaseolus aureus daidzein may be converted to coumestrol, glucosylated, degraded or polymerized [139]. Barz and Hoesel [132] use this and other examples to stress the great sensitivity of phenol metabolism to physiological parameters such as the supply of precursors and hormones. In soybeans if glyceollin biosynthesis is specifically related to disease resistance then the redirection of daidzein metabolism into the glyceollins should be examined as one potentially critical step. Is the change in pathway specifically controlled in resistantly responding plants or does it occur simply as a shunt pathway due to a massive stimulation of phenol metabolism in general? The possibility that phytoalexins may accumulate because of increased production of precursors and insufficient activity or blockage of degradative or other metabolic processes has not been satisfactorily resolved either in soybeans or other systems [102,122,123,140, 141,142].

Elicitors

Bailey [143,144] has argued cogently that the wide range of elicitor molecules and pathogenic forms that stimulate phytoalexin production require a common general mechanism and that this would be provided if they all caused some degree of localized injury. On the basis of earlier work in which tissue was damaged by freezing, it was concluded that cell injury caused the release of a constitutive elicitor that stimulated phytoalexin production in surrounding healthy cells [145]. It is not known how this elicitor might promote phytoalexin production in surrounding cells, although it is possible that it also causes a degree of trauma as may the initial elicitor. The nature of the elicitor was not determined, but evidence for transmission of signals in wounded cells, arising from depolarization and dissolution of membranes suggests areas that might be explored. However, until inhibition of phytoalexin metabolism is ruled out as a mechanism of phytoalexin accumulation, the possibility that the observed effects are due to the release of inhibitors of metabolism rather than elicitors of synthesis should not be overlooked. Evidence for release of a second or constitutive elicitor possibly is supported by the demonstration that *Phaseolus vulgaris* cell suspensions treated with denatured ribonuclease A generated an elicitor that passed through a dialysis membrane and stimulated PAL activity in cells on the other side [91]. However, the possibility that the transmitted elicitor was released from the ribonuclease by the cell suspensions rather than the reverse, together with the observation that an equally effective elicitor from cell walls of the bean pathogen *Colletotrichum lindemuthianum* did not give rise to a constitutive elicitor in parallel experiments suggests that the results are not conclusive. While the hypothesis is attractive direct evidence for the involvement of constitutive phytoalexin elicitors is still lacking.

In related work it has been proposed [146,147] that fragments of pectic cell wall polysaccharides from castor bean, soybeans or other sources are released during microbial invasion and serve as endogenous elicitors of phytoalexins. The evidence to support this proposal is limited. In castor bean, treatment of seeds with an endopolygalacturonase initially isolated from a saprophytic contaminant, *Rhizopus stolonifer*, results in increased casbene synthetase activity in cell-free extracts. Products of endopolygalacturonase activity, presumably pectic fragments also stimulate activity of casbene synthetase [146,148]. Casbene is one of several cyclization products of mevalonic acid or geranylgeranyl pyrophosphate synthesized by cellfree extracts but no evidence that the compound is a phytoalexin or that it accumulates *in vivo* has been provided. In soybeans, preparations from purified cell walls obtained by acid hydrolysis [147] or the activity of endopolygalacturonic acid lyases produced by *Erwinia carotovora* [149], and isolated pectic fragments, one of which was identified as dodeca-1,4-D-galacturonide [150], all caused the accumulation of glyceollin when applied to wounds in cotyledons.

Both these findings are of great interest for they indicate that under the assay conditions used, pectic fragments can stimulate metabolic activity. For consideration as an endogenous elicitor *in vivo* there are however a number of concerns that should be satisfied. Evidence is required for the activity of similar fragments *in vivo* preferably in a legitimate host-pathogen interaction. In the soybean system the activity of the pectic fragments has only one thousandth of the elicitor activity of the enzyme that releases the fragment and of a fungal cell wall elicitor [149,150,151]. It is therefore a relatively inefficient transmitter of information, and possibly activity of the enzyme as an elicitor is not related primarily to its catalytic activity. The time-delay in generation of hydrolysis products with elicitor activity [149] may not be consistent with other evidence from phytoalexin biosynthesis in soybeans [102]. Information is needed as to whether pectic fragments can transmit information from the initially contacted host cell to other cells where enhanced biosynthesis occurs, or whether they merely cause irritation or disruption to the initial cell with other messengers transmitting information to surrounding cells as in wounded tissues. Clearly a different type of assay is required to distinguish between primary and endogenous elicitors.

Bailey [144] summarized evidence that abiotic elicitors primarily were toxic to plant cells in various degrees, but observed that there was little information about biotic elicitors. It has been suggested that biotic elicitors act directly on host plant metabolism to stimulate biosynthetic pathways for phytoalexin production [89,152]. However, as observed by Stoessl [153], the demonstration that the cell-wall glucan elicitor from *Phytophthora megasperma* f. sp. *glycinea* not only elicits phytoalexins in several different plant species, but elicits phytoalexins that are elaborated by entirely different metabolic routes, suggests that at least for this elicitor a more indirect mechanism is more probable. Furthermore although considered primarily as elicitors of phytoalexins, it is becoming increasingly evident that many other processes are affected also. Some of these, such as stimulation of ethylene and hyroxyproline production have been referred to earlier. Evidence has been provided also that arachidonic acid, an elicitor from *Phytophthora infestans*, in addition to being an elicitor of sesquiterpenoid phytoalexins in potatoes, stimulates PAL and lignin formation and significantly affects respiration [154]. Chitosan, an effective elicitor of pisatin in peas [89], stimulates callose formation in soybeans cells, as noted earlier [15]. An elicitor obtained from cell-walls of *Colletotrichum lindemuthianum* stimulates both kievitone formation and the accumulation of cell wall hydroxyproline in cell suspensions of *Phaseolus vulgaris* [104,125].

There is some evidence that elicitors may cause cell death or damage which would thus provide a general rationale for their mode of action [155,156,157,158,159]. However, there is no reason why chemicals should not be found that more directly stimulate pathways of phytoalexin biosynthesis without first causing cell-death just as many other metabolic processes can be manipulated by the application of chemicals. As yet there is no evidence that phytoalexin production in infected plants is regulated in this way. In fact, with the exception provided by De Wit et al. [156], there is little direct evidence that biotic elicitors are involved *in vivo*. The glucan elicitor from *Phytophthora megasperma* f. sp. *glycinea* cell walls for example failed to prevent infection and colonization of soybean tissues when applied together with a compatible race of the fungus [160]. There is clearly a need to demonstrate that elicitors function in host pathogen interactions *in vivo* and are not merely artifacts.

RESISTANCE AND SUSCEPTIBILITY AND RACE-SPECIFICITY

Bailey [143,144] has proposed that the trigger for the initiation of defensive processes is death of initially infected host cells. The hypothesis was discussed for obligate pathogens where resistance was associated with penetration of host-cells subsequent to infection as well as for hemibiotrophs such as *Colletotrichum lindemuthianum* on bean and *Phytophthora megasperma* f. sp. *glycinea* on soybeans where necrosis is associated with resistance and phytoalexin production. A somewhat different interpretation is required for necrotrophs for which cell damage is greatest in susceptible tissues. Either such pathogens must be assumed to kill cells too rapidly or extensively for defence responses to be effective, or they may possess mechanisms to neutralize defence mechanisms for example by phytoalexin degradation.

The association of limited host cell-death or hypersensitivity with disease resistance has been much discussed for many years [see Bailey, 161]. That it occurs in resistant reactions to pathogens of all types suggests that localized cell death is a general non-specific response to contact with foreign organisms. It is presumably generated by any of a variety of structural or biochemical attributes of the pathogen that cause a disruption in the normal functioning of host cells and tissues. It is unlikely to be based only on a single molecule such as the glucan elicitor obtained from cell walls of Phytophthora megasperma f. sp. glycinea [151], for mutation to virulence by a simple structural change [162] presumably would have occurred long ago. The term recognition has been used frequently in relation to resistance but, as discussed by Daly [163] recently, without much supporting evidence. Localized cell death occurs also in physical wounds, and as we have seen, sets in motion a vast array of metabolic activity dedicated to the return of the tissue to its normal state.

These basic resistance and wound-healing processes are very effective. The majority of plants are resistant to the vast majority of potential pathogens and wounds that are not too severe heal. Susceptibility therefore is rare and, as argued elsewhere [164,165,166], the development of compatibility must require great specificity. Successful adaptation to a host appears to require that biotrophic organisms do not cause hypersensitivity in their hosts, or in other words do not stimulate basic resistance mechanisms. This is clearly demonstrated by the development of haustoria of these pathogens. Penetration of host cell-walls must occur without perturbation of the host cell-membrane otherwise hypersensitivity results. At the same time the membrane must be induced to expand greatly as it invaginates to accommodate the invading haustorium. How such specificity has developed is not understood, but attempts have been made to determine the number of steps in successful colonization at which specific adaptations would be required [167,168]. Heath has referred to these as switching points, and it is evident that many such points must be negotiated by rusts and powdery mildews if host-cell injury and death and consequent restriction of the pathogen are to be avoided. The complexity of the system envisaged for the rusts and powdery mildews [167,168] suggests that compatibility requires the activity of many genes (virulence genes?) in the pathogen. Evidence that compatibility is an active function of pathogens is provided possibly by experiments where growth of compatible pathogens in tissues has been interrupted by various treatments and has resulted in the development of host cell necrosis [111,113,114,115,169].

So far resistance has been discussed without reference to the well known phenomena of race-specific avirulence and cultivar specific-resistance. Resistance has been considered as a basic response of plant cells to incompatible organisms, and susceptibility the result of specific adaptation of pathogens that leads to compatibility. This has been referred to as "basic" compatibility by for example Heath [170,171] and Ellingboe [172], although there seems to be no reason why this distinction should be made. Introduction of genes for resistance into the genome of a susceptible partner renders the interaction incompatible. Development of physiologic races and repetition of the process has provided the basis for the elaboration of the gene-for-gene hypothesis in which, apparently, a gene for avirulence in the pathogen interacts with a gene for resistance in the host. These genes usually are dominant. This suggests that specificity is associated with resistance phenomena and not with susceptibility; the converse of the conclusions drawn from the physiological and biochemical evidence that has been summarized in this paper.

However, the following considerations indicate that this contradiction may be more apparent than real. It has been emphasized that gene-for-gene systems develop only in established compatible host-pathogen systems [170,171,172]. In such systems, any changes introduced into the host that interfere with the highly specific processes evolved by the pathogen to establish its compatibility would create an incompatible situation with consequent localized host-cell injury and the initiation of general and non-specific resistance responses. These responses would be fundamentally the same whether controlled by specific genes for resistance or by other factors that induce resistance in otherwise compatible interactions [61,111,115,173]. Thus a gene introduced into a susceptible cultivar that interfered with the specific mechanisms inducing susceptibility would appear to be a gene for resistance. Such genes would not be resistance genes in the sense that they operate a resistance mechanism and perhaps the term is misleading. Instead, they would prevent the establishment or maintenance of a state of compatibility and non-specific resistance mechanisms or incompatibility would develop as the inevitable consequence. The immediate effects of such genes could be at the level of their primary products as required by genetic evidence [172]. Arguments that phytoalexin production or other resistance-related activities cannot operate in gene-for-gene systems because they obviously require the activity of many genes [172] would not apply. Susceptibility in such systems would appear to be passive, but only in relation to the introduced gene. For example, demonstration that deletion of dominant host genes for resistance [174] or dominant pathogen genes for avirulence [175] results in a susceptible reaction-type, equivalent to that conditioned by the recessive alleles of such genes, is not evidence that compatibility occurs passively, but rather that the active processes involved in the generation of compatibility, governed by many other genes, continue to function when not interfered with. Many observations of the physiological and biochemical similarities of defence mechanisms when resistance occurs in the same host-pathogen combination either under the control of single genes for resistance or due to environmental or other manipulations of compatible interactions [61,111,112,114,115,173] are also understandable in terms of interference with mechanisms of compatibility. Thus, both lack of specificity in resistance mechanisms and the high specificity required by genetic analyses would be satisfied by such a system. If recognition is a valid concept in plant pathogen interactions, primary and specific recognition must occur in interactions that are compatible; incompatibility follows when compatibility fails. Recognition would be for likeness e.g. protein copolymerization [165] and hence highly specific. Dissimilarity entails no such precision and would lead to incompatibility.

A rather different interpretation is required for necrotrophic pathogens that kill host cells in both resistant and susceptible plants but do so more extensively in the latter. These fungi cannot be considered to establish a compatible association. Instead virulence must depend on the ability of these fungi to survive or overcome the effects of the host defence responses that are generated by the killing of host cells. It is perhaps significant that among fungal pathogens phytoalexin detoxification mechanisms have been demonstrated primarily in necrotrophs.

CONCLUSIONS

The destruction of plant cells in wounded tissues provides a simpler model for study of the response of plants to trauma than does infectious disease because only one organism is involved. Following wounding there is rapid depolarization of membranes, breakdown in membrane structure and liberation of membrane components, fatty acids and enzymes, especially oxidative enzymes, and evolution of ethylene. These events are immediately communicated to surrounding cells in the tissue and within hours gene activation, synthesis of mRNA and proteins commences. There is rapid increase in PAL activity and pathways of secondary metabolism become highly active. All these processes have been studied also in relation to disease resistance, which suggests that wound responses and disease resistance mechanisms have much in common. They both appear to be generated by cell damage, typified by the hypersensitive response in the case of infection. They both have as their ultimate objective the restoration of the damaged tissue to a state of normal organization and function. Interpretation of these responses in infected tissues is complicated and the responses themselves are more complex because of the involvement of two organisms. However, the evidence suggests that as proposed by Bailey [143,144] the immediate death of cells following infection is the basis for the development of resistance mechanisms. These are non-specific mechanisms that are generated to varying degrees in response to a wide variety of agents including wounding. In contrast the establishment of compatibility by biotrophic fungi requires active and highly specific adaptations that permit invasion to take place with a minimum of damage or disruption of host cells. Cultivar specific resistance in gene-for-gene systems may be visualized as an interference with factors essential for the development or maintenance of compatibility. The consequent inability to maintain actively the compatible condition automatically results in incompatibility and the generation of non-specific resistance phenomena. Genes for resistance then would not be involved in resistance mechanisms but only in modulating compatibility. Resistance to necrotrophs would be based on the same general non-specific mechanisms. These organisms cannot be considered to develop compatibility. Instead for virulence they would depend on their ability to bypass or withstand and overcome the host defence mechanisms that they stimulate.

REFERENCES

1. Daly, J.M., 1976, The carbon balance of diseased plants: changes in respiration, photosynthesis and translocation, in: "Physiological Plant Pathology", R. Heitefuss and P.H. Williams, eds., pp. 450-479, Springer-Verlag, Berlin, Heidelberg and New York.
2. Kuč, J., 1978, Changes in intermediary metabolism caused by disease, in: "Plant Disease, An Advanced Treatise. III. How Plants Suffer from Disease", J.G. Horsfall and E.B. Cowling, eds., pp. 349-374, Academic Press, New York, San Francisco and London.
3. Uritani, I. and Oba, K., 1978, The tissue slice system as a model for studies of host-parasite relationships, in: "Biochemistry of Wounded Plant Tissues", G. Kahl, ed., pp. 287-308, Walter de Gruyter and Co., Berlin and New York.
4. Stewart, A. and Mansfield, J.W., 1984, Fungal development and plant responses in detached onion bulb scales and leaves inoculated with Botrytis allii, B. cinerea, B. fabae and B. squamosa, Plant Pathol., 33: 401-409.
5. Littlefield, L.J. and Heath, M.C., 1979, "Ultrastructure of Rust Fungi", Academic Press, New York, San Francisco and London.
6. Stössel, P., Lazarovits, G. and Ward, E.W.B., 1981, Electron microscope study of race-specific and age-related resistant and susceptible reactions of soybeans to Phytophthora megasperma var. sojae, Phytopathology, 71: 617-623.
7. Tomiyama, K., 1960, Some factors affecting the death of hypersensitive potato plant cells infected by Phytophthora infestans, Phytopath. Z., 39: 134-148.
8. Kahl, G., 1982, Molecular biology of wound healing: the conditioning phenomenon, in: "Molecular Biology of Plant Tumors", G. Kahl and J.S. Schell, eds., pp. 211-267, Academic Press, New York and London.
9. Koopowitz, H., Dhyse, R. and Fosket, D.E., 1975, Cell membrane potentials of higher plants: changes induced by wounding, J. Exp. Bot., 26: 131-137.

10. Wheeler, H.E. and Hanchey, P., 1986, Permeability phenomena in plant disease, Annu. Rev. Phytopathology, 6: 331-350.
11. Hanchey, P. and Wheeler, H.E., 1979, The role of host cell membranes, in: "Recognition and Specificity in Plant Host-Parasite Interactions", J.M. Daly and I. Uritani, eds., pp. 193-210, Japan Scientific Societies Press, Tokyo.
12. Pellizari, E.D., Kuć, J. and Williams, E.B., 1970, The hypersensitive reaction in Malus species: changes in leakage of electrolytes from apples leaves after inoculation with Venturia inaequalis, Phytopathology, 60: 373-376.
13. Elnaghy, M.A. and Heitefuss, R., 1976, Permeability changes and production of antifungal compounds in Phaseolus vulgaris infected with Uromyces phaseoli. I. Role of the spore germination self-inhibitor, Physiol. Plant Pathol., 8: 253-267.
14. Tomiyama, K., Okamoto, H. and Katou, K., 1983, Effect of infection by Phytophthora infestans on the membrane potential of potato cells, Physiol. Plant Pathol., 22: 233-243.
15. Köhle, H., Jeblick, W., Poten, F., Blaschek, W. and Kauss, H., 1985, Chitosan-elicited callose synthesis in soybean cells as a Ca^{2+}-dependent process, Plant Physiol., 77: 544-551.
16. Galliard, T., 1978, Lipolytik and lipoxygenase enzymes in plants and their action in wounded tissues, in: "Biochemistry of Wounded Plant Tissues", G. Kahl, ed., pp. 155-201, Walter de Gruyter, Berlin and New York.
17. Shephard, D.V. and Pitt, D., 1976, Purification of a phospholipase from Botrytis and its effects on plant tissues, Phytochemistry, 15: 1465-1470.
18. Moreau, R.A. and Rawa, D., 1984, Phospholipase activity in cultures of Phytophthora infestans and in infected potato leaves, Physiol. Plant Pathol., 24: 187-199.
19. Bostock, R.M., Laine, R.A. and Kuć, J.A., 1982, Factors affecting the elicitation of sesquiterpenoid phytoalexin accumulation by eicosapentaenoic and arachidonic acids in potato, Plant Physiol., 70: 1417-1424.
20. Lazarovits, G. and Ward, E.W.B., 1982, Polyphenoloxidase activity in soybean hypocotyls at sites inoculated with Phytophthora megasperma f. sp. glycinea, Physiol. Plant Pathol., 21: 227-236.
21. Leatham, G.F., King, V. and Stahmann, M.A., 1980, In vitro protein polymerization by quinones or free radicals generated by plant or fungal oxidative enzymes, Phytopathology, 70: 1134-1140.
22. Lee, T.H., McGlasson, W.B. and Edwards, R.A., 1970, Physiology of disks of irradiated tomato fruit. I. Influence of cutting and infiltration on respiration, ethylene production and ripening, Rad. Bot., 10: 521-529.
23. Imazeki, H., 1970, Induction of peroxidase activity by ethylene in sweet potato, Plant Physiol., 46: 172-174.
24. Chalutz, E., 1973, Ethylene-induced phenylalanine ammonia-lyase activity in carrot roots, Plant Physiol., 51: 1033-1036.
25. Hyodo, H. and Yang, S.F., 1971, Ethylene-enhanced formation of cinnamic acid-4-hydroxylase in excised pea hypocotyl tissue, Arch. Biochem. Biophys., 143: 338-339.
26. Hyodo, H. and Yang, S.F., 1971, Ethylene enhanced synthesis of phenylalanine ammonia-lyase in pea seedlings, Plant Physiol., 47: 765-770.
27. Hyodo, H., Kuroda, H. and Yang, S.F., 1978, Induction of phenylalanine ammonia-lyase and increase in phenolics in lettuce leaves in relation to the development of Russet Spotting caused by ethylene, Plant Physiol., 62: 31-35.
28. Ridge, I. and Osborne, D.J., 1970, Regulation of peroxidase activity by ethylene in Pisum sativum. Requirements for protein and RNA synthesis, J. Exp. Bot., 21: 720-734.

29. Grierson, D., Kear, R.J., Thompson, J.R. and Garcia-Mora, R., 1982, Stimulation of in vitro RNA synthesis by pretreating plants with auxins is due to auxin-induced ethylene production, Z. Pflanzenphysiol., 107: 419-426.
30. Zurfluh, L.L. and Guilfoyle, T.J., 1982, Auxin- and ethylene-induced changes in the population of translatable messenger RNA in basal sections and intact soybean hypocotyls, Plant Physiol., 69: 338-340.
31. Hislop, E.C., Hoad, G.V. and Archer, S.A., 1973, The involvement of ethylene in plant diseases, in: "Fungal Pathogenicity and the Plant's Response", R.J.W. Byrde and C.V. Cutting, eds., pp. 87-113, Academic Press, London and New York.
32. Pegg, G.F., 1976, The involvement of ethylene in plant pathogenesis, in: "Physiological Plant Pathology", R. Heitefuss and P.H. Williams, eds., pp. 582-591, Springer-Verlag, Berlin, Heidelberg and New York.
33. Chalutz, E., DeVay, J.E. and Maxie, E.C., 1969, Ethylene-induced isocoumarin formation in carrot root tissue, Plant Physiol., 44: 235-241.
34. Hislop, E.D. and Stahmann, M.A., 1971, Peroxidase and ethylene production by barley leaves infected with Erysiphe graminis f. sp. hordei, Physiol. Plant Pathol., 1: 297-312.
35. Daly, J.M., Seevers, P.M. and Ludden, P., 1970, Studies on wheat stem rust resistance controlled at the Sr 6 locus. III. Ethylene and disease reaction, Phytopathology, 60: 1648-1652.
36. Montalbini, P. and Elstner, E.F., 1977, Ethylene evolution by rust-infected detached bean (Phaseolus vulgaris L.) leaves susceptible and hypersensitive to Uromyces phaseoli (Pers.) Wint, Planta, 135: 301-306.
37. Hislop, E.C., Archer, S.A. and Hoad, G.V., 1973, Ethylene production by healthy and Sclerotinia fructigena-infected apple peel, Phytochemistry, 12: 2081-2086.
38. Koch, F., Baur, M., Burba, M. and Elstner, E.F., 1980, Ethylene formation by Beta vulgaris leaves during systemic (Beet Mosaic Virus and Beet Mild Yellowing Virus, BMV + BMYV) or necrotic (Cercospora beticola Sacc.) diseases, Phytopath. Z., 98: 40-46.
39. Toppan, A., Roby, D. and Esquerré-Tugayé, M-T., 1982, Cell surfaces in plant microorganism interactions. III. In vivo effect of ethylene on hydroxyproline-rich glycoprotein accumulation in the cell wall of diseased plants, Plant Physiol., 70: 82-86.
40. Pegg, G.F. and Cronshaw, D.K., 1976, Ethylene production in tomato plants infected with Verticillium albo-atrum., Physiol. Plant Pathol., 8: 279-295.
41. Stahmann, M.A., Clare, B.G. and Woodbury, W., 1966, Increased disease resistance and enzyme activity induced by ethylene and ethylene production by black rot-infected sweet potato tissue, Plant Physiol., 41: 1505-1512.
42. Chalutz, E. and DeVay, J.E., 1969, Production of ethylene in vitro and in vivo by Ceratocystis fimbriata in relation to disease development, Phytopathology, 59: 750-755.
43. Brown, G.E., 1978, Hypersensitive response of orange-colored Robinson tangerines to Colletotrichum gloeosporioides after ethylene treatment, Phytopathology, 68: 700-706.
44. Pegg, G.F., 1976, The response of ethylene treated tomato plants to infection by Verticillium albo-atrum, Physiol. Plant Pathol., 9: 215-226.
45. Chalutz, E. and Stahmann, M.A., 1969, Induction of pisatin by ethylene, Phytopathology, 59: 1972-1973.
46. Henfling, J.W.D.M., Lisker, N. and Kuć, J., 1978, Effect of ethylene on phytuberin and phytuberol accumulation in potato tuber slices, Phytopathology, 68: 857-862.
47. Esquerré-Tugayé, M-T., Lafitte, C., Mazau, D., Toppan, A. and Touzé, A., 1979, Cell surfaces in plant-microorganism interactions. II. Evidence

for the accumulation of hydroxyproline-rich glycoproteins in the cell wall of diseased plants as a defense mechanism, Plant Physiol., 64: 320-326.

48. Roby, D., Toppan, A. and Esquerré-Tugayé, M-T., 1985, Cell surfaces in plant-microorganism interactions. V. Elicitors of fungal and of plant origin trigger the synthesis of ethylene and of cell wall hydroxyproline-rich glycoprotein in plants, Plant Physiol., 77: 700-704.
49. Paradies, I., Konze, J.R., Elstner, E.F. and Paxton, J., 1980, Ethylene: indicator but not inducer of phytoalexin synthesis in soybean, Plant Physiol., 66: 1106-1109.
50. Kimpel, J.A. and Kosuge, T., 1985, Metabolic regulation during glyceollin biosynthesis in green soybean hypocotyls, Plant Physiol., 77: 1-7.
51. Laties, G.G., 1978, The development and control of respiratory pathways in slices of plant storage organs, in: "Biochemistry of Wounded Plant Tissues", G. Kahl, ed., pp. 421-466, Walter de Gruyter and Co., Berlin and New York.
52. Kahl, G., 1974, De novo synthesis of glucose-6-phosphate-(E.C.1.1.1.49) and 6-phosphogluconate dehydrogenase (E.C.1.1.1.44) in plant storage tissue slices, Z. Naturforsch. C: Bioscience, 29C: 700-704.
53. Rich, P.R., Boveris, A., Bonner, W.D. and Moore, A.L., 1976, Hydrogen peroxide generation by the alternate oxidase of higher plants, Biochem. Biophys. Res. Commun., 3: 695-703.
54. Kahl, G., 1978, Induction and degradation of enzymes in aging plant storage tissues, in: "Biochemistry of Wounded Plant Tissues", G. Kahl, ed., pp. 347-390, Walter de Gruyter and Co., Berlin and New York.
55. Heath, M.C., 1979, Effects of heat shock, actinomycin D, cycloheximide and blasticidin S on non-host interactions with rust fungi, Physiol. Plant Pathol., 15: 211-218.
56. Nozue, M., Tomiyama, K. and Doke, N., 1977, Effect of blasticidin S on development of potential of potato tuber cells to react hypersensitively to infection by Phytophthora infestans, Physiol. Plant Pathol., 10: 181-189.
57. Tani, T., Yamamoto, H., Onoe, T. and Naito, N., 1975, Initiation of resistance and host cell collapse in the hypersensitive reaction of oat leaves against Puccinia coronata avenae, Physiol. Plant Pathol., 7: 231-242.
58. Yamamoto, H., Tani, T. and Hokin, H., 1976, Protein synthesis linked with resistance of oat leaves to crown rust fungus, Ann. Phytopathol. Soc. Japan, 42: 583-590.
59. Yoshikawa, M., Yamauchi, K. and Masago, H., 1978, De novo messenger RNA and protein synthesis are required for phytoalexin-mediated disease resistance in soybean hypocotyls, Plant Physiol., 61: 314-317.
60. Chamberlain, D.W., 1972, Heat-induced susceptibility to non-pathogens and cross protection against Phytophthora megasperma var. sojae in soybean, Phytopathology, 62: 645-646.
61. Ward, E.W.B., Stossel, P. and Lazarovits, G., 1982, Similarities between age-related and race-specific resistance of soybean hypocotyls to Phytophthora megasperma var. sojae, Phytopathology, 71: 504-508.
62. Andebrhan, T., Coutts, R.H.A., Nagih, E.E. and Wood, R.K.S., 1980, Induced resistance and changes in the soluble protein fraction of cucumber leaves locally infected with Colletotrichum lagenarium or tobacco necrosis virus, Phytopath. Z., 98: 47-52.
63. Hadwiger, L.A. and Wagoner, W., 1983, Electrophoretic patterns of pea and Fusarium solani proteins synthesized in vitro or in vivo which characterize the compatible and incompatible interactions, Physiol. Plant Pathol., 23: 153-162.
64. Hwang, B.K., Wolf, G. and Heitefuss, R., 1982, Soluble proteins and multiple forms of esterases in leaf tissue at first and flag leaf

stages of spring barley plants in relation to their resistance to powdery mildew (Erysiphe graminis f. sp. hordei), Physiol. Plant Pathol., 21: 367-372.
65. Yamamoto, H. and Tani, T., 1982, Two-dimensional analysis of enhanced synthesis of proteins in oat leaves responding to the crown rust infection, Physiol. Plant Pathol., 21: 209-216.
66. Harkin, J.M. and Obst, J.R., 1973, Lignification in trees: indication of exclusive peroxidase participation, Science, 180: 296-298.
67. Meudt, W.J. and Stecher, K.J., 1972, Promotion of peroxidase activity in the cell wall of Nicotiana, Plant Physiol., 50: 157-160.
68. Mapson, L.W. and Wardale, D.A., 1972, Role of indolyl-3-acetic acid in the formation of ethylene from 4-methylmercapto-2-oxo-butyric acid by peroxidase, Phytochemistry, 11: 1371-1387.
69. Rathmell, W.G. and Bendall, D.S., 1972, The peroxidase-catalysed oxidation of a chalcone and its possible physiological significance, Biochem. J., 127: 125-132.
70. Ridge, I. and Osborne, D., 1971, Role of peroxidase when hydroxy-proline-rich protein in plant cell walls is increased by ethylene, Nature New Biology, 229: 205-208.
71. Daly, J.M., Ludden, P. and Seevers, P., 1971, Biochemical comparisons of resistance to wheat stem rust disease controlled by the Sr 6 or Sr 11 alleles, Physiol. Plant Pathol., 1: 397-407.
72. Seevers, P. and Daly, J.M., 1970, Studies on wheat stem rust resistance controlled at the Sr 6 locus, Phytopathology, 60: 1642-1647.
73. Hammerschmidt, R., Nuckels, E.M. and Kuć, J., 1982, Association of enhanced peroxidase activity with induced systemic resistance of cucumber to Colletotrichum lindemuthianum, Physiol. Plant Pathol., 20: 73-82.
74. Hammerschmidt, R., Lamport, D.T.A. and Muldoon, E.P., 1984, Cell wall hydroxyproline enhancement and lignin deposition as an early event in the resistance of cucumber to Cladosporium cucumerinum, Physiol. Plant Pathol., 24: 43-47.
75. Thorpe, J.R. and Hall, J.L., 1984, Chronology and elicitation of changes in peroxidase and phenylalanine ammonia-lyase activities in wounded wheat leaves in response to inoculation by Botrytis cinerea, Physiol. Plant Pathol., 25: 363-379.
76. Partridge, J.E. and Keen, N.T., 1977, Soybean phytoalexins: rates of synthesis are not regulated by activation of initial enzymes in flavonoid biosynthesis, Phytopathology, 67: 50-55.
77. Doke, N., 1983, Involvement of superoxide anion generation in the hypersensitive response of potato tuber tissues to infection with an incompatible race of Phytophthora infestans and to the hyphal wall components, Physiol. Plant Pathol., 23: 345-357.
78. Friend, J., 1981, Plant phenolics, lignification and plant disease, Progress in Phytochemistry, 7: 197-261.
79. Rhodes, J.M. and Wooltorton, L.C.S., 1978, The biosynthesis of phenolic compounds in wounded plant storage tissues, in: Biochemistry of Wounded Plant Tissues", G. Kahl, ed., pp. 243-286, Walter de Gruyter and Co., Berlin and New York.
80. Tanaka, V. and Uritani, I., 1976, Immunochemical studies on fluctuation of phenylalanine ammonia-lyase activity in sweet potato in response to cut injury, J. Biochem., 79: 217-219.
81. Biehn, W.L., Kuć, J. and Williams, E.B., 1968, Accumulation of phenols in resistant plant-fungi interactions, Phytopathology, 58: 1255-1260.
82. Börner, H. and Grisebach, H., 1982, Enzyme induction in soybean infected by Phytophthora megasperma f. sp. glycinea, Arch. Biochem. Biophys., 217: 65-71.
83. Henderson, S.J. and Friend, J., 1979, Increase in PAL and lignin-like compounds as race-specific resistance responses of potato tubers to Phytophthora infestans, Phytopath. Z., 94: 323-334.

84. Rahe, J.E., Kuć, J., Chuang, C.M. and Williams, E.B., 1969, Correlation of phenolic metabolism with histological changes in Phaseolus vulgaris inoculated with fungi, Neth. J. Plant Pathol., 75: 58-71.
85. Maule, A.J. and Ride, J.P., 1982, Ultrastructure and autoradiography of lignifying cells in wheat leaves wound-inoculated with Botrytis cinerea, Physiol. Plant Pathol., 20: 235-241.
86. Vance, C.P. and Sherwood, R.T., 1977, Lignified papilla formation as a mechanism for protection in reed canary grass, Physiol. Plant Pathol., 10: 247-256.
87. Loschke, D.C., Hadwiger, L.A., Schroder, J. and Hahlbrock, K., 1981, Effects of light and of Fusarium solani on synthesis and activity of phenylalanine ammonia-lyase in peas, Plant Physiol., 68: 680-685.
88. Hadwiger, L.A. and Schwochau, M.E., 1971, Ultraviolet light-induced formation of pisatin and phenylalanine ammonia-lyase, Plant Physiol., 47: 588-590.
89. Hadwiger, L.A. and Beckman, J.M., 1980, Chitosan as a component of pea-Fusarium solani interactions, Plant Physiol., 66: 205-211.
90. Zahringer, U., Ebel, J. and Grisebach, H., 1978, Induction of phytoalexin synthesis in soybean: elicitor-induced increase in enzyme activities of flavonoid biosynthesis and incorporation of mevalonate into glyceollin, Arch. Biochem. and Biophys., 188: 450-455.
91. Dixon, R.A., Dey, P.M., Lawton, M.A. and Lamb, C.J., 1983, Phytoalexin induction in French bean. Intercellular transmission of elicitation in cell suspension cultures and hypocotyl sections of Phaseolus vulgaris, Plant Physiol., 71: 251-256.
92. Hadwiger, L.A., Jafri, A., Von Broembsen, S. and Eddy, Jr. R., 1974, Mode of pisatin induction. Increased template activity and dye-binding capacity for chromatin isolated from polypeptide-treated pea pods, Plant Physiol., 53: 52-63.
93. Munn, C.B. and Drysdale, R.B., 1975, Kievitone production and phenylalanine ammonia-lyase activity in cowpea, Phytochemistry, 14: 1303-1307.
94. Schwochau, M.E. and Hadwiger, L.A., 1968, Stimulation of pisatin production in Pisum sativum by actinomycin D and other compounds, Arch. Biochem. Biophys., 126: 731-733.
95. Akazawa, T. and Wada, K., 1961, Analytical study of ipomeamarone and chlorogenic acid alterations in sweet potato roots infected by Ceratocystis fimbriata, Plant Physiol., 36: 139-144.
96. Friend, J., Reynolds, S.B. and Aveyard, M.A., 1973, Phenylalanine ammonia-lyase, chlorogenic acid and lignin in potato tuber tissue inoculated with Phytophthora infestans, Physiol. Plant Pathol., 3: 495-507.
97. Kuć, J., Henze, R.E., Ullstrup, A.J. and Quackenbush, F.W., 1956, Chlorogenic and caffeic acids as fungistatic agents produced by potatoes in response to inoculation with Helminthosporium carbonum, J. Amer. Chem. Soc., 78: 3123-3125.
98. Ride, J.P., 1975, Lignification in wounded wheat leaves in response to fungi and its possible role in resistance, Physiol. Plant Pathol., 5: 125-134.
99. Vance, C.P., Kirk, T.K. and Sherwood, R.T., 1980, Lignification as a mechanism of disease resistance, Annu. Rev. Phytopathology, 18: 259-288.
100. Rahe, J.E. and Arnold, R.M., 1975, Injury-related phaseollin accumulation in Phaseolus vulgaris and its implications with regard to specificity of host-parasite interaction, Can. J. Bot., 53: 921-928.
101. Sakai, S., Tomiyama, K. and Doke, N., 1979, Synthesis of a sesquiterpenoid phytoalexin rishitin in non-infected tissue from various parts of potato plants immediately after slicing, Ann. Phytopathological Soc. Japan, 45: 705-711.

102. Yoshikawa, M., Yamauchi, K. and Masago, H., 1979, Biosynthesis and biodegradation of glyceollin by soybean hypocotyls infected with Phytophthora megasperma var. sojae, Physiol. Plant Pathol., 14: 157-169.
103. Paxton, J.D., Goodchild, D.J. and Cruickshank, I.A.M., 1974, Phaseollin production by live bean endocarp, Physiol. Plant Pathol., 4: 167-171.
104. Robbins, M.P., Bolwell, G.P. and Dixon, R.A., 1985, Metabolic changes in elicitor-treated bean cells. Selectivity of enzyme induction in relation to phytoalexin accumulation, Eur. J. Biochem., 148: 563-569.
105. Müller, K.O. and Börger, H., 1940, Experimentelle Untersuchungen über die Phytophthora-Resistenz der Kartoffel, Arbeitender der Biologischen Reichsanstalt für Land-und Forstwirtschaft (Berlin), 23: 189-231.
106. Deverall, B.J., 1982, Introduction, in: "Phytoalexins", J.A. Bailey and J.W. Mansfield, eds., pp. 1-20, John Wiley and Sons, New York.
107. Paxton, J.D., 1980, A new working definition of the term "Phytoalexin", Plant Dis., 64: 734.
108. Stoessl, A., 1983, Secondary plant metabolites in preinfectional and postinfectional resistance, in: "The Dynamics of Host Defence", J.A. Bailey and B.J. Deverall, eds., pp. 71-122, Academic Press, Sydney, New York and London.
109. Hahn, M.G., Bonhoff, A. and Grisebach, H., 1985, Quantitative localization of the phytoalexin glyceollin I in relation to fungal hyphae in soybean roots infected with Phytophthora megasperma f. sp. glycinea, Plant Physiol., 77: 591-601.
110. Mayama, S. and Tani, T., 1982, Microspectrophotometric analysis of the location of avenalumin accumulation in oat leaves in response to fungal infection, Physiol. Plant Pathol., 21: 141-149.
111. Bailey, J.A., Rowell, P.M. and Arnold, G.M., 1980, The temporal relationship between host cell death, phytoalexin accumulation and fungal inhibition during hypersensitive reactions of Phaseolus vulgaris to Colletotrichum lindemuthianum, Physiol. Plant Pathol., 17: 329-339.
112. Börner, H., Schatz, G. and Grisebach, H., 1983, Influence of the systemic fungicide metalaxyl on glyceollin accumulation in soybean infected with Phytophthora megasperma f. sp. glycinea, Physiol. Plant Pathol., 23: 145-152.
113. Kiraly, Z., Barna, B. and Ersek, T., 1972, Hypersensitivity as a consequence, not the cause, of plant resistance to infection, Nature, 239: 456-458.
114. Ward, E.W.B., 1984, Suppression of metalaxyl activity by glyphosate: Evidence that host defence mechanisms contribute to metalaxyl inhibition of Phytophthora megasperma f. sp. glycinea in soybeans, Physiol. Plant Pathol., 25: 381-386.
115. Ward, E.W.B. and Lazarovits, G., 1982, Temperature-induced changes in specificity in the interaction of soybeans with Phytophthora megasperma f. sp. glycinea, Phytopathology, 72: 826-830.
116. Ward, E.W.B., Lazarovits, G., Stössel, P., Barrie, S. and Unwin, C.H., 1980, Glyceollin production associated with control of Phytophthora rot of soybeans by the systemic fungicide, metalaxyl, Phytopathology, 70: 738-740.
117. Bostock, R.M., Nuckles, E., Henfling, J.W.D.M. and Kuć, J.A., 1983, Effects of potato tuber age and storage on sesquiterpenoid stress metabolite accumulation, steroid glycoalkaloid accumulation and response to abscisic and arachidonic acids, Phytopathology, 73: 435-438.
118. Ward, E.W.B., 1983, Effects of mixed or consecutive inoculations on the interaction of soybeans with races of Phytophthora megasperma f. sp. glycinea, Physiol. Plant Pathol., 23: 281-294.
119. Pueppke, S.G. and Van Etten, H.D., 1976, The relation between pisatin and the development of Aphanomyces euteiches in diseased Pisum sativum, Phytopathology, 66: 1174-1185.

120. Tegtmeier, K.J. and Van Etten, H.D., 1982, The role of pisatin tolerance and degradation in the virulence of Nectria haematococca on peas: a genetic analysis, Phytopathology, 72: 608-612.
121. Van Etten, H.D. and Matthews, P.S., 1984, Naturally occurring variation in the inducibility of pisatin demethylating activity in Nectria haematococca mating population VI, Physiol. Plant Pathol., 25: 149-160.
122. Moesta, P. and Grisebach, H., 1981, Investigation of the mechanism of phytoalexin accumulation in soybean induced by glucan or mercuric chloride, Arch. Biochem. Biophys., 211: 39-43.
123. Moesta, P. and Grisebach, H., 1981, Investigation of glyceollin accumulation in soybean infected by Phytophthora megasperma f. sp. glycinea, Arch. Biochem. Biophys., 212: 462-467.
124. Bell, J.N., Dixon, R.A., Bailey, J.A., Rowell, P.M. and Lamb, C.J., 1984, Differential induction of chalcone synthase mRNA activity at the onset of phytoalexin accumulation in compatible and incompatible plant-pathogen interactions, Proc. Natl. Acad. Sci. USA, 81: 3384-3388.
125. Bolwell, G.P., Robbins, M.P. and Dixon, R.A., 1985, Metabolic changes in elicitor-treated bean cells. Enzymic responses associated with rapid changes in cell wall components, Eur. J. Biochem., 148: 571-578.
126. Chappell, J. and Hahlbrock, K., 1984, Transcription of plant defence genes in response to UV. light or fungal elicitor, Nature, 311: 76-78.
127. Cramer, C.L., Ryder, T.B., Bell, J.N. and Lamb, C.J., 1985, Rapid switching of plant gene expression induced by fungal elicitor, Science 227: 1240-1243.
128. Lawton, M.A., Dixon, R.A., Hahlbrock, K. and Lamb, C., 1983, Rapid induction of the synthesis of phenylalanine ammonia-lyase and of chalcone synthase in elicitor-treated plant cells, Eur. J. Biochem., 129: 593-601.
129. Ryder, T.B., Cramer, D.L., Bell, J.N., Robbins, M.P., Dixon, R.A. and Lamb, C.J., 1984, Elicitor rapidly induces chalcone synthase mRNA in Phaseolus vulgaris cells at the onset of the phytoalexin defense response, Proc. Natl. Acad. Sci. USA., 81: 5724-5728.
130. Tietjen, K.G. and Matern, U., 1983, Differential response of cultured parsley cells to elicitors from two non-pathogenic strains of fungi. I. Identification of induced products as coumarin derivatives, Eur. J. Biochem., 131: 401-407.
131. Dixon, R.A., Dey, P.M., Murphy, D.L. and Whitehead, I.M., 1981, Dose responses for Colletotrichum lindemuthianum elicitor-mediated enzyme induction in French bean cell suspension cultures, Planta, 151: 272-280.
132. Barz, W. and Hoesel, W., 1979, Metabolism and degradation of phenolic compounds in plants, Rec. Adv. Phytochemistry, 12: 339-369.
133. Dixon, R.A. and Fuller, K.W., 1978, Effects of growth substances on non-induced and Botrytis cinerea culture filtrate-induced phaseollin production in Phaseolus vulgaris cell suspension cultures, Physiol. Plant Pathol., 12: 279-288.
134. Rathmell, W.G., 1973, Phenolic compounds and phenylalanine ammonia-lyase activity in relation to phytoalexin biosynthesis in infected hypocotyls of Phaseolus vulgaris, Physiol. Plant Pathol., 3: 259-267.
135. Stoessl, A., Unwin, C.H. and Ward, E.W.B., 1972, Postinfectional inhibitors from plants. I. Capsidiol, an antifungal compound from Capsicum frutescens, Phytopath. Z., 74: 141-152.
136. Ward, E.W.B., Unwin, C.H., Hill, J. and Stoessl, A., 1975, Sesquiterpenoid phytoalexins from fruits of eggplants, Phytopathology, 65: 859-863.
137. Keen, N.T., Zaki, A.I. and Sims, J.J., 1972, Biosynthesis of hydroxy-phaseollin and related isoflavonoids in disease-resistant soybean hypocotyls, Phytochemistry, 11: 1031-1039.
138. Banks, S.W. and Dewick, P.M., 1983, Biosynthesis of glyceollins I, II and III in soybean, Phytochemistry, 22: 2729-2733.

139. Barz, W. and Köster, J., 1981, Turnover and degradation of secondary (natural) products, in: "The Biochemistry of Plants 7. Secondary Plant Products", E.E. Conn, ed., pp. 35-84, Academic Press, New York and London.
140. Ishiguri, Y., Tomiyama, K., Doke, N., Murai, A., Katsui, N., Yagihashi, F. and Masamune, T., 1978, Induction of rishitin-metabolizing activity in potato tuber tissue disks by wounding and identification of rishitin metabolites, Phytopathology, 68: 720-725.
141. Stoessl, A., Robinson, J.R., Rock, G.L. and Ward, E.W.B., 1977, Metabolism of capsidiol by sweet pepper tissue: some possible implications for phytoalexin studies, Phytopathology, 67: 64-66.
142. Ward, E.W.B. and Barrie, S.D., 1982, Evidence that 13-hydroxy-capsidiol is not an intermediate in capsidiol degradation in peppers, Phytopathology 72: 466-468.
143. Bailey, J.A., 1982, Physiological and biochemical events associated with the expression of resistance to disease, in: "Active Defense Mechanisms in Plants", R.K.S. Wood, ed., pp. 39-65, Plenum Press, New York and London.
144. Bailey, J.A., 1982, Mechanisms of phytoalexin accumulation, in: "Phytoalexins", J.A. Bailey and J.W. Mansfield, eds., pp. 289-318, John Wiley and Sons, New York.
145. Hargreaves, J.A. and Bailey, J.A., 1978, Phytoalexin production by hypocotyls of Phaseolus vulgaris in response to constitutive metabolites released by damaged cells, Physiol. Plant Pathol., 13: 89-100.
146. Bruce, R.J. and West, C.A., 1982, Elicitation of casbene synthetase activity in castor bean. The role of pectic fragments of the plant cell wall in elicitation by fungal endopolygalacturonase, Plant Physiol., 69: 1181-1188.
147. Hahn, M.G., Darvill, A.G. and Albersheim, P., 1981, Host-pathogen interactions. IXX. The endogenous elicitor, a fragment of a plant cell wall polysaccharide that elicits phytoalexin accumulation in soybeans, Plant Physiol., 68: 1161-1169.
148. Lee, S.C. and West, C.A., 1981, Polygalacturonase from Rhizopus stolonifer, an elicitor of casbene synthetase activity in castor bean (Ricinus communis L.), Plant Physiol., 67: 633-639.
149. Davis, K.R., Lyon, G.D., Darvill, A.G. and Albersheim, P., 1984, Host-pathogen interactions. XXV. Endopolygalacturonic acid lyase from Erwinia carotovora elicits phytoalexin accumulation by releasing plant cell wall fragments, Plant Physiol., 74: 52-60.
150. Nothnagel, E.A., McNeil, M., Albersheim, P. and Dell, A., 1983, Host-pathogen interactions. XXII. A galacturonic acid oligosaccharide from plant cell walls elicits phytoalexins, Plant Physiol., 71: 916-926.
151. Sharp, J.K., Valent, B. and Albersheim, P., 1984, Purification and partial characterization of a β-glucan fragment that elicits phytoalexin accumulation in soybean, J. Biol. Chem., 259: 11312-11320.
152. Cruickshank, I.A.M., 1980, Defenses triggered by the invader: chemical defenses, in: "Plant Disease V.", J.G. Horsfall and E.C. Cowling, eds., pp. 247-267, Academic Press, New York and London.
153. Stoessl, A., 1980, Phytoalexins - a biogenetic perspective, Phytopath. Z., 99: 251-272.
154. Maina, G., Allen, R.D., Bhatia, S.K. and Stelzig, D.A., 1984, Phenol metabolism, phytoalexins, and respiration in potato tuber tissue treated with fatty acid, Plant Physiol., 76: 735-738.
155. Anderson, J., 1978, Isolation from three species of Colletotrichum of glucan-containing polysaccharides that elicit browning and phytoalexin production in bean, Phytopathology, 68: 189-194.
156. De Wit, P.J.G.M., Hofman, A.E., Velthuis, G.C.M. and Kuć, J.A., 1985, Isolation and characterization of an elicitor of necrosis isolated from intercellular fluids of compatible interactions of Cladosporium fulvum (Syn. Fulvia fulva) and tomato, Plant Physiol., 77: 642-647.

157. Doke, N., Sakai, S. and Tomiyama, K., 1979, Hypersensitive reactivity of various host and non-host plant leaves to cell wall components and soluble glucan isolated from Phytophthora infestans, Ann. Phytopath. Soc. Japan, 45: 386-393.
158. Dow, J.M. and Callow, J.A., 1979, Leakage of electrolytes from isolated leaf mesophyll cells of tomato induced by glycopeptides from culture filtrates of Fulvia fulva (Cooke) Ciferri (syn. Cladosporium fulvum), Physiol. Plant Pathol., 15: 27-34.
159. Lazarovits, G. and Higgins, V.J., 1979, Biological activity and specificity of a toxin produced by Cladosporium fulvum, Phytopathology, 69: 1056-1061.
160. Ayers, A.R., Ebel, J. and Valent, B., 1976, Host-pathogen interactions. X. Fractionation and biological activity of an elicitor isolated from the mycelial walls of Phytophthora megasperma var. sojae, Plant Physiol., 57: 760-765.
161. Bailey, J.A., 1983, Biological perspectives of host-pathogen interactions, in: "The Dynamics of Host Defence", J.A. Bailey and B.J. Deverall, eds., pp. 1-32, Academic Press, Sydney, New York and London.
162. Sharp, J.K., McNeil, M. and Albersheim, 1984, The primary structure of one elicitor-active and seven elicitor-inactive hexa(β -D-gluco-pyranosyl)-D-glucitols isolated from the mycelial walls of Phytophthora megasperma f. sp. glycinea, J. Biol. Chem., 259: 11321-11336.
163. Daly, J.M., 1984, The role of recognition in plant disease, Annu. Rev. Phytopathology, 22: 273-307.
164. Daly, J.M., 1972, The use of near-isogenic lines in biochemical studies of the resistance of wheat to stem rust, Phytopathology, 62: 392-400.
165. Vanderplank, J.E., 1978, "Genetic and Molecular Basis of Plant Pathogenesis", Springer-Verlag, Berlin.
166. Ward, E.W.B. and Stoessl, A., 1976, On the question of 'elicitors' or 'inducers' in incompatible interactions between plants and fungal pathogens, Phytopathology, 66: 940-941.
167. Heath, M.C., 1974, Light and electron microscope studies of the interactions of host and non-host plants with cowpea rust - Uromyces phaseoli var. vignae, Physiol. Plant Pathol., 4: 403-414.
168. Johnson, L.E.B., Bushnell, W.R. and Zeyen, R.J., 1982, Defense patterns in non-host higher plant species against two powdery mildew fungi. 1. Monocotyledonous species, Can. J. Bot., 60: 1068-1083.
169. Dekker, J. and Vanderhoek-Scheuer, R.G., 1964, A microscopic study of the wheat-powdery mildew relationship after application of the systemic compounds procaine griseofulvin and 6-azouracil, Neth. J. Plant Pathol., 70: 142-148.
170. Heath, M.C., 1981, A generalized concept of host-parasite specificity, Phytopathology, 71: 1121-1123.
171. Heath, M.C., 1982, The absence of active defense mechanisms in compatible host-pathogen interactions, in: "Active Defense Mechanisms in Plants", R.K.S. Wood, eds., pp. 143-156, Plenum Press, New York and London.
172. Ellingboe, A.H., 1982, Genetical aspects of active defence, in: "Active Defense Mechanisms in Plants", R.K.S. Wood, ed., pp. 179-192, Plenum Press, New York and London.
173. Ward, E.W.B. and Buzzell, R.I. (1983), Influence of light, temperature and wounding on the expression of soybean genes for resistance to Phytophthora megasperma f. sp. glycinea, Physiol. Plant Pathol., 23: 401-409.
174. Loegering, W.Q. and Sears, E.R., 1981, Genetic control of disease expression in stem rust of wheat, Phytopathology, 71: 425-428.
175. Flor, H.H., 1960, The inheritance of X-ray-induced mutations to virulence in a urediospore culture of race 1 of Melampsora lini, Phytopathology, 50: 603-605.

CELL WALL MODIFICATIONS ASSOCIATED WITH THE RESISTANCE OF CEREALS TO FUNGAL PATHOGENS

J.A. Hargreaves and J.P.R. Keon

University of Bristol, Department of Agricultural Sciences
Long Ashton Research Station
Long Ashton
Bristol BA18 9AF, UK

INTRODUCTION

The inability of many fungal pathogens to infect and colonize some of the more economically important members of the Gramineae is frequently associated with localized structural changes to the cell wall at the site of infection, during the initial penetration event [1,2,3]. These modifications involve the deposition of material on the inner surface of the cell wall directly below the penetration site as a cell wall apposition or papilla and an alteration to the structure of the cell wall surrounding the penetration site to form a halo or disc. Papillae and the surrounding modified cell walls are collectively termed reaction sites [4].

Because of the close association, in both time and space, between the cessation of fungal growth and the appearance of these cell wall reaction sites it is generally believed that structural changes to the cell wall have an important role in preventing the pathogen gaining access to the host protoplast [3,5,6]. These deposits may accomplish this by acting as barriers that either physically prevent further development of the fungal hypha or restrict the movement of nutrients or toxins between the host cell and the infecting hypha of the pathogen [7]. Alternatively, reaction sites may be sites of highly localized phytoalexin accumulation and thus present a fungitoxic barrier [8].

Unlike the inducible chemical defence reaction of plants, i.e. production of phytoalexins, which can be both quantified in plant tissues and assessed for effect on microbial growth in vitro, it is not yet possible either to measure accurately this structural response in infected tissues or assess its effect on fungal development. In the absence of this information the role of cell wall modifications in resistance still remains open to conjecture.

In the present paper we will consider the structure and chemical nature of the cell wall deposits that accumulate at sites of abortive penetrations in cereals and suggest possible mechanisms by which these deposits are formed.

NATO ASI Series, Vol. H1
Biology and Molecular Biology of Plant-Pathogen Interactions
Edited by J. Bailey

THE STRUCTURE OF REACTION SITES

Papillae

Using conventional electron microscope fixation (glutaraldehyde and osmium tetroxide) and staining (lead citrate) techniques, papillae often appear as electron dense aggregates which are clearly distinct from the cell wall. Even within the same plant tissue, papillae exhibit considerable variability in structure: some appear as multilayered deposits, whereas others are relatively unstructured and heterogeneous [9,10]. In many cases, these amorphous deposits contain electron dense material in the form of globules or inclusions and are often surrounded by additional material which appears fibrous or membranous.

The inclusion of ruthenium red during fixation of the tissue results in a more detailed image of the papillae (Fig. 1). An inner core of matrix material appears electron lucid and within this are deposits of electron dense globules and particles. This core deposit is surrounded by material which has been laid down in distinct layers and which may have originated from the successive deposition of material associated with cellular membranes. The matrix of the inner core is probably composed of the β-1,4 glucan polymer, callose [2], within and on which encrusting material is deposited.

Cell Wall Modifications

The changes that occur to the cell wall surrounding the penetration site, during papilla deposition, are similar to some of those associated with papilla formation. These modified cell walls are more electron dense than unmodified cell walls [10] and, like papillae, they autofluoresce under ultraviolet light [11]. Furthermore both papillae and associated altered cell walls show dense staining with ammoniacal silver nitrate indicating the presence of reducing groups (Fig. 2).

Thus, it is possible to envisage at least two processes occurring during the formation of these deposits. First, a rapid accumulation of appositional material beneath the cell wall and second, the deposition of encrusting material within and on the matrix polymer of the papilla and the cellulose microfibrils of the surrounding cell wall. Because in the case of penetration of barley epidermal cells by *Septoria nodorum*, the infection hyphae are restricted to the outer layers of the cell wall (Fig. 1), it is likely that the material deposited during this latter process is responsible for arresting the growth of the pathogen.

HISTOCHEMICAL CHARACTERIZATION OF THE COMPONENTS OF REACTION SITES

The cell wall changes induced in barley, oats and wheat after attempted penetration by *S. nodorum* can be visualized by a wide range of histochemical reagents (Table 1). The reaction sites formed by all these cereals were stained with a number of reagents, ammoniacal silver nitrate, p-nitrobenzene tetrafluoroborate or Coomassie brilliant blue R250. At least some components in reaction sites of the different cereals were similar. However, the material that accumulated at reaction sites in oats stained quite differently from those formed by barley and wheat with the basic dyes, bromophenol blue, nile blue and toluidine blue O, suggesting that some components in the modified cell walls of oats are different to those of barley and wheat. Lignin-like material which accumulates in reaction sites formed by reed canary grass (*Phalaris arundinacea*) [12] and wheat [13] could not be detected in modified walls of oats or barley [11]. Even in these cases where lignin

Table 1. The staining of reaction sites in the outer epidermal cells of coleoptiles of wheat, barley and oats by histochemical reagents, adapted from Hargreaves and Keon [11].

Reagent	Wheat	Barley	Oats
Ammoniacal silver nitrate	+	+	+
p-nitrobenzene tetrafluoroborate	+	+	+
Coomassie brilliant blue R250	+	+	+
Toluidine blue O	±	±	+
Bromophenol blue	+	+	-
Nile blue	-	-	+

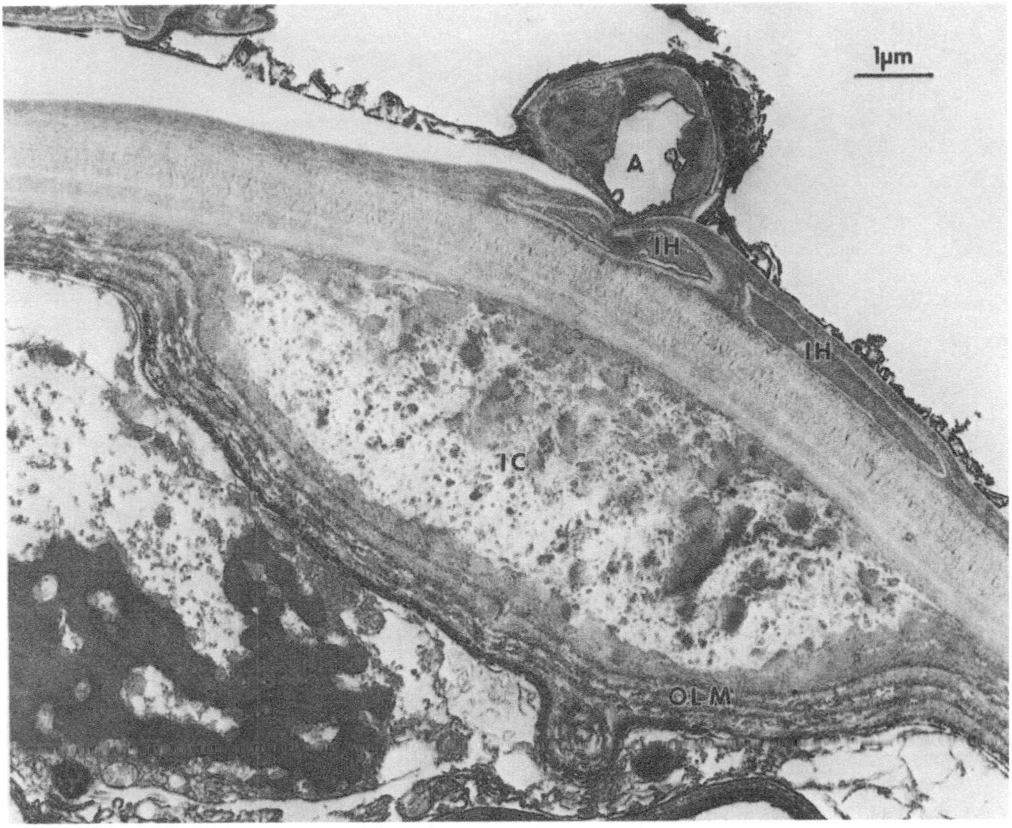

Figure 1. A TEM micrograph of a thin section through a reaction site formed by a barley leaf epidermal cell in response to penetration by Septoria nodorum. The tissue was fixed in glutaraldehyde containing 1% ruthenium red followed by osmium tetroxide containing 1% ruthenium red. Note that the infection hypha (IH) of the pathogen is restricted to the outer layers of the cell wall. A, appressorium; IC, inner core material; IH, infection hypha; OLM, outer layered material.

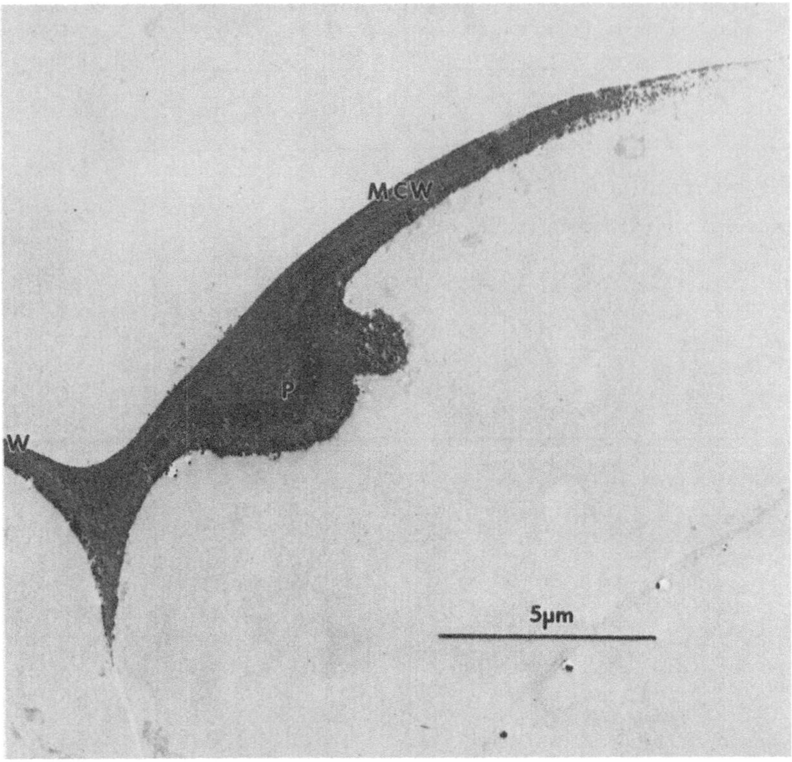

Figure 2. A TEM micrograph of a thin section through a reaction site formed by a barley leaf epidermal cell in response to penetration by *Septoria nodorum*. The tissue was stained with ammoniacal silver nitrate [20]. MCW, modified cell wall; P, papilla.

has been identified, the type of lignin formed appears to be different [12, 13], again indicating the variable nature of at least some components of reaction sites.

Although the detection of components of reaction sites with histochemical reagents is no substitute for direct chemical analysis, it is possible to infer from the staining properties of these deposits that reducing compounds, phenolic compounds and proteins are common to reaction sites formed by cereals.

ISOLATION OF MODIFIED CELL WALLS

A major problem in determining the exact chemical nature of the deposits occurring at failed penetration sites is the inability to detect and analyze the chemical changes occurring in these minute areas of modified wall in intact tissue. To overcome this problem we attempted to isolate the reaction sites formed in barley leaf epidermal cells following infection by *S. nodorum*. The isolation procedure shown in Figure 3 was based on the fact that modified cell walls and papillae are resistant to depolymerization by cell wall degrading enzymes.

Epidermal strips powdered
in liquid N_2

↓

15 mM Pi buffer (pH 7.0) extraction

↓

Hot 2% SDS extraction

↓

H_2O wash

↓

MeOH extraction

↓

$CHCl_3$: MeOH (2 : 1) extraction

↓

Alkaline hydrolysis with
0.1M NaOH for 24 h

↓

Digestion in 2% cellulase
(50mM Ac buffer pH 5.0) for 1 h at 25 °C

↓

Sonicate, layer on 50% glycerol and
centrifuge at 100 g

↓

Collect material at interface, wash and
suspend in 25% glycerol, centrifuge at 100 g.
Reaction sites pelleted at the bottom of the tube

Figure 3. Procedure for the isolation of reaction sites from barley leaves infected by Septoria nodorum [10].

Isolated reaction sites retained their ability to autofluoresce under UV light. Thin sections through the reaction sites showed that they were ultrastructurally similar to those in intact tissue (Fig. 4). The encrusting material had, thus, not been removed during the extraction procedure. Although it was not possible to obtain sufficient pure reaction sites for chemical analysis, these structures were probed with a number of histochemical reagents [10]. Unlike unmodified cell walls, the cell walls in isolated reaction sites did not stain with zinc chlor-iodide, a reagent for cellulose, nor could callose be detected by enhanced fluorescence with aniline blue. These results suggest that encrusting materials prevent these stains from reacting with polymers in the reaction sites. By contrast, strong reactions were obtained with periodic acid-Schiff's reagent, for carbohydrates, and with Coomassie brilliant blue R250, for proteins. These staining properties suggest that the encrusting material deposited at reaction sites is glycoproteinaceous and that it is deposited within the intermicrofibrillar spaces of the cell wall and within the interstitial spaces of the callose polymer laid down in the papillae.

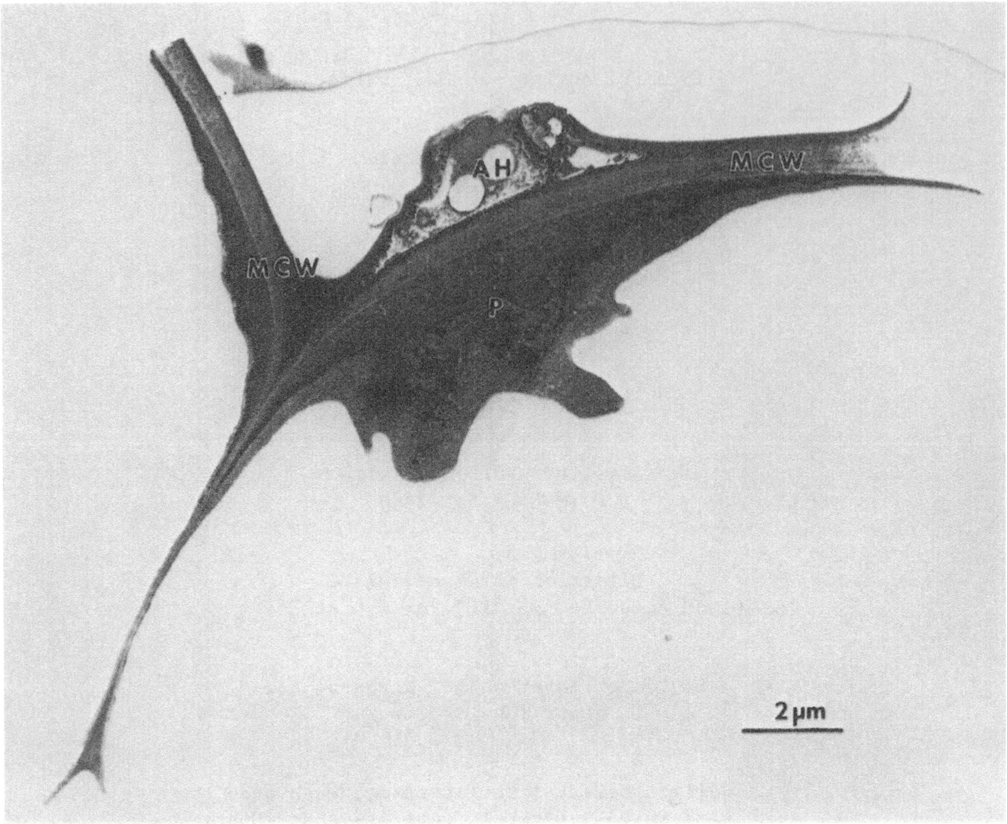

Figure 4. A TEM micrograph of a thin section through an isolated reaction site. AH, area occupied by infection hypha; MCW, modified cell wall; P, papilla.

OXIDATIVE GELATION OF GLYCOPROTEINS AS A MECHANISM OF CELL WALL MODIFICATION

The gelation of glycoproteins within cell walls and papillae could explain the above observations and also a number of other properties of reaction sites (Fig. 5). The addition of small amounts of phenolic acids to wheat flour glycoproteins in the presence of hydrogen peroxide, causes a gel to form [14]. The hardness and elasticity of this gel depends on the proportions of the glycoprotein and the phenolic acids. Reaction sites which can prevent fungal penetration have been shown to be more elastic and harder than those that are breached by infection hyphae [15]. Furthermore, the capacity of the glycoproteins to form a gel is lost when calcium ions are removed. There is evidence to suggest that reaction site formation is also a calcium mediated process [16]. Gelation of these glycoproteins is virtually instantaneous which is also consistent with the rapid reactions seen during this type of resistance mechanism. Such gels could bind to the polymers of the cell wall and papillae through covalent bridges such as diferulic acid [17] or isodityrosine [18] where they could form an impermeable deposit. In addition, such cross-linked glycoproteins may act as a substrate to which other materials such as lignin could adhere [19].

Glycoproteins + Esterified phenolic acids (e.g. Ferulic acid)

$$\xrightarrow[\quad H_2O_2 \quad]{Ca^{2+}}$$

Gel

Figure 5. Oxidative gelation of glycoproteins as a mechanism for the deposition of encrustations in papillae and modified cell walls.

Further advances in understanding the nature of this resistance mechanism could be made by applying some of the techniques of modern molecular biology described in this book. In some respects the response of cereal leaf epidermal cells to fungal penetration offers a number of advantages for this type of approach. It is a very rapid inducible response which occurs in a single cell type (the epidermis). These cells can be isolated from the rest of the leaf with little contamination from underlying mesophyll cells or from the hyphae of the invading pathogen. Methods are now available to produce large numbers of reaction sites in epidermal cells [11] and these should aid further research into their chemical nature.

REFERENCES

1. Aist, J.R., 1976, Papillae and related wound plugs of plant cells, Annu. Rev. Phytopathology, 14: 145-163.
2. Aist, J.R., 1983, Structural responses as resistance mechanisms, in: "The Dynamics of Host Defense", J.A. Bailey and B.J. Deverall, eds., pp. 33-70, Academic Press, Australia.
3. Sherwood, R.T. and Vance, C.P., 1980, Resistance to fungal penetration in Gramineae, Phytopathology, 70: 273-279.
4. Ride, J.P., 1980, The effect of induced lignification on the resistance of wheat cell walls to fungal degradation, Physiol. Plant Pathol., 16: 187-196.
5. Ride, J.P., 1983, Cell walls and other structural barriers in defence, in: "Biochemical Plant Pathology", J.A. Callow, ed., pp. 215-236, John Wiley and Son Ltd., Chichester.
6. Sherwood, R.T. and Vance, C.P., 1982, Initial events in the epidermal layer during penetration, in: "Plant Infection. The Physiological and Biochemical Basis", Y. Asada, W.R. Bushnell, S. Ouchi and C.P. Vance, eds., pp. 27-44, Japan Scientific Societies Press, Tokyo/Springer-Verlag, Berlin.
7. Hargreaves, J.A., 1982, The nature of the resistance of oat leaves to infection by Pyrenophora teres, Physiol. Plant Pathol., 20: 165-171.
8. Mansfield, J.W., 1982, The role of phytoalexins in disease resistance, in: "Phytoalexins", J.A. Bailey and J.W. Mansfield, eds., pp. 253-288. Blackie, Glasgow and London.
9. Keon, J.P.R. and Hargreaves, J.A., 1983, A cytological study of the net blotch disease of barley caused by Pyrenophora teres, Physiol. Plant Pathol., 22: 321-329.
10. Keon, J.P.R. and Hargreaves, J.A. 1984, The response of barley leaf epidermis cells to infection by Septoria nodorum, New Phytologist, 98: 387-398.
11. Hargreaves, J.A. and Keon, J.P.R., 1985, A comparison of the reaction sites associated with the resistance of coleoptiles of barley, oats and wheat to infection by Septoria nodorum, Phytopath. Z., 115: 72-82.

12. Sherwood, R.T. and Vance, C.P., 1976, Histochemistry of papillae formed in reed canary grass leaves in response to non-infecting pathogenic fungi, Phytopathology, 66: 503-510.
13. Ride, J.P. and Pearce, R.B., 1979, Lignification and papilla formation at sites of attempted penetration of wheat leaves by non-pathogenic fungi, Physiol. Plant Pathol., 15: 79-92.
14. Painter, T.J. and Neukom, H., 1968, The mechanism of oxidative gelation of a glycoprotein from wheat flour: Evidence from a model system based upon caffeic acid, Biochim. Biophys. Acta, 158, 363-381.
15. Israel, H.W., Wilson, R.G., Aist, J.R. and Kunoh, H., 1980, Cell wall appositions and plant disease resistance: Acoustic microscopy of papillae that block fungal ingress, Proc. Natl. Acad. Sci. USA, 77: 2046-2049.
16. Marshall, M.R., Smart, M.G., Aist, J.R. and Israel, H.W., 1985, Chlortetracycline and barley papilla formation: localization of calcium and alteration of the response induced by Erysiphe graminis, Can. J. Bot., 63: 876-880.
17. Hartley, R.C. and Jones, E.C., 1976, Diferulic acid as a component of cell walls of Lolium multiflorum, Phytochemistry, 15: 1157-1160.
18. Fry, S.C., 1982, Isodityrosine, a new cross-linking amino acid from plant cell-wall glycoprotein, Biochem. J., 204: 449-455.
19. Roberts, K. and Northcote, D.H., 1972, Hydroxyproline: Observations on its chemical and autoradiographical localization in plant cell wall protein, Planta, 107: 43-51.
20. MacRae, E.K. and Meetz, G.D., 1970, Electron microscopy of the ammoniacal silver reaction for histones in the erythropoietic cells of the chick, J. Cell. Biol., 45: 235-245.

MECHANISM OF RESISTANCE OF WHEAT AGAINST STEM RUST IN THE Sr5/P5 INTERACTION

H.J. Reisener, R. Tiburzy, K.H. Kogel,
B. Moerschbacher and B. Heck

Institut für Biologie III (Pflanzenphysiologie)
RWTH Aachen
D-5100 Aachen
FRG

INTRODUCTION

Some of the wheat/stem rust interactions, e.g. the Sr5/P5 interaction, are characterized by a hypersensitive response of host cells, as first described by Stakman [1]. These cells are easily detected with the aid of an epifluorescence microscope by their bright yellow colour. The fungus can be located following the fluorochromation technique developed by Rohringer and co-workers using the optical brighteners calcofluor or diaethanol [2,3]. In incompatible interactions conditioned by the Sr5 gene the majority of the colonies do not form more than two haustorial mother cells (HMC) [4,5]. Further growth is arrested by the rapid hypersensitive response of the penetrated host cells. Haustorial development is blocked at a very early stage. Hypersensitive host cells are killed rapidly. Fluorescence is not limited to the cell wall but extends over the whole protoplast. There is no hypersensitive response in incompatible interactions; haustoria become rather large in epidermal cells as well as in mesophyll cells [3].

WHAT IS THE NATURE OF THE CONSPICUOUS YELLOW FLUORESCENCE?

UV-fluorescence spectroscopy and solubility characteristics of the substance responsible for the fluorescence, indicated it was lignin or a lignin-like polymer. Histochemical tests for lignin with phloroglucinol/HCl and chlorine/sulphite gave positive results [5,6]. A similar red colour was observed in the vessels which are known to contain lignin. Both epidermal and mesophyll cells showed the hypersensitive response.

The histochemical tests were further substantiated by tracer experiments [5]. If the hypersensitive response is associated with lignification, lignin-precursors (e.g. ^{14}C-cinnamic acid) should accumulate in cells which show the hypersensitive response. Indeed, it was found that the label accumulated only in those cells which had been penetrated by one haustorium at least, and which showed the characteristic yellow fluorescence (Fig. 1). Before applying the photoemulsion the leaves were extensively extracted with hot ethanol, followed by an alkaline hydrolysis to remove unused lignin-precursors and material bound by ester linkages only. The label resisted

NATO ASI Series, Vol. H1
Biology and Molecular Biology of Plant-Pathogen Interactions
Edited by J. Bailey

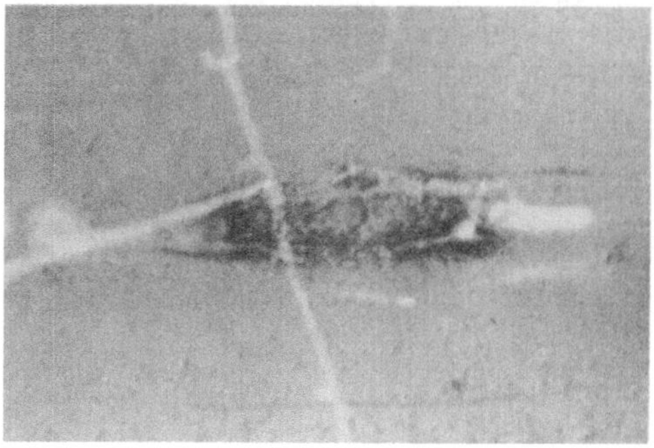

Figure 1. Autoradiogram of an epidermal cell of the resistant cultivar Feldkrone which has been inoculated with an avirulent race of *Puccinia graminis* f.sp. *tritici*. The pathogen has produced a haustorium in the epidermal cell which has responded with a rapid HR as shown by the bright yellow fluorescence. ^{14}C-cinnamic acid was applied 30 h after inoculation. This is the time at which lignification occurs. The label resists extensive extraction procedures.

these different treatments and we conclude that the fluorescent material is lignin or is lignin-like. The resorcinol blue-test indicated that callose was formed in association with lignification.

IS THE FORMATION OF LIGNIN A DETERMINANT OF RESISTANCE?

To answer this question, leaves were treated with the phenylalanine-ammonia-lyase-inhibitor, α-aminooxyacetate (AOA), one day before inoculation with an avirulent race of the fungus. AOA inhibited lignification, especially in epidermal cells. However, it did not completely prevent lignification in the mesophyll cells. Resistance was partially broken, and colony size increased as indicated by the number of haustorial mother cells (HMC) formed per colony. Figure 2 shows three histograms. The different columns indicate the percentage of colonies with 1 to 10, between 10 and 20, and more than 20 HMC. Four days after inoculation most of the colonies of the plants not treated with AOA had formed no more than 2 HMC. Growth of these colonies was arrested at this stage. Most of the colonies induced a hypersensitive response in epidermal cells. In the plants treated with AOA, the majority of colonies formed more than 20 HMC. Penetrated epidermal cells showed no lignification. The haustoria which formed were as large as in compatible interactions. Obviously, there is a close association between the lignification process and colony size.

If lignification is a factor of resistance in the Sr5/P5 gene-interaction, it has to be a fast event and it should occur after haustoria have penetrated but before fungal growth has been arrested. In Figure 3 the course of events is demonstrated. Two near isogenic lines were compared. By 30 h after inoculation 92% of all colonies had formed the first haustorium. However, only 11% of all penetrated epidermal cells showed the hypersensitive response. No significant difference in colony size between the two near isogenic lines was visible at this time. Within the next 2 h, lignification increased to 36.5%. At that time the difference in colony size between the two lines was evident. From then the difference between the two populations increased steadily. We conclude that the sequence of events is: i) penetration of haustorium, ii) induction of lignification, iii) arrest of colony growth.

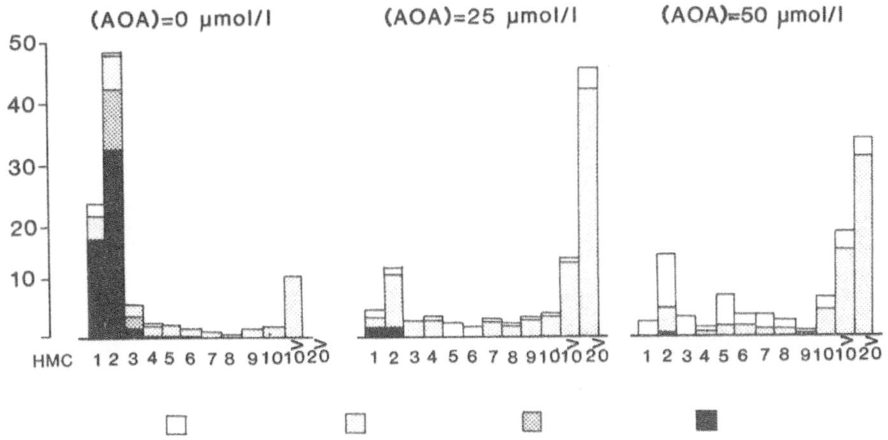

Figure 2. Size of stem rust colonies in primary leaves of the resistant cultivar Feldkrone 4 days post-infection after application of AOA 24 h before inoculation. Control: no AOA applied. The columns represent the percentage of colonies with the indicated number of HMC.

Dotted, crosshatched, black, and white parts of the columns indicate: colonies with HR in epidermal cells only, in mesophyll cells only, in epidermal as well as in mesophyll cells, and without any HR, respectively.

ARE PHYTOALEXINS INVOLVED IN THE ARREST OF THE COLONIES?

The recent discovery of the avenalumins in oat plants by Mayama and co-workers [7] stimulated us to search for similar compounds in wheat plants. Although inhibitory substances in wheat are easily detected, they do not show the characteristics of the avenalumins. They belong to the benzoxazinone-group [H.J. Grambow, personal communication]. At present there is no evidence available that these compounds play a role in the highly resistant or moderately resistant interactions. In accordance with this, we have not been able to demonstrate cross protection [unpublished results]. Experiments in which leaves were preinoculated with a virulent rust race, followed by a second inoculation with an avirulent race 48 h later, showed that lignification of epidermal cells originating from an avirulent rust spore located in the centre or at the edge of a large colony of the virulent rust race did not protect adjacent cells from being successfully penetrated by the haustoria of the virulent colony. Large and healthy looking haustoria were formed. We therefore conclude that in the Sr5/P5 interaction lignification and callose formation in epidermal cells do not arrest the virulent fungus in adjacent cells. Therefore, it is highly unlikely that phytoalexins are involved in the expression of the resistance response.

ACTIVITIES OF ENZYMES INVOLVED IN LIGNIN SYNTHESIS

Activities of enzymes of the general phenylpropanoid pathway: L-phenylalanine ammonia-lyase (PAL) and 4-coumarate: CoA-ligase, as well as one of the specific pathways leading to the formation of lignin: cinnamyl-alcohol-dehydrogenase, were greater in incompatible interactions than in compatible ones. The resistant cultivar Feldkrone and the susceptible cultivar Little Club were used as host plants. Peroxidase activity increased continuously in both interactions [unpublished results].

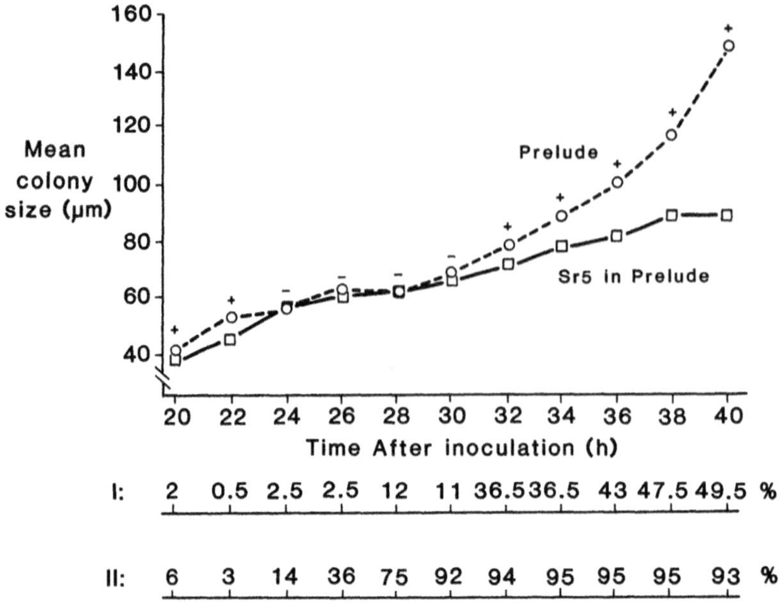

Figure 3. Time course of colony growth of stem rust on resistant and susceptible near isogenic wheat lines. Genetic background is Prelude. Resistance is conditioned by the Sr5 gene. Total length of fungal mycelium (including branches) is plotted against the time after incoulation. "+" indicates significant differences between the two lines.

(Kolmogoroff-Smirnoff, $p = 0.05$).

Scales below: I percentage of colonies which have formed their first haustorium.
II percentage of penetrated epidermal cells which have undergone lignification.

IS IT POSSIBLE TO INTERFERE WITH THE PROPOSED SEQUENCE OF EVENTS?

The interaction we are dealing with fulfils Flor's gene for gene hypothesis. This implies a highly specific interaction between the race of the pathogen and the host cultivar. Following currently discussed models the hypothesis further implies a recognition event between a specific fungal product (an elicitor) and a specific host receptor [8]. It is assumed that the binding of the elicitor to the receptor (presumably at a membrane) releases a signal that leads to a response in the nucleus of the penetrated host cell, causing enhanced rates of transcription and translation of the enzymes which are involved in the expression of resistance.

The experiment with the inhibitor AOA, described above, indicates that it is possible to suppress lignification. Likewise, the lignification er the response can be suppressed by injection of a solution of 2,4-dinitrophenol (DNP) or p-chloromercuryphenylsulphonic acid (PCMPS) into intercellular spaces of the leaves two hours before inoculation. As in the experiment with AOA, large colonies with large haustoria are formed [unpublished results]. DNP blocks ATP-synthesis, PCMPS is a SH-reagent and may interfere with several proteins. These agents can be taken up by the host cells and act presumably at rather late stages in the general scheme.

Of greater relevance is the finding that certain lectins with specificity for galactose, like soybean agglutinin (SBA) and the lectin of Erythrina corallodendron (ECO) cause the incompatible interaction to be shifted to a more compatible one when injected into intercellular spaces before inoculation [9]. Because of their high molecular weight it is very unlikely that lectins are absorbed by the host cytoplasm. Moreover, the effect can be reversed by the simultaneous application of monomeric galactose, indicating that lectins act in a specific way by interfering with events that take place at the outer surface of the plasma membrane. This is in accordance with the finding that all the lectins which are active in suppressing the hypersensitive response agglutinate wheat protoplasts by binding to lectin receptors that are an integral part of the host plasma membrane [9]. The lectin receptors were identified as mono-, and di-galactosyldiglycerides [10]. These results show that the recognition between the two reacting partners takes place at the plasma membrane.

DOES THE PATHOGEN CONTAIN COMPONENTS THAT ARE ABLE TO ELICIT THE CHARACTERISTIC HYPERSENSITIVE RESPONSE?

We used the rust mycelium of axenic cultures and found a fraction with elicitor activity. However, the axenic culture of Puccinia graminis is time-consuming and provides small yields. We therefore extracted germinated uredospores and isolated a fraction that turned out to be highly active. This fraction (Pgt-elicitor) not only elicited the lignification response, but also caused an increase in PAL-activity. In addition to PAL, the activities of all enzymes of the lignin pathway tested were greater than in water-injected tissue [11,12,13].

The isoenzyme pattern of the peroxidases present in elicitor treated leaves is presented in Figure 4. Two near isogenic lines carrying either the Sr5 or the sr5 gene were used. For comparison the profiles of water-injected controls of untreated leaves, and of the natural interaction were added. Only a limited number of rather weak bands were visible in the controls. After inoculation, synthesis (or activation) of peroxidase isoenzymes was found for both interactions. However, different profiles were found in the compatible (lane 9) and in the incompatible (lane 10) responses. The elicitor treated Sr5 and sr5 leaves (lane 6 and 5) showed the characteristic PO pattern of the corresponding natural interactions.

A thorough inspection of the isoenzymes typical of the incompatible interaction demonstrates that the same amount of elicitor causes a greater response in the incompatible Sr5 host than in the compatible sr5 host. The fact that the Pgt-elicitor simulates the natural interaction very closely encourages us to assume that we are not dealing with an artificial phenomenon. This is further substantiated by the third and the fourth lanes of the electrophoretogram. This originated from leaves that were treated with a chitin hydrolysate from crab shells. Though this material induces lignification in wheat leaves, the PO profiles are different from those of the natural interaction and from the Pgt-elicitor treated cells.

The differential effect of the Pgt-elicitor against the two isogenic lines is evident from two more responses: PAL induction and the sensitivity of protoplasts towards the Pgt-elicitor. PAL activities in resistant and susceptible plants were assayed 24 h after application of different elicitor concentrations. The resistant Sr5 line reacted more strongly than the susceptible sr5 line. The threshold concentration was approximately 20 ng glucose equivalents of the Pgt-elicitor per infected leaf.

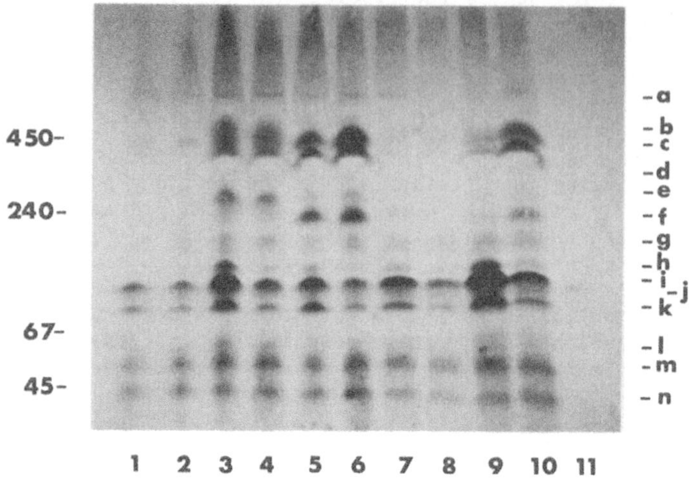

Figure 4. Patterns of isoenzymes of PO extracted from wheat leaves of two near isogenic lines after gradient polyacrylamide gel electrophoresis of identical aliquots (75 µg of protein) under non-denaturing conditions. Lanes designated with even numbers stem from plants carrying the Sr5 gene, those with uneven numbers from plants carrying the sr5 gene.

1 and 2: injection of water into intercellular spaces
3 and 4: injection of an acid hydrolysate of crab chitin
5 and 6: injection of a Pgt-elicitor solution
9 and 10: inoculation with stem rust

Lanes 7 and 8 represent extracts from untreated plants; lane 11 represents a PO bond found in the intercellular washing fluid of Sr5 plants.

The numbers on the left indicate MW in kdalton.

The Pgt-elicitor caused wheat protoplats to burst. Within minutes after addition of the elicitor the cytoplasm aggregated at one side of the protoplast, followed by extrusion of the cytoplasm. The tonoplast was left behind. In the control medium, protoplasts were stable for several days. Again the Sr5 line and sr5 line reacted differently, as demonstrated by the time course of bursting. We conclude that the Pgt-elicitor closely simulates the natural interaction as it shows a differential effect when tested against the two near isogenic lines.

The elicitor is most probably a glycoprotein. The glycopart consists of glucose, mannose and galactose in the ratio 9:1:1. Protein content is approximately 30%. It is heat stable and resists mild acid and mild alkaline treatments. Periodate oxidation destroys the activity completely, whereas pronase digestion has no effect [12]. When the eliciting fraction was subject to a SDS-porogel electrophoresis, silver staining revealed more than 20 bands. After electroblotting approximately 20 bands were detected by staining with ConA-Peroxidase. On a two-dimensional Western blot analysis it became evident that the Pgt-elicitor fraction contained about 50 ConA-binding components. In the same experiment only one component was able to bind to soybean-agglutinin (SBA) [unpublished]. This shows that the Pgt-elicitor is heterogeneous. Before further chemical characterization, efforts have to be made to fractionate the components.

PLASMA MEMBRANES OF HOST CELLS

According to our working hypothesis the elicitor binds to the plasma membrane. As already mentioned lectins can be used as probes for glycoconjugates of the plasma membrane. All galactose-specific lectins, and to a lesser extent glucose- or mannose-specific lectins such as ConA, agglutinate protoplasts of the two isolines equally. More important is the discovery that the lectin of Bandairaea simplicifolia I has a different affinity for protoplasts of the two near isogenic lines (Sr5/sr5), indicating differences in the glycoconjugates of plasmalemma between both isolines [unpublished results]. By means of two-dimensional gel electrophoresis of total membranes approximately 200 spots were detected. From these only two spots showed a different position on the gel when the two near-isogenic lines were compared. The total number of spots decreased to approximately 85 when a more highly purified plasma membrane fraction was analyzed. There was a difference in one spot only between the two near-isogenic lines carrying the Sr5 or the sr5 gene respectively [14].

REFERENCES

1. Stakman, E.C., 1915, Relation between Puccinia graminis and plants highly resistant to its attack, J. Agric. Res., 4: 193-200.
2. Rohringer, R., Kim, W.K., Samborski, D.J. and Howes, N., 1977, Calcofluor: An optical brightener of fluorescence microscopy of fungal parasites in leaves, Phytopathol., 67: 808-810.
3. Kuck, K.H., Tiburzy, R., Hänssler, C. and Reisener, H.J., 1981, Visualization of rust haustoria in wheat leaves by using fluorochroms, Physiol. Plant Pathol., 19: 439-441.
4. Rohringer, R., Kim, W.K. and Samborski, D.J., 1979, A histological study of interaction between avirulent race of stem rust and wheat containing resistance genes Sr5, Sr6, Sr8, or Sr22, Can. J. Bot., 57: 324-331.
5. Tiburzy, R., 1984, Untersuchungen zur Bedeutung der hypersensitiven Reaktion befallener Weizenzellen als Resistenz-faktor gegen Puccinia graminis f.sp. tritici, PhD Thesis, RWTH Aachen.
6. Tiburzy, R. and Reisener, H.J., 1984, Fluoreszenzmikroskopische Beobachtungen am Weizenschwarzrost, Zeiss Information, 27: 54-58.
7. Mayama, S., Matsuura, Y., Iida, H. and Tani, T., 1982, The role of avenalumin in the resistance of oat to crown rust, Puccinia coronata f.sp. avenae., Physiol. Plant Pathol., 20: 189-199.
8. Callow, J.A., 1984, Cellular and molecular recognition between higher plants and fungal pathogens, in: "Encyclopedia of Plant Physiology, New Series, Vol. 17, Cellular Interactions", H.F. Linskens and J. Heslop-Harrison, eds., pp. 212-237, Springer-Verlag, Berlin.
9. Kogel, K.H., Schrenk, F., Sharon, N. and Reisener, H.J., 1984, Suppression of the hypersensitive response in wheat stem rust interaction by reagents with affinity for wheat plasmamembrane galactoconjugates, J. Plant Physiol., 118: 343-352.
10. Kogel, K.H., Ehrlich-Rogozinski, S., Reisener, H.J. and Sharon, N., 1984, Surface galactolipids of wheat protoplats as receptors for soybean agglutinin and their possible relevance to host-parasite interaction, Plant Physiol., 76: 924-928.
11. Kogel, K.H., Heck, B., Kogel, G., Moerschbacher, B. and Reisener, H.J., 1985, A fungal elicitor of the resistance response in wheat, Z. Naturf., 40c: 743-744.
12. Moerschbacher, B., Kogel, K.H., Noll, U. and Reisener, H.J., 1985, An elicitor of the hypersensitive response in wheat leaves isolated from the rust fungus Puccinia graminis f.sp. tritici. I. Isolation and partial characterization, Z. Naturforsch., 41c: in press.

13. Moerschbacher, B., Heck, B., Kogel, K.H., Obst, O. and Reisener, H.J., 1985, An elicitor of the hypersensitive lignification response in wheat leaves isolated from the rust fungus _Puccinia graminis_ f.sp. _tritici_. II. Induction of enzymes correlated with the biosynthesis of lignin, _Z. Naturforsch._, in press.
14. Heck-Wattjes, B., 1986, Gelelektrophoretischer Vergleich Schwarzrost-resistenter und - suszeptibler isogener Weizenlinien (_Triticum aestivum_/ _Puccinia graminis_ f.sp. _tritici_), PhD Thesis, RWTH Aachen.

ELICITATION OF ACTIVE RESISTANCE MECHANISMS

P.J.G.M. de Wit

Department of Plant Pathology
Agricultural University
Wageningen, The Netherlands

INTRODUCTION

In gene-for-gene systems involving biotrophic parasites, active defence mechanisms are usually induced by avirulent but initially not by virulent races of a parasite. The reason why active defence responses do not result from compatible race-cultivar combinations or are ineffective is of considerable interest. Heath [1], among others, has advanced the hypothesis that compatibility could be the result of a suppression of host resistance responses by a virulent parasite (induced susceptibility). If such induced susceptibility was an active plant response rather than a constitutive failure of recognition, metabolic inhibitors might be expected to turn a compatible combination into an incompatible one. This has usually not been found to occur. However, there are many reports of metabolic inhibitors turning the incompatible combination into a compatible one. This indicates that resistance needs to be actively induced [2,3], although it does not preclude the possibility that susceptibility results from a constitutive absence of the induction stimulus in the parasites.

Genetic analyses of gene-for-gene systems support the contention that incompatibility as opposed to compatibility is the specific event [4,5], since the former is uniquely conditioned by the recombination of resistance and avirulence alleles. This in turn implies that avirulent races of a parasite should specifically induce defence mechanisms. Much research has been directed to examination of this prediction, but investigations of specific induction have concentrated on only a few mechanisms, notably hypersensitivity (HR), phytoalexin accumulation and lignification.

Instances where resistance can be attributed solely to a single defence mechanism are very few; several responses may occur simultaneously or consecutively. It is likely therefore that more than one biochemical mechanism may be responsible for the restricted growth of a pathogen. For example, HR is very often associated with the accumulation of phytoalexins [6] or with lignification [7], although very often only the induction of one response has been studied.

Induction of resistance mechanisms has been achieved with molecules that have been isolated from the cell walls, culture filtrates and cytoplasm of

NATO ASI Series, Vol. H1
Biology and Molecular Biology of Plant-Pathogen Interactions
Edited by J. Bailey

various parasitic as well as non-parasitic bacteria and fungi. These molecules have come to be called elicitors [8].

ELICITORS OF THE HYPERSENSITIVE RESPONSE (HR)

Bacterial Parasites

There is little information on the molecules responsible for the elicitation of HR by bacterial parasites [3]. Klement [9] concluded that no compounds had been isolated from bacteria which are able to induce HR, this being a characteristic solely of living bacteria. It is possible that inducers of HR only exist *in vivo* and are highly unstable, so that their HR-inducing activity is lost during isolation and purification. There is a report that pectic enzymes produced by bacteria induce HR [10], but whether the observed symptoms were equivalent to HR has been disputed. Bashan et al. [11] reported that a substance produced by *Pseudomonas syringae* pv. *tomato*, which appeared to be a glycoprotein, induced necrosis on leaves of non-host species such as bean, pepper and cucumber, but not in the host species tomato. In addition, Bashan et al. [11] and Whatley et al. [12] have presented data to show that lipopolysaccharides (LPS) from avirulent strains of *Pseudomonas solanacearum* induce HR in tobacco, while LPS from virulent strains do not.

In compatible combinations, bacteria often become enveloped by fibrillar material. This may play a role in the HR by ensuring cell to cell contact thus facilitating an interaction between bacterial elicitor and host receptor molecules [13]. It is possible that virulent bacteria have evolved a mechanism for preventing envelopment and thus direct host contact. It has been suggested that the production of extracellular polysaccharides (EPS) could prevent recognition of elicitors or induce a water-soaking condition necessary for bacterial multiplication.

Fungal Parasites

Elicitors of HR produced by fungi have usually been studied in association with the accumulation of phytoalexins [6]. Jones and Deverall [14] found indirect evidence for the existence of a *Lr20* gene-specific 'toxin' produced by avirulent races of *Puccinia recondita* f. sp. *tritici*. This 'toxin' induced HR when it came into contact with mesophyll cells of resistant cultivars carrying the *Lr20* gene. De Wit and Spikman [15] and De Wit et al. [16,17] found race-specific elicitors of chlorosis and necrosis in the intercellular fluids extracted from compatible combinations of *Cladosporium fulvum* and tomato. The chlorosis and necrosis induced by these specific elicitors reflected the HR which is typical for incompatible interactions between races of *C. fulvum* and tomato cultivars.

ELICITORS OF PHYTOALEXIN ACCUMULATION

Various types of molecule (glucans, proteins, glycoproteins and unsaturated fatty acids) will elicit phytoalexins. As mentioned above, HR and the accumulation of phytoalexins are often associated. Bailey [6] obtained evidence that phytoalexin formation occurs after infected cells die, suggesting that phytoalexins accumulate as a consequence of cell death (HR). However, in this section, only the phytoalexin-inducing activity of elicitors will be emphasized although it is true that they also often induce necrosis or cell death.

Glucan Elicitors

Glucans have been extensively studied in the relationship between Phytophthora megasperma f. sp. glycinea and soybean [18,19]. The elicitors which have been isolated from culture filtrates and cell walls of the fungus are β-glucans consisting of terminal 3-,6- and 3,6 linked glucosyl residues. Partial acid hydrolysis has led to isolation of the smallest cell wall component able to elicit the accumulation of glyceollin.

The exact structure of the smallest active elicitor molecule has now been reported as a hexa (β-D-glucopyranosyl)D-glucitol [19,20]. Eight of these heptamers have been described but only one isomer appeared to be active as an elicitor of glyceollin accumulation [20]. The structure of the biologically active heptamer has now been confirmed by chemical synthesis [21].

Keen et al. [22] isolated glucomannans enzymically from the cell wall of P. megasperma f. sp. glycinea with a soybean β-1,3-glucanase. These glucomannans were ten times more active as elicitors of glyceollin in soybean cotyledons than the glucan mentioned above. They also showed race-specificity in that they elicited more glyceollin in cultivars with which the isolate concerned is incompatible.

An elicitor from Colletotrichum lindemuthianum found in culture filtrates and cell walls of the fungus was presumed to be a 3- and 4-linked glucan. This compound induced the accumulation of phaseollin, hydroxyphaseollin and other phytoalexins in bean [23]. Hadwiger and Beckman [24] and Hadwiger et al. [25] found that chitosan, a 1,4-linked glucosamine isolated from cell walls of Fusarium solani, induced the accumulation of pisatin in pea.

Protein and Glycoprotein Elicitors

Cruickshank and Perrin [26] isolated a protein from mycelium of Monilinia fructicola, monilicolin A, that induced the accumulation of phaseollin in bean.

Glycoproteins which are potent inducers of phytoalexins have been isolated from a number of fungi. Keen and Legrand [27] found glycoproteins which induced glyceollin in soybean cotyledons from P. megasperma f. sp. glycinea at the mycelial wall surface. However, the concentration of glycoproteins required was considerably higher than with the glucomannan and glucan elicitors from the same fungus discussed previously. Stekoll and West [28] found a glycoprotein elicitor in culture filtrates of Rhizopus stolonifer that induced the accumulation of casbene in castor bean. It also showed an endo-polygalacturonase activity [29]. De Wit and Roseboom [30] and De Wit and Kodde [31] found peptidogalactoglucomannans in culture filtrates and cell walls of C. fulvum which induced rishitin accumulation in tomato fruits. These same elicitors from C. fulvum also induced pisatin in pea pods and glyceollin in soybean cotyledons.

Fatty Acids

Bostock et al. [32] found that arachidonic and eicosapentaenoic acid were effective inducers of sesquiterpenoid accumulation in potato tubers. These acids were released from the lipophylic materials in cell wall preparations of Phytophthora infestans. According to Bostock et al. [32], the lipids alone were sufficient to induce phytoalexin accumulation, but Kurantz and Zacharius [33] only found significant accumulation of sesquiterpenes when carbohydrate and lipid fractions from P. infestans were applied to potato tuber tissue in combination. More recently, Bostock et al. [34] and Maniara et al. [35] reported that the elicitor activity of these fatty acids could be greatly enhanced by the addition of β-1,3- and β-3,6-glucans. Interestingly, combining the carbohydrate fractions with other unsaturated C_{20} fatty acids also led to sesquiterpenoid accumulation (36). Arachidonic and eicosapentaenoic acid also induced capsidiol accumulation in pepper but not, however, rishitin in tomato fruits [37]. Apparently, species of the genus Phytophthora produce at least three types of elicitors: glucans, glycoproteins and unsaturated fatty acids. The first two induce glyceollin accumulation in soybean cotyledons, but not rishitin in potato tubers, while the latter induces sesquiterpene accumulation in potato tubers, capsidiol in pepper fruits but not rishitin in tomato fruits.

Constitutive (Endogenous) Elicitors

It is suggested that these elicitors, of host origin, are released (i) as a consequence of damage to the plant [38,39], for example during HR or the necrotrophic phase of parasitism or (ii) are released by pectic enzymes of pathogen or host origin during early pathogenesis. Pectic or other cell wall fragments released by these enzymes exhibit elicitor activities [19,40,41]. The endogenous elicitor from plant cell walls identified by Nothnagel et al. [42] appeared to be a dodeca α-1,4-D galacturonide. Some cell wall derived fragments also show proteinase inhibitor inducing activity [19,43,44,45] and very small fragments could very well be the proteinase inhibitor inducing factors themselves. Bailey [6] has proposed a hypothesis to explain the mechanisms of phytoalexin synthesis and accumulation. The hypothesis is that cell death gives rise to endogenous elicitors which in turn induce phytoalexin accumulation. This takes account of the facts that: (i) phytoalexins are not induced in compatible interactions while the fungus is in its biotrophic phase; (ii) at the onset of the necrotrophic phase, phytoalexin accumulation has often been reported in susceptible cultivar-isolate combinations, and (iii) in incompatible interactions of the HR-type, phytoalexins accumulate immediately. The hypothesis also suggests that necrotrophic parasites inevitably induce phytoalexin accumulation but may have evolved a way of escaping their deleterious effects. Alternatively, necrotrophic parasites might suppress the accumulation of phytoalexins by killing host cells so quickly that the host synthetic capability is impaired.

Abiotic Elicitors

Many abiotic factors such as heavy metals, detergents or exposure to freezing or heating induce phytoalexin accumulation [6,46]. The reason for this accumulation could be the release of endogenous elicitors caused by the damaging effects of these factors. However, since host tissues are never exposed to these substances or conditions in nature, they will not be discussed further here, other than to point out that abiotic elicitors indicate the non-specific nature of the phytoalexin response.

ELICITORS OF LIGNIFICATION AND PAPILLA FORMATION

There have been few studies of the elicitation of lignification or papilla formation; most have been conducted with living organisms [7]. The induction of lignin in wheat appears to be almost specifically a feature of filamentous fungi (saprophytes or parasites). Yeasts and bacteria are poor elicitors of lignification. Chitin, chitosan and ethylene glycol induce lignification in wounded wheat leaves [47], the first of these being of interest since it is a major constituent of the cell walls of many fungal parasites. As chitin itself is highly insoluble and may be unlikely to come in contact with plant cell membranes during the infection process, degradation products of chitin (N-acetyl-glucosamine, small chitin oligomers and glucosamine) have also been examined. However, all these substances appeared to be inactive as inducers of the lignification. In addition, purified cell walls from Botrytis cinerea and Agaricus bisporus, which do not contain chitin, did nevertheless elicit lignification. Thus it appears that cell wall components other than chitin can elicit the lignification response also. There are no reports of species or race-specific induction of lignification.

ELICITORS AND THEIR SPECIFICITY

HR and the associated accumulation of phytoalexins are frequently associated with incompatibility in gene-for-gene systems. For this reason a great deal of effort has been put into looking for molecules produced by avirulent races of a pathogen which specifically induce HR and the accumulation of phytoalexins: the so-called race-specific elicitors.

Most of the elicitors that have been isolated so far do not exhibit race-specificity; many do not even have species-specificity [30]. Often, specificity would not be expected since no differential relationship between races of the fungus and cultivars of the host plant occurs. The relationships that have been most studied are those for which a reciprocal check arrangement is evident. These are the associations between: potato and Phytophthora infestans, soybean and P. megasperma f. sp. glycinea, and bean and C. lindemuthianum. The conclusions from these studies are contradictory and undoubtedly the procedures used to isolate elicitors are crucial.

Elicitors obtained from cell walls of different fungal races by using crude homogenization procedures appear to be non-specific. These elicitors in addition seem to be rather diverse (unsaturated fatty acids from P. infestans; β-1,3- and β-1,6- branched glucans from P. megasperma f. sp. glycinea; and a peptidogalactoglucomannan from C. fulvum).

When milder extraction methods have been used, some elicitors obtained have indeed appeared to be race-specific [23,27,45]. Keen and Yoshikawa [49] found that a β-1,3-endoglucanase from soybean released potent race-specific elicitors from P. megasperma f. sp. glycinea; they appeared to be glucomannans and were different from the compounds previously isolated from this fungus.

De Wit and Spikman [15] have obtained race-specific elicitors of chlorosis and necrosis from in vivo intercellular fluids extracted from compatible combinations of C. fulvum and tomato. These elicitors are probably of fungal origin [16] and are different from the non-specific glycoproteins isolated from the fungus grown in vitro [31]. The elicitor of necrosis appears to be a small peptide [17] of which the amino acid sequence is currently under research.

SUPPRESSORS AND THEIR SPECIFICITY

In some studies, no evidence for the existence of specific elicitors has been obtained and a contrary hypothesis has emerged. It is suggested that the observed specificity in vivo could be explained by the existence of specific suppressor molecules which eliminate or decrease the effect of non-specific elicitors in compatible interactions. In other words, non-specific elicitors and specific suppressors produced by virulent races only, would work in concert, resulting in the observed in vivo specificity. Garas et al.

[50], Doke et al. [51,52] and Doke and Tomiyama [53] have isolated high molecular weight non-specific phytoalexin elicitors from mycelium of P. infestans. In addition, however, low molecular weight glucan molecules from homogenates and germination fluids of the same pathogen appeared to function as race-specific suppressors of HR and phytoalexin accumulation. These workers have proposed that race-specific suppressors confer gene-for-gene specificity. These suppressor molecules are branched β-1,3 glucans of 17-23 glucose units. However, they did not suppress, but indeed enhanced the phytoalexin-inducing activity of arachidonic and eicosapentaenoic acids [34,35,36]. This indicates that in vivo, these acids are not free but must be esterfied or otherwise linked to other macromolecules.

Ziegler and Pontzen [54] have found specific inhibition of glucan-elicited glyceollin accumulation in soybeans by an extracellular mannan-glycoprotein obtained from virulent races of P. megasperma f. sp. glycinea. The mannan-glycoprotein appeared to be an extracellular invertase. From these results, it was concluded that a susceptible host response (suppression of glyceollin accumulation) and not a resistance response (high glyceollin accumulation) is specifically induced by the fungus.

RECEPTORS FOR ELICITORS OF ACTIVE RESISTANCE MECHANISMS

Elicitors of active defence responses have been studied more extensively than their receptors in host plants. Keen and Bruegger [46] distinguished distinct determinative (recognition) and expressive phases in their model to explain gene-for-gene complementation. Recognition was envisaged as a static process while expression was thought of as dynamic. The contention is that recognition of an elicitor confers resistance by setting in motion a series of steps from, for example, de novo mRNA synthesis, through protein synthesis to the accumulation of phytoalexins (or any other active resistance response).

Recognition factors in plants are envisaged as constitutively produced surface molecules, as are elicitors from the parasite. Their occurrence at host and parasite cell surfaces would facilitate contact during the early stages of host-parasite relationship. The recognition factors in plants that have attracted most attention are the lectins. These are proteins or glycoproteins that have the ability to bind certain carbohydrate structures. Lectins are also known as haemagglutinins or phytohaemagglutinins due to their ability to agglutinate red blood cells. Some lectins also agglutinate bacteria and fungal spores. Their hapten specificity makes lectins plausible candidates for the receptors of elicitors, which are frequently carbohydrates or glycoproteins. Recognition events between two organisms through complementary surface-macromolecules including lectins have been the subject of intensive study and several recent reviews have appeared [55,56].

The Role of Lectins

Bacterial parasites. It has been suggested that lectins present on higher plant cell walls may function as recognition factors for avirulent strains of plant parasitic bacteria [56]. In this situation, it is envisaged that recognition results in incompatibility associated with HR and sometimes phytoalexin accumulation. Sequeira and Graham [57] proposed that in incompatible combinations between tobacco and avirulent strains of P. solanacearum, the LPS portion of the bacterial wall bound with a lectin (potato lectin) from the plant cell wall [58]. It was suggested that virulent strains avoided this binding by their ability to secrete soluble EPS which became bound instead, thus saturating the lectin binding sites. This would prevent the attachment of virulent bacterial cells and allow them to multiply freely in the intercellular spaces. However, contrary to the hypothesis, injection of EPS prior to inoculation with avirulent bacteria did not result in the prevention of HR. Since binding of EPS to the lectin was affected by ionic strength it is questionable whether EPS plays an important role in preventing attachment of bacteria to plant cell walls in vivo. In addition, the hypothesis does not explain why certain isolates of P. solanacearum which produce copious amounts of EPS in culture also induce HR in tobacco [56]. Furthermore, certain avirulent strains of P. solanacearum are neither agglutinated by the lectin nor do they induce HR. It is now believed, however, that other properties of EPS, such as its ability to cause water-soaking, might be more important than its lectin-binding properties in preventing attachment of virulent bacteria to cell walls.

Slusarenko and Wood [59] have extracted factors from green bean leaves that specifically agglutinate cells of P. syringae pv. phaseolicola in vitro. These extracts agglutinated cells of an avirulent strain but agglutination of a virulent strain occurred to a lesser extent. The purified and isolated agglutinins appeared not to be lectins but pectic polysaccharides [60].

In the association between soybean and P. syringae pv. phaseolicola no specific attachment of avirulent strains to host cells was observed in incompatible interactions [61]. It was therefore concluded that attachment of bacterial cells was not a prerequisite for HR and phytoalexin accumulation to occur.

Fett and Sequeira [62] have found a bacterial agglutinin in soybean that agglutinated several strains of Xanthomonas phaseoli var. sojensis but did not agglutinate P. syringae pv. glycinea. There was, however, no correlation between agglutination and pathogenicity.

Romeiro et al. [63] have also isolated a low molecular-weight, heat-stable protein from apple which agglutinates avirulent strains of E. amylovora to a greater extent than virulent strains. Evidence was also obtained for binding of bacterial LPS to this apple agglutinin. However, since no quadratic or reciprocal check exists in this association, no insight into cultivar specificity is provided.

There is therefore some evidence that lectins may be receptors for parasite 'elicitors' but this is by no means certain and no other candidates have been investigated. In host plants from which lectins have been obtained there is some evidence for specific binding at the host-bacterial species level, but there is no good evidence for race-specific binding by lectins. The role for lectins as receptors for elicitors has not so far been examined in associations where a quadratic or reciprocal check arrangement occurs.

Fungal parasites. There are a few indications that lectins could also play a role in relationships between fungal parasites and their hosts. Furuichi et al. [64] suggested that potato lectin is involved in binding the cell wall surfaces of P. infestans to cell membranes. However, no race-specificity was found. Garas and Kuć [65] found that potato lectin precipitated elicitors of terpenoid accumulation which had been extracted from the same fungus, while Nozue et al. [66] found that the induction of HR in incompatible race-cultivar combinations could be inhibited by adding N.N'-diacetyl-chitobiose, a specific hapten for the potato lectin.

Kojima et al. [67] reported that a lectin-like agglutination factor from sweet potato roots can agglutinate non-germinated spores of seven strains of Ceratocystis fimbriata including one parasitic on sweet potato. This factor, however, also showed agglutinating activity with germinated spores of five strains of the fungus parasitic on hosts other than sweet potato, while germinated spores of strains parasitic on sweet potato and almond were not agglutinated. The agglutination factor in this case was thought to be a high molecular-weight polysaccharide. Yoshikawa et al. [12] have recently reported a receptor on soybean membranes for a non-specific fungal elicitor (mycolaminarin). This receptor appears to be a protein or glycoprotein. However, binding of ^{14}C-mycolaminarin was irreversible and could not be displaced with an excess of unlabelled ligand. This finding suggests that the receptor was not likely to be a lectin.

There are other reports of lectins binding fungal elicitors, but often the lectins concerned are unrelated to those found in the host of the parasite in question. Interestingly, Hinch and Clarke [68] reported an inverse binding phenomenon: a fungal lectin interacted with a carbohydrate receptor from the host, but the significance of this is unclear. Little experimental evidence has therefore been accumulated to suggest a role for lectins in the recognition of fungal parasites.

CONVERSION OF ELICITOR 'SIGNAL' INTO METABOLIC EVENTS

In previous sections we have seen that there is preliminary evidence for the existence of receptors of elicitors on plant cell walls or membranes. What do we understand of the cellular machinery located between the primary recognition event and the eventual defence response? Does the primary recognition event cause plant cell damage which in turn induces the release of endogenous elicitors followed by phytoalexin synthesis and other related defence responses?

In a number of cases *de novo* synthesis of mRNA encoding enzymes involved in isoflavonoid derived phytoalexin accumulation have been found [69,70,71, 72] such as chalcone synthase (CHS), phenyl-ammonia-lyase (PAL) and chalcone isomerase (CHI). This demonstration of specific early changes in gene expression after treating plant tissues with elicitors may lead to the elucidation of signal response coupling mechanisms in plant pathogen interactions. However, one should be aware of the fact that changes in mRNA synthesis is yet no proof of concomitant accumulation of phytoalexins. Ebel et al. [73] observed that Xanthan gum, like a glucan elicitor, stimulated *de novo* mRNA synthesis for CHS in soybean cells, but did not induce the subsequent accumulation of glyceollin, indicating that activation of enzyme activity as such does not include accumulation of glyceollin. Unfortunately, in most studies involving rapid switching of plant gene expression non-specific elicitors were used as inducers [69,70,71,73,74].

PHYSIOLOGICAL MODELS TO EXPLAIN GENE-FOR-GENE RELATIONSHIPS

Although speculative, a widely held opinion is that basic compatibility between host and parasite evolved first and that the gene-for-gene relationship has become superimposed upon this [4,5]. The theory that plant parasites evolved from saprophytes raises the question of how many different attributes are required for an organism to be a successful parasite. Clearly, the evolutionary changes required for a saprophyte to become a parasite are greater than for a race of a parasitic species avirulent on a particular cultivar to become virulent to it. Most characteristics conferring parasitic ability are envisaged as positive functions enabling successful host colonization. Some of these functions, for example, could be the production of toxins; pectic enzymes; enzymes that degrade preformed antimicrobial compounds or phytoalexins; substances that mask recognition factors at parasite surfaces and substances that suppress the expression of

HR [1,75]. Undoubtedly, many genes must be involved in the successful development of a parasite in its host and the relationship with host gene functions could be complex.

Gabriel et al. [76] attempted to obtain temperature-sensitive parasite mutants of Phyllosticta maydis to probe which characters were crucial for basic compatibility. A class of mutants was obtained which was temperature sensitive in vivo but not in vitro. The conclusion was therefore, that these mutations affected genes crucial to parasitism. If large numbers of this type of mutant could be obtained, it would be possible both to map the loci concerned and estimate how many genes are necessary for the establishment of basic compatibility. Perhaps more important than the numbers of genes would be knowledge of the relationship between genes controlling parasitic capability and host genes governing susceptibility. Is there a relationship analogous with the gene-for-gene system for race-specific resistance? Since almost nothing is known about the genetics of basic compatibility only speculation is possible. Lack of knowledge is not surprising since such an investigation is equivalent to unravelling the course of the presumed evolution.

Given the existence of basic compatibility, there is likely to be a selection pressure on the plant towards resistance. In the unnatural circumstances of crop species, a plant breeder's role in the selection of resistant cultivars accelerates this process. Once resistant genotypes emerge either naturally or under man's influence, selection pressure is again imposed on the parasite to respond by overcoming the resistance. It has been suggested that this is how the gene-for-gene relationship comes into operation. Only after basic compatibility has been established can specificity at the race-cultivar level develop. Heath [1] has pointed out that mechanisms controlling basic compatibility and race-specific resistance may operate simultaneously and this has a bearing on the interpretation of biochemical and physiological data as well as the design of models to explain gene-for-gene relationships. However, race-specific resistance has generally been studied with little regard to the processes responsible for basic compatibility. Many of the apparent contradictions between studies may be a reflection of this, and paying too little regard to basic compatibility processes may be one reason why mechanisms of race-specific resistance are still poorly understood.

Genetical Implications

Ellingboe [5] has summarized the development of research on the physiology of host resistance since the 1950s. Much of the early work was conducted using two host lines, one susceptible and one resistant to a particular isolate of a parasite (Fig. 1). In most studies the different host lines may have differed greatly for many characters other than resistance. Thus, it is inevitable that any difference between the two lines is absolutely correlated with resistance and susceptibility, and the same argument holds for investigations where a single cultivar or a line of a plant species is inoculated with virulent or avirulent isolates of a parasite (Fig. 2). Any difference between two isolates is absolutely correlated with virulence and avirulence. But establishing 'cause and effect' is almost impossible from this type of study. Rowell et al. [74] influenced thinking by introducing the concept of the quadratic check as a genetical model for physiological studies of resistance to Puccinia graminis (stem rust) in wheat (Fig. 3). Examination of the quadratic check leads to the conclusion that three different genotypical combinations (A_1/r_1; a_1/r_1; a_1/R_1) all lead to a compatible phenotype while only one unique genotype combination (A_1/R_1) leads to incompatibility. This can be interpreted to mean that specificity is controlled by the interaction between the gene products of the

	Host Lines	
	A	B
Parasite	C	I

Figure 1. The possible combinations between one parasite and two host lines (A and B) which vary in resistance. C = compatible or susceptible; I = incompatible or resistant.

	Parasite isolates	
	X	Y
Host	C	I

Figure 2. The possible combinations between two parasite isolates (X and Y) which differ in parasitic capability, and one host line. C and I as in Figure 1.

	Genotypes of host lines	
Genotypes of parasite isolates	R_1R_1	r_1r_1
A_1	I	C
a_1	C	C

Figure 3. The 'quadratic check'. Possible combinations between two host lines and two haploid parasite isolates differing in resistance and parasitic capability respectively. R_1 and A_1 are alleles for resistance and avirulence respectively, and r_1 and a_1 are the opposite alleles for susceptibility and virulence. C and I as in Figure 1.

	Genotypes of host lines	
Genotypes of parasite isolates	$R_1R_1r_2r_2$	$r_1r_1R_2R_2$
A_1a_2	I	C
a_1A_2	C	I

Figure 4. The 'double reciprocal check'. Possible combinations between two host lines and two haploid parasite isolates differing in resistance and parasitic capability respectively. R_1 and R_2 are different alleles for resistance and r_1 and r_2 are the opposite alleles for susceptibility. A_1 and A_2 are different alleles for specific avirulence on R_1 and R_2 respectively; a_1 and a_2 are the opposite alleles for virulence. C and I as in Figure 1.

dominant A and R alleles. In all other cases, it is envisaged that there is no recognition or interaction between gene products and as a consequence, compatibility results. However, Martin and Ellingboe [77] have found small differences between the phenotypes of the three allele combinations leading to compatibility.

Loegering and Harmon [78] introduced the use of near-isogenic lines, more or less identical to one another except for their resistance genes. This made it possible to study differences between resistant and susceptible lines that were more likely to be related to genes for resistance. The arrangement between isolates and cultivars employed in this study was called the (double) reciprocal check (Fig. 4). This arrangement can be interpreted to mean that A_1 interacts specifically with R_1 and not with R_2 and that A_2 interacts specifically with R_2 and not with R_1. There is a strict gene- for-gene relationship. In addition, this arrangement implies that incompatibility is epistatic to compatibility.

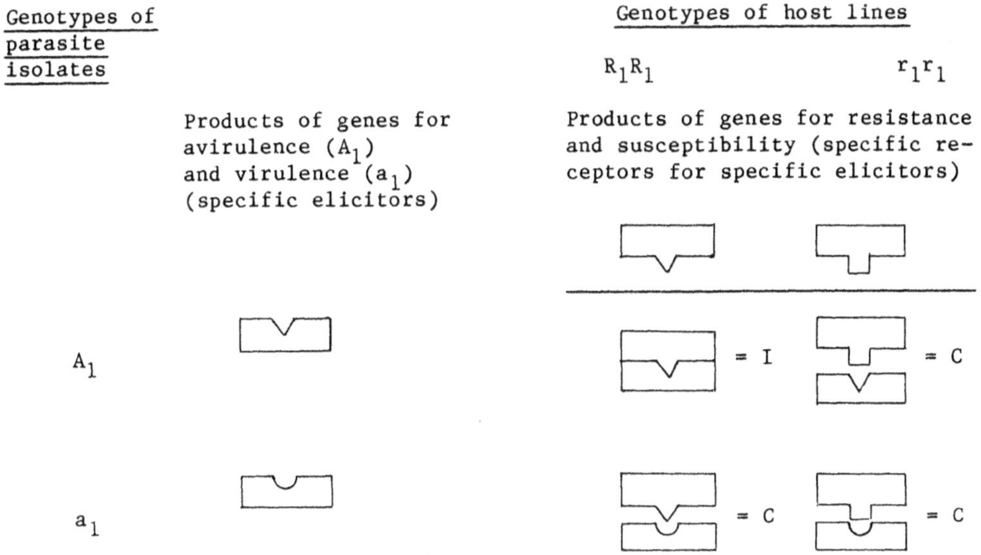

Figure 5. A physiological model for expression of incompatibility after interactions between the product of the gene for resistance (R_1) and the product of the gene for avirulence (A_1). I = incompatible; C = compatible.

Ellingboe [5] and Keen [75] have individually reviewed a number of models which have been proposed to explain the gene-for-gene relationship on a mechanistic basis. These models are briefly elaborated and commented upon here.

Constitutive Model

One model has been advanced by Ellingboe [5] and is derived from genetic studies of the gene-for-gene relationship. It is suggested that the primary product of the dominant allele for avirulence (A_1) interacts directly with the primary product of the dominant gene for resistance (R_1) and that the interaction itself is responsible for incompatibility (Fig. 5). It is suggested that a structural dimer is directly responsible for inhibiting parasite growth.

Lectins might contribute to inhibition of growth on the basis of this model, but there are no reports of lectins with race-specific binding properties, nor are there indications that cultivar or R-gene-specific lectins exist. Plant lectins can inhibit fungal growth [79] but this is probably a non-specific phenomenon. A further objection that has been raised to this model is that compatibility can be induced with metabolic inhibitors, thus suggesting that resistance is not a constitutive character and that the expression of incompatibility requires active processes.

A different type of constitutive model has been observed in relation to the action of host-selective toxins [80]. This model is based on the supposition that a parasite gene controls the production of a toxin which is detected by the product of a host gene for sensitivity to that toxin. Sensitivity to the toxin leads the host to be susceptible to the parasite, for example in the case of some toxin-producing Helminthosporium species. Since there is no double reciprocal check arrangement in diseases involving host-selective toxins, evidence for a gene-for-gene relationship is lacking. However, in other host-parasite combinations which are demonstrably gene-for-gene controlled, it has also been suggested that HR may be a specific response to a selective toxin, the production of which is controlled by an avirulence gene. The hypothetical toxin is thought to cause cellular disruption and the release of pre-formed antimicrobial metabolites.

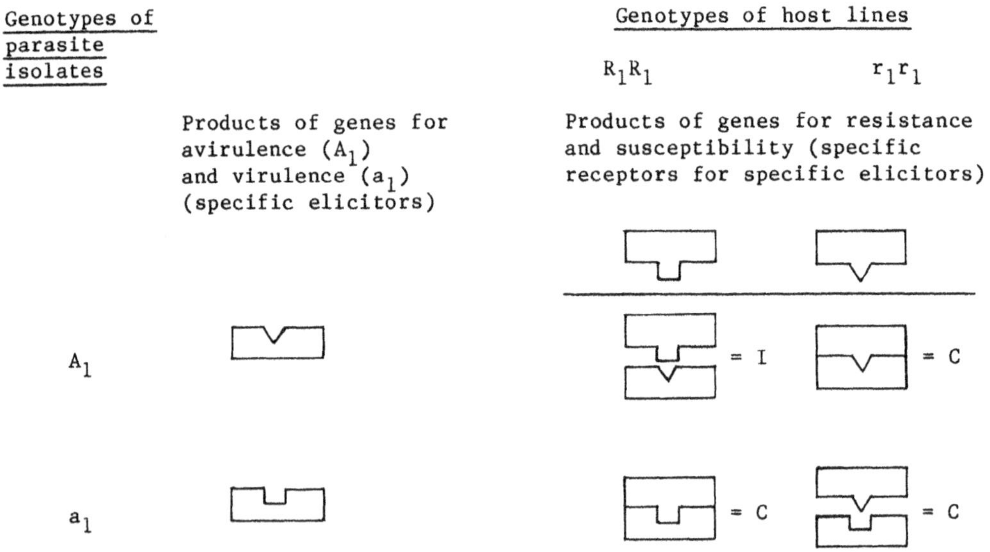

Figure 6. A physiological model for expression of compatibility after interaction between products of genes for resistance (R_1) and susceptibility (r_1) and products of genes for avirulence (A_1) and virulence (r_1). Note that products of a_1 and r_1 do not interact yet a compatible association is observed in practice. C and I as in Figure 5.

The double induction model

This model proposes that there is a host-mediated release of avirulence gene product (specific elicitor) when host-parasite contact is established, which in turn initiates the host defence response [81]. Although this model is hypothetical, some supporting experimental data have recently been forthcoming. Plant enzymes like β-1,3 glucanases and chitinases can rapidly liberate soluble phytoalexin elicitors. Keen and Yoshikawa [49] and Keen et al. [22] found that a β-1,3 glucanase from soybean released glucomannans with elicitor activity from cell walls of the fungus P. megasperma f. sp.

glycinea. These elicitors appeared to be race-specific. Also, De Wit and Spikman [15] found host-mediated release (or production) of race-specific elicitors of necrosis in intercellular fluids of compatible combinations between C. fulvum and tomato. The factor mediating the production of these elicitors in tomato is unknown but it is not dependent upon the R-gene carried by the host in which the fungus is growing [16]. This model is in agreement with genetic studies of gene-for-gene relationships.

The Non-Specific Elicitor and Specific Suppressor Model

This model is based on the premise that non-specific elicitors in concert with race- specific suppressors confer specificity in host-pathogen associations (Fig. 6). It is envisaged that race-specific suppressors produced by virulence genes specifically bind to host receptors (products of genes for resistance). In this way, they either prevent the binding of non-specific elicitors, thus causing the suppression of a resistance response, or alternatively prevent the expression of resistance in another way. Doke et al. [51,52], Garas et al. [50] and Doke and Tomiyama [53] found experimental evidence to support this hypothesis from studies with potato and P. infestans. Ziegler and Pontzen [54] obtained data from studies with soybean and P. megasperma f. sp. glycinea which could also be interpreted on the basis of this hypothesis. The suppressors identified by Doke et al. [51,52] and Garas et al. [50] are active in crude preparations and non-specific elicitors from P. infestans but unaccountably, they enhanced the activity of purified elicitors [33,34,35,36,82]. Hence the role of race-specific suppressors is unclear.

Apart from the confusing experimental evidence, this model is not entirely in agreement with genetic studies of gene-for-gene relationships. In the model, compatibility is thought to be the specific event, yet the combinations A_1/r_1, a_1/R_1 and a_1/r_1 (Fig. 3) all lead to compatibility. As represented in Figure 6 it can be seen that specific recognition between products of A_1 and r_1, and a_1 and R_1 could occur, but not between a_1 and r_1. It seems to be more feasible to consider suppressors to be one of the many characters needed by a parasite to establish basic compatibility.

In this context, Davidse and Boekeloo [83] have found evidence for the production of race non-specific suppressors by P. infestans. These suppressors inhibited the induction of necrosis in potato leaves caused by non-specific elicitors from P. infestans. Heath [84] also presented evidence for general suppressors and has suggested that they may act below the level of race-cultivar specificity. Bushnell and Rowell [85] have proposed a model in which race-specific suppressors may have evolved from general species-specific suppressors. This implies that one macromolecule possesses general as well as race-specific suppressor activity, and that basic compatibility in addition to race-specific compatibility could be determined by one macromolecule. According to this model, both basic compatibility and race-specific compatibility would be inherited as dominant characters. But whilst almost nothing is known about the genetics of basic compatibility, race-specific compatibility in host-parasite combinations involving diploid biotrophic fungi usually involves combinations of recessive alleles.

The specific elicitor-specific receptor model. This model is similar to the first constitutive model described above (Fig. 5). However, it differs in that initial recognition leads to a sequence of events culminating in (locally) induced resistance (HR, accumulation of phytoalexins, lignification etc.). Incompatibility is the specific event. The suggestion is that the primary product of a gene for avirulence (A_1) interacts with the primary product of a gene for resistance (R_1). As a consequence, a second messenger-like substance transfers information to the plant cell nucleus, where defence mechanisms are initiated. These second messengers could be the endogenous elicitors mentioned earlier.

Jones and Deverall [14], Bruegger and Keen [48], Anderson [23], Keen and Legrand [27], De Wit and Spikman [15], Keen and Yoshikawa [49], Keen et al. [22] and De Wit [16,17] have all found evidence for race-specific elicitors of defence responses. The specific elicitor-specific receptor model does not conflict with the idea of there being a basic compatibility since the gene-for-gene system is envisaged as superimposed upon it. The implication is that the general suppressors discussed earlier [1,83,84] are a component of basic compatibility.

Ellingboe [5] has argued that if responses such as phytoalexin accumulation are important in gene-for-gene controlled resistance, one gene-for-one-gene relationships would not be expected. The argument is that many enzymes and hence genes are responsible for phytoalexin synthesis, and that mutations at any of these loci should impair resistance. Similarly, since glycoproteins (as specific elicitors) are also synthesized by many enzymes, it is argued that their production is unlikely to relate to a single avirulence gene. Ellingboe [5] is of the opinion that if this model were to operate, specific resistance and virulence would both be polygenic traits. With respect to a glycoprotein, the glycosyl transferase which provides the final structure is the most important in terms of potential specificity, and this could be coded by the avirulence locus. Mutants lacking enzymes, or having less efficient enzymes for synthesis of phytoalexins have not been studied and may simply be lethal.

There are no indications that gene-for-gene incompatibility is based upon shared genes between host and parasite. Complete correspondence between these two types of relationship would imply that a resistance allele behaves as it does because it codes for the same protein as an allele located in the parasite which thus becomes identifiable as an avirulence allele.

FURTHER CONSIDERATIONS

Models such as those discussed above are important in providing a stimulus for further experimentation, but there is a danger that they channel thinking in a single direction. Allegiance to a particular model is soon likely to inhibit genuine testing of hypotheses.

There seems to be one major point at issue concerning the mechanistic basis of gene-for-gene relationships. To explain specific resistance, is there a need to invoke expressions of incompatibiity such as HR, phytoalexin accumulation and lignification, or are these merely symptoms that inevitably follow incompatibility resulting from some prior event? This is the distinction between constitutive models and those which explain resistance in terms of some actively induced host response.

It cannot be denied that these host responses occur and that they are strongly associated with incompatibility in gene-for-gene relationships. There is also persuasive evidence, discussed previously, which points to their direct involvement in the cessation of parasite development in some cases.

It is nevertheless possible to provide alternative explanations to account for the universal occurrence of some active reaction in plants associated with gene-for-gene incompatibility. For example, if cell death is a necessary consequence to post-penetration incompatibility (as envisaged in at least two of the constitutive models), its occurrence in the absence of some response could be potentially detrimental to the plant. All sorts of substances leak from dead cells, encouraging the growth of micro-organisms in their vicinity. The entry and establishment of unspecialized and damaging necrotrophic parasites would be facilitated, unless a 'wound-healing' process prevented this eventuality. Accumulation of phytoalexins or cell wall modification for example may provide essential non-specific protection against secondary invasion. It is probable that small lesions resulting from penetration by avirulent parasites (particularly those to which the plant is a non-host) abound at plant surfaces under natural circumstances. Some mechanism to restrict access of micro-organisms to the plant via these wounds is to be expected.

CONCLUSION: PERSPECTIVES AND CHALLENGES

Plant pathology, like most fields of endeavour, has evolved by trial and error rather than maturing to a plan. Inevitably, this has led to great leaps forward, gaps and inconsistencies. The study of resistance calls on many disciplines: genetics; plant and microbial physiology; biochemistry and molecular biology. Each has made valuable contributions to the whole, but clearly there is a limit to what can be achieved by any single approach. Thus physiological studies cannot alone explain the underlying genetic controls, and genetic analysis cannot fully explain biochemical mechanisms. A major challenge in studying resistance mechanisms today is to bring together these disparate but complementary disciplines. Another is to translate fundamental knowledge into measures for practical crop protection.

These are easier said than done; there are problems inherent in the systems chosen for study. Near-isogenic lines or races of hosts and pathogens are only rarely available. From earlier sections, it is clear that host-pathogen associations which are amenable to genetic study all too often involve obligate parasites such as rusts and mildews, where certain types of biochemical or physiological study become very difficult. On the other hand, those organisms which can be cultured easily and are more amenable to biochemical study are often fungi imperfecti, making genetic analysis difficult or impossible. Some organisms are renowned for having unstable genomes, making long-term experiments difficult. Pathogens grown in culture can lose their parasitic ability unless they are periodically 'recycled' through the plant.

There are also difficulties with plant material. The conditions of propagation can have a profound influence. The urge is to chose a 'model' host-pathogen system which is most convenient for the type of analysis involved, but even the best systems are prone to artifacts or other factors which confuse the issue. Tissue age and type are among the commonest offenders. Some plant tissues are easier to examine physiologically and biochemically than others, which may be more important in the pathogenic relationship. Basing general assumptions on experiments carried out on specific tissues is a constant temptation. Much work on elicitors of phytoalexin accumulation, for example, has been done with cotyledons, hypocotyls, cultured cells and various storage organs. How representative are they of stems and leaves, and what is the relevance of a result from a laboratory experiment to the crop in the field?

The concentration on model systems also produces a fragmentary picture. There is a temptation to extrapolate findings to other cases, but the enormous diversity of reaction types makes this a dangerous practice, and valid generalities are hard to come by, even for closely related systems.

Having emphasized the difficulties, it is appropriate to balance the picture by marking the progress that has been made. Fortunately, studies on disease resistance mechanisms have moved on from an early fixation with the isolation and characterization of phytoalexins to molecular biology and molecular genetics. Effort is now being devoted to answering more important questions on the mechanisms that lead to their accumulation, and their role in defence. Recombinant DNA-technology gives us the tools to test the different physiological models put forward to explain gene-for-gene systems. Much progress has been made with bacterial plant pathogens. Staskawicz et al. [86] have cloned an avirulence gene of P. syringae pv. glycinea which determines race-specific incompatibility. This avirulence gene was subcloned to obtain a 3 kb. fragment and experiments are underway to isolate the avirulence gene product and assess its phytoalexin-inducing activity. In the C. fulvum - tomato interaction, research has moved on in another direction [17]. A necrosis-inducing peptide which has been suggested to be an avirulence gene product has now been characterized. When the amino acid sequence of this peptide has been determined search for the gene for avirulence from genomic libraries can begin with the help of a synthetic nucleotide probe. Purified elicitors will also permit a search for their receptors, the putative resistance gene products. It is clear that recombinant DNA-technology will enable us to investigate a number of problems which were hitherto difficult to study.

REFERENCES

1. Heath, M.C., 1982, Absence of active defense mechanisms in compatible host pathogen interactions, in: "Active Defense Mechanisms in Plants", R.K.S. Wood, ed., pp. 143-156, Plenum Press, New York and London.
2. Keen, N.T., Ersek, T., Long, M., Bruegger, B. and Holliday, M., 1981, Inhibition of the hypersensitive reaction of soybean leaves to incompatible Pseudomonas spp. by blasticidin S, streptomycin or elevated temperature, Physiol. Plant Pathol., 18: 325-337.
3. Keen, N.T. and Holliday, M.J., 1982, Recognition of bacterial pathogens by plants, in: "Phytopathogenic Prokaryotes", Vol. 2, M.S. Mount and G.H. Lacy, eds., pp. 179-217, Academic Press, New York and London.
4. Ellingboe, A.H., 1981, Changing concepts in host-pathogen genetics, Annu. Rev. Phytopathology, 19: 125-143.
5. Ellingboe, A.H., 1982, Genetic aspects of active defense, in: "Active Defense Mechanisms in Plants", R.K.S. Wood, ed., pp. 179-192, Plenum Press, New York and London.
6. Bailey, J.A., 1982, Mechanisms of phytoalexin accumulation, in: "Phytoalexins", J.A. Bailey and J.A. Mansfield, eds., pp. 289-318, Blackie, Glasgow and London.
7. Vance, T., Kirk, K. and Sherwood, R.T., 1980, Lignification as a mechanism of disease resistance, Annu. Rev. Phytopathology, 18: 259-288.
8. Keen, N.T., 1975, Specific elicitors of plant phytoalexin production: determinants of race specificity in pathogens, Science, 74-75.
9. Klement, Z., 1982, Hypersensitivity, in: "Phytopathogenic Prokaryotes", Vol. 2, M.S. Mount and G.H. Lacy, eds., pp. 149-177, Academic Press, New York and London.
10. Gardner, J.M. and Kado, C.I., 1976, Polygalacturonic acid trans-eliminase in the osmotic shock fluid of Erwinia rubri faciens: characterization of the purified enzyme and its effect on plant cells, J. Bacteriol., 127: 451-460.
11. Bashan, Y., Okon, Y. and Henis, Y., 1982, Detection of a necrosis-inducing factor of non-host plant leaves produced by Pseudomonas syringae pv. tomato, Can. J. Bot., 60: 2453-2460.

12. Yoshikawa, M., Keen, N.T. and Wang, M.C., 1983, A receptor on soybean membranes for a fungal elicitor of phytoalexin accumulation, Plant Physiol., 73: 497-506.
13. Stall, R.E. and Cook, A.A., 1979, Evidence that bacterial contact with the plant cell is necessary for the hypersensitive reaction but not the susceptible reaction, Physiol. Plant Pathol., 14: 77-84.
14. Jones, D.R. and Deverall, B.J., 1978, The use of leaf transplants to study the cause of hypersentitivity to leaf rust, Puccinia recondita in wheat carrying the Lr20 gene, Physiol. Plant Pathol., 12: 311-319.
15. De Wit, P.J.G.M. and Spikman, G., 1982, Evidence for the occurrence of race and cultivar-specific elicitors of necrosis in intercellular fluids of compatible interactions of Cladosporium fulvum and tomato, Physiol. Plant Pathol., 21: 1-11.
16. De Wit, P.J.G.M., Hofman, J.E. and Aarts, J.M.M.J.G., 1984, Origin of specific elicitors of chlorosis and necrosis occurring in intercellular fluids of compatible interactions of Cladosporium fulvum (syn. Fulvia fulva) and tomato, Physiol. Plant Pathol., 24: 17-23.
17. De Wit, P.J.G.M., Hofman, J.E., Velthuis, G.C.M. and Kuć, J.A., 1985, Isolation and characterization of an elicitor of necrosis isolated from intercellular fluids of compatible interactions of Cladosporium fulvum (syn. Fulvia fulva) and tomato, Plant Physiol., 77: 642-647.
18. Albersheim, P. and Valent, B.S., 1978, Host-pathogen interactions in plants. Plants when exposed to oligosaccharides of fungal origin defend themselves by accumulating antibiotics, J. Cell Biol. 78: 627-643.
19. Darvill, A.G. and Albersheim, P., 1984, Phytoalexins and their elicitors. A defense against microbial infection in plants, Annu. Rev. Plant Physiol., 35: 243-276.
20. Sharp, J.K., McNeil, M. and Albersheim, P., 1984, The primary structures of one elicitor-active and seven elicitor-inactive hexa (β-D-glucopyranosyl)-D-glucitols isolated from the mycelial walls of Phytophthora megasperma f. sp. glycinea, J. Biol. Chem., 259: 11321-11326.
21. Ossowski, P., Pilotti, A., Garegg, P.J. and Lindberg, B., 1984, Synthesis of a glucoheptaose and a glucooctaose that elicit phytoalexin accumulation in soybean, J. Biol. Chem., 259: 11337-11340.
22. Keen, N.T., Yoshikawa, M. and Wang, M.C., 1983, Phytoalexin elicitor activity of carbohydrates from Phytophthora megasperma f. sp. glycinea and other sources, Plant Physiol., 71: 466-471.
23. Anderson, A.J., 1980, Differences in the biochemical composition and elicitor activity of extracellular components produced by three different races of a fungal plant pathogen, Colletotrichum lindemuthianum, Can. J. Microbiol., 26: 1473-1479.
24. Hadwiger, L.A., Beckman, J.M. and Adams, M.J., 1980, Chitosan as a component of pea-Fusarium solani interactions, Plant Physiol., 66: 205-211.
25. Hadwiger, L.A., Beckman, J.M. and Adams, M.J., 1981, Localization of fungal compounds in the pea-Fusarium interaction detected immunochemically with anti-chitosan and anti-fungal cell wall antisera, Plant Physiol., 67: 170-175.
26. Cruickshank, I.A.M. and Perrin, D.R., 1968, The isolation and partial characterization of monilicolin A, a polypeptide with phaseollin-inducing activity from Monilinia fructicola, Life Sci., 7: 449-458.
27. Keen, N.T. and Legrand, M., 1980, Surface glycoproteins: evidence that they may function as the race specific phytoalexin elicitors of Phytophthora megasperma f. sp. glycinea, Physiol. Plant Pathol., 17: 175-192.
28. Stekoll, M. and West, C.A., 1978, Purification and properties of an elicitor of castor bean phytoalexin from culture filtrates of the fungus Rhizopus stolonifer, Plant Physiol., 61: 38-45.
29. Lee, S.C. and West, C.A., 1981, Polygalacturonase from Rhizopus stolonifer, an elicitor of casbene synthetase activity in castor bean Ricinus communis seedlings, Plant Physiol., 67: 633-639.

30. De Wit, P.J.G.M. and Roseboom, P.H.M., 1980, Isolation, partial characterization and specificity of glycoprotein elicitors from culture filtrates, mycelium and cell walls of Cladosporium fulvum (syn. Fulvia fulva), Physiol. Plant Pathol., 16: 391-408.
31. De Wit, P.J.G.M. and Kodde, E., 1981, Further characterization and cultivar specificity of glycoprotein elicitors from culture filtrates and cell walls of Cladosporium fulvum (syn. Fulvia fulva), Physiol. Plant Pathol., 18: 297-314.
32. Bostock, R.M., Kuć, J.A. and Laine, R.A., 1981, Eicosapentaenoic and arachidonic acids from Phytophthora infestans elicit fungitoxic sesquiterpenes in the potato, Science., 212: 67-69.
33. Kurantz, M.J. and Zacharius, R.M., 1981, Hypersensitive response in potato tuber: elicitation by combination of non-eliciting components from Phytophthora infestans, Physiol. Plant Pathol., 18: 67-77.
34. Bostock, R.M., Laine, R.A. and Kuč, J.A., 1982, Factors affecting the elicitation of sesquiterpenoid phytoalexin accumulation by eicosapentaenoic and arachidonic acids in potato, Plant Physiol., 70: 1417-1424.
35. Maniara, G., Laine, R. and Kuć, J., 1984, Oligosaccharides from Phytophthora infestans enhance the elicitation of sesquiterpenoid stress metabolites by arachidonic acid in potato, Physiol. Plant Pathol., 24: 177-186.
36. Preisig, C.L. and Kuć, J.A., 1985, Arachidonic acid related elicitors of the hypersensitive response in potato and enhancement of their activities by glucans from Phytophthora infestans, Arch. Biochem. Biophys., 236: 379-389.
37. Bloch, C.B., De Wit, P.J.G.M. and Kuć, J.A., 1984, Elicitating of phytoalexins by arachidonic and eicosapentaenoic acids: a host survey, Physiol. Plant Pathol., 25: 199-208.
38. Hahn, M.G., Darvill, A.G. and Albersheim, P., 1981, Host-pathogen interactions. XIX. The endogenous elicitor, a fragment of a plant cell wall polysaccharide that elicits phytoalexin accumulation in soybeans, Plant Physiol., 68: 1161-1169.
39. Hargreaves, J.A. and Bailey, J.A., 1978, Phytoalexin production by hypocotyls of Phaseolus vulgaris in response to constitutive metabolites released by damaged cells, Physiol. Plant Pathol., 13: 89-100.
40. Bruce, R.J. and West, C.A., 1982, Elicitation of casbene synthetase activity in castor bean Ricinus communis. The role of pectic fragments of the plant cell wall in elicitation by a fungal endopolygalacturonase, Plant Physiol., 69: 1181-1188.
41. Davis, K.R., Lyon, G.D., Albersheim, P. and Darvill, A.G., 1984, Host pathogen interactions. XXV. Endo-polygalacturonic acid lyase EC-4.2.2.2 from Erwinia carotovora elicits phytoalexin accumulation by releasing plant cell wall fragments, Plant Physiol., 74: 52-60.
42. Nothnagel, E.A., McNeil, M., Albersheim, P. and Dell, A., 1983, Host pathogen interactions. XXII. A galacturonic acid oligosaccharide from plant cell walls elicits phytoalexins, Plant Physiol., 71: 916-926.
43. Bishop, P.D., Pearce, G., Bryant, J.E. and Ryan, C.A., 1984, Isolation and characterization of the proteinase inhibitor-inducing factor from tomato leaves. Identity and activity of polygalacturonide and oligogalacturonide fragments, J. Biol. Chem., 259: 13172-13177.
44. Walker-Simmons, M., Jin, D., West, C.A., Hadwiger, L. and Ryan, C.A., 1984, Comparison of proteinase inhibitor-inducing activities and phytoalexin elicitor activities of pure fungal endopolygalacturonase pectic fragments and chitosans, Plant Physiol., 76: 833-836.
45. Walker-Simmons, M., Hadwiger, L. and Ryan, C.A., 1983, Chitosans and pectic polysaccharides both induce the accumulation of the antifungal phytoalexin pisatin in pea pods and anti-nutrient proteinase inhibitors in tomato leaves, Biochem. Biophys. Res. Commun., 110: 194-199.

46. Keen, N.T. and Bruegger, B.B., 1977, Phytoalexins and chemicals that elicit their production in plants, in: "Host Plant Resistance to Pests", P. Hedin, ed., pp. 1-26, American Chemical Society Symposium Series 62.
47. Pearce, R.B. and Ride, J.P., 1982, Chitin and related compounds as elicitors of the lignification in wounded wheat leaves, Physiol. Plant Pathol., 20: 119-123.
48. Bruegger, B.B. and Keen, N.T., 1979, Specific elicitors of glyceollin accumulation in the Pseudomonas glycinea-soybean host-parasite system, Physiol. Plant Pathol., 15: 43-51.
49. Keen, N.T. and Yoshikawa, M., 1983, β-1,3-endoglucanase from soybean releases elicitor-active carbohydrates from fungal cell walls, Plant Physiol., 71: 460-465.
50. Garas, N.A., Doke, N. and Kuć, J., 1979, Suppression of the hypersensitive reaction in potato tubers by mycelial components from Phytophthora infestans, Physiol. Plant Pathol., 15: 117-126.
51. Doke, N., Garas, N.A. and Kuć, J., 1979, Partial characterization and aspects of the mode of action of a hypersensitivity-inhibiting factor (HIF) isolated from Phytophthora infestans, Physiol. Plant Pathol., 15: 127-140.
52. Doke, N., Garas, N.A. and Kuć, J., 1980, Effect on host hypersensitivity of suppressors released during the germination of Phytophthora infestans cytospores, Phytopathology, 70: 35-39.
53. Doke, N. and Tomiyama, K., 1980, Suppression of the hypersensitive response of potato tuber protoplasts to hyphal wall components by water soluble glucans isolated from Phytophthora infestans, Physiol. Plant Pathol., 16: 177-186.
54. Ziegler, E. and Pontzen, R., 1982, Specific inhibition of glucan-elicited glyceollin accumulation in soybeans by an extracellular mannan-glycoprotein of Phytophthora megasperma f. sp. glycinea, Physiol. Plant Pathol., 20: 321-331.
55. Dazzo, F.B. and Truchet, G.L., 1983, Interactions of lectins and their saccharide receptors in the Rhizobium-like symbiosis, J. Membrane Biol., 73: 1-16.
56. Sequeira, L., 1982, Determinants of plant response to bacterial infection. in: "Active Defense Mechanisms in Plants", R.K.S. Wood, ed., pp. 85-102, Plenum Press, New York and London.
57. Sequeira, L. and Graham, T.L., 1977, Agglutination of avirulent strains of Pseudomonas solanacearum by potato lectin, Physiol. Plant Pathol., 11: 43-54.
58. Leach, J.E., Cantrell, M.A. and Sequeira, L., 1982, A hydroxyproline-rich bacterial agglutinin from potato: its localization by immunofluorescence, Physiol. Plant Pathol., 21: 319-325.
59. Slusarenko, A.J. and Wood, R.K.S., 1981, Differential agglutination of races 1 and 2 of Pseudomonas phaseolicola by a fraction from cotyledons of Phaseolus vulgaris cv. Red Mexican, Physiol. Plant Pathol., 18: 187-193.
60. Slusarenko, A.J. and Wood, R.K.S., 1983, Agglutination of Pseudomonas phaseolicola by pectic polysaccharide from leaves of Phaseolus vulgaris, Physiol. Plant Pathol., 23: 217-227.
61. Fett, W.F. and Jones, S.B., 1982, Role of bacterial immobilization in race-specific resistance of soybean to Pseudomonas syringae pv. glycinea, Phytopathology, 72: 488-492.
62. Fett, W.F. and Sequeira, L., 1980, A new bacterial agglutinin from soybean. II. Evidence against a role in determining pathogen specificity, Plant Physiol., 66: 853-858.
63. Romeiro, R., Karr, A. and Goodman, R.N., 1981, Isolation of a factor from apple that agglutinates Erwinia amylovora, Plant Physiol., 69: 772-777.
64. Furuichi, N., Tomiyama, K. and Doke, N., 1980, The role of potato lectin in binding of germ tubes of Phytophthora infestans to potato cell membrane, Physiol. Plant Pathol., 16: 249-256.

65. Garas, N.A. and Kuč, J., 1981, Potato lectin lyses zoospores from Phytophthora infestans and precipitates elicitors of terpenoid accumulation produced by the fungus, Physiol. Plant Pathol., 18: 227-237.
66. Nozue, M., Tomiyama, K. and Doke, N., 1980, Effect of N.N'-diacetyl-D-chitobiose, the potato lectin hapten and other sugars on hypersensitive reaction of potato tuber cells infected by incompatible and compatible races of Phytophthora infestans, Physiol. Plant Pathol., 17: 221-227.
67. Kojima, M., Kawakita, K. and Uritani, I., 1982, Studies on a factor in sweet potato roots which agglutinates spores of Ceratocystis fimbriata black rot fungus, Plant Physiol., 69: 474-478.
68. Hinch, J.M. and Clarke, A.E., 1980, Adhesion of fungal zoospores to root surfaces is mediated by carbohydrate determinants of the root slime, Physiol. Plant Pathol., 16: 303-307.
69. Cramer, C.L., Ryder, T.B., Bell, J.N. and Lamb, C.J., 1985, Rapid switching of plant gene expression induced by fungal elicitor, Science, 227: 1240-1242.
70. Lawton, M.A., Dixon, R.A., Hahlbrock, K. and Lamb, C.J., 1983, Elicitor induction of messenger RNA activity. Rapid effects of elicitor on phenylalanine ammonia lyase EC-4.3.1.5 and chalcone synthase messenger RNA activities in bean Phaseolus vulgaris cells, Eur. J. Biochem., 130: 131-140.
71. Ryder, T.B., Cramer, C.L., Bell, J.N., Robbins, M.P., Dixon, R.A. and Lamb, C.J., 1984, Elicitor rapidly induces chalcone synthase messenger RNA in Phaseolus vulgaris cells at the onset of the phytoalexin defense response, Proc. Natl. Acad. Sci. USA., 81: 5724-5728.
72. Schmelzer, E., Boerner, H., Grisebach, H., Ebel, J. and Hahlbrock, K., 1984, Phytoalexin synthesis in soybean (Glycine max). Similar time courses of messenger RNA induction in hypocotyls infected with a fungal pathogen and in cell cultures treated with fungal elicitor, FEBS Lett., 172: 59-63.
73. Ebel, J., Schmidt, W.E. and Loyal, R., 1984, Phytoalexin synthesis in soybean cells: elicitor induction of PAL and chalcone synthase mRNA's and correlation with phytoalexin accumulation, Arch. Biochem. Biophys., 232: 240-248.
74. Rowell, J.B., Loegering, W.Q. and Powers, H.R., 1963, Genetic model for physiologic studies of mechanisms governing development of infection type in wheat stem rust, Phytopathology 53: 932-937.
75. Keen, N.T., 1982, Specific recognition in gene-for-gene host parasite systems, Adv. Plant Pathol., 1: 35-82.
76. Gabriel, D.W., Ellingboe, A.H. and Rossman, E.C., 1979, Mutations affecting virulence in Phyllosticta maydis, Can. J. Bot., 57: 2639-2643.
77. Martin, T.J. and Ellingboe, A.H., 1976, Differences between compatible parasite/host genotypes involving the Pm4 locus of wheat and the corresponding genes in Erysiphe graminis f. sp. tritici, Phytopathology, 66: 1435-1438.
78. Loegering, W.Q. and Harmon, D.L., 1969, Wheat lines near-isogenic for reaction to Puccinia graminis tritici, Phytopathology, 59: 456-459.
79. Gibson, D.M., Stack, S., Krell, K. and House, J., 1982, A comparison of soybean agglutinin in cultivars resistant and susceptible to Phytophthora megasperma var. sojae (race 1), Plant Physiol., 70: 560-566.
80. Daly, J.M., 1984, The role of recognition in plant disease, Annu. Rev. Phytopathology, 22: 273-308.
81. Cruickshank, I.A.M., 1980, Defenses triggered by the invader: chemical defenses, in: "Plant Disease: An Advanced Treatise", Vol. V, J.G. Horsfall and E.D. Cowling, eds., pp. 247-267, Academic Press, New York, San Francisco and London.

82. Kurantz, M.J. and Osman, S.F., 1983, Class distribution, fatty acid composition and elicitor activity of Phytophthora infestans mycelial lipids, Physiol. Plant Pathol., 22: 363-370.
83. Davidse, L.C. and Boekeloo, M., 1984, Elicitation and suppression of necrosis in potato leaves by culture filtrate compounds of Phytophthora infestans (Mont.) de Bary, Acta Bot. Neerl., 33: 234.
84. Heath, M.C., 1981, A generalized concept of host-parasite specificity, Phytopathology, 71: 1121-1123.
85. Bushnell, W.R. and Rowell, J.B., 1981, Suppressors of defence reactions: a model of roles of specificity, Phytopathology, 71: 1012-1014.
86. Staskawicz, B.J., Dahlbeck, D. and Keen, N.T., 1984, Cloned avirulence gene of Pseudomonas syringae pathovar glycinea determines race-specific incompatibility of Glycine max, Proc. Natl. Acad. Sci. USA, 81: 6024-6028.

EVIDENCE FOR BOTH INDUCED SUSCEPTIBILITY AND INDUCED RESISTANCE IN THE CLADOSPORIUM FULVUM - TOMATO INTERACTION

V.J. Higgins, V. Miao and M. Arlt

Department of Botany
University of Toronto
Toronto, Canada M5S 1A1

Leaf mould of tomato caused by Cladosporium fulvum (syn. Fulvia fulva) has proved to be a useful system for studying mechanisms of host specificity. There are several important advantages to this host-pathogen combination. First, it is a gene-for-gene interaction for which there is available a good collection of C. fulvum races and tomato cultivars with one or a combination of genes for resistance. Some cultivars are near-isogenic, and as well, several races can be considered near-isogenic because of their origin [1,2]. Second, resistance occurs just after stomatal penetration and is rapid and localized [3]. Third, the fungus grows as a biotroph in a strictly intercellular manner [4], a feature which permits the use of intercellular wash fluids in some studies [2,5,6,7] and which also allows for direct contact between races in a doubly inoculated plant. This latter feature was important in this study.

Previous studies on this interaction have shown that, in contrast to a compatible interaction in which there is no apparent initial host cell response to the fungus, the incompatible response is characterized by callose deposits on the host cell wall [3,4], increases in hydroxyproline in cell walls and intercellular fluids [de Wit, personal communication], production of polyacetylene phytoalexins [8] and increases in a pathogenesis-related protein, P14 [de Wit, personal communication]. A glycopeptide molecule of fungal origin was shown to act as a non-specific elicitor of phytoalexins and necrosis [7,9] whereas elicitors of necrosis obtained from intercellular wash fluids were both race and cultivar specific [1,2,5] and appeared to be of fungal origin [5]. The elicitor corresponding to the Cf9 gene for resistance was found to be a low molecular weight basic peptide [6]. To date all information is compatible with the hypothesis that each gene for avirulence in the fungus leads to the production (or release) of a different elicitor which "activates" the appropriate gene for resistance. If this is the case, then it should be possible to induce resistance of a cultivar to a virulent race by prior inoculation with an avirulent race and it should be impossible to induce susceptibility to an avirulent race by prior inoculation with a virulent race. We tested for both possibilities using various in planta interactions between races.

NATO ASI Series, Vol. H1
Biology and Molecular Biology of Plant-Pathogen Interactions
Edited by J. Bailey

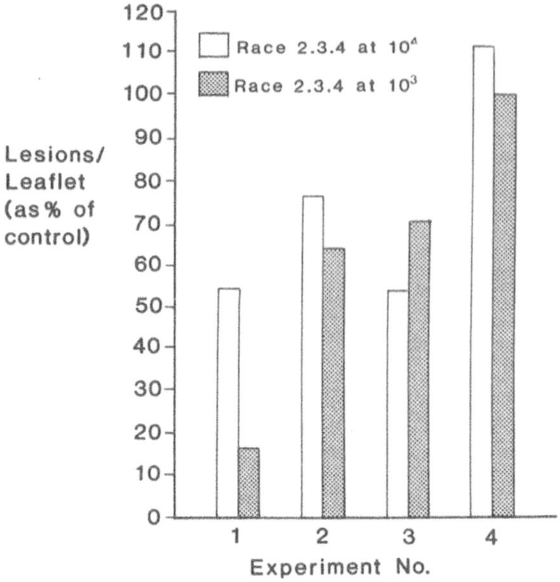

Figure 1. Effect of the presence of high numbers (5×10^5 conidia/ml) of conidia of an avirulent race of *Cladosporium fulvum* on the number of lesions produced by low numbers (10^3 or 10^4 conidia/ml) of conidia of the virulent race 2.3.4 on the cultivar Vinequeen (Cf2 + Cf4 genes for resistance) on simultaneous inoculation. The control was the appropriate concentration of race 2.3.4 alone. The avirulent races used were race 0 for experiment 1 and 3, race 1.2.3 for experiment 2, and race 2.3 for experiment 4.

To test for induced susceptibility, we produced cycloheximide and benomyl tolerant mutants by uv mutagenesis so that a race could be distinguished not only on differential cultivars but also by drug tolerance [10]. A replica leaf plating technique [10] was then used for determining the competitive fitness of the marked mutants and for testing for induced susceptibility. This technique involved holding a sporulating leaf around a foam plug and pressing the sporulating lower leaf area gently on to the surface of rose bengal agar, with and without supplements of cycloheximide and/or benomyl. Tolerant mutants grew both on unsupplemented and supplemented agar whereas wildtype isolates grew only on unsupplemented media. The relative amount of each race was measured by placing a grid under the plates and determining, with a dissecting microscope, the presence of germinated or ungerminated conidia in each quadrant of randomly selected grid squares. In comparing the competitive fitness of the mutants with their wildtype parents considerable variation occurred between experiments (Tables 1 and 2). The cycloheximide tolerant race 2.3 (strain 23C20) was judged to be at least as fit as its wildtype parent on both the universal suscept Potentate and the Cf2 differential, Vetomold (Table 1) but the benomyl tolerant race 4, strain 4B12, definitely showed reduced fitness on Purdue 135, the Cf4 differential (Table 2); however, both races gave good infection when inoculated alone on susceptible cultivars. Both were used to test for induced susceptibility by inoculating Purdue 135 with race 4, waiting one week for infection to become established, and then inoculating with race 2.3. Inoculations were made by either spraying inoculum on leaves or injecting it into interveinal areas [11]. After sporulation had occurred, leaflets were

Table 1. Competition between Cladosporium fulvum race 2.3 strain 23C20, and race 2.3 wildtype, isolate 14-80, on the susceptible cultivars Potentate and Vetomold.

Cultivar	Expt. no.	Quadrants scored*	% 23C20**
Potentate	1	1494	89.3 ± 4.5
	2	1145	65.0 ± 14.2
Vetomold	1	567	64.1 ± 10.8
	2	1069	60.9 ± 17.2

* Leaves were replica printed on cycloheximide-supplemented medium on which the conidia of 23C20, but not the wildtype race 2.3, germinated.
** The percentage of strain 23C20 was calculated by dividing the sum of the number of quadrants with C. fulvum hyphal growth, and half the number of quadrants with both hyphal growth and dead spores, by the total number of quadrants scored in each print.

replica plated and the number of sites with conidia of race 2.3 was determined by the grid method using cycloheximide supplemented media. The results (Table 3) show that race 2.3 was always recovered from race 4 infected plants but not from controls in which water replaced race 4. This phenomena has not yet been explored further.

Table 2. Competition between Cladosporium fulvum race 4, strain 4B12, and wildtype race 4 on the susceptible cultivars Potentate and Purdue 135.

Cultivar	Expt. no.	Quadrants scored*	% 4B12**
Potentate	1	242	57.7 ± 9.5
	2	634	33.4 ± 14.2
	3	340	69.6 ± 10.4
Purdue 135	1	222	27.8 ± 28.3
	2	789	26.4 ± 11.1

* Leaves were printed on benomyl-supplemented medium on which strain 4B12, but not wildtype race 4, could germinate.
** The percentage of strain 4B12 was calculated by dividing the sum of the number of quadrants with C. fulvum hyphal growth, and half the number of quadrants with both hyphal growth and dead spores, by the total number of quadrants scored in each print.

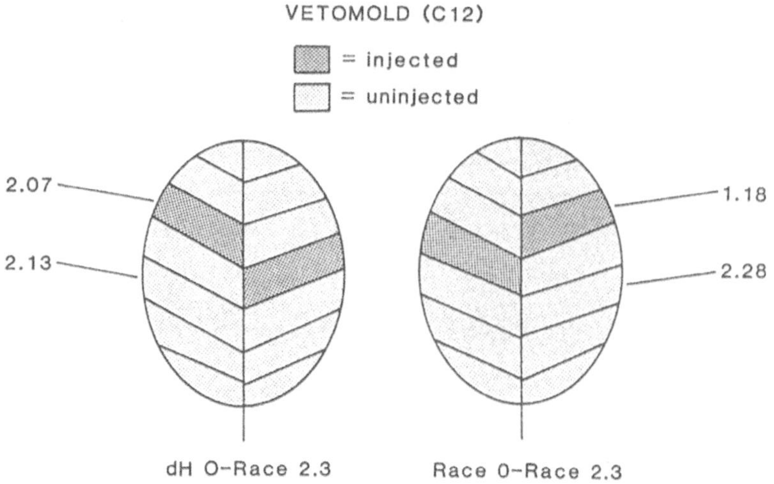

Figure 2. Diagrammatic representation of induced resistance experiments in which the avirulent race 0 (or water) was injected into interveinal panels (hatched areas) of tomato leaflets (cv. Vetomold) and one week later the leaflets were spray inoculated with the virulent race 2.3. The values represent the average sporulation rating (on a scale of 0 to 5) for injected and uninjected areas of race 0 injected (right) and water injected (left) leaflets.

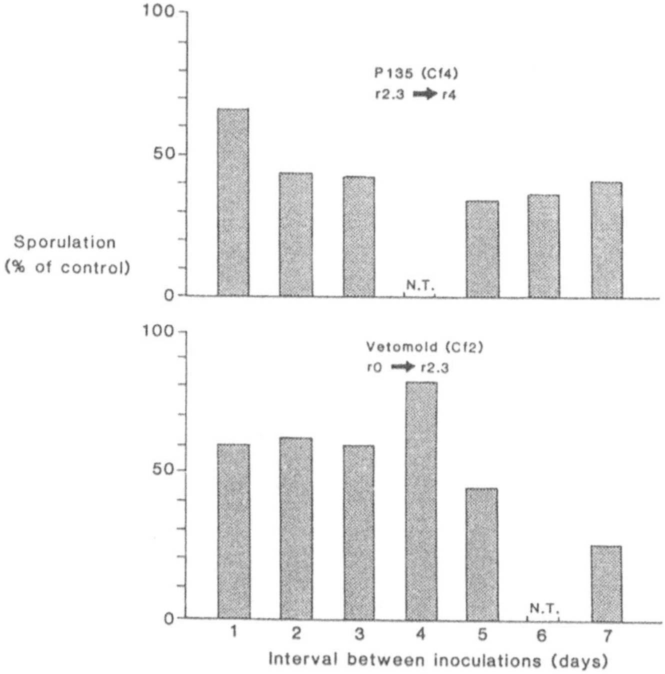

Figure 3. Effect of the time interval between inoculations when conidia of the avirulent race were injected into leaflets and the virulent race was then sprayed on to leaflets. A. Purdue 135 (Cf 4) was inoculated with race 2.3 and then with race 4. B. Vetomold (Cf 2) was inoculated with race 0 and then with race 2.3. The degree of sporulation of race 4 or race 2.3 was rated on a 0-5 scale and compared to that of water injected controls. N.T. indicates not tested.

Table 3. Recovery of race 2.3 of *Cladosporium fulvum* from tomato cv. Purdue 135 by leaf printing after sequential inoculations with race 4 (strain 4B12) and race 2.3 (strain 23C20).

Expt. no.	Inoculation week 1	Inoculation week 2	Number of leaflets or injection sites	No. of leaflets or sites with <1 cycloheximide tolerant colony*
	4B12	23C20		
1	spray	spray	13	9 leaflets
1	inject	spray	60	6 sites (10%)
1	inject	inject	72	22 sites (31%)
2	inject	inject	39	15 sites (38%)
3	inject	inject	133	9 sites (7%)
	water	23C20		
2	inject	inject	24	0 sites
3	inject	inject	216	0 sites

* These isolates were confirmed as race 2.3 by inoculation on standard differentials. The cycloheximide tolerant colonies as a percentage of the total sites scored is given in brackets.

Induced resistance was tested in a similar manner and generally with the same isolates but experiments to date have not required the use of the tolerance markers. The initial tests involved simply using inoculum containing a high number of conidia of an avirulent race and low numbers of a virulent race and comparing numbers of lesions with appropriate control plants. When this was done on the cv. Vinequeen (Cf2 + Cf4) using various avirulent races at 5 x 10^5 conidia/ml and the virulent race 2.3.4 at 10^3 or 10^4 conidia/ml, the number of lesions was reduced in three out of four experiments (Fig. 1). Similar inoculations spaced a week apart and using higher numbers of virulent conidia gave clear-cut protection. This system was not amenable to a number of types of experiments because there was no way of knowing where a lesion was until sporulation occurred and it was difficult to have controls on the same leaf. Hence we explored using the injection inoculation technique for one or both inoculations. Figure 2 illustrates the results of one approach in which sporulation was compared on injected and uninjected interveinal panels of the same leaflet and on control leaflets. Sporulation (rated on a 0 to 5 scale) was reduced only in panels injected with the avirulent race. A time course study was made to determine the optimum time interval between inoculations. Reduced sporulation was evident after a one day interval and there was a general trend for it to be more reduced the longer the interval (Fig. 3) but a two day interval appears a good choice for routine experiments.

We have just begun to explore the mechanism of this induced resistance. Limited histological examination has indicated that the reduced sporulation is indicative of reduced intercellular growth. Based on our knowledge of the resistant reaction, it is probably due to the combination of phytoalexin production, cell wall changes and host cell death. At least some of these same features can be induced with non-specific or specific elicitors and our injection inoculation technique will allow us to test for duplication of the effect of avirulent conidia with either type of elicitor. As well, there are various ways in which one can take advantage of the drug tolerance markers to test mechanisms involved.

The demonstration of induced susceptibility at the sporulation stage is less easily explained. Because the compatible response is well established at the time of inoculation with the avirulent race, it may simply be a shielding type of effect with the avirulent race never sufficiently contacting the host cells to induce a response. In some previous histological work with injected conidia, Higgins [11] proposed that there was evidence for induced susceptibility as both virulent and avirulent conidia initially elicited wall deposits but the virulent race out-grew this response and stopped eliciting it. Such an effect is now difficult to reconcile with the presence of specific elicitors but that, combined with our data in this study, suggests that it warrants further investigation. Nonetheless the sporulation of the avirulent race in the presence of the virulent race may well be of more interest to those studying the development and spread of new races than it is to physiologists.

REFERENCES

1. De Wit, P.J.G.M. and Spikman, G., 1982, Evidence for the occurrence of race and cultivar-specific elicitors of necrosis in intercellular fluids of compatible interactions of Cladosporium fulvum and tomato, Physiol. Plant Pathol., 21: 1-11.
2. Higgins, V.J. and de Wit, P.J.G.M., 1985, Use of race- and cultivar-specific elicitors from intercellular fluids for characterizing races of Cladosporium fulvum and resistant tomato cultivars, Phytopathology, 75: 695-699.
3. Lazarovits, G. and Higgins, V.J., 1976, Histological comparison of Cladosporium fulvum race 1 on immune, resistant, and susceptible tomato varieties, Can. J. Bot., 54: 224-234.
4. Lazarovits, G. and Higgins, V.J., 1976, Ultrastructure of susceptible, resistant and immune reactions of tomato to races of Cladosporium fulvum, Can. J. Bot., 54: 235-249.
5. De Wit, P.J.G.M., Hofman, J.E. and Aarts, J.M.M.J.G., 1984, Origin of specific elicitors of chlorosis and necrosis occurring in intercellular fluids of compatible interactions of Cladosporium fulvum (syn. Fulvia fulva) and tomato, Physiol. Plant Pathol., 24: 17-23.
6. De Wit, P.J.G.M., Hofman, A.E., Velthuis, G.C.M. and Kuč, J.A., 1984, Isolation and characterization of an elicitor of necrosis isolated from intercellular fluids of compatible interactions of Cladosporium fulvum (syn. Fulvia fulva) and tomato, Plant Physiol., 77: 642-647.
7. De Wit, P.J.G.M. and Roseboom, P.H.M., 1980, Isolation, partial characterization and specificity of glycoprotein elicitors from culture filtrates, mycelium and cell walls of Cladosporium fulvum (syn. Fulvia fulva), Physiol. Plant Pathol., 16: 391-408.
8. De Wit, P.J.G.M. and Flach, W., 1979, Differential accumulation of phytoalexins in tomato leaves but not in fruit after inoculation with virulent and avirulent races of Cladosporium fulvum, Physiol. Plant Pathol., 15: 257-267.
9. Lazarovits, G. and Higgins. V.J., 1979, Biological activity and specificity of a toxin produced by Cladosporium fulvum, Phytopathology, 69: 1056-1061.
10. Miao, V., 1984, Interactions between drug tolerant races of Cladosporium fulvum in vitro and in tomato, M. Sc. Thesis, University of Toronto, 102 pp.
11. Higgins, V.J., 1982, Response of tomato to leaf injection with conidia of virulent and avirulent races of Cladosporium fulvum, Physiol. Plant Pathol., 20: 145-155.

DO GALACTURONIC ACID OLIGOSACCHARIDES HAVE A ROLE IN THE RESISTANCE MECHANISM OF CUCUMBER TOWARDS CLADOSPORIUM CUCUMERINUM?

B. Robertsen

Institute of Biology and Geology
University of Tromsø, P O Box 3085 Guleng
N-9001 Tromsø, Norway

INTRODUCTION

Lignification was suggested to be a mechanism of disease resistance in cucumber by Hijwegen [1] who observed that lignin-like material was produced in resistant, but not in susceptible seedlings upon infection with the scab fungus Cladosporium cucumerinum. The lignin-like material was deposited in the walls and cytoplasm of the host tissue [1,2]. Hammerschmidt [3] reported that vanillin was released by cupric oxide oxidation of cell walls from infected resistant cucumber hypocotyls, but not from hypocotyl walls of control seedlings. This further supports the view that lignin is produced during infection of resistant plants. The work of Hammerschmidt and Kuć [4] indicates that lignification also has a role in the resistance mechanism of susceptible cucumber plants which have been systemically immunized against scab disease. Lignification has also been implicated in defence mechanisms of many other plants [5]. The available experimental evidence suggests that lignification may restrict fungal growth either because it makes the plant cell walls resistant to enzymatic degradation [6] or because some of the lignin precursors, like coniferyl alcohol, are toxic to fungi [4].

Genetical analysis of the inheritance of resistance to C. cucumerinum in cucumber by Walker [7] indicated that resistance is controlled by a single dominant gene. If lignification is a component of this resistance mechanism, then induction of it by the fungus may be determined by the presence of one single gene product in cucumber plants. In an attempt to understand how the fungus specifically triggers defence mechanisms in resistant cucumber plants, we are investigating the kinds of molecules from both plant and fungus that are able to elicit the production of lignin-like compounds in cucumber hypocotyls.

BIOASSAY OF ELICITORS OF LIGNIN-LIKE COMPOUNDS IN CUCUMBER HYPOCOTYLS

To detect and purify elicitors it was necessary to develop an assay of their biological activity. There are, however, no colorimetric methods available that give a quick and reliable measure of lignin. An attempt was thus made to develop a bioassay for elicitors of lignin-like compounds in cucumber hypocotyls based upon phloroglucinol/HCl, a stain used to detect lignin in resistant cucumber seedlings infected by C. cucumerinum [2].

NATO ASI Series, Vol. H1
Biology and Molecular Biology of Plant-Pathogen Interactions
Edited by J. Bailey

Table 1. Elicitor activities of α-1,4-D-galacturonide oligomers.

	Galacturonosyl units per oligomer												
	3	4	5	6	7	8	9	10	11	12	13	14	15
Elicitor activity	0	0	0	0	0	2	8	128	128	64	16	1	0

The uppermost 1 cm lengths of the hypocotyls of etiolated cucumber seedlings were excised and sliced lengthwise with a razor blade. The half-segments were washed with water and incubated in test solutions (1 ml) for 20 h at room temperature in darkness. At the end of the incubation period the segments were tested for lignin production by staining with phloroglucinol/HCl [2]. Segments incubated in distilled water were used as controls. The wound surface of the control segments remained colourless after staining, whereas segments incubated with elicitors stained purple red. Oligogalacturonide samples were first adjusted to contain 100 μg uronic acid units ml^{-1}, and then two-fold serial dilutions prepared. The elicitor activity is defined as the reciprocal of the highest dilution which gives visible lignification in the wound surface of the hypocotyl segments.

ELICITORS FROM CUCUMBER CELL WALLS AND POLYGALACTURONIC ACID [2]

Initially it was discovered that elicitors could be extracted from cucumber cell walls by autoclaving the walls for 20 min in distilled water. Autoclaving apparently results in partial hydrolysis of cell wall polysaccharides.

The elicitors bound to a QAE-Sephadex A-25 anion exchange column equilibrated with 0.06 M Tris-HCl buffer at pH 7.9 and could be eluted with a linear gradient of increasing buffer concentration (Fig. 1). The elution profile showed several peaks due to neutral hexose and uronic acid containing material and elicitor activity coeluted with some of these peaks. The material of one of the most active peaks, fractions 128-132, separated into two peaks when subjected to gel filtration on a Biogel P-6 column. The material of the first peak had an elicitor activity of 2 and contained about 41% uronic acid and large amounts of neutral sugars. The material of the second peak had an elicitor activity of 8 and contained about 94% uronic acid. The most active material was thus apparently an oligomer of galacturonic acid and it showed elicitor activity even when diluted down to a concentration of 12.5 μg uronic acid units ml^{-1}.

Purified α-1,4-linked oligogalacturonides were obtained by anion exchange chromatography of autoclaved polygalacturonic acid on the QAE-Sephadex A-25 column using elution conditions as described earlier [2] and by Nothnagel et al. [8]. This resulted in the separation of a series of oligomers of increasing size starting with a trimer. The molecular mass of trimer to nonamer was determined by fast atom bombardment mass spectrometry [2,8]. Oligomers containing 8 to 14 galacturonosyl residues showed elicitor activity with the decamer and the undecamer as the most active (Table 1). The latter oligomers showed activity down to a concentration of 0.8 μg ml^{-1} which is equivalent to about 0.5 μM. The oligomers which were isolated from polygalacturonic acid are apparently considerably more active than the ones obtained from the cucumber cell wall. The cell wall elicitors may thus either be less pure than the oligomers from polygalacturonic acid or they may contain sugars or substituents that inhibit their activity.

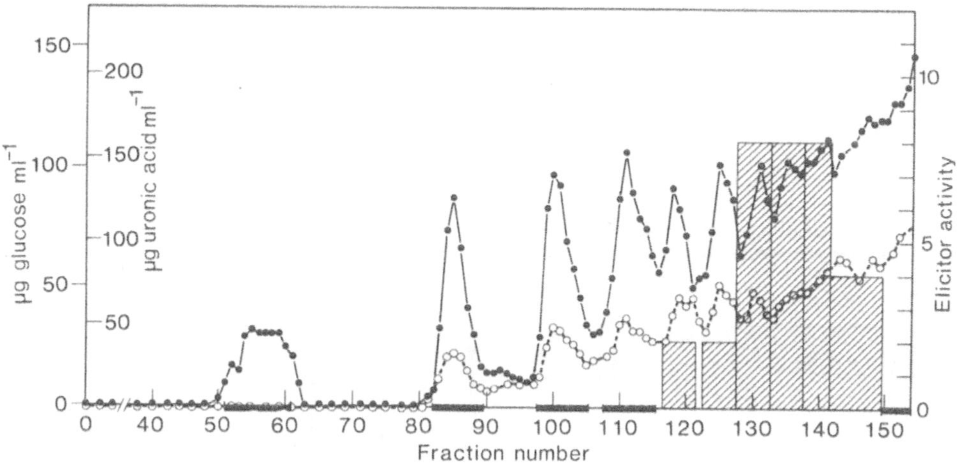

Figure 1. Chromatography of elicitors from cucumber cell walls on QAE-Sephadex A-25.Elicitors released from cucumber cell walls by autoclaving were bound to the column (2.5 x 16 cm) at 0.06 M Tris-HCl, pH 7.9, and then eluted with a linear gradient of 0.06-0.6M Tris-HCl buffer at pH 7.9. Each fraction was assayed for neutral hexoses (o) and uronic acids (●). Fractions were then pooled as indicated and galacturonic acid oligomers isolated and assayed for elicitor activity (hatched bars) as described [2]. Solid lines indicate zero activity.

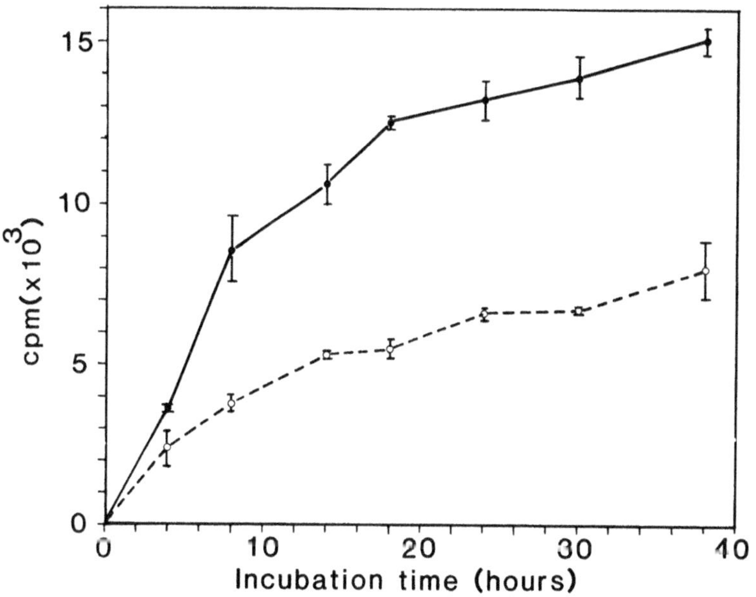

Figure 2. Incorporation of ^{14}C-cinnamic acid into cucumber cell walls during incubation of hypocotyl segments in water (--o--) and a solution of 2.5 μg/ml α-1,4-decagalacturonide (—●—). The segments were prepared and incubated as described for the bioassay of elicitors. At the start of the experiment ^{14}C-cinnamic acid corresponding to 2.6 x 10^4 cpm was added. Cell walls were prepared at different times, collected on filter paper and counted.

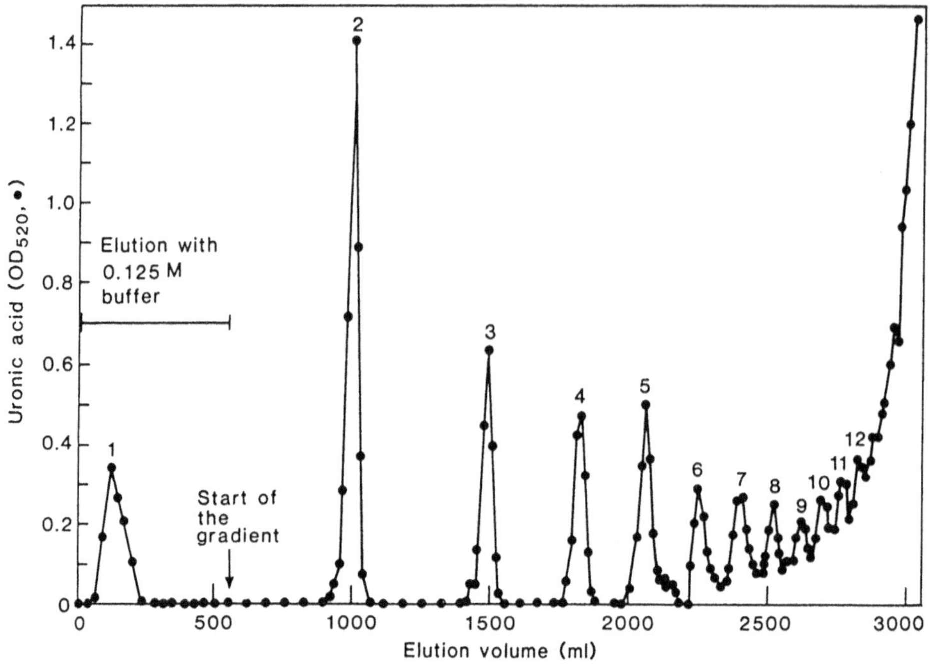

Figure 3. Separation of galacturonic acid oligomers produced during degradation of polygalacturonic acid with endo-polygalacturonase from *C. cucumerinum*. PGA was treated with endo-PG and applied to a QAE-Sephadex column (2.5 x 17.5 cm), equilibrated with 0.05 M Imidazol-HCl buffer at pH 7.0. Galacturonic acid was eluted with 0.125 M buffer whereas the dimer and larger oligomers were eluted by a 2.5 l linear gradient of increasing buffer concentration ranging from 0.125 M to 0.75 M Imidazol-HCl at pH 7.0. The number of galacturonosyl units per oligomer is indicated above each peak.

It is possible, however, that *C. cucumerinum* is able to release more effective elicitors from the wall than those that are obtained by autoclaving.

The deca-galacturonide stimulated incorporation of ^{14}C-labelled cinnamic acid, a lignin precursor, into cucumber hypocotyl cell walls (Fig. 2; Robertsen, unpublished results). This further supports the idea that the lignin-like material really is lignin although it can yet not be ruled out that cinnamic acid derivatives become esterified to cell wall polymers. Wounding itself also stimulates some incorporation of ^{14}C-cinnamic acid into the cell walls. However, as this material was not stained by phloroglucinol/HCl, it must be different to the material which was formed in response to the elicitor.

ENDOPOLYGALACTURONASE RELEASES OLIGOGALACTURONIDE ELICITORS FROM POLYGALACTURONIC ACID

Because oligogalacturonide elicitors could be released from cucumber cell walls, it was essential to investigate the role of endopectic degrading enzymes in the infection process.

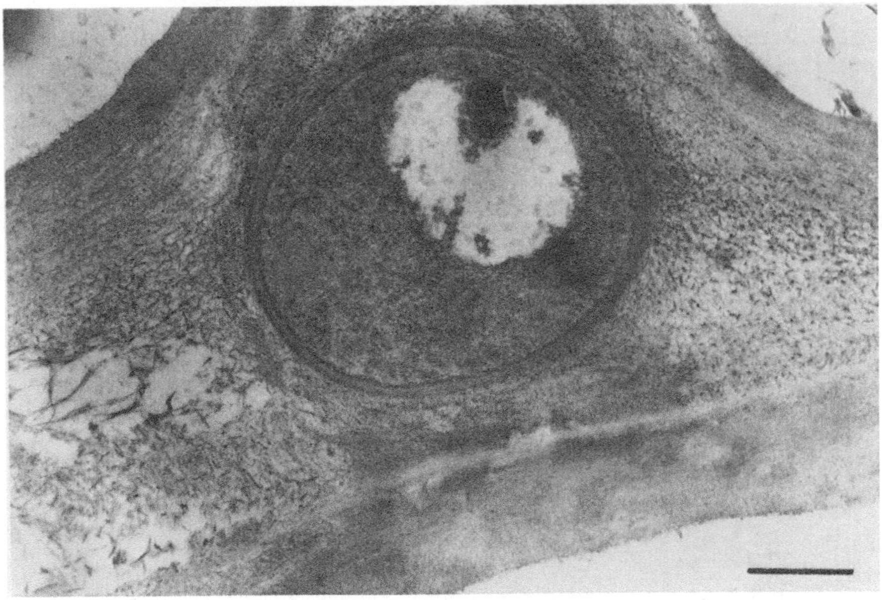

Figure 4. Electron micrograph showing hypha of C. cucumerinum growing in the middle lamellae of hypocotyl tissue of a resistant cucumber seedling 48 h after inoculation (Bar = 0.5 μm).

C. cucumerinum produces endopolygalacturonase when grown in a medium with citrus pectin or cucumber cell walls as the carbon source and ammonium nitrate as the nitrogen source [9]. The endopolygalacturonase produced on the pectin medium has now been partially purified through a series of steps including anion exchange chromatography on DEAE-cellulose, cation exchange chromatography on CM-50 Sephadex and gel filtration on Biogel P-200 [Robertsen, unpublished results]. By chromatography on CM-50 Sephadex it was possible to separate two exopolygalacturonases from the endopolygalacturonase. The endopolygalacturonase, but not the exopolygalacturonase, was able to elicit lignification in the cucumber hypocotyl segments. The partially purified preparation of the enzyme had elicitor activity down to a concentration of 70 ng ml^{-1} equivalent to 0.08 units ml^{-1}. The boiled enzyme showed no elicitor activity.

The enzyme degraded polygalacturonic acid into a series of oligomers starting with galacturonic acid (Fig. 3). Material isolated from the fractions corresponding to nonamer up to dodecamer was shown to have elicitor activity comparable with the oligomers which were isolated from autoclaved polygalacturonic acid.

It was also discovered that the fungus produced polygalacturonic acid lyase in culture when gelatin instead of ammonium nitrate was used as the nitrogen source and pectin as the carbon source. Two endopolygalacturonic acid lyases could be separated by gel filtration of the culture filtrate. Both of the partially purified enzymes were able to elicit lignification in the cucumber hypocotyl segments [Robertsen, unpublished results]. Further purification is needed to ascertain that the elicitor activity is caused by lyase activity and not by other proteins.

ELECTRON MICROSCOPE STUDIES OF THE INFECTION PROCESS

Along with the biochemical studies we have used electron microscopy to find out if it is possible to see differences in cell wall degradation during infection of susceptible and resistant cucumber seedlings [Robertsen and Sandberg, unpublished results]. For this purpose we have used etiolated seedlings because the lignification reaction is much more vigorous in hypocotyls of resistant seedlings grown in darkness than in seedlings grown in light. The hypocotyls were fixed 48 h after inoculation. At this stage lignification was detected in resistant hypocotyls, but disease symptoms were not yet visible in susceptible hypocotyls. These studies have shown that as found by Paus et al. [10] the fungus invades both resistant and susceptible tissue between epidermal cells whereafter it grows intercellularly. The micrographs also indicate that the fungus degrades the middle lamellae of both resistant and susceptible seedlings (Fig. 4). This degradation was definitely most extensive in susceptible seedlings. Because of the lack of specific stains for cell wall polymers it was difficult to estimate if there were other differences in cell wall degradation during infection of susceptible and resistant hypocotyls.

DISCUSSION

α-1,4-linked oligogalacturonides have recently been reported to be elicitors of phytoalexins in plants [8,11]. Lee and West [12] demonstrated that purified endopolygalacturonase from _Rhizopus stolonifer_ is an elicitor of casbene synthetase in castor bean. Davis et al. [13] showed that two purified endopolygalacturonic acid lyases from the bacterium _Erwinia carotovora_ are elicitors of phytoalexin production in soybean. The present report further supports the hypothesis that endo-pectic degrading enzymes have a role as inducers of defence mechanisms in plants. It is, however, still uncertain if these enzymes have a role in the specific resistance mechanism of cucumber to _C. cucumerinum_. Neither the oligogalacturonide elicitors nor the endo-pectic degrading enzymes of the fungus elicit lignification specifically in resistant hypocotyl segments. The ultrastructural studies indicate that the fungus secretes endo-pectic degrading enzymes during infection of both susceptible and resistant cucumber hypocotyls. Oligogalacturonide elicitors may thus be released during infection both of susceptible and resistant seedlings. If these elicitors really are the signal molecules for the lignification reaction in the infected resistant plants, there are a number of possible explanations why they do not work during infection of susceptible plants. The production of pectolytic enzymes may be differently regulated so that the oligogalacturonide elicitors are degraded faster during fungal growth in susceptible than in resistant plants. The fungus produces in culture at least two exopolygalacturonases which are able to destroy the elicitor activity of the decagalacturonide [unpublished results]. It is possible that the fungus macerates susceptible tissue before the host cells are able to lignify or that it secretes compounds that inhibit lignification in susceptible plants.

The oligogalacturonides are not the only elicitors which may be responsible for the lignification response in resistant cucumber plants. Elicitors can be extracted from the mycelium of the fungus [2]. The chemical nature of these elicitors is unknown at the present. Chitosan, a common cell wall component in fungi, is also an elicitor of lignification in cucumber hypocotyls [2]. Hopefully the study of elicitors in extracts from infected plants can give a more definite answer to the kind of molecules which elicit the lignification reaction during infection of resistant plants.

REFERENCES

1. Hijwegen, T., 1963, Lignification, a possible mechanism of active resistance against pathogens, Neth. J. Plant Pathol., 69: 314-317.
2. Robertsen, B., 1986, Elicitors of the production of lignin-like compounds in cucumber hypocotyls, Physiol. Mol. Plant Pathol., in press.
3. Hammerschmidt, R., 1980, Lignification and related phenolic metabolism in the induced systemic resistance of cucumber to Colletotrichum lagenarium and Cladosporium cucumerinum, Ph.D. Dissertation, University of Kentucky, Lexington.
4. Hammerschmidt, R. and Kuč, J., 1982, Lignification as a mechanism of induced systemic resistance in cucumber, Physiol. Plant. Pathol., 20: 61-71.
5. Vance, C.P., Kirk, T.M. and Sherwood, R.T., 1980, Lignification as a mechanism of disease resistance, Annu. Rev. Phytopathology, 18: 259-288.
6. Ride, J.P. and Pearce, R.B., 1979, Lignification and papilla formation at sites of attempted penetration of wheat leaves by non-pathogenic fungi, Physiol. Plant Pathol., 15: 79-92.
7. Walker, J.C., 1950, Environment and host resistance in relation to cucumber scab, Phytopathology, 40: 1094-1102.
8. Nothnagel, E.A., McNeil, M., Albersheim, P. and Dell, A., 1983, Host-pathogen interactions. XXII. A galacturonic acid oligosaccharide from plant cell walls elicits phytoalexins, Plant Physiol., 71: 916-926.
9. Skare, N.H., Paus, F. and Raa, J., 1975, Production of pectinase and cellulase by Cladosporium cucumerinum with dissolved carbohydrates and isolated cell walls of cucumber as carbon sources, Physiol. Plant., 33: 229-233.
10. Paus, F. and Raa, J., 1973, An electron microscope study of infection and disease development in cucumber hypocotyls inoculated with Cladosporium cucumerinum, Physiol. Plant Pathol., 3: 461-464.
11. Jin, D.F. and West, C.A., 1984, Characteristics of galacturonic acid oligomers as elicitors of casbene synthetase activity in castor bean seedlings, Plant Physiol., 74: 989-992.
12. Lee, S.C. and West, C.A., 1981, Polygalacturonase from Rhizopus stolonifer, an elicitor of casbene synthetase activity in castor bean (Ricinus communis L.) seedlings, Plant Physiol., 67: 633-639.
13. Davis, K.R., Lyon, G.D., Darvill, A.G. and Albersheim, P., 1984, Host-pathogen interactions. XXV. Endopolygalacturonic acid lyase from Erwinia carotovora elicits phytoalexin accumulation by releasing plant cell wall fragments, Plant Physiol., 74: 52-60.

CARBOHYDRATE AND LIPID-CONTAINING ELICITORS FROM PHYTOPHTHORA INFESTANS. DO THEY HAVE A COMMON MECHANISM OF ACTION?

P.J. Keenan[a], J.S. Ellis[a], W.G. Rathmell[b] and J. Friend[a]

[a]Department of Plant Biology & Genetics
University of Hull
Hull
HU6 7RX, UK

[b]ICI Plant Protection Division
Jealott's Hill Research Station
Bracknell
Berks
RG12 6EY, UK

INTRODUCTION

The interaction between Phytophthora infestans and potatoes has been considered to be an example of a gene-for-gene interaction. There is a series of major genes for resistance in the potato which are expressed in both tubers and leaves which confer resistance against all races of the fungus except those containing specific virulence genes which overcome the corresponding resistance genes. Major gene resistance or incompatibility is characterized by a hypersensitive response in which the first observable phenomenon is cell death. This is followed in tuber slices or discs by biochemical reactions which include the accumulation of sesquiterpenoid phytoalexins such as rishitin and phytuberin [1], the accumulation of insoluble high molecular weight phenolic compounds [2] and the encasement of haustoria with callose-like material [3]. In compatible reactions between a virulent race of the fungus and tuber discs of a susceptible cultivar hypersensitive cell death does not occur; the other biochemical reactions occur more slowly and to a more limited extent.

When tuber discs are challenged by fungal extracts, rather than intact fungal structures, specificity is lost [1,4]. Elicitors can be isolated from all fungal races which elicit incompatible reactions in tuber discs irrespective of their resistance genotype. Water-soluble glucans have been isolated from the fungus which can suppress the hypersensitive reaction elicited by crude fungal elicitors. The suppressors from virulent races are more effective than those from avirulent races [5,6,7]. It seems to be generally assumed that specificity may be controlled by a combined action of non-specific elicitors and specific suppressors.

In order to test the validity of this hypothesis we have isolated non-specific elicitors from fungal culture filtrates and from mycelial preparations and compared their mode of action with each other and with that

NATO ASI Series, Vol. H1
Biology and Molecular Biology of Plant-Pathogen Interactions
Edited by J. Bailey

of arachidonic acid [8,9]. The results which are presented in this paper indicate that elicitors with a wide variety of chemical composition may have a common mode of action; however, the picture is complicated by apparently different effects of elicitors on tuber and leaf tissue.

MATERIALS AND METHODS

Elicitors were isolated from culture filtrates of P. infestans by the method of Keenan, Bryan and Friend [9] and from mycelium by the method of Bryan, Rathmell and Friend [8]. The activities of the elicitors on tuber discs were determined by their ability to elicit either browning, as measured by a reflectance method [8], or the accumulation of rishitin and lubimin. The other analytical methods are described in the two papers cited.

When potato leaves were used as the test material, activity was determined by the ability of the elicitors to produce visible symptoms comparable with those of a characteristic hypersensitive reaction.

RESULTS AND DISCUSSION

The culture filtrate elicitor was isolated after ultrafiltration of a five-week culture filtrate through an Amicon YM-10 membrane (cut-off 10,000), gel filtration on Sephadex G-50 and preparative flat-bed isoelectric focussing (IEF). The active fractions contained both protein and carbohydrate; they were separated on G-50 from an all-carbohydrate fraction which did not contain any eliciting activity. The active materials focussed in the first six fractions of the acidic region of the gel (pH 3-4.5); the first four fractions also contained polygalacturonase activity. Galactanase and β-1,3 glucanase activities were well separated, appearing in fractions 8-13 and 9-15 respectively. Analytical IEF of combined fractions 1-6 showed the presence of 4 protein bands staining with Coomassie Blue R250, three of which corresponded with carbohydrate bands detected by periodic acid-Schiff's stain. The importance of carbohydrate, rather than protein, in the eliciting activity of combined fractions 1-6 was shown by their resistance to boiling and the proteolytic enzyme pronase (both of which abolished polygalacturonase activity). In addition, elicitor activity was completely destroyed by periodate oxidation, which is presumed to oxidize cis-1,2-diols characteristic of 1,4- and 1,6-linked rather than 1,3-linked monosaccharide residues in oligosaccharides. The monosaccharide composition of the active IEF fractions was mannose 63%, galactose 3% and glucose 34%.

At this stage the role of the individual sugars cannot be assessed; there may well be excess glucose since incubation of the combined fractions 1-6 with a β-1,3-glucanase, either from the culture filtrate itself (fraction 12) or obtained commercially, caused a 2- or 3-fold increase in elicitor activity, respectively. No fatty acids could be detected by GLC after transesterification in methanolic-HCl.

When a preparation of cell walls derived from mycelium of P. infestans was incubated with β-1,3-glucanase, an elicitor fraction was released. It had the same analytical IEF pattern as preparative IEF fractions 1-6 from the culture filtrate. It is assumed that this fraction is probably similar to that obtained from the culture filtrate.

Mycelium was fractionated by step-wise extraction with distilled water and increasing concentrations of phosphate buffer followed by chloroform-methanol then acetone and finally autoclaving in distilled water. The fractions were rotary evaporated and separated into a supernatant and a residue; the residues from the earliest aqueous and buffer extracts were combined. In all, 11 separate fractions were obtained, each of which contained elicitor activity. They were analyzed for esterified fatty acids, carbohydrate and protein. Three fractions appeared to contain fatty acids, with no protein and either no or only a trace of carbohydrate. Two fractions contained fatty acids together with carbohydrate and protein and 6 fractions contained mainly carbohydrate and protein with only a trace of fatty acids. Analytical IEF examination of the soluble elicitors showed that each contained a series of acidic bands which stained for both carbohydrate and protein and which were similar to the IEF bands from fractions 1-6 of the culture filtrate.

Examination of the specific activities based on either unit weight of eliciting fatty acid (arachidonic plus eicosapentaenoic acids) or of carbohydrate, confirmed the classification and emphasized the relative importance of these two fatty acids in these two fractions which contained fatty acids together with carbohydrate and protein.

These results emphasize the potential importance of fatty acid-containing elicitors; recently Kuć and his colleagues [10,11] have shown that the eliciting activities of both arachidonic and eicosapentaenoic acids can be enhanced by the glucans which suppress the eliciting activities of crude mycelial homogenates.

We have therefore investigated the possibility that the metabolism of arachidonic acid could be an important step in the elicitation of phytoalexins in the potato. In mammalian tissue, arachidonic acid can be converted by cyclo-oxygenase to prostaglandins and by lipoxygenase to either leukotrienes or hydroxy C_{20} fatty acids. In plants, the known metabolites of unsaturated fatty acids are those of linolenic acid, namely traumatic acid [12], 12-oxo-transdodecenoic acid [13] and jasmonic acid [14]. However in potato none of the C_{18} fatty acids acts as an elicitor [11], so that it must be considered highly unlikely that any of these compounds is involved in elicitation.

Non-steroidal anti-inflammatory drugs (NSAIDs) are inhibitors of the conversion of arachidonic acid to prostaglandins by cyclo-oxygenase [15]. They also inhibit the activity of both mammalian and soybean lipoxygenase [15,16]. We have therefore examined the effects of NSAIDs on the elicitor activity of arachidonic acid, culture filtrate and mycelial elicitors. In Table 1 the inhibitory effects of two NSAIDs, piroxicam and mefenamic acid, on a range of elicitors are shown.

Table 1. Inhibitory effects of piroxicam and mefenamic acid on elicitation of rishitin and lubimin in Kennebec tuber discs.

Elicitor	% Inhibition with Piroxicam	% Inhibition with Mefenamic acid
Arachidonic acid	100	98
Mycelial fraction (no fatty acid detectable)	96	100
Culture filtrate	100	94
Sporangia (race 4)	100	77

The finding that piroxicam, mefenamic acid and other NSAIDs inhibit not only arachidonic acid-based elicitors but also non-fatty acid-containing elicitors indicates that elicitors of a wide range of chemical composition may act by the production of a common intermediate. Such an intermediate may be superoxide anion since we have shown that the NSAIDs which inhibit elicitation also inhibit superoxide anion production by the xanthine-xanthine oxidase system. Moreover it has already been proposed by Doke [17], that superoxide is an early product of the hypersensitive reaction. One method of production of superoxide anion could be the action of host lipoxygenase on a fatty acid-containing elicitor; such a reaction is known to occur when soybean lipoxygenase acts on linoleate [18]. In order to explain that production of superoxide anion by those elicitors which apparently lack fatty acids we must assume that they act as substrates for a different superoxide anion-generating system. Alternatively it might be possible that they contain low levels of arachidonic acid, which are not converted to methylesters by the transesterification method we have used, and are also acted upon by lipoxygenase.

The results described so far have all used potato tuber tissue as the bioassay material. When leaves are used the measured effects may be different. For example, when P. infestans culture filtrate was fractionated by gel permeation HPLC two elicitors with different effects on tubers and leaves were isolated. Fraction 10 had slight elicitor activity on tuber discs at the highest concentration tested (840 μg glucose equivalents/ml) but caused a strong necrotic reaction on leaves even at 100-fold dilution. On the other hand fractions 17-19 which elicited 7 times the amount of phytoalexin when they were used at 113 μg glucose equivalents/ml had very little reaction on leaves.

Recently we have examined the effects of injecting ultrafiltrate of sporangial washings into leaves and comparing them with the effects of sporangia. On Kennebec leaves (an R_1 genotype) both race 4 sporangia and sporangial washings cause the necrosis characteristic of the hypersensitive response, whereas on King Edward (an r genotype) the race 4 sporangia cause disease symptoms and the ultrafiltrate produced little effect. These are the results to be expected if there is a race-specific elicitor in the ultrafiltrate. However, when the ultrafiltrate from the sporangia of a complex race, which is virulent and gives a compatible reaction on both Kennebec and King Edward, was injected into leaves of both cultivars it gave a necrotic reaction with Kennebec but not with King Edward. These last results indicate that whilst the sporangial washings of the different fungal races gave different reactions on leaves of different potato genotypes, they did not reproduce the race-specific reactions of the sporangia. These findings are reminiscent of earlier ones with tuber slices [9], when it was shown that the magnitude of the phytoalexin response of tuber slices to the culture filtrate elicitor was also cultivar-dependent but unrelated to resistance genotype.

Thus P. infestans seems to contain a range of non-specific elicitors of either phytoalexin accumulation in tuber slices or the hypersensitive reaction in leaves. The relative importance of the different elicitors in the expression of race specificity will probably depend on their interaction with specific suppressors. The recent finding that these suppressors can enhance the eliciting effect of arachidonic acid [10,11] is a complicating factor. The simplest explanation is that arachidonic acid is not an important elicitor in vivo. However a definitive answer will only be possible after further investigation of the interaction of suppressors with all types of elicitor.

We thank the AFRC for a research grant and SERC for the award of a CASE studentship (to JSE).

REFERENCES

1. Varns, J.L, Kuć, J. and Williams, E.B., 1971, Terpenoid accumulation as a biochemical response of the potato tuber to Phytophthora infestans, Phytopathology, 61: 174-177.
2. Henderson, S.J. and Friend, J., 1979, Increase in PAL and lignin-like compounds as race-specific resistance responses of potato tubers to Phytophthora infestans, Phytopath. Z., 94: 323-334.
3. Allen, F.H.E. and Friend, J., 1983, Resistance of potato tubers to infection by Phytophthora infestans: a structural study of haustorial encasement, Physiol. Plant Pathol., 22: 285-292.
4. Varns, J.L., Currier, W.W. and Kuć, J., 1971, Specificity of rishitin and phytuberin accumulation by potato, Phytopathology, 61: 968-971.
5. Garas, N.A., Doke, N. and Kuć, J., 1979, Suppression of the hypersensitive reaction in potato tubers by mycelial components from Phytophthora infestans, Physiol. Plant Pathol. 15: 117-126.
6. Doke, N., Garas, N.A. and Kuć, J., 1979, Partial characterization and aspects of the mode of action of a hypersensitivity-inhibiting factor (HIF) isolated from Phytophthora infestans, Physiol. Plant Pathol. 15: 127-140.
7. Doke, N., Garas, N.A. and Kuć, J., 1980, Effect on host hypersensitivity of suppressors released during the germination of Phytophthora infestans cystospores, Phytopathology, 70: 35-39.
8. Bryan, I.B., Rathmell, W.G. and Friend, J., 1985, The role of lipid and non-lipid components of Phytophthora infestans in the elicitation of the hypersensitive response in potato tuber tissue, Physiol. Plant Pathol. 26: 331-341.
9. Keenan, P.J., Bryan, I.B. and Friend, J., 1985, The elicitation of the hypersensitive response of potato tuber tissue by a component of the culture filtrate of Phytophthora infestans, Physiol. Plant Pathol. 26: 343-355.
10. Maniara, G., Laine, R.A. and Kuć, J., 1984, Oligosaccharides from Phytophthora infestans enhance the elicitation of sesquiterpenoid stress metabolites by arachidonic acid in potato, Physiol. Plant Pathol. 24: 177-186.
11. Preisig, C.L. and Kuć, J., 1985, Arachidonic acid-related elicitors of the hypersensitive response in potato and enhancement of their activities by glucans from Phytophthora infestans (Mont.) de Bary, Arch. Biochem. Biophys., 236: 379-389.
12. English, J., Bonner, J. and Haagen-Smit, A.J., 1939, The wound hormones of plants IV. Structure of a traumatin, J. Am. Chem. Soc., 61: 3434-3436.
13. Vick, B.A. and Zimmerman, D.C., 1976, Lipoxygenase and hydroperoxide lyase in germinating watermelon seedlings, Plant Physiol., 57: 780-788.
14. Vick, B.A. and Zimmerman, D.C., 1983, The biosynthesis of jasmonic acid: a physiological role for plant lipoxygenase, Biochem. Biophys. Res. Comm., 111: 470-477.
15. Higgs, G.A. and Vane, J.R., 1983, Inhibition of cyclo-oxygenase and lipoxygenase, Brit. Med. Bull., 39: 265-270.
16. Sircar, J.C., Schwender, C.F. and Johnson, E.A., 1983, Soybean lipoxygenase inhibition by nonsteroidal anti-inflammatory drugs, Prostaglandins, 25: 393-396.
17. Doke, N., 1983, Involvement of superoxide anion generation in the hypersensitive response of potato tuber tissues to infection with an incompatible race of Phytophthora infestans and to the hyphal wall components, Physiol. Plant Pathol., 23: 345-357.
18. Lynch, D.V. and Thompson, J.E., 1984, Lipoxygenase-mediated production of superoxide anion in senescing plant tissue, FEBS Lett., 173: 251-254.

DIFFERENTIAL ELICITATION ACTIVITIES OF FRACTIONS FROM PHYTOPHTHORA SPP. ON SEVERAL HOST-PLANTS

P. Ricci[a], P. Bonnet[a], P. Abad[a], P.M. Molot[b], P. Mas b
M. Bruneteau[c], I. Fabre[c], O. Lhomme[c] and G. Michel[c]

[a]I.N.R.A., Pathologie végétale
B.P.2078 F-06606 Antibes Cx
France

[b]I.N.R.A., Pathologie végétale
B.P.94 F-84140 Montfavet
France

[c]Université LYON I, Biochimie microbienne
43 Bd du 11 novembre 1918
F-69622 Villeurbanne Cx, France

INTRODUCTION

Early observations suggested that neutral polysaccharides from parasitic fungi could be universal elicitors of plant defence reactions [1]. However, more recent findings have indicated that unsaturated fatty acids and lipoconjugates are the active elicitors of phytoalexin accumulation in potato tubers [2]. But arachidonic and eicosapentaenoic acids are not universal elicitors either: they induce phytoalexin accumulation in potato and pepper, but not in twelve other plant species surveyed, including other Solanaceae, tobacco and tomato [3].

We have examined the elicitor activity of materials isolated from the mycelium or the culture filtrate of several Phytophthora species on three host-plants : carnation, pepper and tobacco. We have defined the elicitor activity as the ability to protect these plants against a subsequent inoculation with corresponding pathogenic Phytophthora species.

In carnation, protection against P. parasitica was correlated with the accumulation of benzoylanthranilic phytoalexins (dianthalexin [4] and diantramide A [5] in cuttings [6]), and was taken as a criterion for elicitation. In pepper, capsidiol is of limited importance in the resistance of cultivars to P. capsici [7] and protection can be induced without capsidiol accumulation [8]. Therefore, elicitation was assessed from a biological assay on surviving leaflets. In tobacco, stem inoculation of detopped plants with P. parasitica var. nicotianae was similarly used as a biological assay. In addition, the induction of b proteins, which have been associated with the hypersensitive reaction of tobacco to several other pathogens [9], was also recorded.

NATO ASI Series, Vol. H1
Biology and Molecular Biology of Plant-Pathogen Interactions
Edited by J. Bailey

Table 1. Phytoalexin elicitation in carnation cuttings.

Fraction		Concentration (μg/ml)	Induction by the fraction	Induction by MWE
P. parasitica (mycelium)	glucans	100	13	246
	glucan I	100	15	149
	glucan II	100	33	149
	glycopeptides	100	34	137
P. capsici (mycelium)	phospholipid	20	2	58
(culture filtrate)	lipoconjugates	100	6	63
		500	23	-
P. cryptogea (culture filtrate)	peptide	3	3	89
		30	23	254
water (control)		-	1 to 6	

Phytoalexins (dianthalexin + dianthramide A) are expressed in μg/g fresh weight. MWE : mycelial wall extract from P. parasitica.

Table 2. Protection of pepper leaflets (cv. Yolo Wonder).

Fraction		Concentration (μg/ml)	Susceptibility rating experiment A	Susceptibility rating experiment B
P. parasitica (mycelium)	MWE	500	96	
	glucans	500	98	
	glycopeptides	500	94	
P. capsici (mycelium)	phospholipid	50		33
		100		16
		200		4.5
(culture filtrate)	lipoconjugates	1000	11	21
P. cryptogea (culture filtrate)	peptide	8	30	
		17	45	
		33	87	
		66	100	
water (control)		-	84	96

Scale from 0 (full protection) to 100 (total invasion) as in [12].

MATERIALS AND METHODS

Fungal Fractions

Lipid-free mycelial walls of P. parasitica were purified and extracted with hot water. The soluble extract (MWE) was fractioned by anion exchange chromatography into glucans and glycopeptides [10]. Further purification of the glucans revealed the two major components (I and II) were β-1,3 glucans with different β-1,6 branches [11].

The mycelium and culture filtrate of P. capsici were produced and fractionated as previously described [12]. From an organic extract of the mycelium, an inositol-containing phospholipid was purified. Differential precipitations of the culture filtrate yielded a complex fraction (G4 in ref. 12). It is not glycosidic, water-insoluble and consists mainly of lipoconjugates. Its structure is being elucidated.

Finally, the culture filtrate of P. cryptogea was fractionated and a toxic activity on tobacco, similar to that described by Csinos and Hendrix [13], was associated with a peptide of about 3500 d [14].

Biological Assays

Carnation. Ten cuttings were dipped for 24 h in 15 ml of each test-solution. The lower 3 mm of the stem, where most of the phytoalexins accumulate, was extracted with acetone. Dianthalexin and dianthramide A were quantitated by HPLC [10]. Because the absolute response of the test varied with the season, MWE at the same concentration was included in each experiment as a positive control: water-treated cuttings were the negative control.

Pepper. The test was similar to that already described [12], except that cotyledons were replaced by 30 detached young leaflets. A 25 µl drop of the test-solution was placed on the non-wounded lower side and 24 h later a 10 µl drop containing 1200 zoospores of P. capsici was placed at the same place. After 3 days, control leaflets showed a discoloured invaded area whereas protected ones showed only small flecks or no symptoms at all.

Tobacco. Two month-old plants of cv. Xanthi were detopped above the 6th expanded leaf and a 20 µl drop of the test-solution was applied to the cut. Two days later a mycelium plug of P. parasitica var. nicotianae was placed on the cut and the subsequent invasion of the stem was recorded. On the 4th day, the 3 upper leaves of non-inoculated plants were extracted and the presence of b proteins examined by PAGE [15].

RESULTS

Elicitor activities of the various fractions and compounds on carnation, pepper and tobacco are presented in Tables 1, 2 and 3 repectively.

The water-soluble extract of lipid-free cell-walls of P. parasitica (MWE) and its fractions were inactive on pepper and tobacco, but induced phytoalexin accumulation in carnation. Among the glucans, only glucan II, with the lower molecular weight (about 9000 d), had any effect. As it represents only 9% of MWE, it cannot contribute much to the total activity. The activity of the mannose-containing glycopeptidic fraction of MWE, as well as that of MWE itself, was sensitive to both pronase and periodate oxidation (results not shown). It seems probably, therefore, that the active component of MWE on carnation is a glycopeptide.

Table 3. Protection and induction of b proteins in tobacco plants (cv. Xanthi).

Fraction		Concent. (µg/ml)	Av. length (mm) of stem invasion after 3 and 7 d.		Presence of proteins
P. parasitica (mycelium)	MWE	500	28	53	0
	glucans	500	20	53	0
	glycopeptides	500	31	65	0
P. capsici (mycelium)	phospholipid	200	25	67	0
		500	31	58	0
(culture filtrate)	lipoconjugates	200	4	13	+
		500	0	0	+
P. cryptogea (culture filtrate)	peptide	1.6	7	23	+
		3	0	0	+
		33	0	0	+
Arachidonic acid		500	24	57	0
water (control)		-	29	62	0

Table 4. Effect of arachidonic acid (AA) on phytoalexin accumulation in carnation.

AA conc. (µg/ml)	Without MWE	In mixture with MWE (100 µg/ml)
0	1.2	63
20	0.5	19
100	0.5	7
500	1.1	1

MWE : Mycelial wall extract from P. parasitica.

The phospholipid from the mycelium of P. capsici protected pepper at 200 µg/ml. It was ineffective on the two other plants. The lipoconjugate fraction of the culture filtrate had limited activity on carnation and pepper, even at high concentration (500-1000 µg/ml). In contrast it protected tobacco stems and induced b proteins in the leaves.

The peptide from P. cryptogea, applied to tobacco detopped stems at 30 µg/ml, which is its concentration in the culture filtrate. This caused extensive systemic leaf necrosis and induced protection of the stem and the accumulation of b proteins in the leaves. Leaf necrosis was less severe at 3 µg/ml, but protection was still observed. This peptide had only limited effect on carnation and pepper.

Arachidonic acid was also tested. Even at 500 µg/ml, it did not behave as an elicitor. In carnation, it inhibited the elicitor activity of MWE when applied as a mixture with it (Table 4).

CONCLUSION

Among the components of fungi belonging to the same genus, Phytophthora, diverse plants "recognize" chemically different molecules as resistance inducing factors : polysaccharides for soybean [1], glycopeptides for carnation, lipids for potato [2], pepper and tobacco (but, apparently, different types of lipids for these Solanaceous plants) and, in addition, a necrosis-inducing peptide for tobacco. As more research is being done on the nature of the elicitors involved in a variety of host-parasite interactions, diversity rather than uniformity emerges as an increasingly common phenomenon.

REFERENCES

1. Albersheim, P. and Valent, B.S., 1978, Plants, when exposed to oligosaccharides of fungal origin, defend themselves by accumulating antibiotics, J. Cell Biol., 78: 627-643.
2. Bostock, R.M., Kuć, J. and Laine, R.A., 1981, Eicosapentaenoic and arachidonic acids from Phytophthora infestans elicit fungitoxic sesquiterpenes in potato, Science, 212: 67-69.
3. Bloch, C.B., De Wit, P.J.G.M. and Kuć, J., 1984, Elicitation of phytoalexins by arachidonic and eicosapentaenoic acids : a host survey, Physiol. Plant Pathol., 25: 199-208.
4. Bouillant, M.L., Favre-Bonvin, J. and Ricci, P., 1983, Dianthalexine, nouvelle phytoalexine, de type benzoxazinone, isolée de l'oeillet Dianthus caryophyllus L. (Caryophyllacées), Tetrahedron Lett., 24: 51-52.
5. Ponchet, M., Martin-Tanguy, J., Marais, A. and Poupet, A., 1984, Dianthramides A and B, two N-benzoylanthranilic acid derivatives from elicited tissues of Dianthus caryophyllus, Phytochemistry, 23: 1901-1903.
6. Ricci, P., Bouillant, M.L. and Bernillon, J., Dianthalexin, a phytoalexin from carnation. Its involvement in induced resistance to Phytophthora parasitica Dastur. (Submitted).
7. Molot, P.M., Mas, P., Conus, M., Ferrière, H. and Ricci, P., 1981, Relations between capsidiol concentration, speed of fungal invasion and level of induced resistance in cultivars of pepper (Capsicum annuum) susceptible or resistant to Phytophthora capsici, Physiol. Plant Pathol., 18: 379-389.
8. Molot, P.M., Staron, T. and Mas, P., 1980, La résistance du piment à Phytophthora capsici. VIII- Induction de résistance et de capsidiol chez

Capsicum annuum avec des fractions obtenues à partir de filtrats de culture et du mycelium de P. capsici, Ann. Phytopathol., 12: 379-387.
9. Gianinazzi, S., 1982, Antiviral agents and inducers of virus resistance : analogies with interferon, in: "Active Defence Mechanisms in Plants", R.K.S. Wood, ed., pp. 275-298, Plenum Press, New York and London.
10. Fabre, I., Bruneteau, M., Ricci, P. and Michel, G., 1986, Isolement, composition et activité élicitrice chez l'oeillet de fractions extraites de la paroi de Phytophthora parasitica, Agronomie, 6: 35-45.
11. Fabre, I., Bruneteau, M., Ricci, P. and Michel, G., 1984, Isolement et étude structurale de glucanes de Phytophthora parasitica, Eur. J. Biochem., 142: 99-103.
12. Molot, P.M., Mas, P., Conus, M. and Ferrière, H., 1980, La résistance du piment à Phytophthora capsici. VII- Protection des organes foliaires après mise en survie sur filtrat de culture du parasite. Caractérisation des conditions d'expression du phénomène, Ann. Phytopathol., 12: 95-107.
13. Csinos, A. and Hendrix, J.W., 1977, Laminar necrosis, growth inhibition, and death of tobacco plants caused by toxic extracts of Phytophthora cryptogea, Phytopathology, 67: 434-438.
14. Bonnet, P., Poupet, A. and Bruneteau, M., 1985, Toxicité vis-à-vis du tabac d'un filtrat de culture de Phytophthora cryptogea Pethyb. & Laff., Agronomie, 5: 275-282.
15. Abad, P., Poupet, A., Ponchet, M., Venard, P. and Bettachini, B., 1985, Separation and quantitative assay of three pathogenesis-related (b)proteins from tobacco mosaic virus hypersensitive Nicotiana tabacum by reverse-phase high-performance liquid chromatography, J. Chromatography, 318: 417-426.

REGULATION OF STEROID GLYCOALKALOID AND SESQUITERPENOID STRESS METABOLITE ACCUMULATION IN POTATO TUBERS BY INHIBITORS OF STEROID SYNTHESIS AND PHYTOHORMONES IN COMBINATION WITH ARACHIDONIC ACID

E.C. Tjamos[a], E. Nuckles[b] and J. Kuć[b]

[a]Benaki Phytopathological Institute
Kifissia, 14561
Athens, Greece

[b]Department of Plant Pathology
University of Kentucky
Lexington, Kentucky 40546, USA

INTRODUCTION

The polyunsaturated fatty acids, arachidonic (AA) and eicosapentaenoic (EPA), both constituents of Phytophthora infestans, reproduce the hypersensitive response when applied to the surface of potato tuber slices [1]. Their effect involves rapid, but localized cell death, browning of the tissue, leakage of electrolytes and foremost, marked inhibition of steroid glycoalkaloid (SGA) accumulation with parallel elicitation of sesquiterpenoid stress metabolites (SSM) accumulation [1,2]. This study refers to the effect of chemicals which inhibit steroid synthesis in fungi and plants such as sterol-inhibiting fungicides [3,4,5] on potato tuber slices. Research concentrated on their effects on SGA and SSM accumulation. The assumption that AA and EPA have a hormonal effect by inhibiting or promoting biochemical processes like phytohormones, justified experimental work on the effect of the main phytohormones indole-3- acetic acid, gibberellic acid and abscisic acid on potato tuber slices. The latter has been reported to affect SSM accumulation in applications with AA [6]. Our aim was to elucidate the mode of action of AA as an elicitor of SSM.

MATERIALS AND METHODS

Certified seed potato tubers cv. Kennebec were used in all experiments. Tubers were stored at 4 °C until 24 h before their removal to room temperature for slicing and treatment. Tuber discs, 0.5 cm thick and 3.0 cm in diameter, were cut from the medullary tissue of the potato. Arachidonic acid (cis-5, 8, 11, 14-eicosatetraenoic acid) was obtained from Sigma Chemicals Co. Stock solutions in chloroform were stored at -20 °C. Suspensions of the acid were prepared by sonication in distilled sterile water and applied under aseptic conditions to slices in moist glass petri dishes. The fungicides triadimenol, b-(4-chlorophenoxyl)-a-(1, 1, dimethylethyl)-1H-1, 2, 4-triazole-1-ethanol), and fenarimol, a(2-chlorophenyl)-a-(4-chlorophenyl-)- 5-pyrimidine-methanol , were initially

NATO ASI Series, Vol. H1
Biology and Molecular Biology of Plant-Pathogen Interactions
Edited by J. Bailey

dissolved in acetone and then olive oil (Sigma Chemicals Co) was added. Acetone was evaporated and the chemicals were suspended by sonication with distilled sterile water. The final concentration of olive oil was 20%. Abscisic acid (ABA) was suspended in 20% olive oil in distilled sterile water as described for fungicides, while indole-3-acetic acid (IAA) and gibberellic acid (GA) were applied as water suspensions. The volume of suspension applied to each potato disc was always 100 μl. Sesquiterpenoid accumulation was determined at various time intervals after treatment. The top mm of each slice was harvested and SSM were quantified by gas liquid chromatography (7). Since rishitin and lubimin are the predominant SSM (> 90%) that accumulate, we considered the amount of rishitin and lubimin that accumulated as representing the elicitor activity. Total steroid glycoalkaloids (SGA) were determined by a spectrophotometric assay [8]. The values are expressed as α-solanine.

RESULTS

Effect of Sterol-Inhibiting Fungicides on SGA and SSM Accumulation

The fungicides triadimenol and fenarimol were applied to potato tuber slices at a concentration of 200 μg/slice. Arachidonic acid (100 μg/slice) was added to study the effect of the fungicides on AA-eliciting activity. The data (Table 1) demonstrate that both fungicides inhibit SGA accumulation without triggering SSM synthesis. It is also evident that consequent application of AA elicits phytoalexin production.

Table 1. The effect of triadimenol, fenarimol and arachidonic acid (AA) on steroid glycoalkaloid (SGA) and sesquiterpenoid stress metabolite (SSM) accumulation in Kennebec potato tuber slices

Triadimenol μg/slice	Fenarimol μg/slice	AA μg/slice	SGA μg/g fresh weight	SSM μg/g fresh weight
200	0	0	18±4	N.D.
0	200	0	15±2	N.D.
0	0	100	13±1	444±57
200	0	100	14±1	547±41
0	200	100	15±1	405±41
olive oil 20% in water			650±109	N.D.
water			818±35	N.D.

Potato tuber slices were aged for 6 h at 19 °C. Each fungicide or a mixed suspension of the fungicide and AA were applied to the upper surface of each slice. The top 1 mm was removed for extraction 120 h after treatment. Values of SGA and SSM are means (±SE) of three replicates containing eight slices per replicate and two determinations for each replicate. SGA are expressed as a-solanine and SSM are the total accumulation of rishitin and lubimin. N.D. = none detected.

Effect of Fenarimol on the Elicitor Activity of Arachidonic Acid

The fungicide fenarimol (200 µg/slice) was applied to potato tuber slices and treatment with 100 µg/slice of AA followed 0, 24, 48, 72 and 96 h after fungicide application. The results, summarized in Table 2, show that AA is capable of shifting the acetate-mevalonate pathway to SSM. This diversion leads to the accumulation of levels of SSM which are proportionally related to the time elapsed after AA application, when SGA inhibition has already been established by fenarimol.

Table 2. The effect of fenarimol and arachidonic acid (AA) on steroid glycoalkaloid (SGA) and sesquiterpenoid stress metabolites (SSM) accumulation in Kennebec potato tuber slices

Fenarimol µg/slice	AA µg/slice	SGA µg/g fresh weight	SSM µg/g fresh weight
0	100 (0 time)	11±1	521±23
200 (0 time)	0	37±3	N.D.
200 (0 time)	100 (0 time)	15±2	417±44
200 (0 time)	100 (24 h)	38±3	410±24
200 (0 time)	100 (48 h)	30±5	274±14
200 (0 time)	100 (72 h)	34±5	121±8
200 (0 time)	100 (96 h)	39±6	20±6
olive oil 20% in water		338±42	N.D.
water		398±65	N.D.

Potato tuber slices were aged for 6 h at 19 °C. Fenarimol was added (0 time) to the upper surface of slices and AA was applied simultaneously (0 time) or 24, 48, 72 and 96 h after the treatment with fenarimol. The top 1 mm of the slices was removed and extracted 120 h after the application of the fungicide. Values of SGA and SSM are means (±SE) of three replicates containing eight slices per replicate and two determinations for each replicate. SGA are expressed as α-solanine and SSM are the total accumulation of rishitin and lubimin. N.D. = none detected.

Effect of Abscisic Acid (ABA) on the Elicitor Activity of Arachidonic Acid

Single applications of various doses of ABA inhibit SGA accumulation without affecting SSM synthesis. At a concentration of 100 µg/slice, SGA production is almost completely blocked. In mixed applications of ABA and AA, SGA accumulation is entirely blocked, while SSM synthesis is gradually restricted depending on the amount of the applied ABA (Table 3). It was also noticed that ABA and AA treated potato tuber slices develop very limited browning, although SSM levels are relatively high.

Table 3. The effect of abscisic acid (ABA) and arachidonic acid (AA) on steroid glycoalkaloid (SGA) and sesquiterpenoid stress metabolite (SSM) accumulation in Kennebec potato tuber slices

ABA µg/slice	AA µg/slice	SGA µg/g fresh weight	SSM µg/g fresh weight
1	0	417±28	N.D.
10	0	236±20	N.D.
50	0	50±7	N.D.
100	0	35±3	N.D.
1	100	21±6	381±22
10	100	13±5	221±34
50	100	10±0	148±37
100	100	10±0	88±9
0	100	10±0	302±23
water		791±16	N.D.

Potato tuber slices were aged for 2 h at 19 °C. Water or ABA was applied to the upper surface of each slice and AA was applied 5-6 h later. The top 1 mm was extracted 96 h after the application of AA. Values of SGA and SSM are means (±SE) of three replicates containing eight slices per replicate and two determinations for each replicate. SGA are expressed as α-solanine and SSM are the total accumulation of rishitin and lubimin. N.D. = none detected.

Effect of Gibberellic Acid (GA) and Abscisic Acid (ABA) on the Elicitor Activity of Arachidonic Acid

The data of Table 4 demonstrate that single GA applications neither affect SGA accumulation nor trigger SSM synthesis.

Table 4. The effect of gibberellic acid (GA), abscisic acid (ABA) and arachidonic acid (AA) on steroid glycoalkaloid (SGA) and sesquiterpenoid stress metabolites (SSM) accumulation in Kennebec potato tuber slices

GA µg/slice	ABA µg/slice	AA µg/slice	SGA µg/g fresh weight	SSM µg/g fresh weight
0	0	50	107±9	270±26
100	0	0	331±18	N.D.
0	50	0	36±4	N.D.
100	50	0	38±2	N.D.
0	50	50	50±3	162±28
100	50	50	40±2	124±15
100	0	50	30±5	105±5
olive oil 20% in water			325±26	N.D.
water			398±50	N.D.

Potato tuber slices were aged for 3 h at 19 °C. Water, 20% olive oil in water or GA were applied to the upper surface of slices and ABA was added 2 h later. AA was applied 6 h after ABA treatment. The top 1 mm of the slices was extracted 96 h after treatment with AA. Values of SGA and SSM are means (±SE) of three replicates containing eight slices per replicate. SGA are expressed as α-solanine and SSM are the total accumulation of rishitin and lubimin. N.D. = none detected.

In contrast however, in mixed applications of GA and AA, however, GA inhibits AA-eliciting activity up to 60%. GA applied prior to treatment with ABA did not affect the inhibition of SGA accumulation by ABA. ABA applications prior to AA treatment resulted in the expected inhibition of SSM accumulation. Restricted SSM accumulation was also evident when GA was applied before treatment with ABA and AA.

Effect of Indole-3-acetic Acid (IAA) and Abscisic Acid (ABA) on the Elicitor Activity of Arachidonic Acid

The results presented in Table 5 show that IAA inhibition of SGA accumulation is not as effective as that caused by AA or ABA. Mixed IAA and ABA applications did not alter the pattern of inhibitory activity of ABA on SGA accumulation. Mixed IAA and AA also did not affect the dual activity of AA. Mixed application of IAA, ABA and AA, however, indicated that IAA restored the eliciting activity of AA restricted by ABA.

Table 5 The effect of indole-3-acetic acid (IAA), abscisic acid (ABA) and arachidonic acid (AA) on steroid glycoalkaloid (SGA) and sesquiterpenoid stress metabolite (SSM) accumulation in Kennebec potato tuber slices

IAA µg/slice	ABA µg/slice	AA µg/slice	SGA µg/g fresh weight	SSM µg/g fresh weight
10	50	50	18±1	112±6
50	50	50	17±1	163±27
100	50	50	16±1	193±60
10	50	0	54±16	N.D.
50	50	0	47±5	N.D.
100	50	0	65±10	N.D.
100	0	50	24±4	255±17
100	0	0	135±15	N.D.
100	0	50	54±4	283±16
water			388+29	N.D.

Potato tuber slices were aged for 3 h at 19 °C. Water or IAA was applied to the upper surface of slices and ABA was added 5-6 h later. AA was applied 12 h after ABA addition. The top 1 mm of the slices was removed and extracted 96 h after the treatment with AA. Values of SGA and SSM are means (±SE) of three replicates containing eight slices per replicate and two determinations for each replicate. SGA are expressed as α-solanine and SSM are the total accumulation of rishitin and lubimin. N.D. = none detected.

DISCUSSION

Earlier work [2] demonstrated that arachidonic and eicosapentaenoic acids exercise a dual biochemical activity when applied to potato tuber slices. This application results in inhibition of steroid glycoalkaloid (SGA) synthesis, a normal wounding reaction of the potato, and activation of sesquiterpenoid stress metabolite accumulation. Sterol-inhibiting fungicides of the triazole or pyrimidine group inhibit SGA accumulation in potato tuber slices without tissue browning. This inhibition, however does not lead to the SSM accumulation. It seems that inhibition of the oxidative dimethylation reactions incited by sterol inhibitors [5] has no similarities with the inhibition exercised by AA and EPA on potato tuber slices.

It has been reported [6,9] that ABA affects the eliciting activity of incompatible races of Phytophthora infestans on potato tuber slices and also inhibits SSM accumulation incited by arachidonic acid. In our experiments we observed the same phenomenon but we also found that ABA inhibited SGA accumulation. This inhibition could be attributed to a general hormonal effect. ABA intervenes with the healing process of the sliced potato tubers restricting the metabolic activities of the tissue. It is also known that ABA inhibits GA-promoted events in aleurone layers such as synthesis of proteases, ribonucleases, endoplasmic reticulum and lecithin [10]. It was observed that combined application of ABA and AA (even 1 μg/slice ABA) resulted in very limited browning of the tissue although substantial quantities of SSM accumulated. This lack of browning constitutes another rather rare example of phytoalexin production without visible necrosis [11]. Neither IAA nor GA affect the intensity of discolouration incited by AA application. IAA but not GA restores SSM accumulation close by the expected levels when applied to the sliced potatoes prior to ABA and AA treatment. IAA alone slightly inhibits SGA accumulation but does not promote SSM synthesis, whereas GA does not intervene with the above processes. GA, however, inhibits SSM accumulation elicited by AA.

The interpretation of the results could involve the following points: treatment of potato tuber slices with AA, may inhibit SGA accumulation by inactivating a key step in the synthesis of SGA or activate a key step in SSM synthesis thus diverting precursors in the mevalonate pathway to SSM synthesis and away from SGA synthesis. The inability of sterol-synthesis inhibiting fungicides to initiate both SGA inhibition and SSM synthesis may be due to the site of their action which is past squalene in the acetate-mevalonate pathway (demethylation of cyclic intermediate). On the contrary the action of AA may be at the site of conversion of 2 farnesyl pyrophosphates (C-15 noncyclic intermediate) to squalene (C-30 noncyclic intermediate) catalyzed by squalene synthetase [12]. An alternative candidate for inhibition could be the step catalyzed by squalene cyclase. The two fungicides tested may cause the accumulation of cyclic methylated derivatives distant from the branch point for sesquiterpene (SSM) synthesis and be unable to activate a key step in SSM synthesis.

Combined treatments (ABA plus AA and GA plus AA) inhibit SSM accumulation in response to AA application. Since ABA inhibits SGA synthesis, its use for studying the mode of action of AA is restricted. GA, however, does not inhibit SGA accumulation. This suggests that the effect of GA is at a step of SSM synthesis in the branch of the pathway leading away from the pathway for SGA synthesis. Before drawing any definite conclusions further work is needed with labelled precursors, and enzyme inhibitors specifically acting on squalene synthetase or cyclase.

REFERENCES

1. Bostock, R.M., Kuć, J.A. and Laine, R.A., 1981, Eicosapentaenoic and arachidonic acids from Phytophthora infestans elicit fungitoxic sesquiterpenoids in the potato, Science, 212: 67-69.
2. Tjamos, E.C. and Kuć, J., 1982, Inhibition of steroid glycoalkaloid accumulation by arachidonic and eicosapentaenoic acids in potato, Science, 217: 542-544.
3. Buchenauer, H. and Rohner, E., 1981, Effect of triadimefon and triadimenol on growth of various plant species as well as on gibberellic content and sterol metabolism in shoots of barley seedlings, Pesticide Biochem. Physiol., 15: 58-70.
4. Schmitt, P. and Benveniste, P., 1979, Effect of fenarimol on sterol biosynthesis in suspension cultures of bramble cells, Phytochemistry, 18: 1659-1665.
5. Siegel, M.R., 1981, Sterol-inhibiting fungicides:Effects on sterol biosynthesis and sites of action, Plant Diseases, 65: 986-989.
6. Bostock, R.M., Nuckels, E., Henfling, J.W.D.M. and Kuć, J., 1983, Effects of potato tuber age and storage on sesquiterpenoid stress metabolite accumulation, steroid glycoalkaloid accumulation and response to abscisic and arachidonic acid, Phytopathology, 73: 435-438.
7. Henfling, J.W.D.M. and Kuć, J., 1979, A semi-micro method for the quantitation of sesquiterpenoid stress metabolites in potato tuber tissue, Phytopathology, 69: 609-612.
8. Cadle, L.S., Stelzig, D.A., Harper, K.L. and Young, R.J., 1978, Thin layer chromatography system for identification and quantitation of potato tuber glycoalkaloids, J. Agric. Food Chem., 26: 1453-1454.
9. Henfling, J.W.D.M., Bostock, R. and Kuć, J., 1979, Effect of abscisic acid on rishitin and lubimin accumulation and resistance to Phytophthora infestans and Cladosporium cucumerinum in potato tuber tissue slices, Phytopathology, 70: 1074-1078.
10. Chrispeels, M.J. and Varner, J.E., 1966, Inhibition of gibberellic acid induced formation of α-amylase by abscisic II, Nature, 212: 1066-67.
11. Darvill, A.G. and Albersheim, P., 1984, Phytoalexins and their elicitors -A defense against microbial infection in plants, Annu. Rev. Plant Physiol., 35: 243-275.
12. Agnew, W.S. and Popjak, G., 1978, Squalene synthetase: Stoichiometry and kinetics of presqualene pyrophosphate and squalene synthesis by yeast microsomes, J. Biol. Chem., 253: 4566-4573.

ELICITATION AND SUPPRESSION OF NECROSIS IN POTATO LEAVES BY CULTURE FILTRATE COMPONENTS OF PHYTOPHTHORA INFESTANS

L.C. Davidse, M. Boekeloo and A.J.M. van Eggermond

Department of Phytopathology
Agricultural University
Wageningen, The Netherlands

Necrosis of potato leaf tissue after invasion by *Phytophthora infestans* is a characteristic of both late blight lesion development and the hypersensitive response. The rate at which colonized tissue becomes necrotic in compatible interactions might be an important factor in determining lesion extension and sporulation. It has been shown that rapid necrosis is often associated with a high level of general resistance [1]. Hence components of *P. infestans* that induce necrosis might be useful in screening programmes for general resistance. On the other hand, the occurrence of necrosis during the hypersensitive response suggests that (some of) these components might be factors that determine race specificity. Purification and characterization of necrosis-inducing factors produced by *P. infestans* *in vivo* and *in vitro* is necessary before any conclusion can be drawn. We therefore initiated a search for these factors using the leaf injection technique [2] as a bioassay.

Necrosis-inducing factors were present in infiltration fluids [2] from diseased leaves but not in healthy potato leaves. The ability to induce necrosis was independent of the compatible combination of cultivar and race used to produce the infiltration fluid or the cultivar tested in the bioassay. Since bacterial contamination of diseased leaves may also give rise to necrosis-inducing factors, the use of infiltration fluids as a source of these factors may cause artefacts. Therefore, filtrates of 10 to 14 day-old mycelial cultures in liquid synthetic medium [3] were utilized as they were considered to be a more reliable source of factors of purely fungal origin [4]. Necrosis-inducing factors from culture filtrates eluted in the void volume fractions of Sephadex G-25 columns, were heat stable and sensitive to treatment with periodate. Gel filtration of dialyzed and freeze dried culture filtrates on Sephadex G-100 columns eluted with water yielded a number of active fractions including the void volume fraction. Fractions eluting close to and at the column volume were inactive. When active fractions were diluted with these latter fractions in a 1:20 to 1:200 (v/v) ratio, the resulting mixture was less active or inactive in inducing necrosis, whereas dilutions of the active fractions with water at the same ratios were highly active. This indicates that elicitation of necrosis can be suppressed by other components of the culture filtrate. At this level of purification necrosis-inducing factors did not show race or cultivar specificity, neither did the suppressing components.

NATO ASI Series, Vol. H1
Biology and Molecular Biology of Plant-Pathogen Interactions
Edited by J. Bailey

Cultivars, however, did differ in sensitivity to the necrosis-inducing factors. Some cultivars with a low level of general resistance were rather insensitive, while some with a high level were highly sensitive.

Necrosis-inducing factors were further purified by DEAE-Sepharose anion exchange chromatography. Weakly anionic substances proved to be the most active. Active fractions contained protein (90%) and polysaccharide (10%). Monomeric units were mannose (70%) and glucose (30%). Quantities as low as 0.1 μg per injection site were able to induce necrosis. The necrosis-suppressing components of the culture filtrates have not been purified further. Sugar analyses revealed glucans to be the major component present.

As yet it is impossible to draw any definite conclusions from these data. Further characterization of the necrosis-inducing factor(s) will be necessary. The high mannose content of the polysaccharide associated with protein in the DEAE-Sepharose preparation, points to actively secreted glycoproteins [5] as the factor(s) responsible for induction of necrosis. Further quantitative testing of the biological activity of purified factor(s) from several races on various potato cultivars will reveal if any cultivar or race specificity is associated with a specific component of the culture filtrate.

REFERENCES

1. Coffey, M.D. and Wilson, U.E., 1983, An ultrastructural study of the late-blight fungus *Phytophthora infestans* and its interaction with the foliage of two potato cultivars possessing different levels of general (field) resistance, *Can. J.Bot.*, 61: 2669-2685.
2. De Wit, P.J.G.M. and Spikman, G., 1982, Evidence for the occurrence of race and cultivar-specific elicitors of necrosis in intercellular fluids of compatible interactions of *Cladosporium fulvum* and tomato, *Physiol. Plant Pathol.*, 21: 1-11.
3. Henniger, H.C., 1959, Versuche zur Kultur verschiedener Rassen von *Phytophthora infestans* (Mont.) de By. auf Künstlichen Nährboden, *Phytopath. Z.*, 34: 285-306.
4. Stolle, K. and Schröber, B., 1984, Wirkung eines Toxins von *Phytophthora infestans* (Mont.) de Bary auf Kartoffelknollengewebe, *Potato Research*, 27: 173-184.
5. Ziegler, E. and Albersheim, P., 1977, Host-pathogen interactions. XIII. Extracellular invertases secreted by three races of a plant pathogen are glycoproteins which possess different carbohydrate structures, *Plant Physiol.*, 59: 1104-1110.

ORGANIZATION, STRUCTURE AND ACTIVATION OF PLANT DEFENCE GENES

T.B. Ryder, J.N. Bell, C.L. Cramer,
S.L. Dildine, C. Grand, S.A. Hedrick,
M.A. Lawton and C.J. Lamb

Plant Biology Laboratory
The Salk Institute for Biological Studies
P.O. Box 85800
San Diego, CA 92138, USA

INTRODUCTION

Plants exhibit natural resistance to disease which has been exploited by breeders to reduce crop losses and hence increase yield. Disease resistance involves not only static protection, but also inducible defence mechanisms including: (i) accumulation of host-synthesized phytoalexins; (ii) deposition of lignin-like material; (iii) accumulation of hydroxyproline-rich glycoproteins and (iv) increases in the activity of certain hydrolytic enzymes such as chitinase and glucanase [1]. Although the genetics, physiology and cytology of plant:pathogen interactions have been extensively studied, until recently relatively little was known at the biochemical level about how plants respond to infection to activate these defence responses.

Over the last few years we have examined the interaction between bean (*Phaseolus vulgaris* L.) and *Colletotrichum lindemuthianum*, causal agent of anthracnose, as a model for analysis of the molecular mechanisms underlying activation of plant defence responses. Our work has involved study of the response of suspension-cultured cells to various elicitor preparations and also the response of intact hypocotyls to infection by *C. lindemuthianum*. Elicitor treatment of cell cultures represents a system of reduced complexity which greatly facilitates initial biochemical analysis, purification of enzymes for antibody production, isolation of mRNA for cDNA cloning etc. Parallel analysis of the response of intact hypocotyls to infection by *C. lindemuthianum* offers a complementary approach which, while less convenient experimentally, provides information on the activation of defence responses during race:cultivar specific interactions and the spatial pattern of the response of the plant to microbial attack under conditions closely resembling the natural infection process. Specific advantages of the bean : *C. lindemuthianum* system for such studies have been discussed in detail elsewhere [2].

Initially our attention was directed towards the regulation of accumulation of phaseollin and other structurally related phytoalexins such as phaseollidin, phaseollinisoflavan and kievitone since there is strong evidence that the differential accumulation of these phenylpropanoid

NATO ASI Series, Vol. H1
Biology and Molecular Biology of Plant-Pathogen Interactions
Edited by J. Bailey

phytoalexins in incompatible and compatible interactions plays a crucial role in the specificity of host resistance [3]. Phytoalexin accumulation is largely a result of increased synthesis from remote precursors [4] and increases in the levels of appropriate biosynthetic enzymes are characteristically observed at the onset of the defence response [5]. Thus in elicitor treated bean cells there are marked, rapid coordinate increases in the activity levels of at least 5 enzymes of phenylpropanoid biosynthesis: the central pathway enzymes phenylalanine ammonia-lyase (PAL), cinnamic acid 4-hydroxylase (C4H), and 4-coumarate: CoA ligase (4CL); and the first two enzymes of a branch pathway specific for flavonoid and isoflavonoid biosynthesis: chalcone synthase (CHS) and chalcone isomerase (CHI), concomitant with the onset of phytoalexin accumulation [6,7,8]. Increases in the activity levels of enzymes of phytoalexin biosynthesis have also been observed in race:cultivar specific interactions between bean hypocotyls and the fungus *C. lindemuthianum* associated with expression of the hypersensitive defence response during incompatible interactions (host resistant) and attempted lesion limitation in compatible interactions (host susceptible) [6,9,10].

ELICITOR AND INFECTION STIMULATE THE SYNTHESIS OF PHYTOALEXIN BIOSYNTHETIC ENZYMES

In vivo labelling studies have shown that elicitor causes rapid, marked, concomitant but transient increases in the rates of synthesis of PAL, 4CL, CHS and CHI [6,11, unpublished data]. These responses are extremely rapid and increases in the rate of enzyme synthesis can be observed within 20 min of elicitor treatment. Maximum rates of enzyme synthesis are attained about 3-4 h after elicitation and each enzyme accounts for between 0.2% and 1.5% of total protein synthesis. These data strongly suggest that stimulation of enzyme synthesis is an important element regulating increased enzyme activity and hence phytoalexin accumulation. However, it should be noted that there also appears to be post-translational control mechanisms regulating the activities of at least some of these enzymes in elicitor-treated cells [8,11,12], and under certain circumstances the enzyme activity levels are regulated by reciprocal changes in the rate of enzyme production and removal [13]. Such a multiple control system provides for rapid, amplified and flexible responses to environmental signals in slowly growing plant cells and may be of widespread biological significance [14].

In vitro translation of mRNA isolated from bean cell cultures indicates that elicitor induction of PAL, CHS and CHI synthesis arises mainly if not exclusively from increases in the activity levels of the corresponding mRNAs encoding these enzymes [6,15]. Marked increases in the activity levels of the mRNAs encoding enzymes of phytoalexin biosynthesis have also been observed in bean hypocotyls infected with *C. lindemuthianum* [6,9]. *In vitro* translation of isolated polysomal RNA allows quantitative analysis of the pattern of protein synthesis in intact tissue without the interpretative problems inherent in an *in vivo* labelling approach which arise from possible changes in the uptake and compartmentalization of exogenous labels as the interaction between host and pathogen develops. There are early increases in the activity levels of mRNAs encoding PAL, CHS and CHI in the incompatible interaction (host resistant) prior to the onset of phytoalexin accumulation and expression of hypersensitive resistance. In contrast, in the compatible interaction there is no induction of PAL, CHS or CHI mRNA activity in the early stages of infection, but rather a delayed widespread response at the onset of lesion formation associated with attempted lesion limitation. Hence there are clear spatial and temporal differences in the pattern of synthesis of phytoalexin biosynthetic enzymes, as judged by translation of isolated mRNA in vitro, between genetically determined compatible and incompatible race:cultivar specific interactions.

TRANSCRIPTIONAL ACTIVATION OF GENES ENCODING PHYTOALEXIN BIOSYNTHETIC ENZYMES

In order to investigate further the molecular mechanisms underlying regulation of phytoalexin accumulation in bean, cloned cDNA libraries containing sequences complementary to mRNA present in elicitor-treated cells have been constructed [16,17, unpublished data]. A number of clones containing CHS sequences were identified by hybridization with cDNA sequences complementary to light-induced CHS mRNA from parsley (*Petroselinum hortense*) [17]. cDNA clones containing PAL sequences were identified by hybrid-select translation using a monospecific antiserum to bean PAL [16].

Using [^{32}P]-labelled PAL and CHS cDNA sequences as probes in RNA blot hybridization experiments, we have demonstrated rapid, marked but transient coordinate increases in PAL and CHS mRNA in response to elicitor treatment [16,17]. There was a close correspondence between the induction kinetics for enzyme synthesis as measured by in vivo labelling techniques, the level of translatable mRNA activity and the amount of hybridizable mRNA for PAL and CHS in both total cellular RNA and polysomal RNA fractions. Thus it can be concluded that the transient accumulation of PAL and CHS mRNA is the major factor governing the rate of enzyme synthesis throughout the phase of rapid increase in enzyme activity.

Marked accumulation of hybridizable PAL and CHS mRNA was also observed in bean hypocotyls infected with *C. lindemuthianum* (unpublished data). In the incompatible interaction with race β, there was an early induction of hybridizable CHS mRNA in polysomal RNA samples isolated from tissue immediately adjacent to the site of spore inoculation. Accumulation was first observed about 52 h after inoculation, and during the phase of phytoalexin accumulation and expression of hypersensitive resistance the level of hybridizable mRNA remained between 80 and 100 fold above that in equivalent, uninfected control hypocotyls. Transient accumulation of CHS mRNA was also observed in polysomal RNA isolated from tissue distant from the site of spore inoculation, although accumulation was less pronounced with a maximum 10 to 20 fold increase over the levels in polysomal RNA from samples of equivalent uninoculated hypocotyls.

In the compatible interaction with race γ, there was no significant increase in hybridizable CHS mRNA above control levels during the early stages of infection equivalent to the phase in incompatible interactions of induction of CHS mRNA and initial expression of hypersensitive resistance. Subsequently however, at the start of lesion development, there was a marked increase of hybridizable mRNA correlated with the onset of phytoalexin accumulation during attempted lesion limitation. Induction of CHS mRNA was more pronounced and occurred slightly earlier in directly infected tissue at the site of spore inoculation but there was also significant induction in tissue distant from the initial site of infection.

Changes in the level of CHS mRNA as a proportion of total cellular RNA followed broadly the same pattern as in polysomal RNA, although in the incompatible interaction increased CHS mRNA levels in directly infected tissue were observed somewhat earlier and elevated levels in tissue distant from the site of infection were maintained for longer in total cellular RNA than in the polysomal RNA fraction. Detailed analysis of the early stages of the incompatible interaction revealed that accumulation of hybridizable CHS mRNA in the total cellular RNA fraction could be observed as early as 35 h after spore inoculation, i.e. about 30 h before the onset of phytoalexin accumulation and the first visible sign of hypersensitive flecking. The pattern of accumulation of hybridizable PAL mRNA in the total cellular RNA fraction followed a very similar pattern to that exhibited by CHS mRNA, with the exception that the relative accumulation in tissue distant from the site of infection in the incompatible interaction was somewhat less in relation to the extent of accumulation in directly infected tissue for PAL mRNA compared to CHS mRNA.

The rapid accumulation of PAL and CHS mRNA from very low, almost undetectable basal levels in unelicited cells and uninfected hypocotyl tissue, strongly suggests stimulation of the transcription of PAL and CHS genes. Transcriptional activation of PAL and CHS genes and induction of mRNA synthesis following elicitor treatment of suspension-cultured cells has been confirmed by in vivo and in vitro labelling of newly-synthesized PAL and CHS transcripts [18, unpublished data]. Newly-synthesized mRNA was purified from pre-existing mRNA by organo-mercurial affinity chromatography following in vivo labelling with 4-thiouridine. This technique was initially used to study developmental regulation of gene expression in yeast with a strain heterotrophic for uridine [19]. During adaptation of this technique to an autotrophic plant system, uptake of 4-thiouridine into suspension-cultured cells was observed with linear incorporation into RNA as a function of time. By double labelling with 4-thiouridine and [^{3}H]-uridine a substantial enrichment for newly-synthesized mRNA was demonstrated following subsequent affinity chromatography. Blot hybridization of newly-synthesized RNA purified in this way from elicitor treated and control cells pulse labelled with 4-thiouridine showed that elicitor markedly stimulated the rates of PAL and CHS mRNA synthesis [16,18].

More recently we have confirmed transcriptional activation of PAL and CHS genes by analysis of in vitro runoff transcripts in nuclei isolated from control and elicitor treated cells [unpublished data]. cDNA clones H1 and H7 contain sequences complementary to highly abundant and rare RNA species respectively which are unaffected by elicitor treatment. As a control it was shown that elicitor had no effect on the constitutive transcription of the genes corresponding to these cDNA sequences. In contrast, elicitor caused a marked and very rapid stimulation of transcription of PAL and CHS genes. Maximum rates of transcription were observed about 80 min after elicitor treatment corresponding to the onset of the phase of most rapid increase in hybridizable mRNA levels. Elicitor stimulation of PAL and CHS gene transcription represents the most rapid molecular event hitherto observed that can be causally related to the expression of a plant defence response.

BIOLOGICAL STRESS CAUSES MASSIVE CHANGES IN THE OVERALL PATTERN OF PROTEIN AND RNA SYNTHESIS

The highly coordinated increases in synthesis of a number of enzymes of phytoalexin biosynthesis in both elicitor treated cells and infected tissue argue strongly that a similar mechanism of induction operates and for many if not all enzymes in the pathway and suggests that rapid and extensive changes in the pattern of gene expression underlie the response of plant cells to biological stress. To examine the extent and complexity of changes in protein synthesis we therefore monitored the pattern of protein synthesis at various times after elicitor treatment and infection, by high resolution 2-dimensional gel electrophoretic analysis of [^{35}S]methionine labelled polypeptides synthesized by in vitro translation of isolated polysomal RNA.

With RNA from untreated control cell cultures, incorporation of [^{35}S]methionine into about 400 polypeptides could be discerned following prolonged fluorographic exposure of 2-dimensional gel electrophoretograms. Treatment of cell cultures with elicitor, heat released from cell walls of C. lindemuthianum, caused a rapid and marked change in the overall pattern of protein synthesis [unpublished data]. While elicitor treatment had relatively little effect on the synthesis of some polypeptides, there was down regulation of the synthesis of a considerable number of other polypeptides. More strikingly, elicitor stimulated the synthesis of at least 60 polypeptides, in many instances from very low or zero basal rates in control cells. While some of these polypeptides represented relatively minor components, the synthesis of other polypeptides was stimulated to the extent that they became major components of the overall pattern of protein synthesis in elicited cells. In most cases stimulation of the rate of polypeptide synthesis could be observed 1 h after elicitor treatment but was considerably more marked 3 h after elicitor treatment. However, the synthesis of a subset of polypeptides was more markedly stimulated 1 h than 3 h after elicitor treatment. Synthesis of a further subset of polypeptides was not stimulated until 3 h after elicitor treatment. Broadly similar changes in the pattern of protein synthesis were observed with the same elicitor preparation at different concentrations and also other elicitor preparations such as chitosan, citrus pectic fragment and proteinase inhibitor inducing factor [20,21]. Interestingly, dilution of cell cultures into new medium also induced very similar changes in the pattern of protein synthesis.

Marked changes in the pattern of polypeptide synthesis were also observed following infection of bean hypocotyls with C. lindemuthianum. In the incompatible interaction with race β, marked but transient changes in the pattern of protein synthesis were observed i.. tissue at the site of spore inoculation. Changes were first observed 52 h after spore inoculation and were most marked in the period 69 to 103 h after inoculation. These changes reflected stimulation of the synthesis of a specific set of polypeptides against a relatively constant background pattern of polypeptide synthesis characteristic of equivalent uninfected hypocotyl tissue. Subsequently the pattern of polypeptide synthesis reverted to that of control uninfected hypocotyls. Certain features of the pattern of induced polypeptide synthesis in directly infected tissue samples could also be discerned in tissue distant from the site of spore inoculation, although the changes were weaker and more transient.

Following inoculation of hypocotyls with spores of the compatible race γ, there was no early change in the pattern of protein synthesis during the period when marked changes in protein synthesis were observed in hypocotyl tissue infected with the incompatible race β. However, changes in the pattern of protein synthesis in the later stages of the compatible interaction were observed, which unlike the incompatible interaction involved not only stimulation of the synthesis of a set of polypeptides, but also down regulation of the synthesis of many polypeptides whose synthesis was characteristic of uninfected hypocotyl tissue, leading to a massive change in the overall pattern of protein synthesis. Changes in the pattern of protein synthesis could first be discerned 126 h after spore inoculation and were most pronounced in the period 150-168 h after inoculation, concomitant with the onset of lesion formation. Similar changes in the pattern of protein synthesis were also observed in tissue distant from the site of fungal infection, but were slightly delayed and somewhat less pronounced than in directly infected tissue.

Although the timing of changes in the pattern of protein synthesis was markedly different in the compatible interaction compared to the incompatible interaction, a number of common features could be discerned, and furthermore, many of the major polypeptide species induced in infected hypocotyl tissue were also induced in elicitor-treated cells.

Interestingly, isolated newly-synthesized RNA in vivo labelled with 4-thiouridine can serve as a template for protein synthesis in in vitro translation systems [18]. By a number of criteria this in vitro translation is of high fidelity and it is possible to compare the pattern of translation products encoded by newly-synthesized mRNA isolated from control and elicitor stimulated cells. By this approach we demonstrated that the rapid and pronounced change in the pattern of protein synthesis following elicitor treatment was the result of a dramatic switch in the pattern of mRNA synthesis involving down regulation of the synthesis of a small number of mRNA species and up regulation often from very low basal levels, of the synthesis of a large number of new mRNA species.

EXPRESSION OF OTHER DISEASE RESISTANCE MECHANISMS ALSO INVOLVES DEFENCE GENE ACTIVATION

The picture that emerges from these studies is of marked, characteristic changes in the patterns of mRNA accumulation and protein synthesis in elicitor-treated cells and infected hypocotyls. At least some of these changes are involved in the induction of phytoalexin accumulation and these observations raise the question of whether expression of other plant disease resistance mechanisms also involves accumulation of specific mRNAs.

Hydroxyproline-rich glycoproteins (HRGPs) are major structural components of plant cell walls and accumulation of cell wall HRGP in response to infection has been observed in a number of systems and is correlated with expression of disease resistance. HRGPs may function in defence as specific agglutinins of microbial pathogens and/or as structural barriers either directly or by providing sites for lignin deposition. As discussed in more detail by Showalter et al. in this volume [22], cloned genomic HRGP sequences have been used to demonstrate, by RNA blot hybridization, that elicitor treatment of suspension-cultured bean cells causes a marked increase in hybridizable HRGP mRNAs. Accumulation of 3 transcripts, size 1.6, 2.7 and about 5.6 kilobases was observed. Interestingly, the response was less rapid but more prolonged than that observed for mRNAs encoding enzymes of phytoalexin biosynthesis, with a lag of 3-4 h followed by a phase of rapid accumulation between 6 and 12 h after elicitation, after which the mRNAs remained at high levels. Accumulation of HRGP mRNAs was also observed in infected hypocotyls with temporal and spatial patterns broadly similar to those previously noted for PAL and CHS mRNAs [22,23].

In addition, recent studies have demonstrated that elicitor stimulates the translatable activities of mRNAs encoding chitinase and cinnamyl alcohol dehydrogenase (CAD) in suspension-cultured bean cells [unpublished data]. Chitinase has lysozymal activity and may function in defence by degradation of the fungal cell wall polymer chitin [24]. CAD is an enzyme of phenylpropanoid metabolism specific to a branch pathway for synthesis of lignin monomers [25]. Interestingly, elicitor stimulation of the level of CAD mRNA activity is more rapid than that previously observed for mRNAs encoding phytoalexin biosynthetic enzymes, with maximum rates of synthesis occurring between 1.5 and 2.0 h after elicitation. More rapid stimulation of CAD compared to the phytoalexin biosynthetic enzymes was also observed at the enzyme activity level. The different kinetics for accumulation of mRNAs encoding (i) CAD, (ii) PAL and CHS and (iii) HRGP may reflect distinct stimuli or a single stimulus leading to either sequential effects or divergent signal pathways.

POLYMORPHISM OF DEFENCE GENES AND PROTEINS

From a cDNA library of about 20,000 clones containing sequences complementary to mRNA in elicitor treated cells, 48 clones were found to contain CHS sequences [17]. Restriction fragment analysis grouped these clones into 5 classes and representatives of each of these classes have been sequenced [unpublished data]. There was strong conservation of sequence in the coding regions of these 5 cDNAs with a small number of base substitutions evenly distributed throughout the coding region. Most of these base substitutions were silent with respect to amino acid sequence. However, within the 3' non-coding region, following a sequence of about 25 nucleotides which are relatively conserved, there was considerable sequence divergence among 4 of the 5 cDNAs.

Southern blotting of genomic DNA from P. vulgaris revealed the presence of a number of fragments containing CHS sequences. Divergent, transcript specific sequences complementary to the 3' untranslated regions of 2 CHS mRNAs were subcloned in the transcription vector Sp64 and used as hybridization probes to correlate CHS genes on 2 different genomic fragments with the respective CHS transcripts. Recently we have isolated and characterized a number of CHS genes from a P. vulgaris genomic library in λ 1059 (kindly provided by Dr. T. C. Hall). Southern blot analysis of these cloned genomic sequences, using CHS cDNA probes showed considerable restriction fragment polymorphism and the 22 different clones analyzed exhibited 6 distinguishable patterns. Taken overall, the data indicate the presence of at least 6 CHS genes within the haploid genome of bean, some of which may be closely clustered and furthermore that at least 5 of these genes are activated by elicitor treatment.

Recently, we have sequenced 2 of the bean CHS genes in their entirety, together with flanking regions. One of these genes corresponds almost exactly to one of the cDNAs, and hence represents an elicitor-induced gene. The other gene apparently represents an isogene distinct from those represented by the cDNAs. Relative to the 3' untranslated region, the 5' untranslated region of these genes are highly conserved. The degree of sequence conservation of the single intron, present in the same position of each gene, is not quite as high as that in the 5' untranslated region but is probably significant in comparison to the divergence of the 3' untranslated sequence. The sequence of the putative promoter region of the elicitor induced gene shows a 15 bp stretch of high homology with the TATA box of the gene for the bean seed storage protein phaseolin [26] but the significance of this observation, if any, is not yet known.

Polymorphism of CHS has also been observed at the protein level. Two-dimensional gel electrophoretic analysis of immunoprecipitated CHS synthesized in vivo in elicitor treated cells or by in vitro translation of mRNA from elicited cells indicates the presence of about 10 different CHS polypeptides with similar M_r but different pI (unpublished data). To correlate polymorphism at the gene, mRNA and polypeptide level, a double hybrid select translation experiment was performed in which CHS mRNA was first purified by hybridization to an immobilized full-length CHS cDNA which therefore contained the highly conserved coding sequences. CHS mRNA purified in this way was then subjected to a second hybrid selection using 2 subclones of 3' noncoding CHS sequences specific to individual transcripts as the targets for hybridization. The selected CHS mRNA was then translated in vitro and analyzed by 2-dimensional gel electrophoresis. By this approach it was demonstrated that these 2 specific CHS mRNA species encoded specific subsets of the set of CHS polypeptides. Northern blot analysis using these transcript specific 3' probes showed that the corresponding mRNAs were somewhat more slowly induced by elicitor with maximum induction about 5 h after elicitor treatment compared to 3 to 4 h for the overall CHS mRNA population. Interestingly, the corresponding CHS polypeptides encoded by these 2 specific mRNAs also became more prevalent relative to the other CHS polypeptides during the later stages of the induction.

Polymorphism of PAL and HRGP has also been observed. At the protein level, PAL exists in a series of 4 forms of the native tetrameric enzyme which exhibit different kinetic properties and which in cell suspension cultures are differentially induced by elicitor [27]. Interestingly, the form predominating in untreated cultures has a high K_m for phenylalanine, whereas elicitor induction of phenylpropanoid synthesis results in the appearance of lower K_m forms, which presumably leads to more efficient channelling of the phenylalanine pool into phenylpropanoid compounds. 2-Dimensional gel electrophoretic analysis of PAL polypeptide subunits synthesized in vitro and in vivo also reveals extensive polymorphism [27]. Genomic DNA clones of PAL show restriction fragment polymorphism and representatives of at least 3 classes have been observed so far [unpublished data]. As noted above, polymorphism of HRGP has been observed at the mRNA level, with at least 3 different size classes of transcript [22,23], and families of HRGP genes have been demonstrated in the tomato, carrot and bean genomes [28,29, unpublished data].

DISCUSSION

The picture beginning to emerge from these studies is that expression of disease resistance mechanisms involves a rapid and selective activation of defence genes and that these systems exhibit extensive polymorphism at the gene, mRNA and protein levels. It seems probable that this polymorphism is of considerable biological importance and may reflect gene amplification to enhance the ability of the plant to respond effectively to infection following perception of an appropriate signal. Furthermore, genetic polymorphism may allow evolution of a family of genes encoding closely related polypeptides with similar but subtly different functional properties attuned to optimal operation in specific biological circumstances. One interesting example already uncovered is the selective induction of elicitor of PAL enzyme forms exhibiting a low K_m for phenylalanine. This arrangement might be seen as a device to assert metabolic priority for phenylpropanoid biosynthesis in the cellular economy of phenylalanine under conditions of biological stress. For CHS, different substrate specificities might account in part for the series of multiple subunit forms induced in response to biological stress, associated with the production of: (i) phaseollin and other 6'-deoxyisoflavonoids and (ii) the 6'-oxyisoflavonoids kievitone [5].

Furthermore genetic polymorphism allows a set of genes encoding identical or very similar polypeptides to be positioned in different regulatory circuits, thereby allowing a flexible response to a variety of environmental conditions. Preliminary evidence indicates that whereas biological stress induces the synthesis of about 10 CHS polypeptide species, illumination of dark grown hypocotyl tissue stimulates the synthesis of a smaller set of CHS polypeptides. It is not yet clear whether illumination and biological stress activate 2 distinct sets of CHS genes or whether illumination activates a specific subset of the genes stimulated by biological stress. Furthermore, even within the syndrome of biological stress, different regulatory circuits may be involved in defence gene activation. Thus the phytoalexin defence response is induced in several rather different biological circumstances inducing: (i) the early stages of an incompatible interaction following a putative molecular recognition event between a disease resistance gene product of the plant and an avirulence gene product of the pathogen and (ii) the later stages of a compatible interaction following a switch to necrotrophic growth and traumatization of the plant tissue at the onset of lesion formation. These two induction systems relate to the roles of phytoalexin accumulation in hypersensitive resistance and attempted lesion limitation respectively. Furthermore the present studies show that the PAL and CHS genes are activated, not only in directly infected tissue but also in hitherto uninfected tissue distant from the site of spore inoculation presumably in response to an endogenous signal [30,31,32]. Thus different regulatory circuits, signal molecules and DNA sequences may be involved in the activation of PAL, CHS and other defence genes in these various stress situations. Similarly the different kinetics for accumulation of different defence related mRNAs in elicitor treated cells implies considerable regulatory complexity in terms of either the nature, pereception or transduction of the signal or the regulatory sequence of the target genes.

The genetic polymorphism of PAL and CHS is of great interest in terms of the evolution of these genes in relation to the biological functions of phenylpropanoid natural products. It is probable that this pathway initially evolved as a means for intracellular excretion following adoption of a terrestrial habit. Subsequent evolution of more sophisticated adaptive functions such as those of lignin as a cell wall structural polymer and defence barrier, and low molecular weight phenolics as pigments and antibiotics, may have been intimately associated with duplication and rearrangement of the genes encoding the biosynthetic enzymes. In this context it is of interest that in antirrhinum and petunia there appear to be only 1 or 2 CHS genes per haploid genome [33,34], in marked contrast to the situation we have uncovered in bean. It is possible to speculate that the additional genes present in the bean genome have evolved in relation to the adaptation of CHS as an enzyme for the synthesis of the isoflavonoid phytoalexin antibiotics characteristic of legumes in addition to the ubiquitous flavonoid pigments.

The observation of extensive polymorphism of defence genes and induction of multiple arrays of genes encoding proteins involved in biochemically and functionally distinct defence responses has considerable implication for the interpretation of the genetics of host:pathogen interactions. In the gene-for-gene hypothesis a specific disease resistance gene in the host can be paired with a specific avirulence gene in the pathogen such that the gene products of the corresponding dominant alleles form a molecular recognition complex leading to incompatibility. Analysis of the segregation of plant disease resistance genes has failed to reveal the presence of other plant genes which modify the effects of the disease resistance genes. This has been interpreted to indicate that the formation of the molecular complex between avirulence gene product and disease resistance gene product is not merely the recognition event but also represents the actual primary resistance response [35]. In this model, recognition would itself be sufficient for expression of hypersensitive resistance and ensuing disease resistance mechanisms including phytoalexin accumulation, etc. would not have a primary role in hypersensitive resistance, but rather some ill-defined secondary function. However, our present observations argue strongly against this interpretation, since a mutation in a specific defence gene, (e.g. 1 of the 6 CHS isogenes) is individually unlikely to have any significant effect on the outcome of the host:pathogen interaction in view of the multiple induction of several isogenes encoding that biochemical function (e.g. CHS activity) and furthermore the multiple induction of several complementary defence responses only 1 of which involves that particular defence gene family (e.g. phytoalexin accumulation).

The magnitude and rapidity of elicitor stimulation of defence gene activation in suspension cultured cells also strongly argues that the resultant defence responses play a primary role in disease resistance. Induction of PAL and CHS mRNA accumulation within 35 h of inoculation of bean hypocotyls with the spores of an incompatible race of C. lindemuthianum likewise represents an extremely early event in relation to the ingress of the pathogen, since following spore inoculation there is a period of 30 to 40 h during which the spores germinate and produce an infection peg before fungal hyphae come in contact with the first host epidermal cell [3,36]. Thus in relation to the timing of this initial cell:cell contact at which the putative molecular recognition event occurs, induction of defence mRNAs in an incompatible interaction is extremely rapid and occurs well before the onset of phytoalexin accumulation, the first signs of hypersensitive flecking and expression of hypersensitive resistance. These observations do not appear to be consistent with the hypothesis that the induction of PAL and CHS mRNAs is causally preceded by hypersensitive cell death and initial expression of hypersensitive resistance. In the light of our recent observations on the organization and activation of defence genes, our working hypothesis is that disease resistance genes encode receptor proteins involved in molecular recognition of pathogens which leads to rapid activation of defence genes and hence expression of hypersensitive resistance.

Induction of defence mRNAs in the later stages of a compatible interaction in hitherto uninfected cells ahead of the invading pathogen provides a plausible mechanism for the process of attempted lesion limitation, which under appropriate physiological conditions can restrict the size of the lesion and prevent complete rotting of the organ and hence plant death, even though the interaction is genetically compatible. Likewise, observation of the preactivation of defence genes in hitherto uninfected cells distant from the initial site of infection by an incompatible pathogen may provide a clue to the molecular basis of induced systemic resistance or immunity [37]. It should be noted that defence gene activation in distant uninfected cells does not necessarily imply that the induced mRNAs are translated into active protein leading to complete activation of the overall defence response (phytoalexin synthesis, HRGP accumulation etc.). Indeed it is possible to speculate that the immunized state reflects accumulation of untranslated mRNAs encoding defence proteins so that following a second, challenge infection these pre-existing mRNAs can be rapidly incorporated into polysomes leading to more rapid and effective activation of phytoalexin accumulation etc. than following infection of equivalent non-immunized tissue with a compatible pathogen. Thus pre-challenge with an avirulent race apparently does not lead to phytoalexin accumulation in distant tissue, but following subsequent challenge inoculation with a normally virulent pathogen, there is a very rapid accumulation of phytoalexins reminiscent of an incompatible interaction [38]. Evidence for the operation of post-transcriptional controls during the infection of bean hypocotyls with an incompatible race of C. lindemuthianum comes from observation of differences in the kinetics of CHS mRNA accumulation in the polysomal RNA fraction compared to the total RNA fraction. Hence although defence mRNAs accumulate in both directly infected tissue and distantly uninfected tissue, efficient and/or prolonged incorporation of these mRNAs accumulate in both directly infected tissue and distantly uninfected tissue, efficient and/or prolonged incorporation of these mRNAs into polysomes might require a second signal localized to directly infected cells and near neighbours in an incompatible interaction. The concomitant accumulation of CHS mRNA in polysomal and total cellular RNA fractions observed in the later stages of a compatible interaction might then reflect simultaneous and widespread deployment of all signals needed for full activation of defence mechanisms in response to extensive traumatization of host tissue at the onset of lesion formation.

In conclusion, application of molecular biological approaches to the study of plant responses to infection has provided some insight into the mechanisms governing disease resistance and focuses attention on the structure, organization and expression of plant defence genes.

ACKNOWLEDGEMENTS

We thank Evelyn Wilson for preparation of the manuscript. Research in the Salk Institute was supported by grants to CJL from ICI (plc), USDA (CRGO CRCR-1-1464) and the Samuel Roberts Noble Foundation. CLC is an NSF Post-doctoral Fellow in Plant Biology.

REFERENCES

1. Sequeira, L., 1983, Mechanisms of induced resistance in plants, Annu. Rev. Microbiol., 37: 51-79.
2. Lamb, C.J., Bell, J.N., Cramer, C.L., Dildine, S.L., Grand, C., Hedrick, S.A., Ryder, T.B. and Showalter, A.M., 1985, Molecular response of plants to infection, in: "Biotechnology for Solving Agricultural Problems", Beltsville Agricultural Research Center Symposia, Vol. X, J.B. St. John, ed., Roman and Allanheld, Totowa, N.J., in press.
3. Bailey, J.A., 1982, Physiological and biochemical events associated with the expression of resistance to disease, in: "Active Defense Mechanisms in Plants", R.K.S. Wood, ed., pp.39-65, Plenum Press, New York.
4. Moesta, P. and Grisebach, H., 1980, Effects of biotic and abiotic elicitors on phytoalexin metabolism in soybean, Nature, 286: 710-711.
5. Dixon, R.A., Dey, P.M. and Lamb, C.J., 1983, Phytoalexins: Enzymology and Molecular Biology, Adv. Enzymol., 55: 1-135.
6. Cramer, C.L., Bell, J.N., Ryder, T.B., Bailey, J.A., Schuch, W., Bolwell, G.P., Robbins, M.P., Dixon, R.A. and Lamb, C.J., 1985, Coordinated synthesis of phytoalexin biosynthetic enzymes in biologically-stressed cells of bean (Phaseolus vulgaris L.), EMBO J., 285-289.
7. Dixon, R.A. and Bendall, D.S., 1978, Changes in the levels of enzymes of phenylpropanoid and flavonoid synthesis during phaseollin production in cell suspension cultures of Phaseolus vulgaris, Physiol. Plant Pathol., 13: 295-306.
8. Robbins, M.P. and Dixon, R.A., 1984, Induction of chalcone isomerase in elicitor-treated bean cells: Comparison of rates of synthesis and appearance of immunodetectable enzyme, Eur. J. Biochem., 145: 195-202.
9. Bell, J.N., Dixon, R.A., Bailey, J.A., Rowell, P.M. and Lamb, C.J., 1984, Differential induction of chalcone synthase mRNA activity at the onset of phytoalexin accumulation in compatible and incompatible plant-pathogen interactions, Proc. Natl. Acad. Sci. USA, 81: 3384-3388.
10. Robbins, M.P., Bolwell, G.P. and Dixon, R.A., 1985, Metabolic changes in elicitor-treated bean cells: Selectivity of enzyme induction in relation to phytoalexin accumulation, Eur. J. Biochem., 148: 563-569.
11. Lawton, M.A., Dixon, R.A., Hahlbrock, K. and Lamb, C.J., 1983, Rapid induction of the synthesis of phenylalanine ammonia-lyase and of chalcone synthase in elicitor-treated plant cells, Eur. J. Biochem., 129: 593-601.
12. Dixon, R.A., Gerrish, C., Lamb, C.J. and Robbins, M.P., 1983, Elicitor-mediated induction of chalcone isomerase in Phaseolus vulgaris cell suspension cultures, Planta, 159: 561-569.
13. Lawton, M.A., Dixon, R.A. and Lamb, C.J., 1980, Elicitor modulation of the turnover of L-phenylalanine ammonia-lyase in French bean cell suspension cultures, Biochem. Biophys. Acta, 633: 162-175.

14. Lamb, C.J., Merritt, T.K. and Butt, V.S., 1979, Synthesis and removal of phenylalanine ammonia-lyase activity in illuminated discs of potato tuber parenchyme, Biochem. Biophys. Acta, 582: 196-212.
15. Lawton, M.A., Dixon, R.A., Hahlbrock, K. and Lamb, C.J., 1983, Elicitor induction of mRNA activity: Rapid effects of elicitor on phenylalanine ammonia-lyase and chalcone synthase mRNA activities in bean cells, Eur. J. Biochem., 130: 131-139.
16. Edwards, K., Cramer, C.L., Bolwell, G.P., Dixon, R.A., Schuch, W. and Lamb, C.J., 1985, Rapid transient induction of phenylalanine ammonia-lyase mRNA in elicitor treated bean cells, Proc. Natl. Acad. Sci. USA, 82: in press.
17. Ryder, T.B., Cramer, C.L., Bell, J.N., Robbins, M.P., Dixon, R.A. and Lamb, C.J., 1984, Elicitor rapidly induces chalcone synthase mRNA in Phaseolus vulgaris cells at the onset of the phytoalexin defense response, Proc. Natl. Acad. Sci. USA, 81: 5724-5728.
18. Cramer, C.L., Ryder, T.B., Bell, J.N. and Lamb, C.J., 1985, Rapid switching of plant gene expression induced by fungal elicitor, Science, 227: 1240-1243.
19. Stetler, G.L. and Thorner, J., 1984, Molecular cloning of the hormone-responsive genes from the yeast Saccharomyces cerevisiae, Proc. Natl. Acad. Sci. USA, 81: 1144-1148.
20. Darvill, A.G. and Albersheim, P., 1984, Phytoalexins and their elicitors: A defense against microbial infection in plants, Annu. Rev. Plant. Physiol., 35: 243-275.
21. Ryan, C.A., 1984, Systemic responses to wounding, in: "Plant-Microbe Interactions: Molecular and Genetic Perspectives", Vol. 1, T. Kosuge and E.W. Nester, eds., pp.307-320, MacMillan, New York.
22. Showalter, A.M., Bell, J.N., Cramer, C.L., Bailey, J.A., Varner, J.E. and Lamb, C.J., 1986, Accumulation of hydroxyproline-rich glycoprotein mRNAs in biologically stressed cell cultures and hypocotyls, in: "Biology and Molecular Biology of Plant-Pathogen Interactions", J.A. Bailey, ed., Plenum Press, New York.
23. Showalter, A.M., Bell, J.N., Cramer, C.L., Bailey, J.A., Varner, J.E. and Lamb, C.J., 1985, Accumulation of hydroxyproline-rich glycoprotein mRNAs in response to fungal elicitor and infection, Proc. Natl. Acad. Sci. USA, 82: 6551-6555.
24. Boller, T., 1985, Induction of hydrolases as a defense reaction against pathogens, in: Cellular and Molecular Biology of Plant Stress", J.L. Key and T. Kosuge, eds., pp.247-262, Alan R. Liss Inc., New York.
25. Sarni, F., Grand, C. and Boudet, A.M., 1984, Purification and properties of cinnamoyl-CoA reductase and cinnamyl alcohol dehydrogenase from poplar stem (Populus euramericana), Eur. J. Biochem., 139: 259-265.
26. Slightom, J.L., Sun, S.M. and Hall, T.C., 1983, Complete nucleotide sequence of a French bean storage protein gene: phaseolin, Proc. Natl. Acad. Sci. USA, 80: 1897-1901.
27. Bolwell, G.P., Bell, J.N., Cramer, C.L., Schuch, W., Lamb, C.J. and Dixon, R.A., 1985, L-Phenylalanine ammonia-lyase from Phaseolus vulgaris: Characterization and differential induction of multiple forms from elicitor-treated cell suspension cultures, Eur. J. Biochem., 149: 411-419.
28. Chen, J. and Varner, J.E., 1985, Isolation and characterization of cDNA clones for carrot extensin and a proline-rich 33-kDa protein, Proc. Natl. Acad. Sci. USA, 82: 4399-4403.
29. Chen, J. and Varner, J.E., 1985, An extracellular matrix protein in plants: characterisation of a genomic clone for carrot extensin, EMBO J., 4: 2145-2152.
30. Hahn, M.G., Darvill, A.G. and Albersheim, P., 1981, Host-pathogen interactions XIX: The endogenous elicitor, a fragment of a plant cell wall polysaccharide that elicits phytoalexin accumulation in soybeans, Plant Physiol., 68: 1161-1169.

31. Hargreaves, J.A. and Bailey, J.A., 1978, Phytoalexin production by hypocotyls of Phaseolus vulgaris in response to constitutive metabolites released by damaged bean cells, Physiol. Plant Pathol., 13: 89-100.
32. Jin, D.F. and West, C.A., 1984, Characteristics of galacturonic acid oligomers as elicitors of casbene synthetase activity in castor bean seedlings, Plant Physiol., 74: 989-992.
33. Reif, H.J., Niesbach, U., Deumling, B. and Saedler, H., 1985, Cloning and analysis of two genes for chalcone synthase from Petunia hybrida, Mol. Gen. Genet., 199: 208-215.
34. Wienand, U., Sommer, H., Schwarz, Z., Shepherd, N., Saedler, H., Kreuzaler, F., Ragg, H., Fautz, E., Hahlbrock, K., Harrison, B.J. and Peterson, P.A., 1982, A general method to identify plant structural genes among genomic DNA clones using transposable element induced mutations, Mol. Gen. Genet., 187: 195-201.
35. Ellingboe, A.H., 1981, Changing concepts in host-pathogen genetics, Annu. Rev. Phytopathology, 19: 125-143.
36. O'Connell, R.J., Bailey, J.A. and Richmond, D.V., 1985, Cytology and physiology of infection of Phaseolus vulgaris by Colletotrichum lindemuthianum, Physiol. Plant Pathol., 27: 75-98.
37. Dean, R.A. and Kuć, J., 1985, Induced systemic resistance protection in plants, Trends in Biotechnol, 3: 125-129.
38. Kuć, J., 1985, Expression of latent genetic information for disease resistance in plants, in: "Cellular and Molecular Biology of Plant Stress", J.L. Key and T. Kosuge, eds., pp.303-318, Alan R. Liss Inc., New York.

MOLECULAR TARGETS FOR ELICITOR MODULATION IN BEAN (PHASEOLUS VULGARIS) CELLS

R.A. Dixon, G.P. Bolwell, M.P. Robbins and M.A.M.S. Hamdan

Department of Biochemistry
Royal Holloway and Bedford New College
Egham
Surrey TW20 OEX, UK

INTRODUCTION

This chapter reviews recent work on the biochemistry of disease resistance expression in relation to the French bean (*Phaseolus vulgaris*)/ *Colletotrichum lindemuthianum* interaction. The basic biological and genetical factors which make this system highly amenable to studies at the biochemical and molecular genetic levels have been reviewed previously [1], and the advantages and disadvantages of using elicitor-treated cell suspension cultures as a model system in initial biochemical studies have also been discussed [2].

Addition of elicitor macromolecules, heat-released from cell walls of the bean pathogen *Colletotrichum lindemuthianum*, to bean cell suspension cultures results in accumulation of isoflavonoid phytoalexins, synthesized *de novo* from L-phenylalanine (Fig. 1). The elicitor fractions from *Colletotrichum* are polysaccharide, composed predominantly of mannose, galactose and glucose, are heterogeneous with respect to M_r, and can be separated as both neutral and anionic molecules [3]. Phytoalexin accumulation is preceded by rapid but transient increases in the extractable activities of L-phenylalanine ammonia-lyase (PAL, EC 4.3.1.5), chalcone synthase (CHS) and chalcone isomerase (CHI, EC 5.5.1.6), enzymes involved in the conversion of L-phenylalanine to C_{15} flavonoid intermediates. Induction of PAL, CHS and CHI is preceded by increased activities of the mRNAs encoding these enzymes [4,5,6]; this, in the case of PAL and CHS, has now been shown to result from elicitor-mediated increases in their specific mRNA levels [7,8; Ryder, this volume].

It is clear from an analysis of 2-dimensional isoelectric focussing : SDS-polyacrylamide gels of polypeptides, synthesized either *in vivo* or *in vitro* from isolated mRNA, that a significant number of proteins and their corresponding mRNAs are induced during the early stages of the phytoalexin response; in addition, synthesis of some polypeptides appears to decline on elicitation [1,9,10,11]. It is important to identify the gene products whose synthesis is rapidly switched on or off in response to elicitors. Enzymes such as PAL, CHS and CHI are not solely involved in isoflavonoid phytoalexin biosynthesis, and other later control points undoubtedly exist as is made clear by the observation that elicitors from *Colletotrichum* culture filtrate induce PAL, CHS and CHI without inducing phytoalexin accumulation [3]. One

NATO ASI Series, Vol. H1
Biology and Molecular Biology of Plant-Pathogen Interactions
Edited by J. Bailey

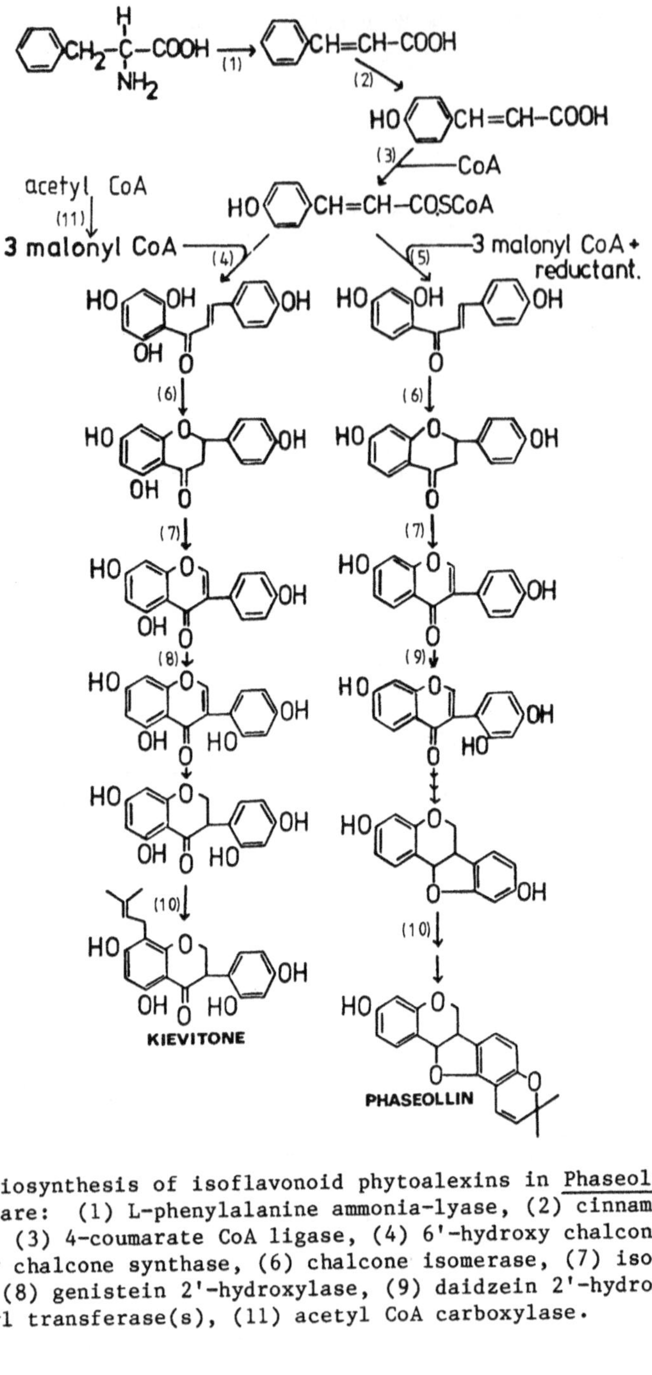

Figure 1. Biosynthesis of isoflavonoid phytoalexins in Phaseolus vulgaris. The enzymes are: (1) L-phenylalanine ammonia-lyase, (2) cinnamic acid 4-hydroxylase, (3) 4-coumarate CoA ligase, (4) 6'-hydroxy chalcone synthase, (5) 6'-deoxy chalcone synthase, (6) chalcone isomerase, (7) isoflavone synthetase, (8) genistein 2'-hydroxylase, (9) daidzein 2'-hydroxylase, (10) dimethylallyl transferase(s), (11) acetyl CoA carboxylase.

of the main goals of present work is the elucidation of gene sequences controlling expression in response to environmental stress; the possible involvement of (a) gene products which may arise by differential expression of one or more genes from a multi-gene family and (b) gene products which may be elicitor-inducible but metabolically unrelated to the phytoalexin pathway, makes the investigation of induced gene expression in this system of particular interest.

This chapter presents evidence which confirms that the changes in protein synthetic patterns observed on elicitation are not limited to enzymes of the phytoalexin pathway; extremely rapid changes which affect cell wall composition are also induced following exposure of bean cells to elicitor. The role of endomembrane-bound proteins in the induced resistance response is stressed, and some new aspects of the induction and coordination of key regulatory enzymes discussed. The reader is referred to earlier reviews for details of previous work on the bean/*Colletotrichum* elicitor system [12,13].

METABOLIC CHANGES IN ELICITOR-TREATED BEAN CELL CULTURES

Studies on the effects of application of biotic and abiotic elicitors to bean hypocotyls have indicated that, within the area of phenylpropanoid metabolism, isoflavonoid accumulation is induced specifically; no significant changes were observed in the levels of related phenylpropanoids such as hydroxycinnamic acids and flavonoids [14]. This also seems generally to be the case in elicitor-treated bean cell cultures [15], although here cell-wall-bound phenolics were increased significantly on elicitation, this increase correlating with the degree of browning of the cultures [15,16]; in addition, changes are also observed in the levels of free and esterified hydroxycinnamic acids [16]. A characteristic feature of phytoalexin induction in bean cultures is the clear temporal distinction between the appearance of the 5-hydroxylated isoflavanone phytoalexin kievitone and the 5-deoxy-isoflavonoid-derived phytoalexin phaseollin [17]; such temporal differentiation of expression of 5-hydroxy and 5-deoxy-isoflavonoid pathways was previously observed in studies on wounded and elicitor-treated bean hypocotyls [18], and probably reflects differential control of two distinct chalcone synthase activities whose mechanisms have been predicted from results of ^{13}C-labelling studies [19]. The chalcone synthase activity involved in the biosynthesis of phaseollin (6-deoxy CHS, Fig. 1), has yet to be measured *in vitro*, but is clearly a key target for modulation, either by elicitor or, in view of its late appearance, possibly by some secondary response coupling factor.

The timings of appearance of phytoalexins, wall-bound phenolics and wall-associated hydroxyproline in elicitor-treated bean cell cultures are summarized in Table 1. It is clear that the increase in wall hydroxyproline is more rapid than the appearance of kievitone, the earliest anti-microbial isoflavonoid to accumulate in this system. No phaseollin is detected in these cells until 24 h after elicitation. Although elicitor-treatment can cause extensive browning of the cultures, cell viability (as assessed by the ability to take up and reduce 2,3,5-triphenyltetrazolium chloride) is not greatly decreased; rather, a reversible decrease in the efficiency of mitochondrial reduction appears to be an early consequence of elicitation [3].

SELECTIVITY OF ENZYME INDUCTION

To date, the activities of five enzymes of the isoflavonoid pathway (PAL, CHS, CHI, cinnamic acid 4-hydroxylase and 4-coumarate : CoA ligase)

Table 1. Accumulation of end-products in response to elicitation in cell suspension cultures of *Phaseolus vulgaris* cv. Immuna.

Compound	Units/g fresh wt.	Level at zero time (units/g fresh wt.)	Maximum level (units/g fresh wt.)	Time of attainment of maximum level (h)
Cell wall-bound hydroxyproline	mg	0.1	0.28	6
Kievitone	nmol	Not detected	42.0	12
Hemicellulose-bound phenolics	A_{310}	4.2	22.8	> 24
Cellulose-bound phenolics	A_{310}	0.6	20.5	> 24
Phaseollin	nmol	Not detected	537.0	48

have been shown to be induced in elicitor-treated bean cell cultures [16,20]. Further, later enzyme activities specific for isoflavonoid phytoalexin biosynthesis have now been demonstrated in cell-free extracts from soybean; these are isoflavone synthetase [21] a pterocarpan 6a-hydroxylase [22] and a pterocarpan dimethylallyl transferase [23]. These enzymes are membrane-associated and are apparently very unstable *in vitro*, but it is clear that their activities are inducible by microbial stress. An isoflavone synthetase(s) and dimethylallyl transferases (probably of differing substrate specificities) are required for the synthesis of the bean phytoalexins, and these components will probably be among those detected by 2-D gel analysis of newly synthesized proteins. However, it is also clear that microbial stress can induce changes in enzymes other than those directly involved in phytoalexin synthesis; in soybean hypocotyls infected with *Phytophthora megasperma* f.sp. *glycinea* (Pmg), or in cell cultures treated with elicitor from this fungus, acetyl CoA carboxylase, glutamate dehydrogenase and glucose 6-phosphate dehydrogenase were strikingly induced along with PAL and CHS [24,25]. In contrast, none of these three enzymes was induced along with PAL, CHS and CHI at the time of phytoalexin accumulation in elicitor-treated bean cell cultures or infected bean hypocotyls [17], nor were changes observed in the extractable activities of caffeic acid O-methyltransferase, peroxidase, phospho-2-keto-3-deoxy-heptonate aldolase, shikimate dehydrogenase, glutamine synthetase, glutamate synthase, β-D-glucosidase, α-D-mannosidase or endo-1,3 β-D-glucanase. Several of these enzymes lie on pathways peripheral to, but metabolically interrelated with, the isoflavonoid phytoalexin pathway. Interestingly, the activities of glutamate dehydrogenase [17] and 4-coumarate hydroxylase [16], which were increasing in untreated control cultures, showed no increase following addition of elicitor. These enzymes do not play a direct role in isoflavonoid synthesis, and their "suppression" by elicitor reflects similar effects of Pmg elicitor on light-induced CHS and acetyl CoA carboxylase activities in parsley cell suspension cultures, these latter enzymes being involved in flavonoid, but not phytoalexin, biosynthesis in this system [26]. Detailed investigations into the molecular basis of such "suppressive" effects of elicitor molecules are clearly now required.

Although the above screening experiments were by no means exhaustive, the results suggest a highly selective induction of key biosynthetic enzymes in response to elicitor in bean cells, with major changes so far being seen in the enzymes of the phenylalanine ⟶ isoflavonoid pathway and membrane-bound enzymes involved in hydroxyproline protein synthesis (see below).

MEMBRANE-BOUND PROTEINS AS IMPORTANT TARGETS FOR ELICITOR MODULATION

A number of membrane-bound enzymes/proteins are clearly induced by elicitor in bean cells (Table 2). With respect to phytoalexin synthesis, cinnamic acid 4-hydroxylase(CA4H), which catalyzes the first hydroxylation reaction in the biosynthesis of kievitone and phaseollin from L-phenylalanine, is located in the microsomal fraction and is a cytochrome P450-dependent mixed function monooxygenase. Both CA4H activity and cytochrome P450 levels rapidly increased in bean microsomes following elicitation [16]. In this work, cytochrome P450 was measured by ^{125}I-protein A binding in dot blot assays of bean microsomal extracts previously incubated with an anti-(rat P450) monoclonal antibody. This antibody, which appears to be specific for the highly conserved heme binding site of P450, partially inhibits bean CA4H activity and recognizes a polypeptide of M_r 48 000 from bean extracts in immuno-precipitations of newly synthesized polypeptides synthesized *in vivo* and *in vitro*, and on Western blots [27]. Further microsomal P450-linked enzyme systems involved in the phytoalexin pathway may catalyze aryl migration (isoflavone synthetase) and the

Table 2. Changes in membrane-bound enzymes and protein components following elicitation of bean cell suspension cultures.

		Bean cultivar			
		Immuna		Canadian Wonder	
	Time after elicitation	0 h	6 h	0 h	6 h
Components	Units				
Prolyl hydroxylase	μ kat/kg protein	0.76	18.0	0.41	12.76
Membrane-bound hydroxyproline	μ g/mg membrane protein	4.0	7.0	ND[a]	ND
Protein arabinosyl transferase	μ kat/kg protein	4.8	14.5	4.12	9.34
M_r 42 500 arabinose acceptor protein	relative level	1.0	4.2	1.0	2.8
Cytochrome P-450	relative level	1.0	2.1	ND	ND
Cinnamic acid 4-hydroxylase	μ kat/kg protein	5.0	14.6	ND	ND

(a) ND = Not determined

2'-hydroxylations of the isoflavones daidzein and genistein (Fig. 1). These are at present under investigation.

A most striking change observed among the membrane-bound components following elicitation is the rapid but very transient induction of prolyl hydroxylase (peptide proline : 2-oxoglutarate dioxygenase) activity [16, Table 2]. This induction was concomitant with the appearance of increased hydroxyproline in the cell walls (Table 1) and endomembrane system (Table 2). A varietal difference was observed here, with relatively low levels of hydroxyproline in walls from unelicited cells of cv. Immuna compared with cv. Canadian Wonder so that the relative transient increase in wall hydroxyproline corresponding to increased hydroxylase activity is apparently much greater. Little or no transient increase in wall hydroxyproline in cv. Canadian Wonder was also observed by Showalter et al. [28], and differences may also exist in the capacities for cross-linked immobilization [29,30] within the wall. A major hydroxyproline rich glycoprotein produced by membranes during elicitation has been possibly identified as an M_r 42 500 arabinosylated glycoprotein [31]. The levels of this acceptor and a protein arabinosyl transferase activity increase along with the induced prolyl hydroxylase [16, Table 2]. Interestingly, in unelicited cells undergoing primary cell wall synthesis arabinose residues are incorporated from UDP arabinose into polysaccharide by an arabinan synthase which is immunologically distinct from the protein : arabinosyl transferase; elicitation, however, leads to induction of the transferase and arabinose is now transferred solely to the M_r 42 500 protein [31,32]. Arabinosylation of this protein appears to involve a lipid oligosaccharide intermediate formed from UDP arabinose and the glycoprotein binds in a specific manner to thyroglobulin - and fetuin - Sepharose [33] and thus shares a specificity with phytohaemagglutinin (PHA) [34]. However, it does not react with anti-(PHA) IgG [33] and may therefore be more related to the potato lectin-type arabinosylated glycoproteins [35,36,37]. Finally, an as yet unidentified M_r 58 000 polypeptide, which shares a common epitope with the arabinan synthase and may belong to a family of glycosyl transferases [38] is one of the most highly elicited membrane bound proteins.

FURTHER STUDIES ON THE PROPERTIES AND INDUCTION OF BEAN PHENYLALANINE AMMONIA-LYASE AND PROLYL HYDROXYLASE

Chromatofocussing analysis of bean PAL has revealed the presence of four iso-forms of the active tetrameric enzyme in elicitor-treated cell cultures; furthermore, up to 11 subunit forms, differing in pI value but of identical M_r are seen on 2-D gels of immunoprecipitates from in vivo labelled cells [39]. The M_r of the native PAL subunits is 77 000, although this subunit is inherently unstable in vitro [40]; the pI values of the tetrameric form vary from 5.4 to 4.95. The multiple forms individually exhibit Michaelis-Menten kinetics (K_m values 0.072, 0.122, 0.256 and 0.302 mM in order of decreasing pI value), although apparent negative rate cooperativity, as reported for PAL from a number of sources [13], is observed in preparations up to the chromatofocussing stage. Of great interest is the observation that elicitation results in differential increased appearance of the two PAL forms of highest pI value (lowest K_m) (Fig. 2). PAL multiplicity is also seen at the level of subunits synthesized in vitro from polysomal mRNA [39], and the development of cDNA clones for bean PAL [8] should provide the basis for analysis of the differential expression of the PAL iso-forms at the molecular genetic level.

Prolyl hydroxylase has now been purified to apparent homogeneity from elicitor-treated bean cell cultures by a procedure involving solubilization from microsomes, ammonium sulphate precipitation, ion-exchange chromatography and affinity chromatography on poly-L-proline-Sepharose 4B [41]. The enzyme,

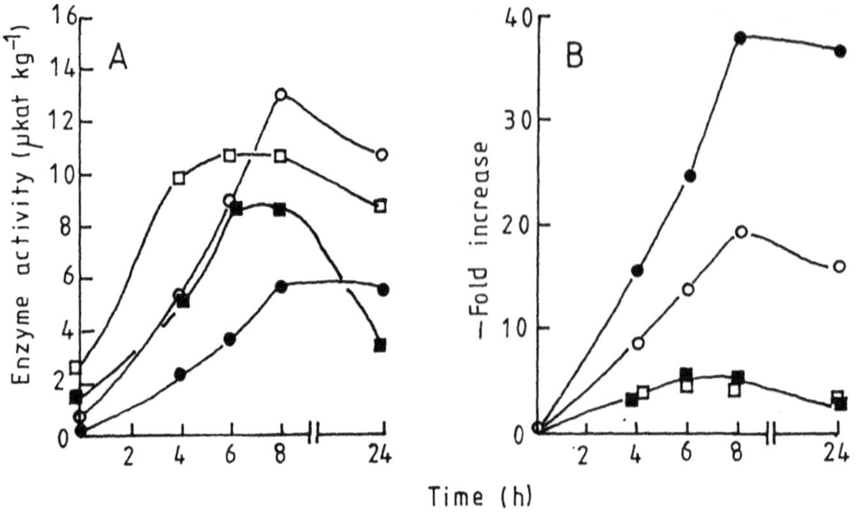

Figure 2. Differential induction of PAL multiple forms in elicitor-treated bean cell cultures. The data are calculated by integration of peak areas from chromatofocussing elution profiles, and it has therefore been assumed that the stabilities of the different forms are identical during partial purification. The multiple forms are of pI value 5.4 (● —— ●), 5.2 (○ —— ○), 5.05 (■ —— ■) and 4.85 (□ —— □).

which is a loosely-associated, minor, but highly active, component of smooth endomembranes, has been assayed by measurement of the poly-L-proline-dependent stoichiometric decarboxylation of 2-oxoglutarate. In many respects, its properties are very similar to those of the mammalian enzyme (subunit M_r 65 000, requirement for Fe^{++} and O_2), suggesting high conservation, although the mammalian enzyme does not hydroxylate poly-L-proline. The enzyme activity undergoes a very rapid 20-30 fold induction within 6 h after exposure of cells to elicitor; however, the activity then declines rapidly before increasing again at a slower rate [16]. This latter increase may be related to the recently reported increase in extensin mRNA levels in elicitor-treated bean cells [28].

A polyclonal antiserum against the purified bean prolyl hydroxylase has been recently produced [27]. This immunoprecipitates an *in vivo* labelled protein exhibiting a subunit doublet of M_r around 65 000, although it only inhibits the enzyme at low titre values and does not immunoprecipitate M_r 65 000 polypeptides from proteins synthesized *in vitro* from total or specifically enriched membrane-bound polysomal mRNA. Although this may be due to low abundance of prolyl hydroxylase mRNA, it is also possible that prolyl hydroxylase may undergo post-translational glycosylation in a similar manner to the vertebrate enzyme, and the antiserum may not be totally specific for the peptide portion of the molecule.

COORDINATION OF ENZYME SYNTHESIS AND TURNOVER IN INDUCED TISSUES

Under certain conditions, the induction of PAL, CHS and CHI in elicitor-treated cell cultures appears to be highly coordinated with respect to the relative timings of changes in enzyme activities, rates of synthesis *in vivo*, and mRNA activities measured in the total and polysomal mRNA fractions [6]. This is suggestive of a similar mechanism of induction for these three enzymes, presumably involving rapid increased gene transcription [6,7,8,11].

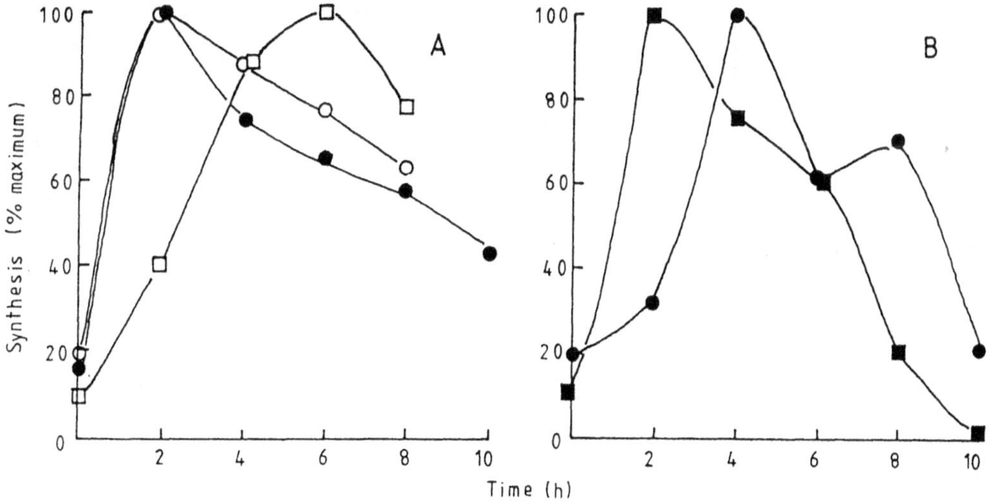

Figure 3. Enzyme synthesis in elicitor-treated bean cell suspension cultures. The data in A and B are from separate batches of bean culture. A, synthesis *in vitro* (polysomal mRNA activities); B, synthesis *in vivo* (measured by ^{35}S-methionine pulse labelling). ●——● = PAL, ○——○ = CHS, ■——■ = CHI, □——□ = prolyl hydroxylase.

In our laboratory, we have always observed that CHI activity is induced more slowly, and over a longer time period, than is that of PAL or CHS in both elicitor-treated cell cultures and infected bean hypocotyls, although increased synthesis is nevertheless rapidly induced; this is confirmed at the level of mRNA activity in the data in Figure 3A. An unidentified polypeptide of M_r 47 000 co-induces, at the protein level, with CHI as assessed by Western blot analysis [42]. Changes in the rate of prolyl hydroxylase synthesis *in vivo* closely follow, or even precede, those of PAL and CHS (Fig. 3B). The rapidity of induction of prolyl hydroxylase in the bean cells suggests that this, like the induction of PAL, CHS and CHI, is an early event directly linked to the elicitor stimulus. In contrast, the accumulation of hydroxyproline-rich glycoproteins in elicitor-treated melon tissues is less rapid and may require the intermediacy of ethylene formation [43].

It is unlikely that a common mechanism exists in bean whereby induction of PAL, CHS and CHI is obligately linked. Fractionation of *Colletotrichum* culture filtrate elicitor by gel-filtration followed by chromatography on concanavalin A-Sepharose yields fractions which will induce CHS activity without inducing PAL or CHI activities (Table 3) [3].

Although increased synthesis, reflecting mRNA changes, is involved in the induction of all the enzymes so far studied in the elicitor/bean cell culture sytem, post-translational changes are also involved in determining the characteristic shapes of the enzyme activity profiles. Density labelling with ^{2}H from ^2H_2O followed by analysis of enzyme activity distribution on CsCl density gradients has clearly demonstrated that the induction of CHI is characterized by a significant degree of activation of pre-existing, unlabelled enzyme [44]; at the same time, *de novo* synthesis of both activated and inactive enzyme subunits is observed [42]. The mechanism of CHI activation remains unclear, although it is interesting to note that exogenous

Table 3. Enzyme induction in bean, cv. Immuna, cell suspension cultures by crude and fractionated elicitor preparations from *Colletotrichum lindemuthianum*.

Elicitor fraction	Induced enzyme activity[(a)] at 6 h at elicitor concentrations (μg glucose equiv. ml^{-1}) shown					
	PAL		CHS		CHI	
	20	100	20	100	20	100
Crude cell wall elicitor	126	360	230	40	6.0	9.0
Culture filtrate elicitor:						
Crude	140	290	170	70	2.5	5.1
Sephacryl S-300 eluate	235	320	280	300	0.0	5.9
Concanavalin A-Sepharose fractions:						
(i) Unbound	0	0	350	0	0.0	0.0
(ii) Eluted with α-methyl mannoside	6	4	560	60	0.0	0.4

(a) Units/kg protein above basal level. Basal activities were PAL, 80 μ kat/kg; CHS, 60 nkat/kg; CHI, 2.5 mkat/kg. The culture filtrate elicitor was first fractionated by gel filtration on Sephacryl S-300. The major carbohydrate-containing peak was then further resolved by chromatography on concanavalin A-Sepharose.

addition of cinnamic acid can induce CHI extractable activity [45], while this treatment causes rapid removal of CHI polysomal mRNA activity (Robbins, Bolwell and Dixon, unpublished). Coordination of enzyme changes may therefore operate by multiple mechanisms including (i) rapid switching of gene expression in response to an intracellular response coupler related to initial elicitor binding at the plant cell surface, (ii) cessation of synthesis via multiple or common mechanisms, possibly involving pathway intermediates as 'sensors' of metabolic flux and (iii) post-translational effects on enzyme turnover and activation, again possibly involving pathway intermediates.

CONCLUSIONS

At the molecular level, the elicitation response in bean cell cultures exhibits many phenomena which may be important in intact plant-pathogen interactions. It is a useful model system for studies on the nature, selectivity, regulation and co-ordination of gene expression associated with induced resistance. A number of molecular targets have now been identified which, although not all metabolically related, may share functional similarities (i.e. a role in defence) and are modulated by the same elicitor preparations. Work now in progress is aimed at investigating (i) further details of the enzymology of the pathways leading to kievitone and phaseollin, (ii) the structure, function and expression of elicitor-modulated genes encoding both soluble and membrane-bound proteins, (iii) the role of phenylpropanoid pathway intermediates in the regulation of enzyme synthesis and turnover and (iv) the structures of, and molecular targets modulated by, defined elicitor fractions from Colletotrichum and endogenous elicitor fractions from bean cells.

ACKNOWLEDGEMENTS

RAD acknowledges support from the Agricultural and Food Research Council, the Science and Engineering Research Council, Imperial Chemical Industries plc and the University of London Central Research Fund.

REFERENCES

1. Lamb, C.J., Bell, J., Norman, P., Lawton, M.A., Dixon, R.A., Rowell, P. and Bailey, J.A., 1983, Early molecular events in the phytoalexin defense response, in: "Structure and Function of Plant Genomes", O. Cifferi, and L. Dure, eds., pp. 313-327, Plenum Press, New York.
2. Dixon, R.A., 1980, Plant tissue culture methods in the study of phytoalexin induction, in: "Tissue Culture Methods for Plant Pathologists", D.S. Ingram, and J.P. Helgeson, eds., pp. 185-196, Blackwell, Oxford.
3. Hamdan, M.A.M.S. and Dixon, R.A., 1986, Differential biochemical effects of elicitor preparations from Colletotrichum lindemuthianum, Physiol. Plant Pathol., in press.
4. Lawton, M.A., Dixon, R.A., Hahlbrock, K. and Lamb, C.J., 1983, Elicitor induction of mRNA activity. Rapid effects of elicitor on phenylalanine ammonia-lyase and chalcone synthase mRNA activities in bean cells, Eur. J. Biochem., 130: 131-139.
5. Bell, J.N., Dixon, R.A., Bailey, J.A., Rowell, P.M. and Lamb, C.J., 1984, Differential induction of chalcone synthase mRNA activity at the onset of phytoalexin accumulation in compatible and incompatible plant-pathogen interactions Proc. Natl. Acad. Sci. USA, 81: 3384-3388.
6. Cramer, C.L., Bell, J.N., Ryder, T.B., Bailey, J.A., Schuch, W., Bolwell, G.P., Robbins, M.P., Dixon, R.A. and Lamb, C.J., 1985, Co-

ordinated synthesis of phytoalexin biosynthetic enzymes in biologically-stressed cells of bean (Phaseolus vulgaris L.), EMBO J., 4: 285-289.

7. Ryder, T.B., Cramer, C.L., Bell, J.N., Robbins, M.P., Dixon, R.A. and Lamb, C.J., 1984, Elicitor rapidly induces chalcone synthase mRNA in Phaseolus vulgaris cells at the onset of the phytoalexin defense response, Proc. Natl. Acad. Sci. USA, 81: 5724-5728.
8. Edwards, K., Cramer, C.L., Bolwell, G.P., Dixon, R.A., Schuch, W. and Lamb, C.J., 1985, Rapid transient induction of phenylalanine ammonia-lyase mRNA in elicitor-treated bean cells, Proc. Natl. Acad. Sci. USA, 82: 6731-6735.
9. Hadwiger, L.A. and Wagoner, W., 1983, Electrophoretic patterns of pea and Fusarium solani proteins synthesized in vitro or in vivo which characterize the compatible and incompatible interactions, Physiol. Plant Pathol., 23: 153-162.
10. Loschke, D.C., Hadwiger, L.A. and Wagoner, W., 1983, Comparison of mRNA populations coding for phenylalanine ammonia-lyase and other peptides from pea tissue treated with biotic and abiotic phytoalexin inducers, Physiol. Plant Pathol., 23: 163-173.
11. Cramer, C.L., Ryder, T.B., Bell, J.N. and Lamb, C.J., 1985, Rapid switching of plant gene expression induced by fungal elicitor, Science, 227: 1240-1243.
12. Lamb, C.J., Lawton, M.A., Taylor, S.J. and Dixon, R.A., 1980, Elicitor modulation of phenylalanine ammonia-lyase in Phaseolus vulgaris, Ann. Phytopathol., 12: 423-433.
13. Dixon, R.A., Dey, P.M. and Lamb, C.J., 1983, Phytoalexins: enzymology and molecular biology, Adv. Enzymol., 55: 1-136.
14. Rathmell, W.G. and Bendall, D.S., 1971, Phenolic compounds in relation to phytoalexin biosynthesis in hypocotyls of Phaseolus vulgaris, Physiol. Plant Pathol., 1: 351-362.
15. Dixon, R.A. and Bendall, D.S., 1978, Changes in phenolic compounds associated with phaseollin production in cell suspension cultures of Phaseolus vulgaris, Physiol. Plant Pathol., 13: 283-294.
16. Bolwell, G.P., Robbins, M.P. and Dixon, R.A., 1985, Metabolic changes in elicitor-treated bean cells. Enzymic responses associated with rapid changes in cell wall components, Eur. J. Biochem., 148: 571-578.
17. Robbins, M.P., Bolwell, G.P. and Dixon, R.A., 1985, Metabolic changes in elicitor-treated bean cells. Selectivity of enzyme induction in relation to phytoalexin accumulation, Eur. J. Biochem., 148: 563-569.
18. Whitehead, I.M., Dey, P.M. and Dixon, R.A., 1982, Differential patterns of phytoalexin accumulation and enzyme induction in wounded and elicitor-treated tissues of Phaseolus vulgaris, Planta, 154: 156-164.
19. Dewick, P.M., Steele, M.J., Dixon, R.A. and Whitehead, I.M., 1982, Biosynthesis of isoflavonoid phytoalexins. Incorporation of sodium [1,2-$^{13}C_2$] acetate into phaseollin and kievitone, Z. Naturforsch., 37c: 363-368.
20. Dixon, R.A. and Bendall, D.S., 1978, Changes in the levels of enzymes of phenylpropanoid and flavonoid synthesis during phaseollin production in cell suspension cultures of Phaseolus vulgaris, Physiol. Plant Pathol., 13: 295-306.
21. Hagmann, M.L. and Grisebach, H., 1984, Enzymatic rearrangement of flavanone to isoflavone, FEBS Lett., 175: 199-202.
22. Hagmann, M.L., Heller, W. and Grisebach, H., 1984, Induction of phytoalexin synthesis in soybean. Stereospecific 3,9-dihydroxy-pterocarpan 6a-hydroxylase from elicitor-induced soybean cell cultures, Eur. J. Biochem., 142: 127-131.
23. Zähringer, U., Ebel, J., Mulheirn, L.J., Lyne, R. and Grisebach, H., 1979, Induction of phytoalexin synthesis in soybean. Dimethylallyl pyrophosphate : trihydroxypterocarpan dimethylallyl transferase from elicitor-induced cotyledons, FEBS Lett., 101: 90-92.

24. Borner, H. and Grisebach, H., 1982, Enzyme induction in soybean infected by Phytophthora megasperma f.sp. glycinea, Arch. Biochem. Biophys., 217: 65-71.
25. Ebel, J., Schmidt, W.E. and Loyal, R., 1984, Phytoalexin synthesis in soybean cells: Elicitor induction of phenylalanine ammonia-lyase and chalcone synthase mRNAs and correlation with phytoalexin accumulation, Arch. Biochem. Biophys., 232: 240-248.
26. Hahlbrock, K., Lamb, C.J., Purwin, C., Ebel, J., Fautz, E. and Schäfer, E., 1981, Rapid response of suspension-cultured parsley cells to the elicitor from Phytophthora megasperma var. sojae. Induction of the enzymes of general phenylpropanoid metabolism, Plant Physiol., 67: 768-773.
27. Bolwell, G.P. and Dixon, R.A., 1985, Elicitor induction of membrane-bound hydroxylases in bean cells, submitted for publication.
28. Showalter, A.M., Varner, J.E., Cramer, C.L., Bell, J.N. and Lamb, C.J., 1984, Regulation of extensin gene expression by fungal elicitor, in: "Abstracts of the 6th John Innes Symposium. The Cell Surface in Plant Growth and Development", p. 53, John Innes Institute, Norwich.
29. Fry, S., 1982, Isodityrosine, a new cross-linking amino-acid from plant cell wall glycoprotein, Biochem. J., 204: 449-455.
30. Cooper, J.B. and Varner, J.E., 1983, Insolubilization of hydroxyproline-rich cell wall glycoprotein in aerated carrot root slices, Biochem. Biophys. Res. Commun., 112: 161-167.
31. Bolwell, G.P., 1984, Differential patterns of arabinosylation by membranes of suspension-cultured cells of Phaseolus vulgaris (French bean) after subculture or elicitation, Biochem. J., 222: 427-435.
32. Bolwell, G.P. and Northcote, D.H., 1984, Demonstration of a common antigenic site on endomembrane proteins of Phaseolus vulgaris by a rat monoclonal antibody; tentative identification of arabinan synthase and consequences for its regulation, Planta, 162: 139-146.
33. Bolwell, G.P., 1986, An elicitor-induced carbohydrate-binding glyco-protein from suspension cultured bean cells: mechanism of post-translational arabinosylation, Phytochemistry, in press.
34. Vitale, A., Ceriotti, A., Bollini, R. and Chrispeels, M.J., 1984, Biosynthesis and processing of phytohaemagglutinin in developing bean cotyledons, Eur. J. Biochem., 141: 97-104.
35. Allen, A.K., Desai, N.N., Neuberger, A. and Creeth, J.M., 1978, Properties of potato lectin and the nature of its glycoprotein linkages, Biochem. J., 171: 665-674.
36. Leach, J.E., Cantrell, M.A. and Sequeira, L., 1982, A hydroxyproline-rich bacterial agglutinin from potato : extraction, purification and characterisation, Plant Physiol., 70: 1353-1358.
37. Leach, J.E., Cantrell, M.A. and Sequeira, L., 1982, A hydroxyproline-rich bacterial agglutinin from potato : its localization by immunofluorescence, Physiol. Plant Pathol., 21: 319-325.
38. Bolwell, G.P., 1985, Significance of a common epitope on endomembranes of plant and animal cells, submitted for publication.
39. Bolwell, G.P., Bell, J.N., Cramer, C.L., Schuch, W., Lamb, C.J. and Dixon, R.A., 1985, L-Phenylalanine ammonia-lyase from Phaseolus vulgaris. Characterisation and differential induction of multiple forms from elicitor-treated cell suspension cultures, Eur. J. Biochem., 149: 411-419.
40. Bolwell, G.P., Sap, J., Cramer, C.L., Lamb, C.J., Schuch, W. and Dixon, R.A., 1986, L-Phenylalanine ammonia-lyase from Phaseolus vulgaris: Partial degradation of enzyme subunits in vitro and in vivo, Biochim. Biophys. Acta, in press.
41. Bolwell, G.P., Robbins, M.P. and Dixon, R.A., 1985, Elicitor-induced prolyl hydroxylase from Phaseolus vulgaris: Localisation, purification and properties, Biochem. J., 229: 693-699.

42. Robbins, M.P. and Dixon, R.A., 1984, Induction of chalcone isomerase in elicitor-treated bean cells. Comparison of rates of synthesis and appearance of immunodetectable enzyme, _Eur. J. Biochem._, 145: 195-202.
43. Roby, D., Toppan, A. and Esquerré-Tugayé, M-T., 1985, Cell surfaces in plant microorganism interactions. V. Elicitors of fungal and of plant origin trigger the synthesis of ethylene and of cell wall hydroxyproline-rich glycoprotein in plants, _Plant Physiol._, 77: 700-704.
44. Dixon, R.A., Gerrish, C., Lamb, C.J. and Robbins, M.P., 1983, Elicitor-mediated induction of chalcone isomerase in _Phaseolus_ _vulgaris_ cell suspension cultures, _Planta_, 159: 561-569.
45. Gerrish, C., Robbins, M.P. and Dixon, R.A., 1985, _Trans_-cinnamic acid as a modulator of chalcone isomerase in bean cell suspension cultures, _Plant Sci. Lett._, 38: 23-27.

ACCUMULATION OF HYDROXYPROLINE-RICH GLYCOPROTEIN mRNAs IN BIOLOGICALLY STRESSED CELL CULTURES AND HYPOCOTYLS

A.M. Showalter[a], J.N. Bell[b], C.L. Cramer[b]
J.A. Bailey[c], J.E. Varner[a] and C.J. Lamb[b]

[a]Dept. of Biology, Washington University
St. Louis, MO, USA 63130
[b]Plant Biology Laboratory, Salk Institute for Biological Studies, P O Box 85800, San Diego, CA, USA 92138
[c]Long Ashton Research Station, University of Bristol, Long Ashton, Bristol BS18 9AF, UK

INTRODUCTION

Hydroxyproline-rich glycoproteins (HRGPs) are structural components of plant cell walls and may also function in the processes of plant development, growth, and disease resistance [reviewed in 1,2]. In higher plants, these cell wall HRGPs consist of 35-45% hydroxyproline (Hyp) and are also relatively rich in serine, valine, tyrosine and lysine. A repeating pentapeptide sequence, Ser-$(Hyp)_4$, further characterizes these unusual glycoproteins which are often referred to as "extensins". Most of the hydroxyproline residues are O-glycosidically attached to short oligoarabinosides, while some of the serine residues are O-glycosidically linked to galactose.

Several studies support a role for HRGPs in plant defence. First, these HRGPs are known to accumulate in melon plants infected with the anthracnose fungus *Colletotrichum lagenarium* [3]. Second, artificial enhancement or suppression of HRGP levels in melon plants results respectively in increased or decreased resistance to *Colletotrichum lagenarium* [4]. Third, cell wall hydroxyproline levels, indicative of HRGP levels, increase more rapidly in resistant cultivars than in susceptible cultivars of cucumber infected with the fungus *Cladosporium cucumerinum* [5]. Fourth, elicitor treatment of melon and soybean hypocotyls results in the stimulation of HRGP synthesis [6]. Similarly, elicitor treatment of bean cell cultures results in the accumulation of cell wall hydroxyproline [7]. Although the precise role(s) which these glycoproteins play in plant defence is unknown, it is likely that they are acting as specific agglutinins of microbial pathogens [8-10] and/or as structural barriers, either directly or indirectly by providing sites for lignin deposition [11].

Recently, cDNA and genomic clones encoding a carrot cell wall HRGP were isolated and sequenced [12,13]. The coding region of this carrot gene is 306 amino acids in size and specifies a signal peptide sequence, presumably essential for transport of the HRGP to the cell wall, and 25 Ser-$(Pro)_4$ repeats. In the HRGP precursor, the proline residues in these repeat units are post-translationally hydroxylated to form the Ser-$(Hyp)_4$ repeats

NATO ASI Series, Vol. H1
Biology and Molecular Biology of Plant-Pathogen Interactions
Edited by J. Bailey

A.

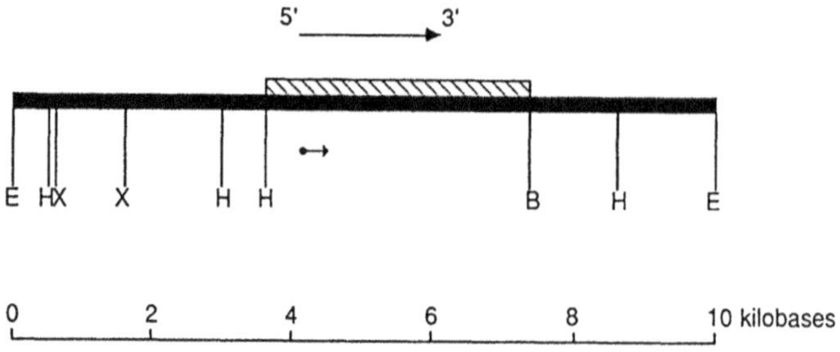

B.

```
TCTCCACCACCACCAAAAACCTTGCCTCCACCACCACCAAAAACCTCGCCTCCACCT
SerProProProProLysThrLeuProProProProProLysThrSerProProPro

CCTGTCCACTCACCACCACCACCACCGGTAGCATCACCTCCCCCCCCCGTGCACTCA
ProValHisSerProProProProProValAlaSerProProProProValHisSer

CCACCACCACCAGTAGCATCACCTCCACCTCCCGTCCACTCACCACCACCACCACCA
ProProProProValAlaSerProProProProValHisSerProProProProPro

GTAGCATCACCTCCACCTCCTGTCCACTCACCACCACCACCGGTAGCATCACCTCCC
ValAlaSerProProProProValHisSerProProProProValAlaSerProPro

CCTCCCGTCCACTCACCACCACCTCCAGTTCACTCACCACCACCACCAGTA
ProProValHisSerProProProProValHisSerProProProProVal
```

Figure 1. Restriction map (A) and partial DNA sequence (B) of the 10 kb EcoRI fragment of the tomato genomic HRGP clone Tom 5. The region of hybridization of this restriction fragment to carrot cDNA and genomic HRGP sequences is indicated by ▨. The direction of transcription is as shown, and the region of DNA sequence analysis is indicated by ●→ . The DNA was cut and labelled at an internal Hinfl site and read in a 5' to 3' direction off the coding strand. B, BamHI; E, EcoRI; H, HindIII; X, XbaI.

characteristic of cell wall HRGPs. Using these carrot HRGP clones as probes, we have isolated and partially characterized a tomato genomic HRGP clone [14]. We report here and elsewhere [15] on the molecular characterization of this tomato HRGP gene clone and its use as a probe in RNA blot hybridizations to investigate HRGP mRNA levels in elicitor-treated bean suspension cell cultures and in the race:cultivar-specific interactions between bean hypocotyls and the partially biotrophic fungus Colletotrichum lindemuthianum, the causal agent of anthracnose.

ISOLATION AND CHARACTERIZATION OF A TOMATO HRGP GENE CLONE

In order to study the structure and expression of the HRGP genes in plants, we have isolated and partially characterized a tomato genomic clone encoding a cell wall HRGP [14,15]. This clone was isolated by screening a tomato genomic DNA library with ^{32}P-labelled carrot cDNA and genomic HRGP sequences and subsequently named Tom 5. By Southern blot hybridization analysis, a 10 kilobase (kb) EcoRI restriction fragment of Tom 5 was observed to hybridize to the carrot HRGP sequences. This 10 kb DNA fragment was subjected to further restriction enzyme and Southern blot analyses and subsequently examined by DNA sequence analysis (Fig. 1). Such sequence analysis elucidated numerous Ser-$(Pro)_4$ units encoded by the gene, consistent with its identification as a cell wall HRGP gene clone. In the experiments described below, we have radioactively labelled this 10 kb EcoRI restriction fragment of Tom 5 with ^{32}P and used it as a probe to monitor HRGP mRNA levels in biologically stressed bean cell cultures and hypocotyls.

HRGP mRNA ACCUMULATION IN ELICITOR-TREATED BEAN CELL CULTURES

Bean (*Phaseolus vulgaris* L. cv. Canadian Wonder) suspension-cultured cells were treated with elicitor for varying lengths of time ranging from 0 to 36 h. The elicitor used here was isolated from mycelial cell walls of *Colletotrichum lindemuthianum* as a high molecular weight, heat-released fraction [16]. Following elicitor treatment, polysomal and total cellular RNA was isolated from the cells, electrophoresed on denaturing agarose gels, and blotted on to nitrocellulose paper [15]. These RNA or northern blots were subsequently hybridized to the ^{32}P-labelled tomato HRGP gene probe [15], and resulted in the hybridization of three HRGP mRNA species of approximately 5.6, 2.7, and 1.6 kb in size in the total cellular RNA pool (Fig. 2). Identically migrating RNA species were also observed in the polysomal RNA pool (data not shown). Furthermore, carrot cDNA and genomic HRGP sequences were used as probes in these same experiments and showed identical hybridization patterns to those detected with the tomato HRGP gene probe (data not shown).

Elicitor treatment of these cell cultures resulted in the marked accumulation of all three HRGP mRNA species (Fig. 2A and B). This accumulation was detected as early as 4 h after elicitor treatment and was most obvious in the case of the 2.7 kb transcript which represented the most intensely hybridizing RNA species. In mock-treated, unelicited control cells, these RNA species did not accumulate appreciably over time but remained at a low basal level (Fig. 2A). These data are consistent with previous studies [3-7] reporting the accumulation of the cell wall HRGPs in response to elicitor treatment and fungal infection of plant cells.

The accumulation pattern of the 2.7 kb HRGP mRNA species in the polysomal and total cellular RNA pools of elicitor-treated cells was compared to that of the mRNA encoding the phytoalexin biosynthetic enzyme, chalcone synthase [17]. Clearly, both the HRGP and chalcone synthase mRNAs accumulate in response to elicitor treatment, but the kinetics of their accumulation differ (Fig. 2C). Elicitor treatment causes a rapid but transient increase in chalcone synthase mRNA. In contrast, the HRGP mRNAs, both in the polysomal and total cellular RNA pools, display a less rapid but prolonged increase in mRNA levels. Other phytoalexin biosynthetic enzymes (phenylalanine ammonia-lyase and chalcone isomerase) also show mRNA accumulation patterns closely resembling those of chalcone synthase [18-20; unpublished observations]. The differences between the patterns of mRNA accumulation for HRGP and the phytoalexin biosynthetic enzymes may reflect the specific distinct roles which HRGPs and phytoalexins have as apparent defence molecules.

In collaboration with Dr. J. Ebel (Albert-Ludwigs-Universität, Freiburg) we have also demonstrated the accumulation of HRGP mRNA in soybean cell cultures treated with *Phytophthora megasperma* glucan elicitor (unpublished results). The accumulation kinetics in these soybean cell cultures were very similar to those described above.

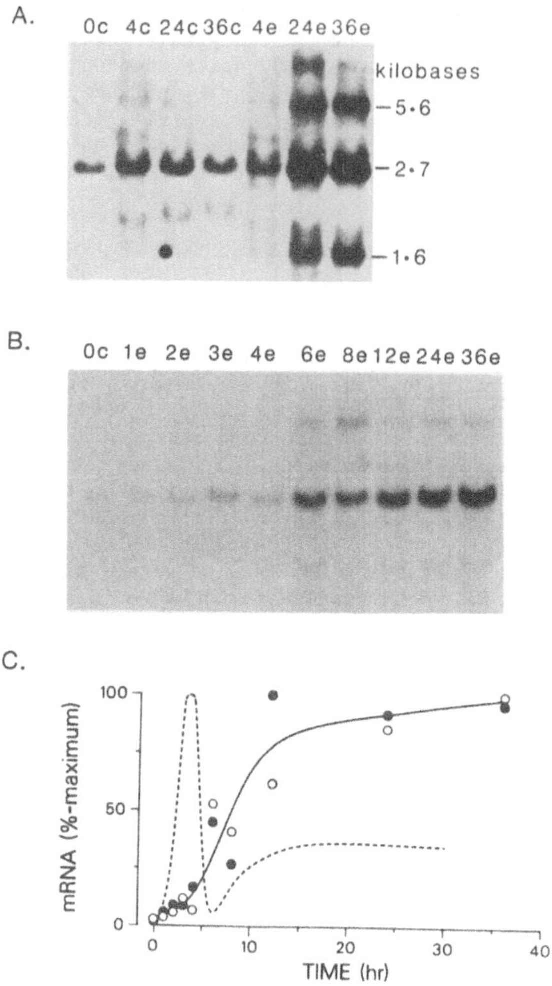

Figure 2. Accumulation of HRGP mRNA in elicitor-treated bean cells. RNA was blot hybridized with ^{32}P-labelled tomato HRGP genomic sequences. (A and B) Total cellular RNA isolated from cells at the times (in hours) indicated after elicitor treatment (e) or from equivalent mock-treated unelicited control cells (c). (C) Kinetics of elicitor-induced accumulation of the HRGP mRNA species size 2.7 kb in total cellular RNA (O) and polysomal RNA fractions (●). Dotted line denotes the kinetics previously observed [17] for the accumulation of mRNA encoding the phytoalexin biosynthetic enzyme chalcone synthase in the same set of RNAs.

HRGP mRNA ACCUMULATION IN INFECTED BEAN HYPOCOTYLS

In the experiments reported here, we chose to investigate the race: cultivar-specific interactions betwen *Phaseolus vulgaris* cv. Kievitsboon Koekoek hypocotyls and two races of *Colletotrichum lindemuthianum*, β and γ, which lead to incompatible (host resistant) and compatible (host susceptible) reactions respectively. The physiology and cytology of these interactions have been well characterized [21,22]. Following application of fungal spores on the unwounded surface of hypocotyls (a natural site of infection for this

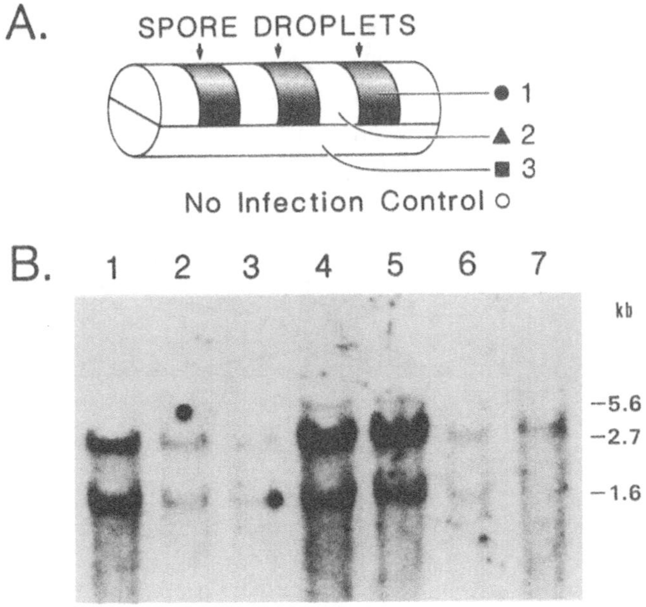

Figure 3. Pattern of accumulation of HRGP mRNAs in hypocotyls of bean cv. Kievitsboon Koekoek during race:cultivar-specific interactions with physiological races of Colletotrichum lindemuthianum. (A) Dissection of hypocotyl tissue. (B) Autoradiograph of RNA blot hybridized with ^{32}P-labelled HRGP genomic sequences. Lanes: 1-3, RNA from hypocotyls 79 h after inoculation with spores of the incompatible race β; 4-6, RNA from hypocotyls 150 h after inoculation with spores of the compatible race γ. RNA was isolated from site 1(lanes 1 and 4), site 2 (lanes 2 and 5), site 3 (lanes 3 and 6). For comparison, lane 7 contains RNA isolated from suspension-cultured cells 7 h after elicitor treatment.

fungus), there is a period of 2-3 days during which time the spores germinate and the fungus enters the cell. It is at this time when differences between the two interactions become evident. In the incompatible interaction, localized host cell death occurs and appears as scattered flecks on the hypocotyl surface. This so-called hypersensitive response leads to restriction of further fungal growth and is correlated with the localized accumulation of phytoalexins. In contrast, in the compatible interaction there is no visible host response and the fungus continues to grow biotrophically for an additional 4-5 days, extensively colonizing the hypocotyl tissue. After this period of extensive biotrophic growth (approximately 6-7 days after spore inoculation), widespread host cell death occurs and brown, watery, spreading anthracnose lesions are produced. Phytoalexins also accumulate at this time correlated with attempted lesion limitation. Depending upon the physiological status of the plant and environmental conditions, considerable necrotrophic fungal growth can continue eventually leading to hypocotyl rotting and plant death or restriction of lesion development can occur preventing extensive hypocotyl rotting and plant death [23].

Our studies here were accomplished by inoculating hypocotyls with spores of the two fungal races, isolating RNA from the infected hypocotyl tissues, and then examining this RNA by blot hybridization to ^{32}P-labelled tomato

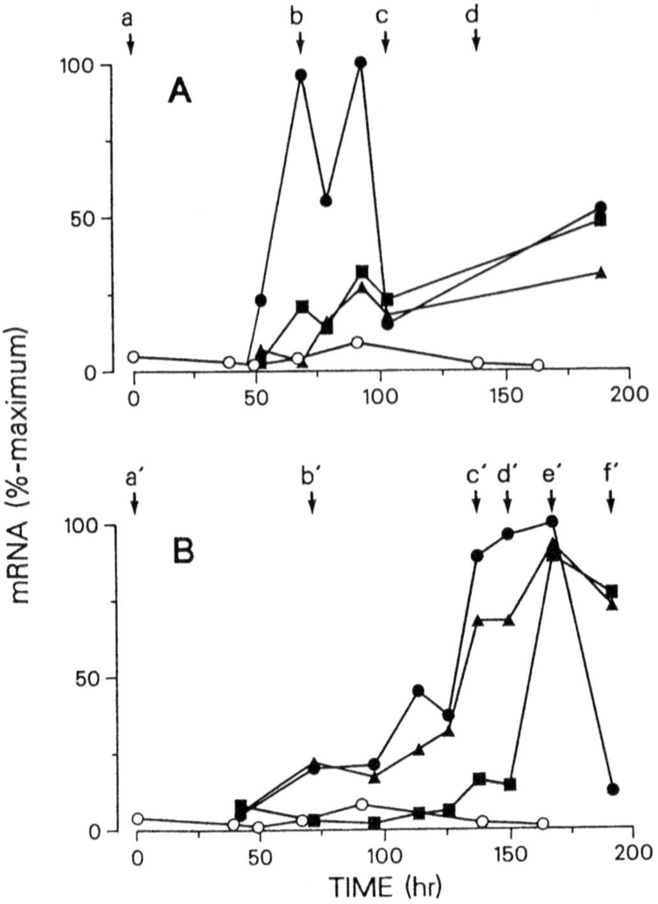

Figure 4. Kinetics of accumulation of the HRGP mRNA 2.7 kb species in hypocotyls of bean cv. Kievitsboon Koekoek during race:cultivar-specific interactions with physiological races of Colletotrichum lindemuthianum. (A) Incompatible interaction with race β. (B) Compatible interaction with race γ. RNA was isolated from site 1 (●), site 2 (▲), site 3 (■), and equivalent uninfected hypocotyls (○). Arrows in A denote events in the expression of hypersensitive resistance at site 1: a, spore inoculation; b, onset of hypersensitive flecking in a few sites; c, hypersensitive flecking apparent at most sites; d, very dense brown flecking at all sites. No visible changes occurred in sites 2 and 3 and control hypocotyls throughout the time course. Arrows in B denote events in lesion development at site 1: a', spore inoculation; b', no visible symptoms (cf. incompatible interaction); c', onset of symptom development at a few sites; d', pale to medium brown lesions apparent at most sites; e', onset of water soaking and development of spreading lesions; f', extensive water soaking and spreading of lesions from site 1, some browning at site 2.

HRGP gene sequences [15]. Three specific regions of the infected hypocotyl tissue were used for RNA isolation; site 1: tissue at and directly underlying the infection site; site 2: tissue laterally adjacent to the infection site; and site 3: tissue directly beneath sites 1 and 2 (Fig. 3A).

In the incompatible interaction (host resistant) with Colletotrichum lindemuthianum, race β, HRGP mRNA markedly accumulated early in the infection, concomitant with the expression of the hypersensitive response. This accumulation involved the 1.6 and 2.7 kb HRGP mRNA transcripts, previously observed in the cell culture RNAs; however, the 5.6 kb species was not detected (Fig. 3B). This accumulation was first observed 52 h after inoculation and was most pronounced in tissue at the infection site (site 1) (10-20 fold more HRGP mRNA hybridization than in equivalent, uninfected control hypocotyls) (Fig. 4A). Adjacent tissue (sites 2 and 3), although removed from the localized region of fungal growth [22], also responded to the infection albeit to a lesser extent (5-6 fold more HRGP mRNA hybridization than in equivalent, uninfected control hypocotyls) (Fig. 4A). This observation indicates that some transmissable signal(s) is involved with the systemic activation of molecular responses to fungal infection. Such a signal may be involved with the phenomenon of induced systemic resistance [24,25].

In the compatible interaction (host susceptible) with Colletotrichum lindemuthianum, race γ, HRGP mRNA also accumulated but with different kinetics from the incompatible interaction. Specifically, marked accumulation of HRGP mRNA occurred somewhat later in the compatible interaction than in the incompatible interaction and was correlated with the onset of lesion formation (Fig. 4). This accumulation involved increases in the 2.7 kb species, the major hybridizing HRGP mRNA species, the 1.6 kb species, and weak but detectable increases in the 5.6 kb species. Although HRGP mRNA accumulation was greatest at the infection site (site 1), considerable accumulation was also detected in sites 2 and 3. Maximum accumulation of HRGP mRNA in the incompatible interaction (site 1, 93 h) was approximately 80% of that found in the compatible interaction (site 1, 168 h). Interestingly, the 3 distinct HRGP mRNA species accumulate to different degrees in the incompatible and compatible interactions; the significance of these differences, however, is unknown.

As with the cell cultures, HRGP mRNA accumulation in both the incompatible and compatible interactions occurred later than the accumulation of mRNAs encoding several phytoalexin biosynthetic enzymes (phenylalanine ammonia-lyase, chalcone synthase, and chalcone isomerase) [19; unpublished results]. Moreover, the accumulation of mRNA in the incompatible interaction relative to the compatible interaction was more pronounced for HRGPs than for the phytoalexin biosynthetic enzymes. Again, these differences may reflect the specific distinct roles which phytoalexins and HRGPs have in plant defence.

CONCLUSIONS

HRGP mRNAs markedly accumulate in elicitor-treated bean suspension cell cultures and in race:cultivar-specific interactions between bean hypocotyls and the partially biotrophic fungus Colletotrichum lindemuthianum, causal agent of anthracnose. Moreover, these HRGP mRNAs accumulate with different kinetics in the race:cultivar-specific interactions. Specifically, in an incompatible interaction (host resistant) there is an early increase in HRGP mRNAs correlated with hypersensitive resistance; whereas, in a compatible interaction (host susceptible) marked accumulation of HRGP mRNAs occurs as a delayed response at the onset of lesion formation, possibly reflecting an attempt to limit lesion development. These data are consistent with previous studies [3-7] demonstrating the accumulation of HRGPs in elicitor-treated and fungally infected plant cells.

The pattern of accumulation for HRGP mRNAs is broadly similar to that observed for mRNAs encoding several phytoalexin biosynthetic enzymes, including phenylalanine ammonia-lyase, chalcone synthase, and chalcone isomerase which have been analyzed in these same systems [15, 17-20; unpublished results]. Some differences in pattern detail between these two responses, however, are apparent. For example, the phytoalexin biosynthetic enzyme mRNAs accumulate more rapidly in both the elicitor-treated cell cultures and in the race:cultivar-specific interactions than the HRGP mRNAs. Also, a more pronounced accumulation of HRGP mRNAs in the incompatible interaction relative to the compatible interaction is observed than in the case of the phytoalexin biosynthetic enzyme mRNAs. Such differences may reflect the specific distinct roles which phytoalexins and HRGPs have in plant defence associated with their functional properties as toxic natural products and structural cell wall glycoproteins respectively.

In our studies, we have also observed that HRGP mRNAs and phytoalexin biosynthetic enzyme mRNAs accumulate in uninfected tissue distant from the site of fungal inoculation, indicating the existence of some intercellular transmission signal(s). Such a signal apparently can activate defence responses in advance of the infection zone and hence may be involved with the establishment and expression of induced systemic resistance [24,25].

Finally, from this and other molecular studies on plant-pathogen interactions [reviewed in 26,27], it is becoming increasingly clear that rapid selected alterations in the pattern of plant gene expression occur in response to pathogen attack. At least some of these alterations are brought about by inducing the transcription of specific defence response genes [18,28]. Thus, it should be possible to isolate, manipulate, and reintroduce these specific, but modified, genes into plants to ascertain the relative importance of such defence gene products _in vivo_ and to determine the efficacy of this approach in genetically engineering plants with superior disease resistance qualities.

ACKNOWLEDGEMENTS

We gratefully acknowledge the generous gifts of carrot HRGP cDNA and genomic clones from J.A. Chen and the tomato genomic DNA library from R.W. Breidenbach. This work was supported by research grants to J.E.V. from the National Science Foundation (PCM8104516), Department of Energy (DE-FG 02-84ER13255), and an unrestricted grant from the Monsanto Company, and by a grant to C.J.L. from the Samuel Robert Noble Foundation. C.L.C. is a National Science Foundation Research Fellow in Plant Biology.

REFERENCES

1. Cooper, J.B., Chen, J.A. and Varner, J.E., 1984, The glycoprotein component in plant cell walls, _in_: "Structure, Function and Biosynthesis of Plant Cell Walls", W.M. Dugger and S. Bartnicki-Garcia, eds., pp. 75-88, American Society of Plant Physiologists, Rockville, MD.
2. Lamport, D.T.A. and Catt, J.W., 1981, Glycoproteins and enzymes of the cell wall, _in_: "Encyclopedia of Plant Physiology", Vol. 13B, W. Tanner and F.A. Loewus, eds., pp. 133-165, Springer-Verlag, New York.
3. Esquerré-Tugayé, M.T. and Lamport, D.T.A., 1979, Cell surfaces in plant-microorganism interactions. I. A structural investigation of cell wall hydroxyproline-rich glycoproteins which accumulate in fungus-infected plants, _Plant Physiol._, 64: 314-319.
4. Esquerré-Tugayé, M.T., Lafitte, C., Mazau, D., Toppan, A. and Touzé, A., 1979, Cell surfaces in plant-micro-organism interactions. II. Evidence for the accumulation of hydroxyproline-rich glycoproteins in the cell wall of diseased plants as a defense mechanism, _Plant Physiol._, 64: 320-326.

5. Hammerschmidt, R., Lamport, D.T.A. and Muldoon, E.P., 1984, Cell wall hydroxyproline enhancement and lignin deposition as an early event in the resistance of cucumber to Cladosporium cucumerinum, Physiol. Plant Pathol., 24: 43-47.
6. Roby, D., Toppan, A. and Esquerré-Tugayé, M.T., 1985, Cell surfaces in plant-microorganism interactions. V. Elicitors of fungal and of plant origin trigger the synthesis of ethylene and of cell wall hydroxyproline-rich glycoprotein in plants, Plant Physiol., 77: 700-704.
7. Bolwell, G.P., Robbins, M.P. and Dixon, R.A., 1985, Metabolic changes in elicitor-treated bean cells. Enzymic responses associated with rapid changes in cell wall components, Eur. J. Biochem., 148: 571-578.
8. Van Holst, G.J. and Varner, J.E., 1984, Reinforced polyproline II conformation in a hydroxyproline-rich cell wall glycoprotein from carrot root, Plant Physiol., 74: 247-251.
9. Leach, J.E., Cantrell, M.A. and Sequeira, L., 1982, A hydroxyproline-rich bacterial agglutinin from potato : extraction, purification and characterization, Plant Physiol., 70: 1353-1358.
10. Mellon, J.E. and Helgeson, J.P., 1982, Interaction of a hydroxyproline-rich glycoprotein from tobacco callus with potential pathogens, Plant Physiol., 70: 401-405.
11. Whitmore, F.W., 1978, Lignin-protein complex catalysed by peroxidase, Plant Sci. Lett., 13: 241-245.
12. Chen, J. and Varner, J.E., 1985, Isolation and characterization of cDNA clones for carrot extensin and a proline-rich 33-kDa protein, Proc. Natl. Acad. Sci. USA., 82: 4399-4403.
13. Chen, J. and Varner, J.E., 1985, An extracellular matrix protein in plants: characterization of a genomic clone for carrot extensin, EMBO J. 4: 2145-2152.
14. Showalter, A.M. and Varner, J.E., 1986, Molecular details of plant cell wall hydroxyproline-rich glycoprotein expression during wounding and infection, in: "Molecular Strategies for Crop Protection", C.J. Arntzen and C.A. Ryan, eds., Alan R. Liss Inc., New York.
15. Showalter, A.M., Bell, J.N., Cramer, C.L., Bailey, J.A., Varner, J.E. and Lamb, C.J., 1985, Accumulation of hydroxyproline-rich glycoprotein mRNAs in response to fungal elicitor and infection, Proc. Natl. Acad. Sci. USA., 82: 6551-6555.
16. Anderson-Prouty, A.J. and Albersheim, P., 1975, Host-pathogen interactions. VIII. Isolation of a pathogen-synthesized fraction rich in glucan that elicits a defense response in the pathogen's host, Plant Physiol., 56: 286-291.
17. Ryder, T.B., Cramer, C.L., Bell, J.N., Robbins, M.P., Dixon, R.A. and Lamb, C.J., 1984, Elicitor rapidly induces chalcone synthase mRNA in Phaseolus vulgaris cells at the onset of phytoalexin defense response, Proc. Natl. Acad. Sci. USA., 81: 5724-5728.
18. Cramer, C.L., Ryder, T.B., Bell, J.N. and Lamb, C.J., 1985, Rapid switching of plant gene expression induced by fungal elicitor, Science, 227: 1240-1243.
19. Cramer, C.L., Bell, J.N., Ryder, T.B., Bailey, J.A., Schuch, W., Bolwell, G.P., Robbins, M.P., Dixon, R.A. and Lamb, C.J., 1985, Coordinated synthesis of phytoalexin biosynthetic enzymes in biologically-stressed cells of bean (Phaseolus vulgaris L.), EMBO.J. 4: 285-289.
20. Edwards, K., Cramer, C.L., Bolwell, G.P., Dixon, R.A., Schuch, W. and Lamb, C.J., 1985, Rapid transient induction of phenylalanine ammonia-lyase mRNA in elicitor treated bean cells, Proc. Natl. Acad. Sci. USA., 82: 6731-6735.
21. O'Connell, R.J., Bailey, J.A. and Richmond, D.V., 1985, Cytology and physiology of infection of Phaseolus vulgaris by Colletotrichum lindemuthianum, Physiol. Plant Pathol., 27: 75-98.

22. Bailey, J.A., 1982, Physiological and biochemical events associated with the expression of resistance to disease, in: "Active Defense Mechanisms in Plants", R.K.S. Wood, ed., pp. 39-65, Plenum Press, New York.
23. Rowell, P.M. and Bailey, J.A., 1983, The influence of cotyledons, roots and leaves on the susceptibility of hypocotyls of bean (Phaseolus vulgaris) to compatible races of Colletotrichum lindemuthianum, Physiol. Plant Pathol., 23: 245-256.
24. Sequeira, L., 1983, Mechanisms of induced resistance on plants, Annu. Rev. Microbiol., 37: 51-79.
25. Dean, R.A. and Kuć, J., 1985, Induced systemic resistance protection in plants, Trends in Biotechnol., 3: 125-129.
26. Lamb, C.J., Bell, J.N., Cramer, C.L., Dildine, S.L., Grand, C., Hedrick, S.A., Ryder, T.B. and Showalter, A.M., 1985, Molecular response of plants to infection, in: "Biotechnology for Solving Agricultural Problems", P.C. Augustine, H.D. Danforth and M.R. Bakst, eds., Rowman and Allanheld, Totowa, New Jersey.
27. Hahlbrock, K., Chappell, J., Johnen, W. and Walter, M., 1985, Early defense reactions of plants to pathogens, in: "Molecular Form and Function of the Plant Genome", L. van Vloten-Doting, G.S.P. Groot and T.C. Hall, eds., pp. 129-140, Plenum Press, New York.
28. Chappell, J. and Hahlbrock, K., 1984, Transcription of plant defence genes in response to U.V. light or fungal elicitor, Nature, 311: 76-78.

BIOCHEMICAL STUDY OF HYDROXYPROLINE-RICH GLYCOPROTEINS IN PLANT-PATHOGEN INTERACTIONS

D. Mazau, D. Rumeau and M.T. Esquerré-Tugayé

Université Paul Sabatier
Centre de Physiologie Végétale - UA 241 CNRS
118, route de Narbonne
31062 Toulouse, France

INTRODUCTION

The cell surfaces of a host plant and of a pathogen play an important and complex role during pathogenesis. Firstly, recognition events are supposed to involve specific sites of the cell wall or plasmalemma of the two partners. Secondly, intense degradation of host plant cell walls by fungal bacterial hydrolases has often been reported during the course of infection; in view of the occurrence of chitinase and β-1,3 glucanase in plants [1,2, 3], there are no reasons to believe a priori that chitin or glucan containing pathogen cell surfaces are not degraded as well, at least to some extent. Thirdly, active defence responses take place at the cell wall level of host plants [4]. It is likely, although not much documented yet, that pathogens have the ability to develop their own protection.

Despite their obvious importance, our knowledge of these events remains scanty, mainly because the biochemistry of the cell surfaces themselves e.g. structure, porosity, pH, charge density, is very complex.

Our approaches to understanding at the cellular and molecular levels a cell surface response of host plants and its elicitation by cell surface components of fungal pathogens is reported in this paper. We will consider the enrichment of plant cell walls in hydroxyproline-rich glycoproteins.

HYDROXYPROLINE-RICH GLYCOPROTEINS OF HIGHER PLANTS

The accumulation of hydroxyproline rich glycoproteins (HRGPs) was first demonstrated in the cell wall of melon seedlings inoculated by Colletotrichum lagenarium [5]. A clear correlation was found between this response and an increased defence of melons against the fungus [6].

Two distinct HRGPs : $HRGP_1$ and $HRGP_2$ have been isolated from melon callus and seedlings by ion exchange chromatography of CM-Sepharose [7]. The two glycoproteins are different from each other by their chemical properties: sugar to protein ratio, chemical composition and structural markers. The sugar moiety of $HRGP_1$ represents 94% of the molecule and is composed of arabinose (66%) and galactose (34%). In the protein moiety (6%), hydroxyproline, serine, alanine and glycine are the most abundant amino acid

NATO ASI Series, Vol. H1
Biology and Molecular Biology of Plant-Pathogen Interactions
Edited by J. Bailey

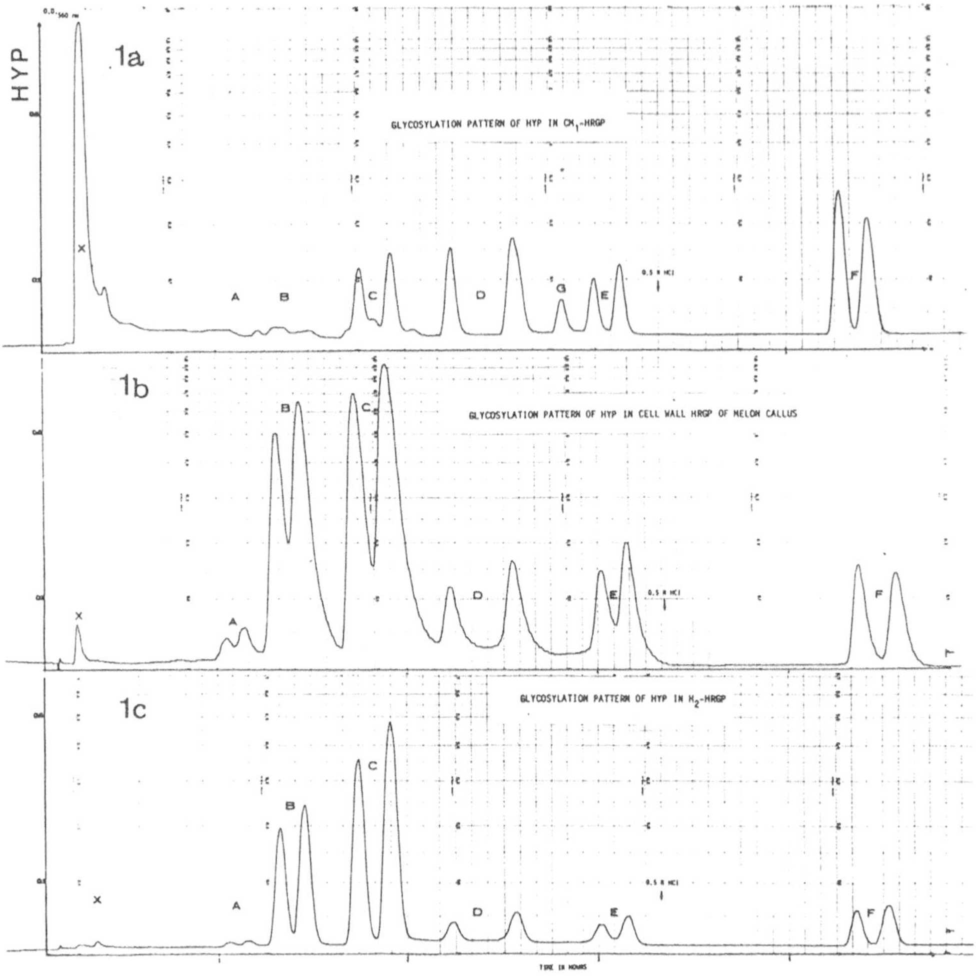

Figure 1. Glycosylation pattern of Hyp in $HRGP_1$ (Fig. 1a), cell wall HRGP (1b) and $HRGP_2$ (1c) : X = Hyp-oligosaccharide, A= Hyp-Ara_5, B = Hyp-Ara_4, C = Hyp-Ara_3, D = Hyp-Ara_2, E = Hyp-Ara, F = Hyp (cis and trans), G = undetermined glycosylated Hyp.

residues (45, 12, 10 and 7% respectively). As shown in Figure 1a, a major hyp-oligosaccharide and hyp-arabinosides are found in the molecule. The overall composition of $HRGP_1$ is typical of an arabinogalactan protein.

The sugar to protein ratio of $HRGP_2$ is 64:36, with arabinose, galactose and glucose as the recovered sugar residues (respectively 77, 22 and 1% of the sugar moiety). The protein moiety is characterized by its high amounts of hydroxyproline (39%), lysine (15%) and tyrosine (7%). These three amino acids can be considered as structural markers since they confer on the molecule its high glycosylation, high basicity and an assumed high crosslinkage. As such, this glycoprotein resembles the HRGP isolated from carrot root discs by Stuart and Varner [8], the P_2-HRGP isolated from tomato cells by Smith et al. [9], and HRGP agglutinins isolated from potato

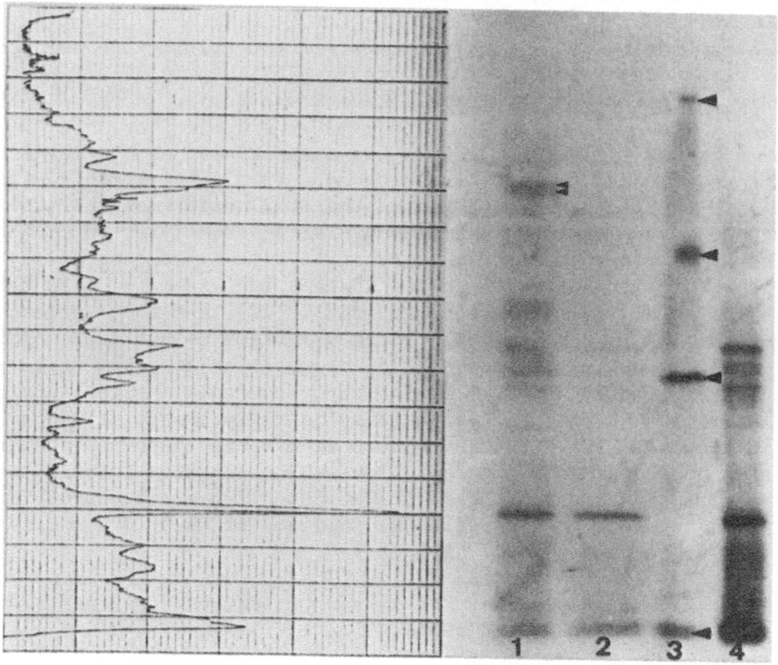

Figure 2. Gel electrophoresis of in vitro translation products of Poly A^+RNA fractions eluted from an oligo(dG) cellulose column. Lanes 1,2 and 4 represent poly A^+RNA which hybridize (=cytosine-rich mRNA), weakly hybridize, or did not hybridize to the column, respectively; lane 3 = MW markers. A densitometric scan of lane 1 is shown on the left part of the figure.

[10] and tobacco [11]. The high basicity of HRGP allowed us to further separate it into three components [12] by affinity chromatography on a heparin-ultrogel column. Although different from each other, the three glycoproteins $HRGP_{2b}$ and $_{2c}$ exhibit the same glycosylation profile as the wall, extensin like, HRGP (Fig. 1b). In particular $HRGP_{2b}$ (Fig. 1c) which is the most abundant, contains the Hyp arabinosides Hyp-$(Ara)_4$ to 1 which typify extensin, and the newly discovered Hyp-$(Ara)_5$. After deglycosylation [13], $HRGP_{2b}$ has an apparent MW comprising between 52 000 and 56 000 as estimated by SDS-PAGE electrophoresis. $HRGP_{2b}$ is also very antigenic.

The high amounts of tyrosine could be responsible for the crosslinkage of the molecule within itself, and to other wall components. So far, intramolecular isodityrosine bonds have been found in tomato and carrot HRGP [14,15,16]. Such tight bonds could account for the strong binding and the insolubility of melon HRGPs once they are in the cell wall.

Thus, four distinct HRGP precursors, pertaining to two different families are presently identified in melons. The HRGP which accumulates in the cell wall of infected melons is thought to be of the $HRGP_2$, extensin like, family since it has the same glycosylation profile [17]. This allows us to calculate that the amount of HRGP increases from 0.5% in healthy tissues to about 5 to 15% of the wall dry weight in infected or stressed melon tissues.

mRNAs OF HYDROXYPROLINE RICH GLYCOPROTEINS

The accumulation of HRGP in the cell wall upon infection results from an increased de novo synthesis of the molecule [18]. It is likely that prolyl hydroxylase and glycosyltransferase activities are also increased [19].

The increased synthesis of HRGPs has been studied at the level of their mRNAs. We have looked for cytosine rich mRNA in view of the high amounts of hyp (40%) in the protein moiety and of the CC"X" nature of the proline (hyp precursor) codons. Total RNA has been extracted from infected melons and healthy controls, four days after inoculation with C. lagenarium, by the phenol-chloroform method [20]. Poly A^+RNA has been isolated from this extract by oligo(dT) cellulose chromatography [21], and then enriched in cytosine-rich mRNA by hybridization to an oligo(dG) column [22]. This procedure has allowed us to recover 0.5% of cytosine rich mRNA from poly A^+ RNA in infected plants, and 0.1% from healthy controls. After in vitro translation with a wheat germ system (BRL) in the presence of ^{3}H-proline, only the C-rich mRNA from infected plants direct the synthesis of a major proline rich peptide. This peptide which was detected by SDS-PAGE and fluorography, has an apparent MW of 56 000 (Fig. 2); a densitometric scan of the corresponding lane even reveals two peptides of MW 56 000 and 54 500 within this peptide band.

Other proline rich peptides are also detected, but they are not unique to infected plants or to C-rich mRNAs.

These data indicate that an increased HRGP level in the cell wall is accompanied by an increase of translatable mRNAs coding for proline-rich peptides. Modification of the genome activity which leads to this increase in melons is presently under study.

HYDROXYPROLINE RICH GLYCOPROTEINS IN THE CELL WALL OF INFECTED PLANTS

It is obvious, when the level of HRGP reaches 5 to 15% of the wall upon infection (versus 0.5% in controls), that such an enrichment leads to dramatic changes in cell wall properties. As a wall polymer, HRGP reinforces the strength of the cell surface, and as a polycation, it modifies the net charge of the wall. It can also act as an agglutinin towards negatively charged pathogens [10,11]. These three properties would be sufficient to account for an active defence reaction of infected plants against pathogens. Indeed, a good correlation was first observed in melons where early enrichment of the cell wall in HRGP was associated with an induced resistance against C. lagenarium, while inhibition of its synthesis was followed by an increased susceptibility [6].

From the more recent analysis of several plant pathogen systems, it appears that an enrichment of the cell wall in HRGP occurs in several plants infected by different pathogens, either fungi, bacteria or viruses [12]. Among the 25 different combinations which have been analyzed so far, significant increases are only found in infected dicotyledons. In systems presenting race-cultivar specificity, this response resembles the phytoalexin response in that it appears earlier and to a greater extent in resistant cultivars than in susceptible ones [12,23,24]. As in the case of phytoalexins, HRGP synthesis is also increased under stress conditions [25].

The precise location of HRGP within the wall of healthy plants, and the sites where it accumulates upon infection remain to be determined. It is hoped that the availability of antibodies will permit the visualization of the molecule and the extent to which it accumulates around lesions.

ELICITATION OF HYDROXYPROLINE RICH GLYCOPROTEINS

The regulatory events which lead to an increased synthesis of HRGP are not yet known. We have found that this synthesis can be triggered *in vitro* by elicitors of fungal and plant origins.

The elicitors which are isolated from the cell wall can be defined as sugar containing molecules. Within this class of molecules, it seems that elicitors of different origin, and with different structures, like the hepta-β-glucoside of *Phytophthora megasperma* var. *glycinea* [26], and the pectic plant endogenous elicitor [27,28] may have the same activity. It was recognized, several years ago, that the activity of fungal elicitors is not restricted to the elicitation of phytoalexins [29].

The elicitors of *C. lagenarium* are glycopeptides with phosphate residues [30]. At μg amounts, these elicitors and elicitors isolated from *Phytophthora megasperma* elicit in their respective hosts (melon, soybean) the synthesis of HRGP [29,31]. At least 12 h are necessary for elicitation of HRGP to occur. Melon endogenous elicitors isolated from the cell wall of healthy melons also elicit the glycoprotein.

The messages involved in the elicitation of HRGP or of other molecules (phytoalexins, chitinases, proteolytic inhibitors ...) are not known yet, because few studies have been devoted to understanding the effect of elicitors at the cellular level. Ethylene might play a role, since elicitors stimulate the production of ethylene [30], and ethylene alone promotes the synthesis of HRGP [6]. But elicitors have many other properties, notably they depolarize plant membranes [25] and, in the melon-*C. lagenarium* system, they interact with melon sap lectins [32]. Whether a link does exist between these different events is not yet known. Elicitation is probably a cascade of complex events.

CONCLUSION

The accumulation of HRGP is an example which shows that a host plant modifies its surface upon interaction with the pathogen cell surface. The cellular and molecular engineering which is now available in plant science will help our understanding of this interaction.

REFERENCES

1. Abeles, F.B., Bosshart, R.P., Forrence, L.E. and Habig, W.H., 1970, Preparation and purification of glucanase and chitinase from bean leaves, *Plant Physiol.*, 47: 129-134.
2. Nichols, J.E., Beckman, J.M. and Hadwiger, L.A., 1980, Glycosidic enzyme activity in pea tissue and pea-*Fusarium solani* interactions, *Plant Physiol.*, 66: 199-204.
3. Toppan, A. and Roby, D., 1982, Chitinase activity in melon seedlings inoculated with *Colletotrichum lagenarium* (Pass.) Ell. et Halst, or treated with ethylene, *Agronomie*, 2: 829-834.
4. Bell, A., 1981, Biochemical mechanisms of disease resistance, *Annu. Rev. Plant Physiol.*, 32: 21-81.
5. Esquerré-Tugayé, M.T., 1973, Effect of a parasitic disease on the content of hydroxyproline in cell walls of melon plant epicotyls and petioles, *C.R Acad. Sci.*, Paris, 276: 525-528.
6. Esquerré-Tugayé, M.T., Laffite, C., Mazau, D., Toppan, A. and Touzé, A., 1979, Cell surfaces in plant-microorganism interactions. II. Evidence

for the accumulation of hydroxyproline-rich glycoproteins in the cell wall of diseased plants as a defense mechanism, Plant Physiol., 64: 320-326.

7. Mazau, D. and Esquerré-Tugayé, M.T., 1986, Physiol. Plant Pathol. (accepted for publication).
8. Stuart, D.A. and Varner, J.E., 1980, Purification and characterization of a salt-extractable hydroxyproline-rich glycoprotein from aerated carrot discs, Plant Physiol., 66: 787-792.
9. Smith, J.J., Muldoon, E.P. and Lamport, D.T.A., 1984, Isolation of extensin precursors by direct elution of intact tomato cell suspension cultures, Phytochemistry, 23: 1233-1239.
10. Leach, J.E., Cantrell, M.A. and Sequeira, L., 1982, Hydroxyproline-rich bacterial agglutinin from potato. Extraction, purification and characterization, Plant Physiol., 70: 1353-1358.
11. Mellon, J.E. and Helgeson, J.P., 1982, Interaction of a hydroxyproline-rich glycoprotein from tobacco callus with potential pathogens, Plant Physiol., 70: 401-405.
12. Mazau, D. and Esquerré-Tugayé, M.T., 1986, (manuscript submitted).
13. Edge, A.S.B., Faltynck, C.R., Hof, L., Reichert, L.E. and Weber, P., 1981, Deglycosylation of glycoproteins by trifluoromethanesulfonic acid, Anal. Biochem., 118: 131-137.
14. Fry, S.C., 1982, Isodityrosine, a new cross-linking amino acid from plant cell-wall glycoprotein, Biochem. J., 204: 449-455.
15. Epstein, L. and Lamport, D.T.A., 1984, An intramolecular linkage involving isodityrosine in extensin, Phytochemistry, 23: 1241-1246.
16. Cooper, J.B. and Varner, J.E., 1984, Cross-linking of soluble extensin in isolated cell walls, Plant Physiol., 76: 414-417.
17. Esquerré-Tugayé, M.T. and Mazau, D., 1981, Les glycoproteines à hydroxyproline de la paroi végétale, Physiol. Veg., 19: 415-426.
18. Toppan, A., Roby, D. and Esquerré-Tugayé, M.T., 1982, Cell surfaces in plant-microorganism interactions. III. In vivo effect of ethylene on hydroxyproline-rich glycoprotein accumulation in the cell wall of diseased plants, Plant Physiol., 70: 82-86.
19. Bolwell, G.P., 1984, Different patterns of arabinosylation by membranes of suspension-cultured cells of Phaseolus vulgaris (French bean) after subculture or elicitation, Biochem. J., 222: 427-435.
20. Mozer, T.J., 1980, Partial purification and characterization of the mRNA for α-amylase from barley aleurone layers, Plant Physiol., 65: 834-837.
21. Aviv, H. and Leder, P., 1972, Purification of biologically active globin messenger RNA by chromatography on oligothymidylic acid-cellulose, Proc. Natl. Acad. Sci. USA., 69: 1408-1412.
22. Rumeau, D., Mazau, M. and Esquerré-Tugayé, M.T., 1986, (manuscript submitted).
23. Hammerschmidt, R., Lamport, D.T.A. and Muldoon, E.P., 1984, Cell wall hydroxyproline enhancement and lignin deposition as an early event in the resistance of cucumber to Cladosporium cucumerinum, Physiol. Plant Pathol., 24: 43.47.
24. Showalter, A.M., Bell, J.N., Cramer, C.L., Bailey, J.A., Varner, J.E. and Lamb, C.J., 1985, Accumulation of hydroxyproline-rich glycoprotein mRNAs in response to fungal elicitor and infection, Proc. Natl. Acad. Sci. USA., 82: 6551-6555.
25. Esquerré-Tugayé, M.T., Mazau, D., Pelissier, B., Roby, D., Rumeau, D. and Toppan, A., 1985, in: "Cellular and Molecular Biology of Plant Stress", J. Key and T. Kosuge, eds., pp. 459-473, Alan R. Liss, Inc.
26. Darvill, A.G. and Albersheim, P., 1983, Phytoalexins and their elicitors - a defense against microbial infection in plants, Annu. Rev. Plant Physiol., 35: 243-275.
27. Nothnagel, E.A., McNeil, M., Albersheim, P. and Dell, A., 1983, Host-pathogen interactions. XXII. A galacturonic acid oligosaccharide from plant cell walls elicits phytoalexins, Plant Physiol., 71: 916-926.

28. Ryan, C.A., Bishop, P., Pearce, G., Darvill, A.G. and Albersheim, P., 1981, A sycamore cell wall polysaccharide and a chemically related tomato leaf polysaccharide possess similar proteinase inhibitor-inducing activities, Plant Physiol., 68: 616-618.
29. Esquerré-Tugayé, M.T., Mazau, D., Toppan, D. and Roby, D., 1980, Ethylene elicitation of cell wall glycoproteins associated with plant defense, Ann. Phytopathology, 12: 403-411.
30. Toppan, A. and Esquerré-Tugayé, M.T., 1984, Cell surfaces in plant-microorganism interactions. IV. Fungal glycopeptides which elicit the synthesis of ethylene in plants, Plant Physiol., 75: 1133-1138.
31. Roby, D., Toppan, A. and Esquerré-Tugayé, M.T., 1985, Cell surfaces in plant-microorganism interactions. V. Elicitors of fungal and of plant origin trigger the synthesis of ethylene and of cell wall hydroxyproline-rich glycoprotein in plants, Plant Physiol., 77: 700-704.
32. Pi, L., 1984, Thèse Doctorat Spécialité, Université Paul Sabatier, Toulouse.

BIOCHEMICAL RESPONSES OF NON-HOST PLANT CELLS TO FUNGI AND FUNGAL ELICITORS

E. Kombrink, J. Bollmann, K.D. Hauffe, W. Knogge, D. Scheel, E. Schmelzer, I. Somssich and K. Hahlbrock

Max-Planck-Institut für Züchtungsforschung
D-5000 Köln 30
FRG

INTRODUCTION

The most frequent interactions of plants with microorganisms are those of non-host plants with microorganisms which are either non-pathogenic to plants at all or pathogenic only to genetically corresponding cultivars of one or a few plant species. We are especially interested in the molecular mechanisms of non-host resistance reactions of plants and have chosen the interaction of parsley (Petroselinum crispum) with the potential soybean pathogen, Phytophthora megasperma f. sp. glycinea (Pmg) as a model system.

Parsley plants respond to attempted infections by Pmg with the rapid formation of small localized lesions, the typical hypersensitive response [1]. These lesions are too small to be visible with the naked eye, but are easily detected under a microscope. They appear as brown necrotic spots in visible light and fluoresce brightly in UV light in various colours, depending on the wavelengths used.

Several lesions can occur in close proximity to one another in Pmg-infected parsley cotyledons, although single infections are much more frequent. Most infection sites comprise approximately 10-20 cells at the final stage of necrosis, but smaller or larger numbers can occur, at least under experimental conditions in the laboratory. The following indications are taken from recent, so far unpublished, results of studies of Pmg-infected parsley plants. The average number of cells responding per site is small, the frequency of lesion formation per cotyledon is low, there is no synchronism of lesion formation, and the concentrations of defence-related materials and enzyme activities already present in uninfected tissue are high (see below). Taken together, these limiting factors make it very difficult to measure individual biochemical reactions of the defence response in intact plants, especially when conventional assay procedures with extracts from relatively large pieces of tissue are applied.

To circumvent these difficulties and provide the necessary materials for the development of new sensitive assays in whole-plant tissues, we have studied some of the reactions of interest in cultured parsley cells. Many of the biochemical reactions which are usually associated with the hypersensitive response in infected plant tissue can be triggered in parsley cell suspension cultures by treatment with Pmg elicitor.

NATO ASI Series, Vol. H1
Biology and Molecular Biology of Plant-Pathogen Interactions
Edited by J. Bailey

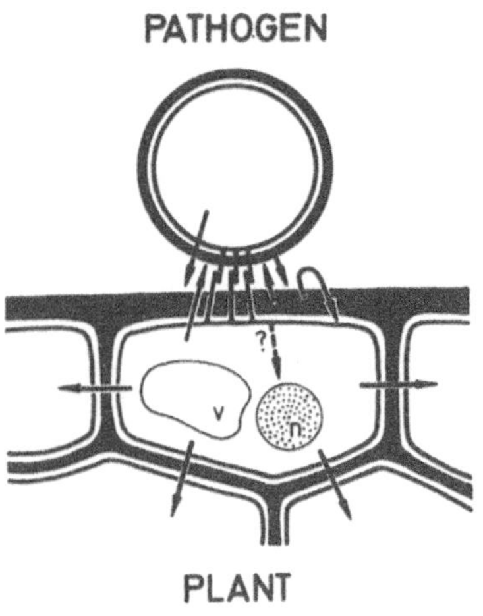

Figure 1. Updated scheme [1] summarizing the most frequently observed biochemical mechanisms of interaction between a fungal or bacterial pathogen and a challenged plant cell. Materials and reaction so far studied in parsley plants or cell suspension cultures are marked with asterisks.

BIOCHEMICAL RESPONSES OF CULTURED PARSLEY CELLS TO *Pmg* ELICITOR

The scheme depicted in Figure 1 includes an up-dated list [1,2], of the most commonly observed microbial and plant constituents which are involved in plant-pathogen interactions. A crude cell-wall preparation from *Pmg* mycelium was one of the most potent elicitors tested in cell suspension cultures of parsley [3] and was used in all experiments described here. Biochemically identified, elicitor-induced materials and activated pathways in elicitor-treated parsley cells are marked with asterisks in Figure 1.

Drastic changes in the metabolism of elicitor-treated parsley cells are reflected by rapid and extensive changes in the labelling patterns of proteins, both *in vivo* (Fig. 2) and *in vitro* (J. Bollmann and K. Hahlbrock, unpublished results). Two-dimensional gel analysis of the *in vivo* labelled proteins revealed that phenylalanine ammonia-lyase (PAL) and 'pathogenesis-related proteins' (PR proteins) were among the most prominent radioactive spots appearing between the second and third hour after elicitor treatment of the cells (Fig. 2B). Many other proteins were also synthesized at greatly increased rates, whereas the synthesis of several others was drastically reduced.

VARIATION OF RESPONSES WITH ELICITOR CONCENTRATION AND GROWTH STAGE OF THE CELL CULTURE

Magnitude and timing of the observed effects were strongly dose dependent [3]. Examples are given in Figure 3 for two enzyme activities and one class of products, the furanocoumarin phytoalexins, which are accumulated in the culture fluid. The two enzymes, PAL and chitinase, represent different defence mechanisms. PAL is the first enzyme of general phenylpropanoid metabolism which converts in three steps, phenylalanine, a product of primary metabolism, to 4-coumaroyl-CoA, the key substrate of various specific phenylpropanoid pathways. These phenylpropanoid pathways

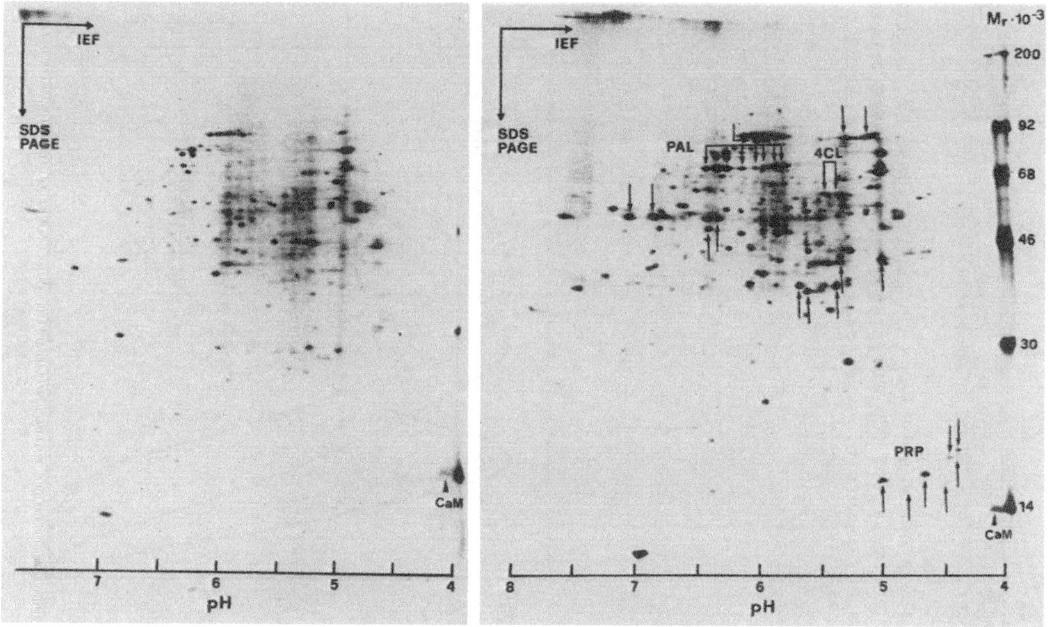

Figure 2. Two-dimensional gel analysis of [^{35}S]methionine-labelled proteins from control cells (left) and from cells labelled during the second and third hour after treatment with elicitor (right). Arrows mark the positions of some of the most prominent, elicitor-induced proteins including two enzymes of general phenylpropanoid metabolism (PAL, 4CL) and several 'pathogenesis-related' proteins (PRP). In addition to standard molecular weight markers, calmodulin (CaM) was used as an internal reference for small, acidic proteins.

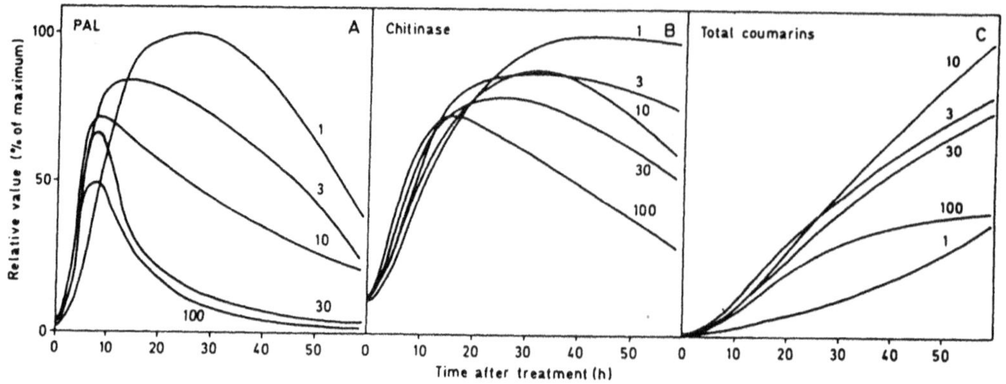

Figure 3. Relative changes in PAL (A) and chitinase (B) activities and furanocoumarin accumulation (C) in parsley cell cultures treated at time 0 with the indicated amounts of elicitor (µg/ml culture fluid). Data taken in part from [3].

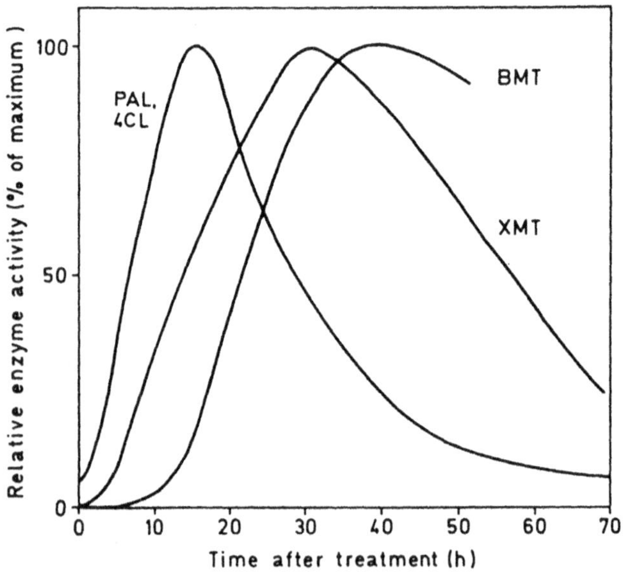

Figure 4. Relative changes in the activities of PAL, 4CL, XMT and BMT in elicitor-treated parsley cells [6].

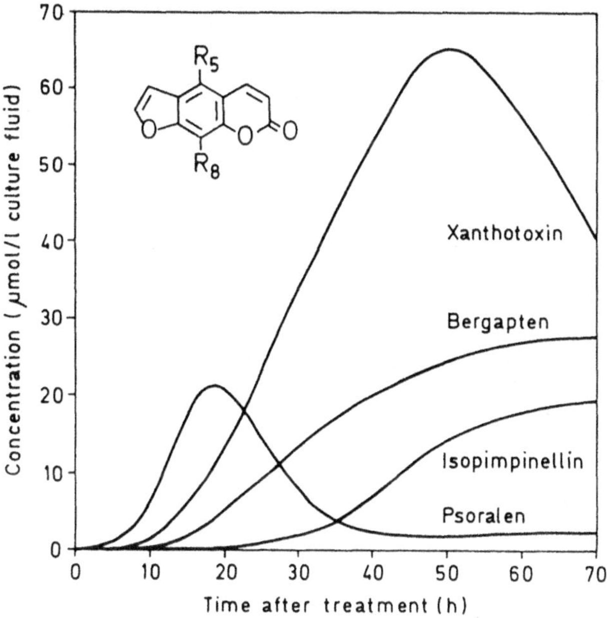

Figure 5. Changes in furanocoumarin concentrations in the culture fluid of elicitor-treated parsley cells [6]. Psoralen: $R_5 = R_8 = H$; xanthotoxin: $R_5 = H$, $R_8 = OCH_3$; bergapten: $R_5 = OCH_3$, $R_8 = H$; isopimpinellin: $R_5 = R_8 = OCH_3$.

include the formation of coumarin derivatives and several cell-wall components, such as lignin, suberin and related phenylpropanoid-derived materials (Fig. 1). By contrast, chitinase is thought to have a direct role in the degradation of fungal and bacterial cell walls. This dual function is possible because the enzyme has both chitinase and lysozyme activities [4].

Figure 3 (A and B) shows that maximal activities of PAL (as well as a closely related enzyme, 4-coumarate:CoA ligase, 4CL [3]) and chitinase were obtained with a crude-elicitor concentration of about 1 μg per ml of culture fluid, with a rather sharp peak for efficient induction around this concentration [3]. Although a similar-shaped dose-response curve was obtained for the elicitor-induced accumulation of furanocoumarins, the peak was shifted to a ten-fold higher elicitor concentration (Fig. 3C). Enzymes of the furanocoumarin pathway proper have not been investigated in this respect, but their induction behaviour is expected to be similar to that of the furanocoumarins.

The observed elicitor effects on parsley cells are also strongly dependent on the growth stage of the culture. While the induction of furanocoumarin accumulation was high from the time of subculturing until cell mass reached a maximum, the induction of PAL and 4CL activities showed a sharp maximum preceding the stage of maximal cell mass [5]. Subsequent to this stage, no effect at all was caused by the addition of elicitor. However, an endogenous mechanism of furanocoumarin and enzyme induction was observed at late culture stages, probably indicating the generation of 'endogenous' elicitor(s) during cell lysis [5].

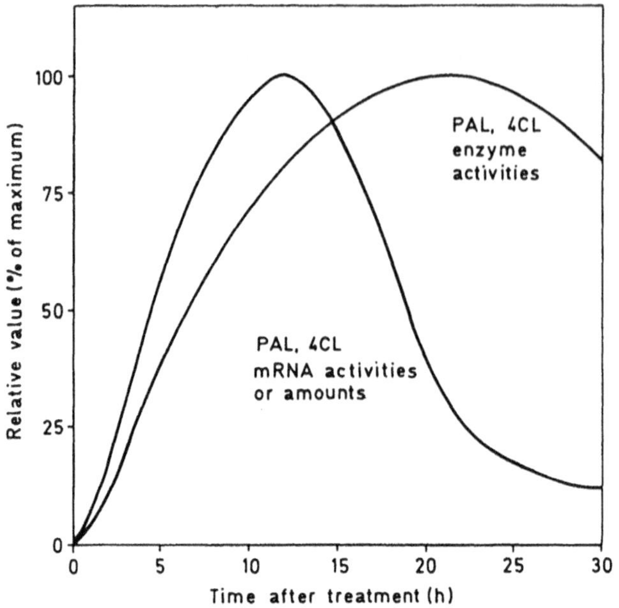

Figure 6. Relative changes in hybridizable amounts and translational activities of PAL and 4CL mRNAs and in the corresponding enzyme activities in elicitor-treated parsley cells (cell culture 2) [9].

RELATIVE TIMING OF CHANGES IN ENZYME ACTIVITIES

Under appropriate conditions with respect to both elicitor concentration and growth stage of the culture, the kinetics of enzyme induction and of accumulation of the individual furanocoumarins have been determined [6]. Some of the essential results are shown in Figures 4 and 5. Two specific enzymes of the furanocoumarin pathway, SAM:xanthotoxol O-methyltransferase (XMT) and SAM:bergaptol O-methyltransferase (BMT), were included in these studies and were found to increase sequentially in activity a few hours later than PAL and 4CL, which showed the usual rapid and highly coordinated activity changes (Fig. 4). In agreement with the sequential changes of these four enzyme activities, psoralen, the unsubstituted, putative early intermediate in the synthesis of the various furanocoumarin derivatives, appeared only transiently prior to the accumulation of the major products. As expected from the sequence of XMT and BMT induction, the respective methylation products, xanthotoxin and bergapten, appeared in the same order as the enzyme activities (Fig. 5).

ELICITOR-INDUCED CHANGES IN GENE TRANSCRIPTION RATES

The two metabolically related enzymes, PAL and 4CL, as well as several PR proteins have been shown to be induced by transient gene activation in elicitor-treated parsley cells [7,8]. Unpublished results (J. Bollmann, I. Somssich and K. Hahlbrock) indicate that this applies to many more proteins. Two recent observations are especially noteworthy in this connection.

Earlier results had suggested a biphasic behaviour of PAL gene transcription under the experimental conditions used [1,7]. This finding was

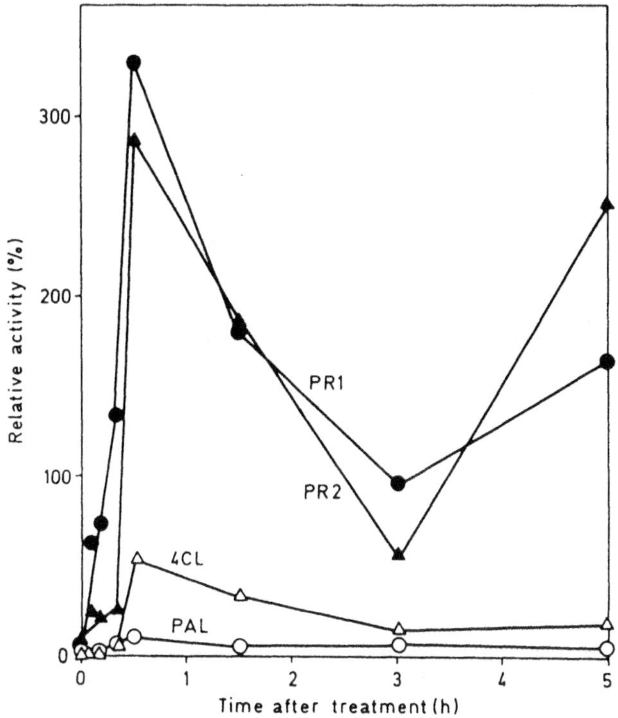

Figure 7. Early changes in the transcription rates of PAL, 4CL, PR1 and PR2 genes in nuclei isolated from elicitor-treated parsley cells.

difficult to explain, because the apparent second peak in the transcription rate was neither reflected in the time course of the changes in PAL mRNA and enzyme activities, nor was it in agreement with the otherwise high degree of coordination in PAL and 4CL induction. Recently, changes in the experimental conditions have eliminated the second peak. Complete coordination was observed at all levels of PAL and 4CL induction, when either a partially purified elicitor preparation or a newly established parsley cell culture (designated as cell culture 2) was used [9]. As an example, the time courses of changes in PAL and 4CL mRNA amounts and translational activities, as well as the resulting changes in the two enzyme activities in elicitor-treated cell culture 2 are shown in Figure 6.

Another recent observation was a very rapid activation of all four genes so far investigated in this regard. As shown in Figure 7, a peak in the transcription rates for the genes encoding PAL, 4CL and two groups of PR proteins was obtained as early as approximately 30 min after the addition of elicitor to cell culture 2, and significant increases were already detectable, at least for the PR protein genes, after a few minutes [8].

LEVELS OF DEFENCE-RELATED MATERIALS IN UNINFECTED PLANT TISSUE

Cultured parsley cells are valuable experimental tools not only because of the rapid synchronous and extensive induction by elicitor of many of the materials associated with pathogen defence in intact plant tissue (Fig. 1),

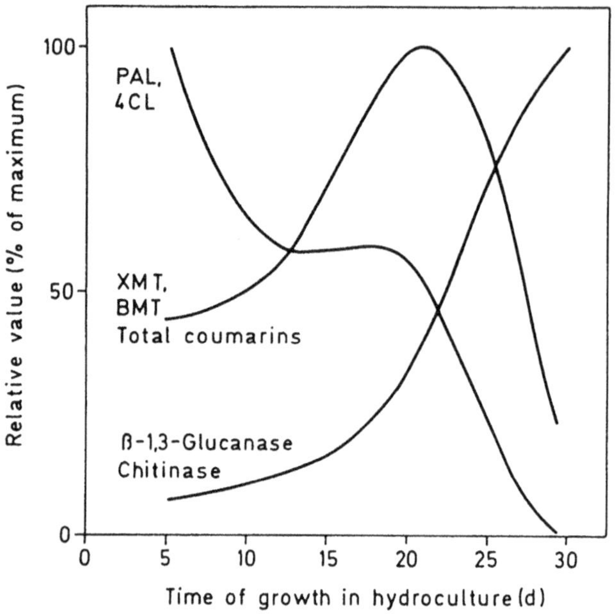

Figure 8. Developmental changes in various enzyme activities and furanocoumarin concentrations in cotyledons of parsley plants grown under sterile conditions.

but also because of the low levels at which most of these materials occur in untreated control cultures. We had expected to find a similar situation in previously uninfected parsley plants prior to the hypersensitive response to an attempted invasion by Pmg. However, the following results were surprisingly different from our expectations.

All of the putative defence-related materials tested were already present in parsley plants which had been grown under sterile conditions. Among these were all of the major funanocoumarins found in elicitor-treated cell cultures (cf. Fig. 5), the four enzymes related to furanocoumarin biosynthesis which have been discussed above (PAL, 4CL, XMT, BMT), the two hydrolytic enzymes studied (chitinase and β-1,3-glucanase), and at least some of the PR proteins and the corresponding mRNAs. Large, differential changes during cotyledon development were observed for all of these enzymes, proteins and phytoalexins, possibly suggesting that at least some of them have additional functions besides their involvement in pathogen defence (W. Knogge, E. Kombrink, E. Schmelzer and K. Hahlbrock, unpublished results). An outline of these results is given in Figure 8.

Possible, rapid increases in the concentrations of these enzymes and products, which might occur in the localized areas of Pmg-infected tissues, could not be detected above such high background levels.

CONCLUSIONS AND OUTLOOK

Thus, the large reduction in complexity of both interacting organisms (from the fungal pathogen to a cell-wall preparation with elicitor activity, and from the diversity of differentiated cells in intact plant tissue to a relatively homogeneous population of cells in suspension culture) has enabled us to study some of the induction mechanisms which are not easily detectable under natural conditions. Although it is important to realize that there are limitations to the interpretation of results obtained with the artificial cell culture-elicitor system in relation to the natural plant-fungus interaction, such model studies are extremely valuable, for the following reasons.

The biochemical reactions comprising the hypersensitive response in plant-pathogen interactions (including the mechanisms of recognition, binding, signal formation and transduction, gene activation, subsequent metabolic regulation from transcript processing to product localization and turnover, confinement of the necrotic lesions, as well as systemic reactions) might be similar in principle in many cell types of many plant species, but differ in detail not only between species, but also between different cell types in the same organ of one species. Moreover, it has long been established that physiological and environmental conditions greatly influence the responsiveness of a given cell or tissue to a certain microorganism or its putative signal molecules. Such differential responses of cells in intact plants might be mimicked, at least in principle, by the observed dose dependence of elicitor-induced reactions [3], by the dependence of these reactions on the growth stage of the cell culture [4], and by an apparent hierarchy [1] governing the responses of a given cell type to combinations of different stresses. On the whole, plant cell suspension cultures are therefore convenient tools for biochemical and physiological studies, even though no selected cell or cell type, whether cultured or from intact plants, can be more than a model for general conclusions.

As a tool, cultured parsley cells have also been very valuable for the induction, isolation and purification of the various enzymes discussed above, and for the generation of specific antisera and of cDNAs to the corresponding mRNAs.

In recently initiated studies with intact plants, these antisera and cDNAs are being used as probes for measurements of the induction kinetics and the localization of defence-related materials in individual cells, or intercellular spaces, of Pmg-infection sites (W. Jahnen, E. Schmelzer and K. Hahlbrock, unpublished results). The cDNA probes have also been employed in the isolation and characterization of genomic DNA clones as a basis for the elucidation of the mechanisms of gene activation in pathogen defence. One preliminary, surprising result has been that the copy number of PAL and 4CL genes in parsley appears to be small (probably two genes each per haploid genome), when compared with the large number of protein spots, at least for PAL, on two-dimensional gels (Fig. 2; [10]; and unpublished results by W. Schulz, C. Douglas and K. Hahlbrock).

REFERENCES

1. Hahlbrock, K., Chappell, J., Jahnen, W. and Walter, M., 1985, Early defense reactions of plants to pathogens, in: "Molecular Form and Function of the Plant Genome", L. van Vloten-Doting, G.S.P. Groot and T.C. Hall, eds., pp. 129-140, Plenum Publishing Corporation.
2. Hahlbrock, K. and Scheel, D., 1986, Biochemical responses of plants to pathogens, in: "Non-Conventional Approaches to Plant Disease Control", I. Chet, ed., John Wiley and Sons, New York, in press.
3. Kombrink, E. and Hahlbrock, K., 1986, Responses of cultured parsley cells to elicitors from phytopathogenic fungi, Plant Physiol., 81: 216-221.

4. Boller, T., Gehri, A., Mauch, F. and Vögeli, U., 1983, Chitinase in bean leaves: induction by ethylene, purification, properties, and possible function, Planta, 157: 22-31.
5. Kombrink, E. and Hahlbrock, K., 1985, Dependence of the level of phytoalexin and enzyme induction by fungal elicitor on the growth stage of Petroselinum crispum cell cultures, Plant Cell Reports, 4: 277-280.
6. Hauffe, K.D., Hahlbrock, K. and Scheel, D., 1986, Elicitor-stimulated furanocoumarin biosynthesis in cultured parsley cells: S-Adenosyl-L-Methionine:Bergaptol and S-Adenosyl-L-Methionine:Xanthotoxol O-Methyltransferases, Z. Naturforsch, 41c: 228-239.
7. Chappell, J. and Hahlbrock, K., 1984, Transcription of plant defense genes in response to UV light or fungal elicitor, Nature, 311: 76-78.
8. Somssich, I., Schmelzer, E., Bollmann, J. and Hahlbrock, K., 1986, Rapid activation by fungal elicitor of genes incoding 'pathogenesis-related' proteins in cultured parsley cells, Proc. Natl. Acad. Sci. USA, 83: 2427-2430.
9. Schmelzer, E., Somssich, I. and Hahlbrock, K., 1985, Coordinated changes in transcription and translation rates of phenylalanine ammonia-lyase and 4-coumarate:CoA ligase mRNAs in elicitor-treated Petroselinum crispum cells, Plant Cell Reports, 4: 293-296.
10. Kuhn, D.N., Chappell, J., Boudet, A. and Hahlbrock, K., 1984, Induction of phenylalanine ammonia-lyase and 4-coumarate:CoA ligase mRNAs in cultured plant cells by UV light or fungal elicitor, Proc. Natl. Acad. Sci. USA, 81: 1102-1106.

PEA GENES ASSOCIATED WITH THE NON-HOST RESISTANCE TO FUSARIUM SOLANI ARE ALSO INDUCED BY CHITOSAN AND IN RACE-SPECIFIC RESISTANCE BY PSEUDOMONAS SYRINGAE

L.A. Hadwiger, C. Daniels, B.W. Fristensky,
D.F. Kendra and W. Wagoner

Molecular Biology of Disease Resistance Laboratory
Department of Plant Pathology
Washington State University
Pullman
WA, 99164-6430, USA

INTRODUCTION

Many of the disease resistance traits in plants can be transferred within the species using conventional cross-breeding techniques. These specific traits are often inherited as single mendelian genes and often occur as alleles at a given locus, suggesting that a single enzyme or protein product is associated with each gene. We have proposed that the "gene for gene" interaction associated with a given disease reaction involves the interaction of a product (inducer) coded by an avirulence gene in the pathogen with a resistance gene associated with a host response [1,2].

"Non-specific" disease resistance traits also exist in plants enabling them to express "non-host" resistance [3] to plant pathogens normally pathogenic on other plant species. The difficulty of interspecies crossing prevents a mendelian analysis of these traits.

We have cloned genes [4] which are activated in association with the non-host disease resistance in peas expressed against the fungus *Fusarium solani* f. sp. *phaseoli*, a pathogen of beans. These same genes become suppressed in the susceptibility reaction of peas to *F. solani* f. sp. *pisi*, a pea pathogen [5]. Recently race-specific resistance in peas [6] against three races of *Pseudomonas syringae* pv. *pisi*, has been described; therefore we have evaluated the response of these non-host resistance pea genes in this race-specific interaction.

In this report we will discuss the regulatory components in the non-host resistance response of peas and the common features of this response to that of race-specific resistance.

CHITOSAN: A COMPONENT OF NON-HOST RESISTANCE IN PEAS

The endocarp tissue of immature pea pods accurately distinguishes between races of pathogenic and nonpathogenic organisms [6,7] in a manner comparable to that of the subsoil-associated pea tissues infected in nature.

NATO ASI Series, Vol. H1
Biology and Molecular Biology of Plant-Pathogen Interactions
Edited by J. Bailey

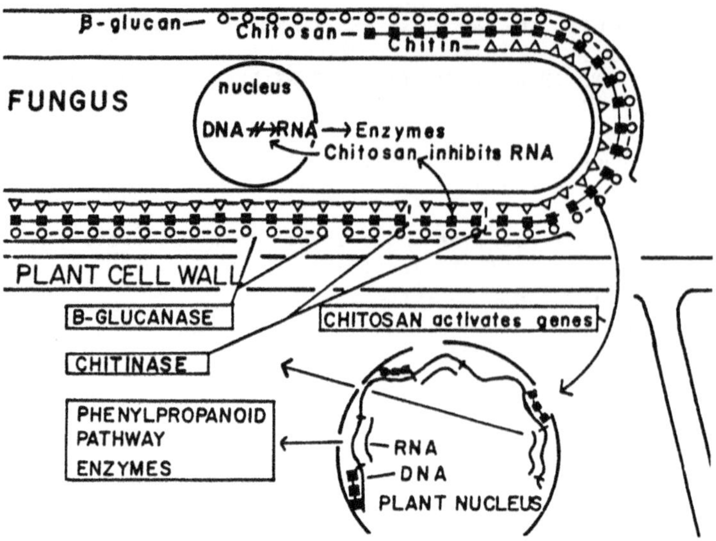

Figure 1. A schematic summarization of the functions attributed to chitosan in the resistance of pea tissue to F. solani f. sp. phaseoli, a pathogen of beans. Chitosan composes 1% of the fungal wall. Pea tissue contains β 1,3-glucanase and β 1,4-chitinase capable of digesting the two major fungal wall components. The activities of these two enzymes increase within 4 h after inoculation. Chitosan accumulates within the fungal cell. Chitosan inhibits RNA synthesis in the fungus and, therefore fungal growth. Chitosan is detected inside the pea cytoplasm within 20 min. A substantial portion of the molecule localizes within the nucleus and because of its high affinity for DNA is believed to complex with this target molecule. Chitosan (like the inoculum of F. solani f. sp. phaseoli) activates a specific group of plant genes which are temporally associated with disease resistance. Also there is an accumulation of mRNA coding for the enzyme, phenylalanine ammonia lyase, a key enzyme in the phenylpropanoid pathway of plants. End products of this pathway include pisatin and lignin.

In the early minutes (20 min) of the interaction between F. solani f. sp. phaseoli and endocarp tissue, chitosan, a hexosamine-rich polymer is released from fungal cells [8] and moves into adjoining epidermal cells of pea endocarp tissue localizing somewhat selectively in the pea nucleus. Chitosan also begins accumulating in the fungal cell itself. The transport into pea tissue and localization of chitosan within the pea cell was monitored by immunochemical, histochemical and radioisotopic techniques [9].

The presence of chitosan both in the plant and fungal tissue is important because of the biological functions possessed by this polymer. Pure chitosan applied to pea tissue mimics in great detail the activation of the same characteristic resistance associated responses induced by F. solani f. sp. phaseoli. Chitosan applied to macroconidial suspensions in liquid media at 7 μg/ml inhibits both germination and growth of F. solani speciales [8].

Chitosan is composed of β-1,4 linked glucosamine residues and its biological activity may be related to the alternate arrangement of the amino groups along the length of the polymer. Chitosan polymers with seven or more glucosamine residues [10] are maximally active in inhibiting fungal growth

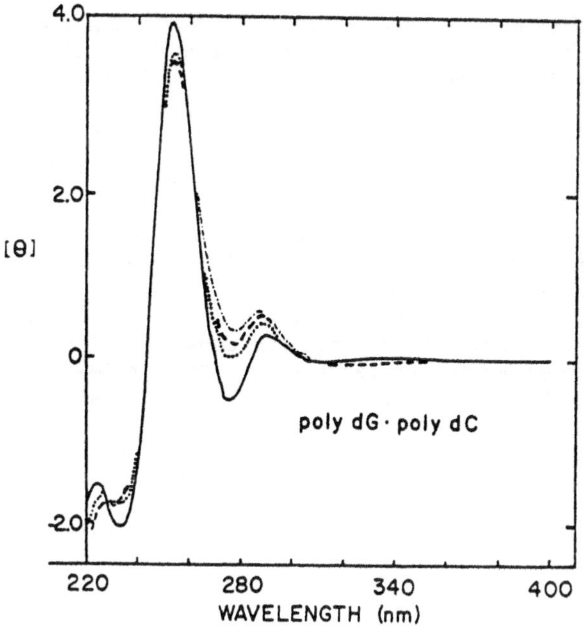

Figure 2. Circular dichroism spectra of a poly dG-poly dC sequence of DNA (50 μg in 2.6 ml of 0.1% SSC, solid line) altered by the addition of 12 μg (...), 25 μg (---) or 50 μg (• - •) of chitosan.

and pre-inducing (at 10 μg/ml) the pathway associated with the synthesis of the phytoalexin, pisatin. The proposed role of chitosan in the pea/*F. solani* f. sp. *phaseoli* interaction is described in Figure 1 [11].

INTERACTION OF CHITOSAN WITH DNA

Chitosan as a polycationic molecule has a potential to interact with macro-molecules possessing anionic groups. Much of the ^{3}H-chitosan taken up by the pea tissue localizes in the nucleus [Hadwiger, unpublished data]. Thus its target molecule may be nuclear DNA. Chitosan and DNA complex readily *in vitro* [9]. The effect of chitosan on the DNA molecule can be detected (i) by a change in the melting temperature curve of DNA [8], (ii) the circular dichroism spectra of C-G-rich segments of DNA (Fig. 2) and (iii) the coprecipitation of DNA and chitosan at near-equivalent concentrations in aqueous solution. Preliminary evidence suggests that chitosan may show some sequence preference in its interaction with DNA. At low concentrations (e.g., 75 μg/ml) chitosan appears to enhance the digestion of pUC12 DNA by the restriction endo nucleases *Taq* 1 and *Hae* 3 [Fristensky and Kendra, unpublished data].

CHITOSAN AND *F. SOLANI* f. sp. *PHASEOLI*-INDUCED RESPONSES.

The early changes in structure and function within the challenged pea cell are important since the resistance induced either by chitosan or *F. solani* f. sp. *phaseoli* develops within 6 h. Chitosan incited visible changes in plant nuclei [12] within 30 min following its application to pea endocarp tissue. RNA species associated with the resistance response are rapidly synthesized and accumulate selectively within 2 h. These species remain predominant well after resistance has been secured [5]. The RNAs were detected with DNA probes from cloned genes selected from a cDNA library of genes identified as "resistance response genes". Such genes are induced in peas challenged with F. solani f. sp. phaseoli [4]. mRNA populations in pea tissue following chitosan treatment, or F. solani f. sp. phaseoli inoculation, have also been translated in vitro into their respective proteins [13]. An accelerated rate of synthesis was observed for more than 20 major pea proteins. The specific mRNAs coding these proteins are prevalent only when resistance can be detected cytologically [14]. These proteins appear briefly in the susceptible reaction in temporal association with a brief suppression of pathogen growth and begin to disappear as growth of the pathogen is renewed. The message for these proteins remains prevalent in the resistance response for at least 48 h. These mRNAs do not accumulate in F. solani f. sp. phaseoli-challenged pea tissue following heat shock [15], a stress which renders the plant temporarily susceptible. After 9 h of recovery following heat shock, the pea tissue resumes both the ability to resist and to accumulate selectively these mRNAs.

Chitosan enhances the accumulation of mRNA for phenylalanine ammonia lyase [16] a key enzyme in the metabolic pathways for pisatin, lignin and other phenylpropanoid synthesis. Within 6 h chitosan can also enhance the activities [17,18] of chitinase and β-1,3 glucanase; pea enzymes possessing the potential to degrade the two major carbohydrate polymers of the fungal cell wall.

ANTIFUNGAL ACTION OF CHITOSAN

As indicated above, low concentration of chitosan added to liquid cultures of F. solani f. sp. phaseoli completely inhibits growth. However, fungi inhibited by as much as 40 μg chitosan/ml can still resume growth if transferred to chitosan-free media within 24 h. Thus chitosan does not completely destroy cell viability [Kendra, unpublished]. The mode by which chitosan suppresses growth is not clear. Its inhibitory action may be related to RNA synthesis since the incorporation of ^{3}H-uridine into RNA is dramatically inhibited within 1-3 h after chitosan treatment [19].

RESISTANCE RESPONSES IN NON-HOST AND RACE-SPECIFIC RESISTANCE

The reactions of endocarp tissues of 5 differential pea varieties to 3 races of Pseudomonas syringae pv. pisi have been characterized [6]. This additional host/parasite system can now be utilized to compare non-host resistance response with race-specific resistance responses within the same host species.

Using DNA probes from the cDNA library of induced genes described above, we found that some of the pea genes which are temporally activated in correlation with the expression of non-host resistance also become more active in race-specific resistance [Daniels, Fristensky and Hadwiger, unpublished] (Fig. 3). The positive correlation with resistance was evident in the interactions of 4 varieties with 3 races when probed with labelled DNA from 5 clones. At this time it is not possible to determine unequivocally if these cloned pea genes actually function to develop non-host or race-specific resistance. The evidence suggests that multiple gene responses are associated with race-specific resistance responses. The gene product interactions between host and race may not be as simple as previously conceived.

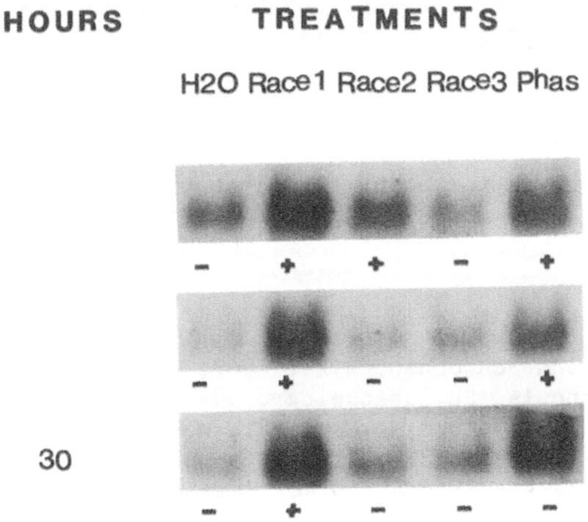

Figure 3. The accumulation of RNA, homologous with cDNA from the clone pI 176 in the differential pea variety, Ceras, which is resistant to Race 1 and susceptible to Races 2 and 3 of Pseudomonas syringae pv. pisi. This cDNA clone was originally selected from a cDNA library of pea genes whose activity is temporally associated with the expression of resistance (R) to F. solani f. sp. phaseoli, a fungal pathogen of beans. See the far right hand lane (Phas) of the Northern excerpts. There is also a selective accumulation (+) of RNA homologous to pI 176 in the race-specific resistance of pea tissue challenged by P. syringae pv. pisi Race 1. This RNA is either absent or present in much smaller quantities (-) in challenges with H_2O, race 2 or race 3 to which the tissue is susceptible (s). RNA homologous to five such clones have been shown to accumulate in correlation with the expression of resistance of 3 pea varieties to these 3 races of P. syringae pv. pisi [Daniels, Fristensky and Hadwiger, unpublished].

We have demonstrated a technique for identifying disease resistance-response genes. New techniques continue to be developed for transforming genomic clones of such plant genes into a number of plant species. Virulence of a given pathogen to its host is usually unique to that host species. However, other plant species often have the genetic potential to resist this designated pathogen. Thus, the potential exists for transforming these "resistance associated" pea genes into other plant species which may be attacked by a different, but related, Fusarium species. A positive increase in the resistance of the transformant would provide evidence that this gene functions as a component of disease resistance in the pea species from which it was derived.

SUMMARY

The chitosan from Fusarium solani f. sp. phaseoli cell walls is released and both enters the host plant cells and accumulates in the fungal mycelium within 30 min. Chitosan is inhibitory to F. solani growth at 4 μg/ml, induces resistance to the virulent pathogenic F. solani f. sp. pisi at 10 μg/ml and induces a complex host response at concentrations <50 μg/ml. Chitosan of F. solani f. sp. phaseoli activate a host response involving >20 plant proteins, some of which degrade the fungal wall carbohydrates, β-1,3 glucan and chitin. Five pea genes selected from a cDNA library of F. solani f. sp. phaseoli-induced genes are activated in chitosan-induced resistance, non-host resistance to the *F. solani* and race-specific resistance to three races of *Pseudomonas syringae* pv. *pisi*.

ACKNOWLEDGEMENT

This research was supported by NSF grant DMB 8414870 and Washington Sea Grant R/X-13.

REFERENCES

1. Hadwiger, L.A. and Schwochau, M.E., 1969, Host resistance — an induction hypothesis, *Phytopathology*, 59: 223-227.
2. Hadwiger, L.A. and Loschke, D.C., 1981, Molecular communication in host-parasite interactions. Hexosamine polymers (chitosan) as regulator compounds in race-specific and other interactions, *Phytopathology*, 71: 756-761.
3. Heath, M.C., 1980, Reactions of non-suscepts to fungal pathogens, *Annu. Rev. Phytopathology*, 18: 211-236.
4. Riggleman, R.C., Fristensky, B.W. and Hadwiger, L.A., 1985, The disease resistance response in pea is associated with increased transcription of specific mRNAs, *Plant Molec. Biol.*, 4: 81-86.
5. Fristensky, B., Riggleman, R.C., Wagoner, W. and Hadwiger, L.A., 1985, Gene expression in susceptible and disease resistant interactions of peas induced with *Fusarium solani* pathogens and chitosan, *Physiol. Plant Pathology*, in press.
6. Hadwiger, L.A. and Webster, D.M., 1984, Phytoalexin production in five cultivars of peas differentially resistant to three races of *Pseudomonas syringae* pv. *pisi*, *Phytopathology*, 74: 1312-1314.
7. Christenson, J.A. and Hadwiger, L.A., 1972, Pisatin induction in the foot region of pea seedlings by pathogenic and nonpathogenic clones of *Fusarium solani*, *Phytopathology*, 62: 1101-1102.
8. Hadwiger, L.A. and Beckman, J.M., 1980, Chitosan as a component of pea-*F. solani* interactions, *Plant Physiol.*, 66: 205-211.
9. Hadwiger, L.A., Beckman, J.M. and Adams, M .J., 1981, Localization of fungal components in the pea-*Fusarium* interaction detected immunochemically with antichitosan and antifungal cell wall antisera, *Plant Physiol.*, 67: 170-175.
10. Kendra, D.F. and Hadwiger, L.A., 1984, Characterization of the smallest chitosan oligomer that is maximally antifungal to *Fusarium solani* and elicits pisatin formation in *Pisum sativum*, *Exp. Mycol.*, 8: 276-281.
11. Hadwiger, L.A., Fristensky, B.W. and Riggleman, R., 1984, Chitosan, a natural regulator in plant-fungal pathogen interactions increases crop yields, *in*: "Advances in Chitin, Chitosan and Related Enzymes", J.P. Zikakis, ed., Academic Press.
12. Hadwiger, L.A. and Adams, M.J., 1977, Nuclear changes associated with the host-parasite interaction between *Fusarium solani* and peas, *Physiol. Plant Pathol.*, 12: 63-72.
13. Wagoner, W., Loschke, D.C. and Hadwiger, L.A., 1982, Two-dimensional electrophoretic analysis of *in vivo* and *in vitro* protein synthesis in peas inoculated with compatible and incompatible *Fusarium solani*, *Physiol. Plant Pathology*, 20: 99-107.
14. Hadwiger, L.A. and Wagoner, W., 1983, Electrophoretic patterns of pea and *Fusarium solani* proteins synthesized *in vitro* which characterize the compatible and incompatible interactions, *Physiol. Plant Pathol.*, 23: 153-162.
15. Hadwiger, L.A. and Wagoner, W., 1983, Effect of heat shock on the mRNA directed disease resistance response of peas, *Plant Physiol.*, 72: 553-556.

16. Loschke, D.C., Hadwiger, L.A. and Wagoner, W., 1983, Comparison of mRNA populations coding for phenylalanine ammonia lyase and other peptides from pea tissue treated with biotic and abiotic phytoalexin inducers, Physiol. Plant Pathol., 23: 163-173.
17. Nichols, J., Beckman, J.M. and Hadwiger, L.A., 1980, Glycosidic enzyme activity in pea tissue and pea-*Fusarium solani* interactions, Plant Physiol., 66: 199-204.
18. Mauch, F., Hadwiger, L.A. and Boller, T., 1984, Ethylene: symptom not signal for the induction of chitinase and β-1,3-glucanase in pea pods by pathogens and elicitors, Plant Physiol., 76: 607-611.
19. Hadwiger, L.A., Kendra, D.F., Fristensky, B.W. and Wagoner, W., 1985, A proposed mechanism for how chitosan both activates genes in plants and inhibits RNA synthesis in fungi, in: "Proceedings of 3rd International Conference on Chitin/Chitosan", R. Muzzarelli, ed.

RACE VARIABLE MACROMOLECULES IDENTIFIED IN *PUCCINIA GRAMINIS* F.SP. *TRITICI*; CORRELATIONS WITH AVIRULENCE/VIRULENCE TO HOST GENES FOR RESISTANCE

N.K. Howes and W.K. Kim
Agriculture Canada Research Station
Winnipeg
Manitoba R3T 2MG
Canada

Many biotrophic parasites have been shown to have differences in spore polypeptides between physiologic races [1,2,3,4,5]. In wheat stem rust (*Puccinia graminis* f.sp. *tritici*) some of these race variable polypeptides were correlated with virulence to host genes for resistance [6], but in most cases the correspondence between races avirulent to a specific host gene and the presence of a particular polypeptide was not complete.

In order to determine the range of polypeptide variation both between races and within different isolates of the same race, and to determine if any polypeptides are always associated with particular virulence genes, 25 field isolates of *P. graminis* f.sp. *tritici* were examined. These isolates represent 6 'Stakman' race groups [7], composed of 21 physiologic races and an additional 4 field isolates of two races collected in separate years.

In all, 28 polypeptides (designated numerically with the prefix pgt) varied between the 21 races, and one additional polypeptide varied between three field isolates of the same race. The location of the polypeptides after 2-dimensional IEF-PAGE and staining with Coomassie blue is shown in Figure 1.

The number of polypeptide differences closely corresponded to the difference in virulence between the races. Races that differed by many genes for virulence, differed by up to 14 polypeptides, while races that only differed by one known virulence gene differed by 0 to 3 polypeptides. Furthermore, cultures of the same race collected in different years usually had identical polypeptides. These 21 races differed in reaction type to 16 of the single-gene-line host differentials used in the Canadian stem rust race survey [7]. Comparing the 28 race variable polypeptides and 16 virulence genes, there are 8 polypeptides associated with 6 genes for avirulence (Table 1).

Polypeptide pgt*4* was always present in races avirulent to *Sr9e* and always absent in races virulent to *Sr9e*. Since the expression of avirulence to *Sr9e* has been shown to be dominant, some of these avirulent races could be heterozygous for the *P9e* gene or may be heterozygous for the gene controlling the synthesis of polypeptide pgt*4*. None of the other polypeptide/avirulence associations has an exact correspondence, but where the polypeptide was

NATO ASI Series, Vol. H1
Biology and Molecular Biology of Plant-Pathogen Interactions
Edited by J. Bailey

Table 1. Correlations between the presence of specific uredospore polypeptides in stem rust races and avirulence to specific host genes for resistance.

Effective host gene	Uredospore polypeptide		Number of races phenotypically Avirulent (A)	Virulent (V)	P*
Sr9e	pgt4	(+)	16	0	< .0001
		(-)	0	5	
Sr9e	pgt21	(+)	17	1	< .005
		(-)	0	3	
Sr7a	pgt2	(+)	8	4	< .005
		(-)	0	9	
Sr10	pgt2, 10**	(+)	8	4	< .005
		(-)	0	9	
Sr15	pgt1	(+)	4	3	< .01
		(-)	0	14	
	pgt9	(+)	8	3	< .005
		(-)	0	10	
Sr17	pgt9	(+)	10	1	< .0001
		(-)	0	10	
	pgt10**	(+)	10	2	< .005
		(-)	0	9	

* Probability of virulence and polypeptide being independent.
** Polypeptides pgt11 and pgt12 have same distribution as pgt10.

absent, the race was always virulent. Thus, these associations are consistent with avirulence and the polypeptide being controlled by the same gene.

Seven polypeptides were also associated with 3 genes for virulence (Table 2). With polypeptide pgt5 and virulence to Sr9d there was an exact correspondence. Virulence to Sr9d has been shown to be inherited as a dominant gene [8].

Two polypeptides pgt13 and pgt18 were associated with virulence to Sr5 and Sr36 respectively. Where these polypeptides were absent, all of the races were avirulent, but where present some races were avirulent and others virulent. Polypeptide pgt13 was located close to another race variable polypeptide (pgt14) on 2-dimensional electrophoreograms. Both pgt13 and pgt14 had identical molecular weights but slightly different isoelectric points. Their Coomassie blue-stained spots had similar sizes, shapes and intensities when present together in one race and are possibly both controlled by alleles of the same gene. If this is the case, races lacking pgt13 would be homozygous for pgt14. All such races were avirulent to Sr5, while races virulent to Sr5 either lacked pgt14 or were heterozygous for pgt14.

Table 2. Correlations between the presence of specific uredospore polypeptides in stem rust races and avirulence to specific host genes for resistance.

Effective host gene	Uredospore polypeptide	Number of races phenotypically Avirulent (A)	Virulent (V)	P*
Sr9d	pgt5 (+)	0	19	< .01
	(-)	2	0	
Sr5	pgt13 (+)	1	16	< .005
	(-)	4	0	
Sr36	pgt3, 18 (+)	4	11	< .01
	(-)	6	0	
	pgt10** (+)	1***	11	< .0001
	(-)	9	0	

* Probability of virulence and polypeptide being independent.
** Polypeptides pgt11 and pgt12 have same distribution as pgt10.
*** Virulence to Sr36 intermediate (2+ to 3- reaction type) in race C24.

Polypeptide pgt18 was also associated with another race variable polypeptide; pgt4. Since pgt4 was associated with avirulence to Sr9e, this polypeptide may be controlled by the same gene that determines avirulence to both Sr36 and Sr9e, with avirulence dominant to Sr9e and recessive to Sr36.

Some polypeptides such as pgt2, pgt9 and pgt10 were associated with more than one gene for virulence (eg. Sr7a, Sr10). Since virulence to Sr7a and Sr10 shows genetic linkage [8] we are assuming that, if they are not products of the same genes, these polypeptides are also controlled by genes located on the same chromosome region as P7a and P10.

Other explanations for the correlations between virulence and polypeptides are that the races examined were drawn from a limited number of asexually reproducing populations. Associations between virulence and polypeptides could then be determined by chance association in common ancestors. Mutations of existing avirulent races, selection for virulent progeny and their examination for changes in polypeptides should show if both avirulence and the polypeptides are controlled by the same gene.

Since race variable polypeptides were present in both dormant uredospores and differentiated uredosporelings [9], three races, C1(17), C17(56) and C36(48) were examined to determine if any race variable polypeptides were located in the uredosporeling cell walls and if any were glycosylated. Polypeptides pgt2, 5, 7, 13 and 14 were present in cell wall preparations. Other race variable polypeptides may also be located in the cell walls, but were washed off during cell wall purification. Protein blots made from 2-dimensional electrophoreograms of these three races were probed with Concanavalin A/peroxidase to detect ConA binding glycoproteins. While approximately 100 glycoproteins were detected, only two were coincident with race variable polypeptides (pgt7 and pgt27) and neither of these was correlated with any known genes for virulence. Thus, if fungal glycoproteins

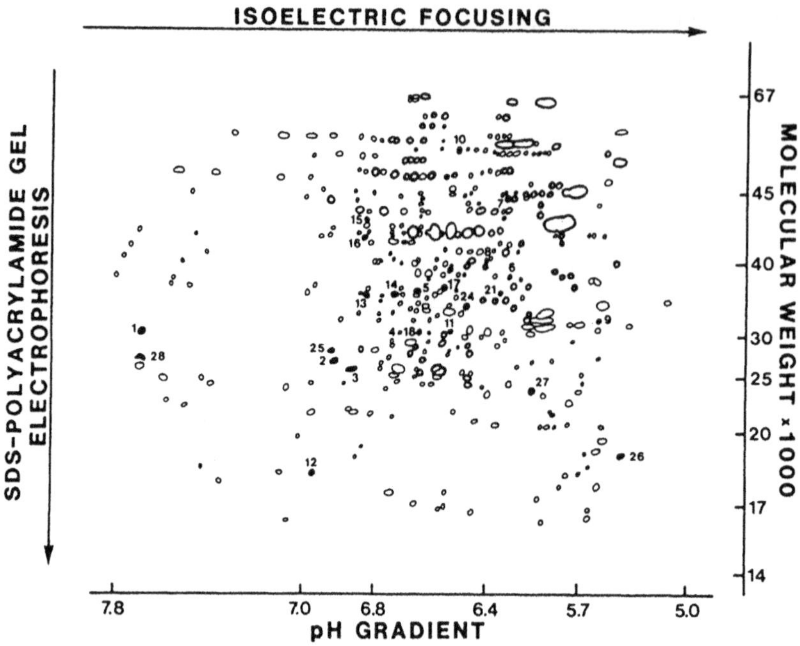

Figure 1. Composite diagram of 2-dimensional IEF-PAGE separation of detergent soluble polypeptides extracted from uredospores from 21 races of P. graminis f. sp. tritici. Open spots, positions of polypeptides present in all races; blacked-in spots, positions of polypeptides not present in all races.

are involved in determining race specific virulence, they must be present at very low concentrations, be absent from uredosporelings, or contain sugars which do not bind to Concanavalin A.

SUMMARY

Polypeptides extracted from dormant uredospores of 21 physiologic races of wheat stem rust (Puccinia graminis f. sp. tritici) were separated by 2-dimensional IEF-PAGE. Of the polypeptides detected by staining with Coomassie blue, 270 were present in all races, while 28 varied between the 21 races. The number of polypeptide differences closely corresponds to the difference in virulence between the races. Races with many virulence differences differ by up to 14 polypeptides, while races with only one virulence gene difference differ by 0-3 polypeptides. Field collections of the same race, collected in different years, usually have identical polypeptides.

There is an exact correspondence between races avirulent to Sr9e and the presence of polypeptide pgt4 and also races virulent to Sr9d and the presence of polypeptide pgt5. An additional 10 polypeptides are associated with 7 additional genes for virulence, such that the polypeptides and avirulence or virulence could be determined by products of the same genes.

Five race variable polypeptides are present in purified uredospore cell walls, including three which were correlated with race specific avirulence/virulence. Although uredosporelings contained approximately 100 polypeptides that bind to Concanavalin A (Con A), only two are coincident with any of the 28 race variable polypeptides and neither are correlated with virulence. No additional race variable Con A binding macromolecules were detected in detergent soluble extracts of either uredosporelings or purified uredosporeling cell walls.

REFERENCES

1. Gabriel, D.W. and Ellingboe, A.H., 1981, Use of mutagenesis and two-dimensional electrophoresis in a search for the primary product of a p gene, Phytopathology, 71: 217.
2. Gabriel, D.W. and Ellinboe, A.H., 1982, Polypeptide mapping by two-dimensional electrophoresis and pathogenic variation in field isolates and induced mutants of Erysiphe graminis f.sp. tritici, Phytopathology, 72: 1496-1499.
3. Howes, N.K., Kim, W.K. and Rohringer, R., 1982, Detergent-soluble polypeptides extracted from uredorspores of four physiologic races of Puccinia graminis f.sp. tritici, Physiol. Plant Pathol., 21: 361-366.
4. Shipton, W.A. and Fleischmann, G., 1969, Disc electrophoresis of proteins from uredospores of races of Puccinia coronata f.sp. avenae, Phytopathology, 59: 883.
5. Torp, J. and Andersen, B., 1982, Two-dimensional electrophoresis of proteins from cultures of Erysiphe graminis f.sp. hordei, Physiol. Plant Pathol., 21: 151-160.
6. Kim, W.K., Rohringer, R., Howes, N.K. and Chong, J., 1983, in: "Plant Biochemistry", T. Akazawa, T. Asahi and H. Imaseki, eds., pp 227-236, Japan Scientific Societies Press, Tokyo.
7. Stakman, E.C., Stewart, D.M. and Loegering, W.Q., 1962, U.S. Dept. of Agric., ARS Bull. E. 617 (Rev. 1962).
8. Green, G.J., 1966, Selfing studies with races 10 and 11 of wheat stem rust, Can. J. Bot., 44: 1255-1260.
9. Kim, W.K., Howes, N.K. and Rohringer, R., 1982, Detergent-soluble polypeptides in germinated uredospores and differentiated uredosporelings of wheat stem rust, Can. J. Plant Pathol., 4: 328-333.

THE MOLECULAR BIOLOGY OF PECTIC ENZYME PRODUCTION AND BACTERIAL SOFT ROT PATHOGENESIS

A. Collmer

Department of Botany
University of Maryland
MD, 20742, USA

Pectic enzymes have been implicated as pathogenicity factors in a broad range of plant diseases caused by necrotrophic pathogens. The role of pectic enzymes in the soft rots caused by highly pectolytic *Erwinia* spp. has received particular attention. Experiments with pectate lyases purified from *Erwinia* cultures have demonstrated that these enzymes can act alone in causing the maceration and host cell killing characteristic of bacterial soft rots [1,2,3,4]. The erwinias are now amenable to a wide array of molecular biological techniques which facilitate exploration of the regulation, export, and role of pectic enzymes in pathogenesis.

E. chrysanthemi and *E. carotovora* (subspp. *carotovora* and *atroseptica*) cause, in addition to soft rots, various other parenchymatal necroses and vascular wilts in a wide array of plants [5,6]. While strains of *E. chrysanthemi* differ in their ability to attack certain hosts, thereby revealing a degree of host specificity, soft rot erwinias, in general, can attack potato tuber tissues, at least in laboratory assays employing tuber slices or injected and anaerobically incubated whole tubers [6,7,8,9]. The majority of research to date on the molecular basis of pathogenicity in the erwinias involves potato tubers as the host tissue. Erwinias are major pathogens of potatoes, causing rot of tubers in the field and storage as well as black leg (or aerial rot) of stems [6]. The purpose of this paper is to explore the relationship between soft rot erwinias, the pectic enzymes they produce, and their ability to cause disease.

ERWINIA PECTIC ENZYMES AND THEIR EFFECTS ON PLANT TISSUES

The soft rot erwinias export at least four different types of pectic enzymes to their environment [10,11,12,13,14]. Table 1 lists the enzymes and intermediates involved in the early steps of polygalacturonate degradation by *E. chrysanthemi*. Pectate lyase (PL) attacks internal glycosidic linkages in pectate (polygalacturonate) by β-elimination. Pectin lyase (PNL) has the same action pattern and reaction mechanism but differs in its substrate specificity: PNL attacks highly methylesterified pectin. Polygalacturonase (PG) cleaves internal glycosidic linkages in pectate by hydrolysis. Exo-poly-α-D-galacturonosidase (exoPG) hydrolytically nibbles dimers from the non-reducing end of pectic polymers. Characterization of pectic enzyme complexes has been facilitated by the development of activity stains

NATO ASI Series, Vol. H1
Biology and Molecular Biology of Plant-Pathogen Interactions
Edited by J. Bailey

Table 1. Initial steps in the degradation of polygalacturonate by *E. chrysanthemi*.

Enzyme (abbreviation; gene symbol)	Intermediate
	Polygalacturonate
Pectate lyase (PL;*pel*) Exo-poly-α-D-galacturonosidase (exoPG;*peh*)	↓
	Digalacturonates[a]
Oligogalacturonide lyase (OGL;*ogl*)	↓
	4-Deoxy-L-*threo*-5-hexos-eulose uronate (DTH)
Ketodeoxyuronate isomerase[b] (KDUI;*kduI*)	↓
	3-Deoxy-D-*glycero*-2,5-hexodiulosonate (DGH)
Ketodeoxyuronate dehydrogenase[c] (KDUD;*kduD*)	↓
	2-Keto-3-deoxy-D-gluc-onate (KDG)

[a]PL degrades polygalacturonate to unsaturated digalacturonate and larger unsaturated oligogalacturonates.
[b]The recommended name for KDUI is 4-deoxy-L-*threo*-5-hexoseulose-uronate ketol isomerase [17].
[c]The recommended name for KDUD is 2-keto-3-deoxy-D-gluconate dehydrogenase [17].

employing pectate agarose overlays for use with high resolution electrophoretic procedures, e.g., ultrathin-layer polyacrylamide gel isoelectric focusing [15,16].

A survey of representative strains of *E. chrysanthemi* and *E. carotovora* (including both subspecies) using these techniques indicates that *E. chrysanthemi* and *E. carotovora* have a substantially different complement of pectic enzymes [Ried and Collmer, in preparation]. *E. chrysanthemi* produces multiple isozymes of PL with isoelectric points ranging from ca. 4.5 to 10. These will be discussed in more detail later. Many strains of *E. chrysanthemi* also produce an exoPG with a pI around 8.3. In contrast, *E. carotovora* produces only PL isozymes (at least two) with isoelectric points above 8.5. In further contrast to *E. chrysanthemi*, both subspecies of *E. carotovora* appear to produce a single species of PG (instead of exoPG) with a very alkaline pI. PNL production can be demonstrated in various strains of *E. chrysanthemi* and *E. carotovora* [14,18,19].

Most, but not all, PL isozymes isolated from erwinia culture supernatants will macerate and kill potato tuber tissues [2,20]. While the interconnecting linkages and arrangement of the complex polysaccharides in the dicot primary cell wall await elucidation [21], the destructive effect of pectic enzymes establishes two features of dicot cell wall structure: (i) α-1,4-galacturonosyl linkages occur in structurally vital polymers in both the middle lamella and the primary cell wall, and (ii) these linkages are accessible to pectic enzymes with molecular weights around 30-45 kd without prior or concerted modification of the cell wall by other microbial polysaccharidases. Thus, the simplest explanation for the killing of plant tissues by PL is that the cells are rendered osmotically fragile by weakening of the cell walls [22]. This is supported by experiments which demonstrate that tissues treated with PL remain viable in the presence of an osmoticum, but undergo massive loss of electrolytes (even after removal of the enzyme and soluble reaction products) when the osmoticum is removed [4,23].

The recent discovery that pectic enzymes or pectic fragments can elicit a variety of plant defence reactions [24,25,26] raises the possibility that additional processes are occurring during the interaction of pectolytic erwinias with their hosts. Two reactions of plants to Erwinia PL isozymes in particular are being explored. First, PL isozymes from E. carotovora can elicit phytoalexin synthesis in soybean cotyledons [27]. This may be relevant to soft rot pathogenesis, since E. carotovora has been shown to be inhibited by potato phytoalexins in culture [28]. Second, at least one E. chrysanthemi PL isozyme (PLc from strain CUCPB 1237) causes an exchange of K^+ and H^+ ions in suspension-cultured tobacco cells [29]. This specific ion flux is rapidly induced and is transient. Upon resumption of normal rates of net K^+ influx and H^+ efflux, tobacco suspension cultures no longer respond to further addition of the enzyme. The K^+/H^+ exchange is triggered by levels of PLc 100-fold lower than that required to cause general electrolyte leakage due to osmotic fragility.

Thus, while PL has been shown unambiguously to be able to cause lethal structural damage to plant cell walls, the possibility remains that the enzymes or biologically active pectic fragments are affecting the cell before the walls have been substantially weakened. These observations further suggest that pectic enzymes can have complex effects on host-parasite interactions, functioning alternately as pathogenicity factors or as elicitors of defence reactions. The amount of pectic enzyme produced in the interaction thus may be critical in determining the outcome.

ERWINIA PECTIC ENZYME REGULATION

The manner in which soft rot erwinias regulate and deploy their pectic enzymes, particularly PL, has long been suspected to be important to their pathogenic abilities [30]. Many saprophytic bacteria also produce pectic enzymes, and comparative studies have revealed three major differences between these bacteria and the soft rot erwinias: (i) the level of PL production (basal or induced) is higher in the erwinias; (ii) Pl synthesis in the erwinias can be rapidly induced by exogenous pectic substrates; and (iii) the erwinias efficiently export PL to the medium [31,32,33]. These observations, on their own, should be interpreted cautiously since they also reveal substantial quantitative differences between E. chrysanthemi and E. carotovora; the former produces several-fold higher levels of extracellular PL and more efficiently exports the enzyme [31].

Several features of pectic enzyme regulation in E. chrysanthemi have been determined by observing changes in PL synthesis following culture shift from a non-inducing growth medium to a medium containing pectic compounds or isolated plant cell walls. (i) PL is induced by low levels (less than 1 μg/ml) of saturated and unsaturated digalacturonate (the major end products of exoPG and PL, respectively) [34,35]. (ii) Polygalacturonate is an effective inducer of PL but only at concentrations above 100 μg/ml, i.e., around the Km of the extracellular enzymes [35]. (iii) Digalacturonates stimulate a 50-fold higher level of PL synthesis than does monomer galacturonate [35]. (iv) PL synthesis can be induced by isolated and washed plant cell walls [36]. (v) Induction of PL by plant cell walls, but not by soluble polygalacturonate, is inhibited by EDTA, a reversible inhibitor of PL activity [10]. (vi) Glucose or concentrations of unsaturated digalacturonate sufficiently high to support substantial bacterial growth inhibit maximum PL production unless cAMP is added [34,35]. (vii) Export of PL to the medium occurs rapidly following induction of PL synthesis without accumulation of an intracellular pool of the enzyme [31].

Based on these observations and analysis of the products of the extracellular enzymes produced by E. chrysanthemi [10], the following model can be evoked to explain how insoluble pectic polymers in the environment can trigger the bacterial cell to synthesize more PL [30]. The two extracellular enzymes, PL and exoPG, are produced at rather substantial basal levels by uninduced cultures. Working together, these enzymes efficiently release digalacturonates from pectic polymers. If conditions in the medium are optimum for PL activity (high pH and sufficient divalent cation) then unsaturated digalacturonate will be the predominant product; otherwise digalacturonate predominates. PL must initiate the attack on insoluble pectic polymers in plant cell walls, but exoPG can release digalacturonate from soluble polygalacturonate in the absence of PL activity. These digalacturonates are taken up by the cell and after further catabolism, to be discussed below, induce the synthesis of more PL and exoPG. If the medium supports high levels of PL activity, and consequently accumulation of unsaturated digalacturonate, or if a preferred carbon source is available, then cAMP-mediated (self) catabolite repression results in reduced levels of PL synthesis. Although monomer galacturonate is rapidly utilized by E. chrysanthemi it is not a product of the extracellular pectic enzymes, nor is it an effective inducer.

That PL is induced so much better by galacturonate dimers than by the monomer suggests that the actual inducer is an intermediate in digalacturonate catabolism. 4-Deoxy-L-threo-5-hexoseulose uronate (DTH) and 3-deoxy-D-glycero-2,5-hexodiulosonate (DGH), occur in the pathway of digalacturonate but not monomer galacturonate catabolism [35,37,38,39,40,41]. These ketodeoxyuronates are formed after cleavage of the digalacturonates by the intracellular enzyme oligogalacturonide lyase (OGL) produces either one or two molecules of unsaturated galacturonate. This unsaturated monomer rearranges by spontaneous enol-keto shift to form DTH which is isomerized by ketodeoxyuronate isomerase to DGH and then reduced by ketodeoxyuronate dehydrogenase (KDUD) in the presence of NADH to 2-keto-3-deoxy-D-gluconate. The latter is also a product of saturated monomer galacturonate catabolism [39].

E. chrysanthemi mutants deficient in OGL provide direct evidence for the role of at least one of these ketodeoxyuronates in PL induction. The mutants are uninducible by plant cell walls, polygalacturonate or digalacturonates, but, like the wild type strains, can be induced by exogenous ketodeoxyuronate [35,37]. Furthermore, a mutant deficient in KDUD (and expected to accumulate the putative inducers) is hyperinduced by digalacturonates [37].

Although the role of ketodeoxyuronates in PL induction in E. carotovora has not been explored, there is substantial evidence that PL products are involved in the induction process and that PL synthesis is under control of cAMP-mediated self-catabolite and glucose repression [42,43,44,45]. The patterns of PL regulation in the soft rot erwinias that have been observed so far seem well suited for efficient and aggressive utilization of pectic polymers as a carbon source during saprophytic growth. The phenotype of pectic enzyme regulatory mutants in planta is puzzling. OGL-deficient mutants of E. chrysanthemi and a cAMP-deficient mutant of E. carotovora retain the wild type ability to macerate potato tuber tissues even though the OGL mutants are uninducible and unable to utilize pectic compounds and the cAMP-deficient mutant produces only reduced levels of PL in culture [30,35,37; M.S. Mount, personal communication].

Even more puzzling is the regulation of PNL in certain strains of E. carotovora and E. chrysanthemi. The enzyme is not induced by pectic compounds, but rather by ultraviolet light irradiation and agents that damage or inhibit DNA synthesis [14,46,47]. PNL induction thus seems to be part of the SOS response in these bacteria. The SOS regulatory system has been determined in Escherichia coli to control a global set of cellular responses to the disruption of DNA metabolism. Such responses include enhanced DNA repair capacity, increased mutagenesis, and prophage induction [48]. PNL is coordinately induced with bacteriocins in E. carotovora [49] and a temperate phage in E. chrysanthemi [14]. Since PNL from fungal plant pathogens has been shown to be very effective at macerating and killing plant tissues [50,51], it is possible that the enzyme has a role in bacterial soft rot pathogenesis, although SOS inducibility of the enzyme is not a feature possessed by all strains of E. carotovora and E. chrysanthemi [14].

The erwinia pectic enzymes must be exported out of the bacterial cell before they can reach their substrates in plant tissues. Consequently, protein export would be expected to be an important process in pathogenesis. The ability to export specific proteins beyond the outer membrane is not a general property of Gram-negative bacteria [52], and there appear to be significant differences in the process of pectic enzyme export in the soft rot erwinias. In E. chrysanthemi, synthesis and export of PL seem closely coordinated, whereas in E. carotovora there is substantial accumulation of cell-bound PL before the enzyme is released to the medium [31]. Export-deficient (Out^-) mutants and PL-producing E. coli clones, discussed below, provide some insight into the export process in E. chrysanthemi. As a Gram-negative bacterium, E. chrysanthemi can potentially localize a protein to any of five cellular compartments: cytoplasm, inner membrane, periplasmic space, outer membrane, and extracellular medium. Andro et al. [53] isolated Out^- E. chrysanthemi mutants which synthesize PL normally but accumulate the enzyme in the periplasmic space. This suggests that PL export occurs in two steps. In the first step, presumably mediated by the Sec machinery which is involved in the export of other periplasmic and outer membrane proteins, PL is exported through the inner membrane to the periplasmic space. In the second step, mediated by the out gene products, PL is exported through the outer membrane to the medium. Mutations in any one of three loci on the E. chrysanthemi chromosome result in the Out^- phenotype [53]. Out^- mutants are readily isolated when mutagenized E. chrysanthemi cells are screened for inability to degrade the pectate around colonies on indicator plates. Although Out^- mutants are nonpathogenic, their use in analysis of the function of PL in pathogenesis is complicated by the fact that these mutants are unable to export cellulase and possibly other proteins having a role in the host-parasite interaction [53].

MOLECULAR CLONING OF ERWINIA PECTIC ENZYME STRUCTURAL GENES

Recent reports from eight laboratories have described the cloning of PL isozyme structural (pel) genes from E. chrysanthemi and E. carotovora [54,55,56,57,58,59,60,61]. The rapid appearance of these publications, beginning with the report by Keen et al. [55] of the cloning of two pel genes from E. chrysanthemi EC16, indicates both the vigour with which molecular biological techniques are being applied to the soft rot erwinias and the ease with which pel genes can be detected in E. coli recombinant DNA libraries. I will describe the cloning of the pelB and pelC genes from E. chrysanthemi CUCPB 1237 as an example [54].

A library of strain 1237 DNA was constructed by inserting partial Sau3A-digested chromosomal DNA fragments larger than 4 kb into the BamH1 site of plasmid pBR322. The recombinant plasmids were transformed into E. coli strain HB101 and screened for their ability to sink into pectate semisolid agar. Eight out of 4,000 colonies were found to sink during several days incubation. Restriction mapping of the cloned DNA in these pectolytic transformants revealed overlapping portions of a common 9.8 kb region of the E. chrysanthemi chromosome. Activity-stained ultrathin-layer isoelectric focusing gels were used to analyze the PL isozyme profiles of E. chrysanthemi and the pectolytic clones. E. chrysanthemi CUCPB 1237 culture supernatants contain five major bands of PL activity. According to the recommendations of a European Molecular Biology Organization Workshop on Soft-Rot Erwiniae (Marseille-Luminy, France, 23-27 July 1984) these isozymes are designated PLa through PLe, beginning with the most acidic form, and the corresponding genes are designated pelA through pelE. The pectolytic E. coli clones were found to produce the "neutral" pI isozymes, PLb and PLc. Further restriction mapping and subcloning indicated that the two isozymes are encoded by separated genes and that these genes are expressed from independent native promoters [54; Schoedel and Collmer, in preparation].

Similar and more extensive results have been obtained with other E. chrysanthemi strains. Kotoujansky et al. [56] cloned all five pel genes from strain 3937 in the lambda vector L47-1; van Gijsegem et al. [60] used the in vivo cloning vector, pULB113 (RP4::mini-Mu), to clone five pel genes from strain B374; and Reverchon et al. [58] isolated a pMMB33 cosmid derivative from a library of strain B374 DNA that contained pelA, pelD and pelE. The pattern of PL isozyme production by these various clones indicates that the pel genes in E. chrysanthemi occur in two separate clusters, one containing pelA, pelD and pelE and the other containing pelB and pelC.

As would be expected from the difference in the extracellular pectic enzyme profiles, E. carotovora DNA libraries have yielded fewer PL isozyme genes. Zink and Chatterjee [61] isolated genes encoding single forms of PL, exoPL and PG from strain 71; Lei et al. [57] isolated three pel genes from strain EC; and Roberts et al. [59] isolated two pel genes from strain EC14. The pel genes in E. carotovora appear to be in a single cluster.

MANIPULATION OF CLONED ERWINIA PECTIC ENZYME STRUCTURAL GENES

The pel genes cloned in E. coli are useful tools for three general experimental purposes. (i) They facilitate biochemical analysis of the PL isozymes; the isozymes in Erwinia culture supernatants are not easily resolved from each other, whereas subcloned pel genes produce individual isozymes which are readily purified from E. coli periplasmic shock fluids. (ii) E. coli provides a convenient cellular environment for analyzing the interactions of genes that act in trans in a complex manner on the production of PL in E. chrysanthemi. For example, the accumulation of PL in the periplasmic space of E. coli provides further evidence for a two-step process of PL export in E. chrysanthemi. Thus, E. coli, which apparently lacks the out genes required for export of PL through the outer membrane, will provide a useful host for the cloning and analysis of out genes. (iii) The cloned pel genes can be readily mutated in E. coli and then substituted for the wild type allele in the Erwinia chromosome by marker-exchange. This process, which permits definitive analysis of the function of the individial isozymes in pathogenesis, will be discussed in more detail.

There are three general steps in the marker-exchange mutagenesis procedure. (i) The cloned gene is inactivated by insertion of an antibiotic resistance gene; (ii) the plasmid carrying the marker-inactivated gene is transferred to the wild type organism and finally; (iii) the mutation is introduced into the chromosome (via homologous recombination) by selecting for antibiotic resistant cells that have lost the delivery plasmid [62]. The E. chrysanthemi CUCPB 1237 pelC gene was inactivated by a simple adaptation of this procedure which should be generally applicable to other Erwinia genes carried on the universal plasmid cloning vector pBR322 [63]. First, a pBR322 derivative carrying the pelC gene was partially digested with Sau3A. Cut plasmids of near full-length were isolated from an agarose gel and ligated with a pUC4K BamH1 fragment which encodes kanamycin resistance [64]. Kanamycin-resistant transformants containing these recombinant plasmids were then analyzed for loss of their ability to make PLc. Second, the plasmid in one such transformant was conjugationally transferred to E. chrysanthemi from an E. coli donor containing the helper plasmids R64drd11 and pLVC9 [65]. Third, the pelC::kan mutation was introduced into the E. chrysanthemi chromosome by selecting for kanamycin resistance in transconjugants cultured in phosphate-limited medium (which renders pBR322 unstable). The E. chrysanthemi mutant was Kan^r and Amp^s, indicating that the delivery plasmid had been lost. Southern blot analysis of the E. chrysanthemi mutant chromosomal DNA confirmed the insertion of the pelC::kan mutation into the chromosome. Activity-stained ultrathin-layer polyacrylamide gels demonstrated that the mutant had lost the ability to make PLc but not other PL isozymes. The same procedure was used to mutate the pelB gene in E. chrysanthemi CUCPB 1237 [66]. Diolez and Coleno [67] mutated the pelC gene in E. chrysanthemi 3937 by a similar procedure, but there was a fundamental difference in the initial mutagenesis step: the cloned pelC gene was inactivated by a Mu-lac insertion which produced a hybrid protein under control of the pelC promoter. Consequently, pelC expression in the E. chrysanthemi mutant can be monitored by following changes in the level of β-galactosidase activity.

The E. chrysanthemi CUCPB 1237 mutants deficient in pelC or pelB were analyzed for PL inducibility, ability to utilize polygalacturonate and virulence. Mutant and wild type cultures were grown in a glycerol-mineral salts medium, harvested and resuspended in the same medium containing polygalacturonate as the sole carbon source. The rates of bacterial growth and PL induction did not differ significantly between either of the mutants and the wild type. Virulence assays (Table 2) were performed by injecting whole potato tubers and incubating them anaerobically according to the procedure of De Boer and Kelman [68]. Again, there was no significant difference between the mutants and the wild type [63,66]. The simplest interpretation for the lack of a phenotype for mutations in pelB or pelC is that the encoded isozymes are redundant, and possibly that the inability to produce one of the isozymes is compensated for by increased production of the others. The alternative explanation is that these isozymes play a necessary role in some other ecological niche of the erwinias that is poorly represented by the assays used (discussed below). Ironically, E. coli clones producing PLb and PLc are able to cause limited maceration when injected in high numbers into anaerobically incubated potato tubers [54]. Thus, production of the neutral PL isozymes is sufficient but not necessary for maceration of potato tuber tissues by enterobacteria under these experimental conditions.

ERWINIAS, PECTIC ENZYMES, AND PATHOGENICITY: FURTHER PERSPECTIVES

Biochemical data indicate that the E. chrysanthemi PL isozymes with high isoelectric points (PLd and PLe) are more effective at macerating and killing

Table 2. Maceration of potato tubers by *E. chrysanthemi* CUCPB 1237 and mutants 1237-K3 (*pelC*) and 1237-K5 (*pelB*).

	Wet wt (g) of macerated tissue from injection site[a]			
	Experiment I		Experiment II	
CFU injected[b]	1237	1237-K3	1237	1237-K5
3×10^6	0.81±0.36	0.87±0.34	0.67±0.16	0.69±0.19
3×10^5	0.71±0.25	0.76±0.31	0.53±0.24	0.55±0.23
3×10^4	0.59±0.23	0.53±0.23	0.48±0.19	0.47±0.17

[a] Tubers were injected with 50 μl bacterial suspensions and incubated anaerobically at 30 °C for 2 days before being assayed. Values represent mean and standard deviation from at least 18 sites per bacterial concentration [63,66].
[b] Colony forming units.

plant tissues than the isozymes with more neutral isoelectric points (PLb and PLc) [2]. Definitive genetic evidence for the necessity of the PL isozymes in pathogenesis may eventually be obtained when mutations in the *pelD* and *pelE* genes or in certain combinations of *pel* genes are constructed. Recent physiological studies, however, have revealed another line of evidence for the importance of pectic enzymes in pathogenesis. Maher and Kelman [69] have observed that potato tubers are more susceptible to maceration by pectic enzymes when they are incubated anaerobically. Since conditions that lower oxygen tensions in tubers, e.g., a film of surface water, also render the tubers more susceptible to *Erwinia* soft rot, this suggests that the oxygen-dependent resistance against the action of pectic enzymes may be important in determining whether soft rot develops [68,69,70].

The interaction of soft rot erwinias and potato tuber tissues provides a simple and agriculturally important model host-parasite system. Unfortunately, the use of this system to analyze the virulence of *Erwinia* mutants has revealed little about the function in pathogenicity of various genes involved in pectic enzyme production. Mutants deficient in the production of OGL, KDUD, cAMP, PLb, or PLc retain the ability to macerate [30,35,37,43,63,66]. Only mutants impaired in their ability to export pectic enzymes (and other proteins) lose macerative ability [53,71]. This suggests that the pectolytic capacity of the erwinias exceeds the requirements of the bacteria for successful pathogenicity, at least when they are acting as opportunistic pathogens on compromised host tissues. The selective advantage of a high pectolytic capacity may reside in other habitats of the erwinias, e.g., the rhizosphere [72], and assays based on the ability of erwinias to multiply competitively in complex, plant-associated, microbial communities may eventually be required to evaluate mutant phenotypes.

Thus, an understanding of the functional basis for the peculiarities of the erwinia pectic enzyme system, e.g., production of five PL isozymes by *E. chrysanthemi*, may require joint exploration of both the ecology and molecular genetics of the erwinias. A precedent for a relationship between molecular mechanisms of catabolic enzyme regulation and the ecological niche of a bacterium has been reported for *E. coli* [73,74]: negative (repressor) regulation is selected when the substrate is rare and the demand for operon expression is accordingly low; positive (activator) regulation is selected when the substrate is common and the demand for operon expression is high.

CONCLUSION

The relationship between soft rot erwinias, the pectic enzymes they produce, and their ability to cause disease has been a classic topic of research in disease physiology. The nature of this relationship can now be evaluated by molecular biological techniques which enable separation of the erwinias and their pectic enzyme structural genes. The *pel* genes from several *Erwinia* strains have been cloned into *E. coli*, and the ability of clones expressing *pel* genes at high levels both to macerate and multiply in host tissues may reveal that PL production is sufficient for enterobacterial phytopathogenicity. On the other hand, *Erwinia* mutants lacking all of the pectic enzyme structural genes may still retain the ability to multiply and cause some symptoms in host tissues. There is already evidence that genes additional to those required for pectic enzyme production are necessary for pathogenicity in *E. chrysanthemi* [75]. It will be particularly interesting to see if nonpectolytic mutants of certain *Erwinia* strains retain their ability to cause vascular wilts and other diseases of growing plant tissues. Pectic enzymes may be the key to a close association of erwinias with higher plants which has enabled the evolution of a variety of pathogenic relationships.

REFERENCES

1. Basham, H.G. and Bateman, D.F., 1975, Killing of plant cells by pectic enzymes: the lack of direct injurious interaction between pectic enzymes or their soluble reaction products and plant cells, *Phytopathology*, 65: 141-153.
2. Garibaldi, A. and Bateman, D.F., 1971, Pectic enzymes produced by *Erwinia chrysanthemi* and their effects on plant tissues, *Physiol. Plant Pathol.*, 1: 25-40.
3. Mount, M.S., Bateman, D.F. and Basham, H.G., 1970, Induction of electrolyte loss, tissue maceration, and cellular death of potato tissue by an endopolygalacturonate *trans*-eliminase, *Phytopathology*, 60: 924-931.
4. Stephens, G.J. and Wood, R.K.S., 1975, Killing of protoplasts by soft-rot bacteria, *Physiol. Plant Pathol.*, 5: 165-181.
5. Pérombelon, M.C.M., 1982, The impaired host and soft rot bacteria, *in*: "Phytopathogenic Prokaryotes", Vol. 2., M.S. Mount and G.H. Lacy, *eds.*, pp. 55-69, Academic Press, New York.
6. Pérombelon, M.C.M. and Kelman, A., 1980, Ecology of the soft rot erwinias, *Annu. Rev. Phytopathol.*, 18: 361-387.
7. Dickey, R.S., 1981, *Erwinia chrysanthemi*: Reaction of eight plant species to strains from several hosts and to strains of other *Erwinia* species, *Phytopathology*, 71: 23-29.
8. Lacy, G.H., Hirano, S.S., Victoria, J.I., Kelman, A. and Upper, C.D., 1979, Inhibition of soft-rotting *Erwinia* spp. strains by 2,4-dihydroxy-7-methoxy-2H-1,4-benzoxazin-3(4H)-one in relation to their pathogenicity on *Zea mays*, *Phytopathology*, 69: 757-763.
9. Samson, R. and Nassan-Agha, N., 1978, Biovars and serovars among 129 strains of *Erwinia chrysanthemi*, *in*: "Proceedings of the Fourth International Conference on Plant Pathogenic Bacteria, Angers, France, 1978", (Station de Pathologie Végétale et Phytobactériologie, Ed.), pp. 547-553, Institut National de la Récherche Agronomique, Angers, France.

10. Collmer, A., Whalen, C.H., Beer, S.V. and Bateman, D.F., 1982, An exo-poly-α-D-galacturonosidase implicated in the regulation of extracellular pectate lyase production in Erwinia chrysanthemi, J. Bacteriol, 149: 626-634.
11. Nasuno, S. and Starr, M.P., 1966, Polygalacturonase of Erwinia carotovora, J. Biol. Chem., 241: 5298-5306.
12. Stack, J.P., Mount, M.S., Berman, P.M. and Hubbard, J.P., 1980, Pectic enzyme complex from Erwinia carotovora: a model for degradation and assimilation of host pectic fractions, Phytopathology, 70: 267-272.
13. Starr, M.P. and Moran, F., 1962, Eliminative split of pectic substances by phytopathogenic soft-rot bacteria, Science, 135: 920-921.
14. Tsuyumu, S. and Chatterjee, A.K., 1984, Pectin lyase production in Erwinia chrysanthemi and other soft-rot Erwinia species, Physiol. Plant Pathol., 24: 291-302.
15. Bertheau, Y., Madgidi-Hervan, E., Kotoujansky, A., Nguyen-The, C., Andro, T. and Coleno, A., 1984, Detection of depolymerase isoenzymes after electrophoresis or electrofocusing, or in titration curves, Anal. Biochem., 139: 383-389.
16. Ried, J.L. and Collmer, A., 1985, An activity stain for the rapid characterization of pectic enzymes in isoelectric focusing and sodium dodecyl sulfate polyacrylamide gels, Appl. Environ. Microbiol., 50: 615-622.
17. International Union of Biochemists. Nomenclature Committee, (1979). Enzyme Nomenclature 1978. Academic Press, New York, 606 pp.
18. Itoh, Y., Sugiura, J., Izaki, K. and Takahashi, H., 1982, Enzymological and immunological properties of pectin lyases from bacteriocinogenic strains of Erwinia carotovora, Agric. Biol. Chem., 46: 199-205.
19. Kamimiya, S., Nishiya, T., Izaki, K. and Takahashi, H., 1974, Purification and properties of a pectin trans-eliminase in Erwinia aroideae formed in the presence of nalidixic acid, Agric. Biol. Chem., 38: 1071-1078.
20. Quantick, P., Cervone, F. and Wood, R.K.S., 1983, Isoenzymes of a polygalacturonate trans-eliminase produced by Erwinia atroseptica in potato tissue and in liquid culture, Physiol. Plant Pathol., 22: 77-86.
21. McNeil, M., Darvill, A.G., Fry, S.C. and Albersheim, P., 1984, Structure and function of the primary cell walls of plants, Annu. Rev. Biochem., 53: 625-663.
22. Bateman, D.F. and Basham, H.G., 1976, Degradation of plant cell walls and membranes by microbial enzymes, in: "Encyclopedia of Plant Physiology New Series, Physiological Plant Pathology", Vol. 4., R. Heitefuss and P.H. Williams, eds., pp. 316-355, Springer-Verlag, New York.
23. Basham, H.G. and Bateman, D.F., 1975, Relationship of cell death in plant tissue treated with a homogeneous endopectate lyase to cell wall degradation, Physiol. Plant Pathol., 5: 249-261.
24. Bishop, P.D., Makus, D.J., Pearce, G. and Ryan, C.A., 1981, Proteinase inhibitor-inducing factor activity in tomato leaves resides in oligosaccharides enzymically released from cell walls, Proc. Natl. Acad. Sci. USA., 78: 3536-3540.
25. Hahn, M.G., Darvill, A.G. and Albersheim, P., 1981, Host-pathogen interactions. XIX. The endogenous elicitor, a fragment of a plant cell wall polysaccharide that elicits phytoalexin accumulation in soybeans, Plant Physiol., 68: 1161-1169.
26. Lee, S.C. and West, C.A., 1981, Polygalacturonase from Rhizopus stolonifer, an elicitor of casbene synthetase activity in castor bean (Ricinus communis L.) seedlings, Plant Physiol., 67: 633-639.
27. Davis, K.R., Lyon, G.D., Darvill, A.G. and Albersheim, P., 1984, Host-pathogen interactions. XXV. Endopolygalacturonic acid lyase from Erwinia carotovora elicits phytoalexin accumulation by releasing plant cell wall fragments, Plant Physiol., 74: 52-60.

28. Lyon, G.D. and Bayliss, C.E., 1975, Effect of rishitin on Erwinia carotovora var. atroseptica and other bacteria, Physiol. Plant Pathol., 6: 177-186.
29. Atkinson, M.M., Baker, C.J. and Collmer, A., 1985, The effect of pectate lyase on K^+ and H^+ transport in tobacco, Plant Physiol., 77s: 106 (abstr.).
30. Collmer, A., Berman, P. and Mount, M.S., 1982, Pectate lyase regulation and bacterial soft-rot pathogenesis, in: "Phytopathogenic Prokaryotes", Vol. 1., M.S. Mount and G.H. Lacy, eds., pp. 395-422, Academic Press, New York.
31. Chatterjee, A.K., Buchanan, G.E., Behrens, M.K. and Starr, M.P., 1979, Synthesis and excretion of polygalacturonic acid trans-eliminase in Erwinia, Yersinia and Klebsiella species, Can. J. Microbiol., 25: 94-102.
32. Zucker, M. and Hankin, L., 1970, Regulation of pectate lyase synthesis in Pseudomonas fluorescens and Erwinia carotovora, J. Bacteriol., 104: 13-18.
33. Zucker, M., Hankin, L. and Sands, D., 1972, Factors governing pectate lyase synthesis in soft rot and non-soft rot bacteria, Physiol. Plant Pathol., 2: 59-67.
34. Chatterjee, A.K., Thurn, K.K. and Tyrell, D.J., 1981, Regulation of pectolytic enzymes in soft rot Erwinia, in: "Proceedings of the Fifth International Conference on Plant Pathogenic Bacteria, Cali, Colombia", J.C. Lozano, ed., pp. 252-262, Centro Internacional de Agriculture Tropical, Cali, Columbia.
35. Collmer, A. and Bateman, D.F., 1981, Impaired induction and self-catabolite repression of extracellular pectate lyase in Erwinia chrysanthemi mutants deficient in oligogalacturonide lyase, Proc. Natl. Acad. Sci. USA., 78: 3920-3924.
36. Collmer, A. and Bateman, D.F., 1982, Regulation of extracellular pectate lyase in Erwinia chrysanthemi: evidence that reaction products of pectate lyase and exo-poly-α-D galacturonosidase mediate induction on D-galacturonan, Physiol. Plant Pathol., 21: 127-139.
37. Chatterjee, A.K., Thurn, K.K. and Tyrell, D.J., 1985, Isolation and characteriztion of Tn5 insertion mutants of Erwinia chrysanthemi that are deficient in polygalacturonate catabolic enzymes oligogalacturonate lyase and 3-deoxy-D-glycero-2,5-hexodiulosonate dehydrogenase, J. Bacteriol., 162: 708-714.
38. Condemine, G., Hugouvieux-Cotte-Pattat, N. and Robert-Baudouy, J., 1984, An enzyme in the pectinolytic pathway of Erwinia chrysanthemi: 2-Keto-3-deoxygluconate oxidoreductase, J. Gen. Microbiol., 130: 2839-44.
39. Kilgore, W.W. and Starr, M.P., 1959, Catabolism of galacturonic and glucuronic acids by Erwinia carotovora, J. Biol. Chem., 234: 2227-2235.
40. Preiss, J. and Ashwell, G., 1963, Polygalacturonic acid metabolism in bacteria. I. Enzymatic formation of 4-deoxy L-threo-5-hexoseulose uronic acid, J. Biol. Chem., 238: 1571-1576.
41. Preiss, J. and Ashwell, G., 1963, Polygalacturonic acid metabolism in bacteria. II. Formation and metabolism of 3-deoxy-D-glycero-2,5-hexodiulosonic acid, J. Biol. Chem., 238: 1577-1583.
42. Hubbard, J.P., Williams, J.D., Niles, R.M. and Mount, M.S., 1978, The relation between glucose repression of endopolygalacturonate trans-eliminase and adenosine 3'5'-cyclic monophosphate levels in Erwinia carotovora, Phytopathology, 68: 95-99.
43. Mount, M.S., Berman, P.M., Mortlock, R.P. and Hubbard, J.P., 1979, Regulation of endopolygalacturonate trans-eliminase in an adenosine 3', 5'-cyclic monophosphate deficient mutant of Erwinia carotovora, Phytopathology, 69: 117-120.
44. Tsuyumu, S., 1977, Inducer of pectic acid lyase in Erwinia carotovora, Nature, 269: 237-238.

45. Tsuyumu, S., 1979, "Self-catabolite repression" of pectate lyase in Erwinia carotovora, J. Bacteriol., 137: 1035-1036.
46. Kamimiya, S., Izaki, K. and Takahashi, H., 1972, A new pectolytic enzyme in Erwinia aroideae formed in the presence of nalidixic acid, Agric. Biol. Chem., 36: 2367-2372.
47. Tomizawa, H. and Takahashi, H., 1971, Stimulation of pectolytic enzyme formation of Erwinia aroideae by nalidixic acid, mitomycin C and bleomycin, Agric. Biol. Chem., 35: 191-200.
48. Little, J.W. and Mount, D.W., 1982, The SOS regulatory system of Escherichia coli, Cell, 29: 11-22.
49. Itoh, Y., Izaki, K. and Takahashi, H., 1980, Simultaneous synthesis of pectin lyase and carotovoricin induced by mitomycin C, nalidixic acid or ultraviolet light irradiation in Erwinia carotovora., Agric. Biol. Chem., 44: 1135-1140.
50. Byrde, R.J. and Fielding, A.H., 1968, Pectin methyl transeliminase as the macerating factor of Sclerotinia fructigena and its significance in brown rot of apple, J. Gen. Microbiol., 52: 287-297.
51. Hislop, E.C., Keon, J.P.R. and Fielding, A.H., 1979, Effects of pectin lyase from Monilinia fructigena on viability, ultrastructure and localization of acid phosphatase of cultured apple cells, Physiol. Plant Pathol., 14: 371-381.
52. Pugsley, A.P. and Schwartz, M., 1985, Export and secretion of proteins by bacteria, FEMS Microbiol. Rev., 32: 3-38.
53. Andro, T., Chambost, J.P., Kotoujansky, A., Cattaneo, J., Bertheau, Y., Barras, F., Van Gijsegem, F. and Coleno, A., 1984, Mutants of Erwinia chrysanthemi defective in secretion of pectinase and cellulase, J. Bacteriol., 160: 1199-1203.
54. Collmer, A., Schoedel, C., Roeder, D.L., Ried, J.L. and Rissler, J.F., 1985, Molecular cloning in Escherichia coli of Erwinia chrysanthemi genes encoding multiple forms of pectate lyase, J. Bacteriol., 161: 913-920.
55. Keen, N.T., Dahlbeck, D., Staskawicz, B. and Belser, W., 1984, Molecular cloning of pectate lyase genes from Erwinia chrysanthemi and their expression in Escherichia coli, J. Bacteriol., 159: 825-831.
56. Kotoujansky, A., Diolez, A., Boccara, M., Bertheau, Y., Andro, T. and Coleno, A., 1985, Molecular cloning of Erwinia chrysanthemi pectinase and cellulase structural genes, EMBO J., 4: 781-785.
57. Lei, S.P., Lin, H.C., Heffernan, L. and Wilcox, G., 1985, Cloning of the pectate lyase genes from Erwinia carotovora and their expression in Escherichia coli, Gene, 35: 63-70.
58. Reverchon, S., Hugouvieux-Cotte-Pattat, N. and Robert-Baudouy, J., 1985, Cloning of genes encoding pectolytic enzymes from a genomic library of the phytopathogenic bacterium, Erwinia chrysanthemi, Gene, 35: 121-130.
59. Roberts, D.P., Berman, P.M., Lacy, G.H., Mount, M.S. and Allen, C., 1984, Erwinia carotovora subsp. carotovora DNA encoding pectate lyases cloned into plasmid pBR322, Phytopathology, 74: 797 (abstr.).
60. Van Gijsegem, F., Toussaint, A. and Schoonejans, E., 1985, In vivo cloning of the pectate lyase and cellulase genes of Erwinia chrysanthemi, EMBO J., 4: 787-792.
61. Zink, R.T. and Chatterjee, A.K., 1985, Cloning and expression in Escherichia coli of pectinase genes of Erwinia carotovora subsp. carotovora, Appl. Environ. Microbiol., 49: 714-717.
62. Ruvkun, G.B. and Ausubel, F.M., 1981, A general method for site-directed mutagenesis in prokaryotes, Nature, 289: 85-88.
63. Roeder, D.L. and Collmer, A., 1985, Marker-exchange mutagenesis of a pectate lyase isozyme gene in Erwinia chrysanthemi, J. Bacteriol., 164: 51-56.
64. Vieira, J. and Messing, J., 1982, The pUC plasmids, an M13mp7-derived system for insertion mutagenesis and sequencing with synthetic universal primers, Gene, 19: 259-268.

65. V Haute, E., Joos, H., Maes, M., Warren, G., Van Montagu, M. and S ell, J., 1983, Intergeneric transfer and exchange recombination of r triction fragments cloned in pBR322: a novel strategy for the r ersed genetics of the Ti plasmids of Agrobacterium tumefaciens, EMBO J 2: 411-417.
66. Roeder, D.L. and Collmer, A., 1986, Marker-exchange mutagenesis of the pelB gene in Erwinia chrysanthemi CUCPB 1237, in: "Proceedings of the Sixth International Conference on Plant Pathogenic Bacteria, June 2-6, 1985, College Park, Maryland, U.S.A.", in press.
67. Diolez, A. and Coleno, A., 1985, Mu-lac insertion-directed mutagenesis in a pectate lyase gene of Erwinia chrysanthemi, J. Bacteriol., 163: 913-917.
68. De Boer, S.H. and Kelman, A., 1978, Influence of oxygen concentration and storage factors on susceptibility of potato tubers to bacterial soft rot (Erwinia carotovora), Potato Res., 21: 65-80.
69. Maher, E.A. and Kelman, A., 1983, Oxygen status of potato tuber tissue in relation to maceration by pectic enzymes of Erwinia carotovora, Phytopathology, 73: 536-539.
70. Burton, W. and Wigginton, M.J., 1970, The effect of a film of water upon the oxygen status of a potato tuber, Potato Res., 13: 180-186.
71. Chatterjee, A.K. and Starr, M.P., 1977, Donor strains of the soft-rot bacterium Erwinia chrysanthemi and conjugational transfer of the pectolytic capacity, J. Bacteriol., 132: 862-869.
72. Stanghellini, M.E., 1982, Soft-rotting bacteria in the rhizosphere, in: "Phytopathogenic Prokaryotes", Vol. 1., M.S. Mount and G.H. Lacy, eds., pp. 249-261, Academic Press, New York.
73. Savageau, M.A., 1974, Genetic regulatory mechanisms and the ecological niche of Escherichia coli, Proc. Natl. Acad. Sci. USA., 71: 2453-2455.
74. Savageau, M.A., 1977, Design of molecular control mechanisms and the demand for gene expression, Proc. Natl. Acad. Sci. USA., 74: 5647-5651.
75. Expert, D. and Toussaint, A., 1985, Bacteriocin-resistant mutants of Erwinia chrysanthemi: possible involvement of iron acquisition in phytopathogenicity, J. Bacteriol., 163: 221-227.

MOLECULAR GENETICS OF PATHOGENICITY OF *XANTHOMONAS CAMPESTRIS*

M.J. Daniels

John Innes Institute
Colney Lane
Norwich, NR4 7UH, UK

INTRODUCTION

Although genetical methods have been used to analyze microbial physiology with great success for several decades, complex phenomena such as pathogenicity present technical difficulties which have acted as a deterrent until recently, when the development of recombinant DNA procedures advanced to the point where novel approaches were feasible. An important milestone was the development of some broad host range gene cloning vectors [e.g. 1,2,3,4,5,6,7,] which are either known or presumed to be suitable for use with members of the several genera of Gram-negative bacterial plant pathogens (*Erwinia*, *Pseudomonas* and *Xanthomonas*). The choice of vector for a particular organism is usually empirical. Factors such as the stability of plasmids *in vivo* are of some importance, and variations may be apparent even between different isolates of the same bacterial species.

This paper outlines a strategy that has been used to clone genes involved in the pathogenicity to plants of a typical bacterial pathogen, *Xanthomonas campestris* pv. *campestris*, the causal agent of black rot of crucifers [8]. This particular model pathogen was chosen because it is possible to screen large numbers of colonies for pathogenicity in a short time. Consequently mutants altered in pathogenicity (but unaffected in growth and behaviour *ex planta*) can be readily obtained [9]. These mutations are assumed to define pathogenicity genes, which are cloned by seeking complementing recombinant plasmids from a library of wild-type *X. c. campestris* DNA [10]. The DNA in the recombinant plasmids has been analyzed by transposon mutagenesis [11]

PATHOGENICITY MUTANTS

Pathogenicity mutants were sought by direct screening of mutagenized bacteria on plants in the hope that this would yield mutations in a wide variety of genes which affect pathogenicity. A simple procedure was developed [9] using aseptically-grown seedlings of turnip (*Brassica campestris* cv. "Just Right"). Seeds were surface sterilized and planted individually in compartments of Replidishes (Sterilin Ltd., Teddington, UK). After 3-4 days at 25 °C, seedlings were ready for use (height about 4 cm) and they were inoculated by touching a sterile needle on a bacterial

NATO ASI Series, Vol. H1
Biology and Molecular Biology of Plant-Pathogen Interactions
Edited by J. Bailey

colony and stabbing it into the stem of a single seedling. The response was scored after a further 3-4 days. Seedlings first showed darkening at the inoculation point, then a translucent zone appeared which gradually spread away from the region and rotted. The cotyledons became yellow and finally the rot engulfed the whole seedling. Any bacterial colonies which gave an abnormal response in the first screen were extensively retested. Since a single operator can easily screen 500 colonies per day it is practicable to use this inoculation technique to look for mutants altered in pathogenicity.

It must be stressed that this pathogenicity test is regarded as a screening test and not an experimental model of this disease. In particular as the bacteria are introduced directly into parenchyma and vascular tissue, mutations in genes which affect the initial entry into intact plants would not be recovered. Outbreaks of black rot in the field originate from contaminated seed which leads to epiphytic colonization of the plant. The bacteria normally enter the plant through hydathodes [12]. To mimic this process we have adapted a procedure used by Bain [13] for screening _Brassica_ lines for resistance to black rot. Seeds are soaked in bacterial suspensions and planted out, and under suitable conditions the seedlings rapidly become diseased [Daniels and Milligan, unpublished]. We hope that the use of two complementary screening tests for mutants will enlarge the range of classes which can be recovered. Mutants which have been isolated to date using the stabbing test fall into a number of classes based on their behaviour in seedlings and mature leaves, their ability to grow in seedlings and the effect of inoculum dose on subsequent events [9]. Only a small proportion of mutants show alterations in levels of known enzymes which may be pathogenicity factors (e.g. pectinases) [10 and unpublished]. These results and the fact that "pathogenicity mutants" as we define the term occur at about 10% of the frequency of auxotrophs under our conditions of mutagenesis suggest that a considerable number of genes - perhaps 50 - are involved in pathogenicity. Similar proportions of avirulent to auxotrophic mutants were found in independent studies of _Pseudomonas solanacearum_ [14] and _Pseudomonas syringae_ pv. _phaseolicola_ [15].

CLONING OF PATHOGENICITY GENES

A library of wild-type _X. c. campestris_ genes was constructed in the broad host range cosmid vector pLARF1 [2] using DNA partially cut with the restriction enzyme _Eco_RI. With this vector DNA inserts have an average size of 26 kb so that only a few hundred clones are necessary to be confident of including any gene. The library was transferred en masse into mutants, and the transconjugants (selected by resistance to tetracycline) were scored for restoration of wild-type phenotype [10]. Appropriate controls verified that recombinant plasmids were responsible for the phenotypic change i.e. that the mutations were complemented by wild-type alleles on the plasmid. So far two recombinant plasmids carrying pathogenicity genes, pIJ3000 and pIJ3020, have been studied in detail. Both appear to have some role in extracellular enzyme production [10 and unpublished].

The pathogenicity genes of pIJ3000 have been extensively analyzed by mutagenesis with the transposon Tn_5_ [11]. pIJ3000 was introduced by transformation into _Escherichia coli_ PCT800 which carries a chromosomal Tn_5_ insertion and by selecting for simultaneous conjugal transfer of resistance to tetracycline and kanamycin to another _E. coli_ strain, a collection of Tn_5_ insertions in pIJ3000 was obtained. The positions of the insertions were mapped with restriction enzymes and a subset was chosen for further work including mutations at intervals no greater than 1 kb. Two basic experiments were performed with each Tn_5_ mutant plasmid: (a) each was transferred into the original _X. c. campestris_ mutant strain 8288 which was used to isolate pIJ3000 by complementation and (b) following transfer into the wild-type *X. c. campestris* strain 8004, a marker-exchange technique [16] was used to transfer the mutation from the plasmid to the corresponding site in the chromosome. When the strains thus constructed are tested for pathogenicity experiment (a) locates the original pathogenicity gene mutated in strain 8288, whilst experiment (b) locates any pathogenicity genes carried on the 26 kb of *Xanthomonas* DNA in pIJ3000. This approach revealed the presence of a linked cluster of pathogenicity genes. A central region of more than 10 kb of DNA is required for pathogenicity [11]. Indirect genetic criteria point to at least 5 genes in this region. Mutations in the regions flanking the 10 kb zone have no effect on pathogenicity, and "genome walking" has now extended to "non-pathogenic" region a further 23 kb to one side of the cluster. Recent experiments suggest that the gene cluster in pIJ3000 is involved in the production of pectate lyase and protease and the function of the genes is now under investigation using both molecular genetic and biochemical techniques.

Analogous experiments with pIJ3020 have not so far indicated the presence of linked pathogenicity genes in the *X. c. campestris* DNA carried on this plasmid.

HOST SPECIFICITY

The problem of the specificity of plant pathogens for their hosts, which may be expressed at various levels from cultivar to family, has been one of the most important and intractable aspects of plant pathology for many years. Recent results using recombinant DNA techniques promise a breakthrough in the near future. Staskawicz et al. [17] were able to clone race-specific incompatibility (*rsi*) genes (or avirulence genes) from certain races of *P. syringae* pv. *glycinea*. Races of *P. s. glycinea* are distinguished by their reaction on a set of soybean differentials. When *rsi* genes were transferred from one race to another, the effect was to confer upon the recipient the inability to cause disease on those differentials with which the donor race was incompatible. In addition to functioning in other races of *P. s. glycinea* the *rsi* genes could express specific incompatibility in a strain of *X. campestris* pv. *glycines* which was previously virulent on all soybean lines [17]. Very similar findings have been reported by Gabriel [18] with the system *X. campestris* pv. *malvacearum* and cotton. Again, genes could be cloned from certain races of *X. c. malvacearum* which conferred on recipient strains the specific patterns of avirulence or incompatibility to cotton differentials shown by the donor race. This system is potentially attractive because a large body of data on the genetics of resistance of cotton to *X. c. malvacearum* has been accumulated by breeders [19].

X. campestris pv. *oryzae* causes bacterial blight of rice and is one of the most destructive bacterial diseases of plants. Many races of the pathogen have been discovered with characteristic virulence/avirulence patterns on rice differentials. In collaboration with G. Todd and J. Callow, we are attempting to clone specificity genes from *X. c. oryzae* races.

It should be noted that in both cases where incompatibility genes have been cloned, the incompatibility reaction takes the form of a rapid hypersensitive response of the host tissue. We have recently attempted to clone genes from *X. campestris* pathovars which influence behaviour on hosts and non-hosts. The pathovars and plants chosen were *X. c. campestris* (homologous host, turnip) and *X. campestris* pv. *translucens* (host, wheat). DNA libraries were prepared from each pathovar and transferred reciprocally into the other. The resulting transconjugant populations i.e. *X. c. campestris* containing *X. c. translucens* DNA, and *X. c. translucens* containing

X. c. campestris DNA were tested both on wheat and turnip seedlings in order to detect either restriction of pathogenicity on the homologous host (equivalent to transferring incompatibility genes) or enhanced ability to attack the heterologous host. In tests of up to 2000 clones for each combination no reproducible effects were observed [Sawczyc and Daniels, unpublished]. For both these pathovars, inoculation of the heterologous host gives a null reaction, not a hypersensitive response. Our negative results may therefore imply that this type of non-host incompatibility is not determined by a single gene (or cluster of genes carried on a single cosmid clone). However the pathogenicity mechanisms of *X. c. campestris* and *X. c. translucens* are at least in part closely related, because it has been possible to complement certain *X. c. campestris* pathogenicity mutations with clones from the *X. c. translucens* DNA library, and genetic analysis has shown that the complementing genes also encode pathogenicity determinants in *X. c. translucens* for wheat.

DISCUSSION

Our aim has been to use genetic methods to build up a comprehensive picture of the disease process of black rot. Although the strategy outlined in this paper has successfully revealed a number of pathogenicity genes one must recognize its limitations. The isolation of pathogenicity mutations may be limited by the "resolving power" of the pathogenicity test. The need to screen large numbers of clones when isolating mutants forces the adoption of simplified tests which may not be realistic models of the naturally diseased plant and consequently the range of mutants recovered may be biassed towards certain classes. I have earlier discussed the use of alternative tests which may reveal additional classes of mutant. The second area of possible difficulty concerns the cloning of pathogenicity genes which has hitherto depended on complementing mutations with wild-type plasmid-borne sequences. Even if the DNA libraries are complete, it is possible that certain mutations cannot be complemented, for example the mutant allele may be dominant. Thus a range of strategies, depending on different assumptions, are necessary in order to pursue the goal of gaining a complete understanding of pathogenicity. Finally, this discussion has been devoted exclusively to pathogen genes, but no picture of disease can be constructed without taking into account the host plant. We are now attempting to identify *Brassica* genes which show changes in expression following infection with different *X. c. campestris* genotypes.

ACKNOWLEDGEMENT

This work has been supported by the John Innes Foundation and the Gatsby Foundation.

REFERENCES

1. Ditta, G., Stanfield, S., Corbin, D. and Helinski, D.R., 1980, Broad host range DNA cloning system for Gram-negative bacteria: Construction of a gene bank of *Rhizobium meliloti*, *Proc. Natl. Acad. Sci. USA.*, 77: 7347-7351.
2. Friedman, A.M., Long, S.R., Brown, S.E., Buikema, W.J. and Ausubel, F.M., 1982, Construction of a broad host range cosmid cloning vector and its use in the genetic analysis of *Rhizobium* mutants, *Gene*, 18: 289-296.
3. Knauf, V.C. and Nester, E.W., 1982, Wide host range cloning vectors: a cosmid clone bank of an *Agrobacterium* Ti plasmid, *Plasmid*, 8: 45-54.

4. Bagdasarian, M., Lurz, R., Rückert, B., Franklin, F.C.H., Bagdasarian M.M., Frey, J. and Timmis, K.N., 1981, Specific-purpose plasmid cloning vectors. II. Broad host range, high copy number RSF1010-derived vectors, and a host-vector system for gene cloning in Pseudomonas, Gene, 16: 237-247.
5. Frey, J., Bagdasarian, M., Feiss, D., Franklin, F.C.H. and Deshusses, J., 1983, Stable cosmid vectors that enable the introduction of cloned fragments into a wide range of Gram-negative bacteria, Gene, 24: 299-308.
6. Sharpe, G.S., 1984, Broad-host-range cloning vectors for Gram-negative bacteria, Gene, 29: 93-102.
7. Tait, R.C., Close, T.J., Lundquist, R.C., Hagiya, M., Rodriguez, R.I. and Kado, C.I., 1983, Construction and characterization of a versatile broad host range DNA cloning system for Gram-negative bacteria, Bio Technology, 1: 269-275.
8. Williams, P.H., 1980, Black rot : A continuing threat to world crucifers, Plant Disease, 64: 736-742.
9. Daniels, M.J., Barber, C.E., Turner, P.C., Cleary, W.G. and Sawczyc, M.K., 1984, Isolation of mutants of Xanthomonas campestris pv. campestris showing altered pathogenicity, J. Gen. Microbiol., 130: 2447-2455.
10. Daniels, M.J. Barber C.E., Turner, P.C., Sawczyc, M.K., Byrde, R.J.W. and Fielding, A.H., 1984, Cloning of genes involved in pathogenicity of Xanthomonas campestris pv. campestris using the broad host range cosmid pLAFRI, EMBO J., 3: 3323-3328.
11. Turner, P., Barber, C. and Daniels, M., 1985, Evidence for clustered pathogenicity genes in Xanthomonas campestris pv. campestris, Molec. Gen. Genet., 199: 338-343.
12. Cook, A.A., Walker, J.C. and Larson, R.H., 1952, Studies on the disease cycle of black rot of crucifers, Phytopathology, 42: 162-167.
13. Bain, D.C., 1952, Reaction of Brassica seedlings to blackrot, Phytopathology, 42: 497-500.
14. Boucher, C., Barberis, P., Demery, D. and Trigalet, A., 1985, in: "Advances in Molecular Genetics of the Bacteria - Plant Interaction", A.A. Szalay, and R.P. Legocki, eds., pp. 165-168, Cornell University, Ithaca, NY.
15. Anderson, D.M. and Mills, D., 1985, The use of transposon mutagenesis in the isolation of nutritional and virulence mutants in two pathovars of Pseudomonas syringae, Phytopathology, 75: 104-108.
16. Ruvkun, G.B. and Ausubel, F.M., 1981, A general method for site-directed mutagenesis in prokaryotes, Nature, 289: 85-88.
17. Staskawicz, B.J., Dahlbeck, D. and Keen, N.T., 1984, Cloned avirulence gene of Pseudomonas syringae pv. glycinea determines race-specific incompatibility on Glycine max (L.) Merr., Proc. Natl. Acad. Sci. USA., 81: 6024-6028.
18 Gabriel, D.W., 1985, in: "Advances in Molecular Genetics of the Bacteria - Plant Interaction", A.A. Szalay and R.P. Logocki, eds., pp. 202-203, Cornell University, Ithaca, NY.
19. Brinkerhoff, L.A., 1970, Variation of Xanthomonas malvacearum and its relation to control, Annu. Rev. Phytopathol., 8: 85-110.

GENETIC CONTROL OF HYPERSENSITIVITY AND PATHOGENICITY IN *PSEUDOMONAS SYRINGAE* PV. *PHASEOLICOLA*

P.B. Lindgren and N.J. Panopoulos

Department of Plant Pathology
University of California
Berkeley, USA

INTRODUCTION

Pseudomonas syringae pathovars are ideal organisms to use in studies of host-pathogen interactions. These organisms are related taxonomically, but are different from each other with respect to pathology, host species, genus or cultivar specificity, type of disease they produce, and the production of chemically distinct phytotoxins [1,2]. Some members of this group share common hosts and virtually all elicit hypersensitive responses in plants they do not normally infect [3]. Furthermore, the high degree of genetic homology among many of them [2] makes it likely that the organisms themselves or the genes which determine their similar properties have common evolutionary origins.

Our present understanding of the virulence system of *Agrobacterium tumefaciens*, e.g. [4] or the battery of plant polymer degradating enzyme systems of *Erwinia* spp. [see Collmer, this volume] at the molecular level is much more advanced than that of the pathogenicity or virulence mechanism of any *P. syringae* pathovars. Examples of pathogenicity, virulence, or host injury inducing factors in these bacteria whose role in disease is understood in some detail include the toxins (phaseolotoxin, produced by *P. s. phaseolicola*; coronatine produced by *P. s. syringae* pathovars *atropurpurea*, *glycinea*, and *tomato*; syringomycin and syringotoxin produced by *P. s. syringae*, [5,6], indoleacetic acid, produced by *P. s. savastanoi* [7], and the ice nucleating protein of *P. s. syringae* [8]. With the development of recombinant techniques and other genetic tools, several pathovars within the group are increasingly becoming model pathogens for the study of biochemical components and gene systems involved in pathogenicity, virulence, host reaction type, and race/cultivar specificity.

POSITIVE AND NEGATIVE GENE FUNCTIONS

Genetic models of bacterial pathogenicity on plants draw distinctions between genes or groups of genes that appear to play different roles in the bacterium-plant interaction [9,10]. One group is made up of genes which actively promote pathogenicity in that the products of the genes are required in a positive sense in order for disease to occur. Genes which code for such things as pectolytic enzymes, toxins, growth hormones, and extracellular

NATO ASI Series, Vol. H1
Biology and Molecular Biology of Plant-Pathogen Interactions
Edited by J. Bailey

polysaccharides (EPS) fall into this class. These gene products may function as pathogenicity or as virulence factors [11], although this distinction is not always clear. For example, are pectolytic enzymes of Erwinia pathogenicity factors or virulence factors? It is known, however, that the chlorosis inducing toxins produced by P. syringae pathovars are virulence factors [12]. The fact that the Tox$^-$ mutants of these bacteria are still pathogenic suggests that other genes whose functions are unknown at this time are required for pathogenicity.

The second group of genes does not actively promote pathogenicity, but has a negative function in that they block infection on a particular host. In recent molecular genetic studies with three bacterial pathogens, a group of genes designated "P" (or rsi) was described which specifies host/pathogen incompatibility at the race/cultivar level in conformance with the gene-for-gene hypothesis [9,13]. According to this hypothesis, P genes act in concert with functionally complementary host genes, designated "R", so that individual P-R gene pairs determine whether disease will occur with a particular race/cultivar combination. Incompatible reactions (no disease occurs, pathogen is avirulent, host is resistant) occur when genetically dominant alleles are present at both P and R loci, and compatible reactions (disease occurs, pathogen is virulent, host is susceptible) when genetically recessive alleles are present at either locus. It appears, therefore, that the P genes and their products actively promote avirulence, rather than virulence/pathogenicity, either by preventing the establishment of cell-cell compatibility or by destroying a compatible cellular relationship once it is established [9]. Genes controlling race-specific avirulence on host differentials have been cloned from P. s. glycinea [14], Xanthomonas campestris pv. vesicatoria [J. Swanson and B. J. Staskawicz, in preparation], and X. c. pv. malvacearum [15].

Exactly what determines host-pathogen specificity at the host species level is not known. It is conceivable that positively acting functions determine some part of this specificity, although no positive factors analogous to the host specific toxins of fungal pathogens have been identified in P. syringae pathovars [7,12]. At one time, it was believed that EPS and membrane lipopolysaccharides played a major role in this process. However, conclusive evidence as to their role in specificity has not been obtained. Negative functions may also determine host species specificity [16]. However, it is probable that this specificity results from the combination of both positive and negative functions.

The term "incompatible" has been used both in a narrow sense to describe pathogen avirulence and host resistance reactions governed by Mendelian gene pairs (e.g. P/R) in race/cultivar systems [9] or in a broader sense to also describe interactions between heterologous pathogens with non-host plants [17]. For the remainder of this paper, the term will be used in the latter sense for convenience.

BACTERIAL HYPERSENSITIVITY AND PATHOGENICITY

A common reaction of plants to incompatible pathogens is the hypersensitive response (HR). HR is considered an expression of plant resistance although its exact role as a plant defence mechanism is unclear at present [3]. Hypersensitivity is a rapid localized necrosis of one or a small number of cells directly contacting the pathogen and is associated with the limited multiplication and spread of the pathogen to surrounding areas. HR is a complex physiological process and we are far from understanding its various steps. These presumably include an early recognition of the incompatible pathogen by the plant, induction and development of necrosis, biosynthesis of antimicrobial substances, i.e. phytoalexins, localization of the pathogen and restriction of its multiplication. HR is a universal phenomenon in the plant kingdom and can occur with fungal, viral, or bacterial pathogens [17].

Pseudomonas syringae pathovars produce a wide array of disease symptoms including galls, cankers, and leaf blights [1]. Generally speaking two types of plant reactions are seen when a plant is inoculated with foliar pathogens such as P. s. syringae, P. s. phaseolicola, P. s. tabaci or P. s. glycinea. Compatible reactions, seen on susceptible host plants, are characterized by the production of water-soaked lesions followed by necrosis of the infected tissue at late stages of infection. The development of the water-soaking usually takes at least 72 h. Incompatible reactions, seen on non-host plants or on resistant host cultivars, are characterized by a hypersensitive response which occurs much more rapidly than compatible reactions (generally within 24 h). It is noteworthy that saprophytic Pseudomonas species which do not cause disease on any host plant are also unable to elicit a HR on any plant [3].

Bacterial HR can be demonstrated most readily by injection or infiltration of leaves of a resistant host cultivar or non-host with bacterial suspensions at approximately 5×10^6 cells/ml or greater. The initial water congestion in the infected or infiltrated area rapidly disappears by evapotranspiration, absorption and translocation. A few hours later, the affected area dries and collapses. No spread of the pathogen occurs outside this area. With more dilute bacterial suspensions (below 10^6 cells/ml), confluent collapse of the infiltrated tissue is not observed, but appropriate histochemical staining reveals individual "dead" cells (i.e., cells no longer able to exclude non-permanent stains) in the leaf mesophyll [18]. For this reason it is believed that HR is in fact a response to infection in natural situations and not just a result of an artificial laboratory manipulation. The ability to produce a hypersensitive response on one non-host plant, tobacco, is in fact used as a standard identification test for P. syringae pathovars. Virtually all pathovars, except P. s. tabaci and P. s. angulata which infect tobacco, produce a HR on this host.

An important question which remains unanswered is how are pathogenicity and hypersensitivity related to one another. The fact that all P. syringae pathovars elicit a HR with "incompatible" host plants suggests a relationship between this property and the ability to cause disease on a "compatible" host plant. More specifically, are there genes whose products are necessary for both virulence and the ability to elicit a hypersensitive response? Since the biochemical basis of pathogenicity and hypersensitivity is not well established or entirely obvious at present, a molecular genetic approach to the study of these processes seemed appropriate. Our basic experimental approach has been to generate Tn5 insertion mutants and to analyze these genetically in the hope that the nature and physiological role of their products might later be clarified.

GENETIC ANALYSIS OF PSEUDOMONAS SYRINGAE pv. PHASEOLICOLA-INTERACTIONS

P. s. phaseolicola (PSP) is the causal agent of halo blight of beans. The bacterium can infect both leaf and pod tissue. Reactions on resistant bean cultivars are often described as hypersensitive and are characterized by a localized necrosis at the site of inoculation and limited multiplication of the pathogen (up to 100 fold over initial inoculum levels). Susceptible reactions on the other hand are characterized by development of typical

Table 1. Representative interactions of P. s. phaseolicola NPS3121 of three classes of mutants on various plants. +: degree of virulence; -:HR; NR:no reaction;(-): variable or inconsistent HR.

Strain	Phenotype	Bean Red Kidney	Bean Red Mexican	Tobacco	Other Heterologous Hosts
3121	W.T.	+++	-	-	-
4000	Class I	NR	NR	NR	NR
4003	Class II	++	-	NR	(-)
4005	Class III	+	(-)	NR	(-)

water-soaked lesions and bacterial multiplication of up to 10^6 fold over initial inoculum levels. The bacterium also produces a chlorosis-inducing toxin, phaseolotoxin, which is responsible for the local and systemic chlorotic symptoms associated with the disease [19]. We have been working with a Race 1 isolate (NPS3121) of the pathogen which, in our present terminology, gives a compatible reaction on susceptible bean cultivars such as Red Kidney, incompatible reactions on resistant bean cultivars such as Red Mexican and on a number of "heterologous" hosts including tobacco, tomato, soybean and cowpea on which the bacterium is normally non-pathogenic. Resistance on Red Mexican and on the heterologous hosts is expressed as a hypersensitive response.

Transposon Mutagenesis

The suicide plasmid pUW964 [20] carrying the transposon Tn5, was introduced into NPS3121 by conjugation with E. coli HB101 (pUW964) [21]. The plasmid contains the wide host range transfer functions of RK2 which allow it to transfer readily into NPS3121, but depends on the narrow host range ColE1 replicon for its stable maintenance. The inability to replicate in P. s. phaseolicola facilitates the selection of mutants having Tn5 insertions in the genome. Approximately 3,200 such mutants were screened for their ability to produce altered host responses. Initially, 800 of the survivors were screened simultaneously for virulence on the susceptible bean cultivar Red Kidney and for ability to elicit HR on tobacco. Of these, three mutants were found which were no longer pathogenic on bean and could not induce HR on tobacco. Because HR is much more rapid than the susceptible reaction on bean, the remaining 2,400 were first screened for their HR phenotype on tobacco, and those found to cause HR were subsequently screened on bean to determine whether they could still produce a susceptible reaction. A total of eight mutants with altered host and/or non-host responses were identified by this means. All eight were also screened for their ability to elicit a hypersensitive response on the resistant bean cultivar Red Mexican and on the heterologous hosts tomato, cowpea and three cultivars of soybean (Acme, Chippewa and Flambeau).

The eight mutants fell into three phenotypic classes (Table 1). All eight were HR^- on tobacco. Class I consisted of six mutants (NPS4000, 4001, 4002, 4004, 4006, 4007) which were characterized by loss of pathogenicity on Red Kidney and loss of HR-inducing ability on Red Mexican and on all heterologous hosts tested. In other words, all six mutants caused no visible reaction on any of the plants tested. No HR was seen where normally a susceptible reaction occurs (i.e., on Red Kidney) and no

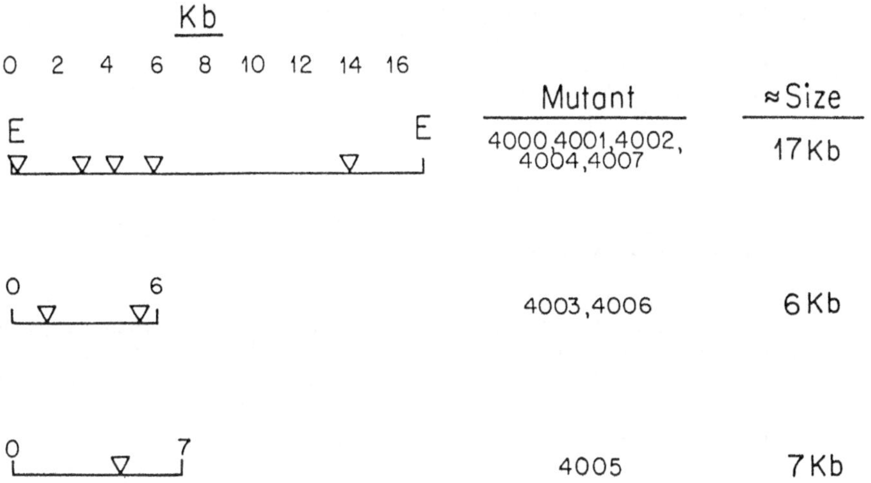

Figure 1. Diagramatic representation of Tn5 insertions associated with mutant phenotypes in *P. s. phaseolicola* NPS3121. EcoRI target fragment for each mutant and the size of each fragment is given in kilobases (kb).

susceptible reaction was seen on the resistant hosts in which a HR normally occurs. Class II consisted of one mutant (NPS4003) which had reduced virulence on Red Kidney bean when compared to the wild-type strain, and still elicited HR on Red Mexican, cowpea, soybean and tomato, although it did not consistently elicit a HR on soybean. Class III also consisted of one mutant (NPS4005). This mutant was further reduced in virulence on Red Kidney when compared to wild-type or NPS4003 and had altered ability to elicit a HR on Red Mexican and the heterologous hosts in that on certain hosts a HR was never seen, while on others a HR was seen but was not consistently reproduced. All eight mutants were prototrophic, produced phaseolotoxin as determined by an *in vitro* bioassay, and had a similar colony morphology to that of the wild-type strains.

The sites of insertion of Tn5 in each mutant were mapped by Southern blot analysis. Genomic DNA of each was digested to completion with EcoRI and probed with ^{32}P-labelled λ::Tn5. This analysis revealed that Tn5 was associated with three EcoRI fragments, approximately 17, 6 and 7 kb in size (Fig. 1). Five mutants with Class I phenotype had insertions in the 17 kb fragment (NPS4000, 4001, 4002, 4004, 4007). These insertions were scattered across the entire length of the fragment. Two mutants had Tn5 insertions in the 6.0 kb fragment (NPS4003, 4006). One mutant is a Class I mutant (NPS4006) while the other is the Class II mutant and each insertion was found at opposite ends of the fragment. The Tn5 insertion in the remaining mutant (NPS4005) was localized in the 7 kb fragment.

Total DNA from mutant NPS4000 was partially digested with Sau3A and cloned into the BamHI site in the broad host range cosmid vector pLAFR3 [B. Staskawicz, unpublished]. Selection for kanamycin resistant transductants yielded one which contained a plasmid with 23 kb of insert DNA including the Tn5 element. The plasmid was designated pPL1. A HindIII fragment from pPL1 containing ca. 6.8 kb of PSP DNA and 1.2 kb of Tn5 was subsequently eluted from a gel and used as a ^{32}P-labelled probe in Southern blot analysis of EcoRI digested genomic DNA from wild-type PSP and the eight mutants. The probe showed homology to a single EcoRI fragment of approximately 22.7 kb in NPS4000, 4001, 4002, 4004, and 4007 which confirmed our earlier conclusion that these five mutants represented Tn5 insertions in the same 17 kb EcoRI fragment. With each of the remaining three mutants the above probe showed homology to two EcoRI fragments, one having the same size (17 kb) as in wild-type NPS3121 and the other corresponding in size to those detected with the λ::Tn5 probe.

Cloning

A genomic library was constructed by cloning partial Sau3A digested DNA fragments from the wild-type strain NSP3121 into the cosmid vector pLAFR3 [22]. The library was screened by colony hybridization for homology to the ^{32}P-labelled 8 kb HindIII fragment from pPL1 containing ca. 6.8 kb of DNA from the wild-type 17 kb EcoRI fragment. Six different recombinant plasmids with homology to the above probe were identified. These plamids were transferred individually from E. coli by pRK2013 assisted mobilization, to all eight mutants. All six plasmids were found to complement at least one of the mutants with insertion of Tn5 in a 17 kb EcoRI fragment (i.e., NPS 4000, 4001, 4002, 4004, 4007). In all cases where a clone restored wild-type response to a mutant, the wild-type phenotypic response was seen for all hosts mentioned above.

One recombinant plasmid, designated pPL6, restored the wild-type phenotype to seven of the eight mutants which had Tn5 insertions in either the 17 or the 6 kb target fragments detected in the mutants. The plasmid has repeatedly been transferred from transformed E. coli HB101 by pRK2013 assisted mobilization to each of the seven mutants which it had originally restored. As expected, this plasmid consistently restored each mutant to the wild-type phenotype. No clone was identified which could restore the phenotype to mutant NPS4005. It appears that there is a large gene cluster which is involved with the ability to cause disease on compatible hosts and the ability to elicit a HR on incompatible hosts. Further studies with this plasmid are currently being carried out in our laboratory to determine the structure and organization of these genes.

Conservation of DNA Sequences

The 8 kb EcoRI fragment from pPL1 used to probe the library was also used to probe a number of P. syringae pathovars to determine if they contain any DNA sequences homologous to P. s. phaseolicola. Pathovars which contain homologous sequences to this probe included isolates of P. s. syringae, 5 races of P. s. glycinea, P. s. tabaci, P. s. angulata, P. s. lachrymans, and P. s. tomato. Xanthomonas campestris pv. phaseoli, X. c. pv. glycine, Erwinia carotovora subsp. betavasculorum, Agrobacterium tumefaciens and two Rhizobium spp. had no homology to the probe. In addition, two saprophytic Pseudomonas species, P. putida and P. fluorescens, as well as E. coli did not contain homologous sequences. Thus, it appears that a region of DNA involved with virulence/hypersensitivity from P. s. phaseolicola is specific to a number of P. syringae pathovars. The functional significance of this conservation for all pathovars remains to be determined.

To determine whether homologous DNA sequences in different pathovars might have the same functional role, pPL6 was transferred into a Tn5 mutant of P. s. glycinea. The mutant had similar phenotype to the Class I PSP mutants described above as the mutant causes no disease reaction on either homologous or heterologous hosts [B. Staskawicz, unpublished data]. Its parental strain (Race 0) gave a compatible reaction on the soybean cultivar Centennial, but a HR on the soybean cultivars Harosoy and Norchief as well as on tobacco. Wild-type phenotypes were restored in the mutant on all plants tested when pPL6 was transferred by conjugation. Thus, there appears to be functional homology between a gene(s) on the P. s. phaseolicola gene cluster and a gene(s) controlling virulence and HR in the soybean pathogen P. s. glycinea.

DISCUSSION

From the data discussed here and elsewhere [21,22, Lindgren et al., in preparation], the following conclusions may be drawn. First, there exists a group of genes which appear to exert dual control on the interaction between P. s. phaseolicola and its host and non-host plants. These genes appear to be required for pathogenicity on susceptible cultivars of the homologous host, bean, and for the elicitation of the hypersensitive reaction on resistant cultivars of the homologous host as well as on non-host plants. These genes are organized into a large gene cluster which appears to be conserved, at least in part, among several P. syringae pathovars whose host range includes legume and non-legume species. Such homology appears to be absent from distantly related phytopathogenic species, saprophytic pseudomonads and E. coli and from legume symbiotic Rhizobium spp.. In the case of X. campestris pv. phaseoli, a bean pathogen, the absence of detectable homology might be interpreted to mean that its pathogenicity requires components that are distinct from those involved in bean halo blight. However, the nucleotide sequence divergence between the two bacteria may have precluded hybrid formation under the conditions of normal stringency used in our experiments. This consideration also applies with the other organisms that were examined. However, the lack of homology in these cases is not surprising since these do not share common hosts or non-host reactions with P. s. phaseolicola.

Tn5 mutants of P. s. phaseolicola [23,24] and P. s. syringae pathogenic on bean [23,25] have been isolated by others. Some of these had similar phenotypes to the Class I mutants described here. It is not clear at present whether the Tn5 insertions in all mutants described thus far were localized in the same genes. Other work in this laboratory indicates that another class of genes may control pathogenicity, but not hypersensitivity, in P. s. phaseolicola. A Tn5 mutant of a P. s. syringae isolate pathogenic on bean has been identified which did not cause brown spot symptoms on bean, but still elicited a hypersensitive reaction on tobacco [26]. The Tn5 target fragment from this mutant (6.6 kb) contained homologous sequences to a DNA segment of equal length in P. s. phaseolicola NPS3121. These results suggest, but do not prove, that genes outside the DNA regions described here may play a role in the pathogenicity of P. s. phaseolicola.

The ability of plasmid pPL6 to restore the Class I mutant phenotype in P. s. glycinea suggests that the DNA sequences which are homologous to the gene cluster above have functional significance as well. The extent of this functional complementation is not presently known. One might expect that HR-inducing and/or bean pathogenic strains must have genes with similar or analogous functions with those present in the gene cluster cloned in pPL6, unless the underlying mechanism for their interaction with plants differs from that of P. s. phaseolicola. A special case is that of P. s. tabaci and P. s. angulata which normally cause disease and do not induce HR on tobacco although like P. s. phaseolicola, they still elicit the latter reaction on the other non-host plants. Similar considerations also apply to those pathogens which elicit HR on the same non-hosts as P. s. phaseolicola but which infect different host plants. It would be interesting to know which part(s) of the gene cluster in question is present or absent from the genomes of these bacteria and how their expression may be modulated by other as yet undiscovered genes.

The experiments discussed here and other recent work on race/cultivar specific interactions in P. s. glycinea [14] and heterospecific DNA transfer experiments between P. s. tomato and P. s. glycinea [16] suggest that the genetic control of host and non-host reactions in phytopathogenic pseudomonads involves the interplay between positively and negatively acting functions and perhaps genes that have not yet been identified in these pathogens. Thus, genes which determine race-specific avirulence of P. s. glycinea on soybean differentials do not play a role in the induction of hypersensitivity on the organism's non-host plants, such as tobacco, since all known races of this pathogen elicit the reaction on this host. Furthermore, the gene cluster of P. s. phaseolicola described here does not impart HR-inducing ability to E. coli or P. fluorescens which do not normally elicit the reaction. Several hypotheses can explain these results. First, expression of genes carried on pPL6 may not occur in these bacteria. Second, additional genes may be required for the elicitation of HR which were not present on pPL6. Third, elicitation of HR may require the combined interplay of genes exemplified by the rsi and the genes controlling pathogenicity and hypersensitivity. Hopefully, these questions will be clarified once the structure, organization and products of the respective coding regions and of any other relevant regions become known. Finally, the conservation of sequence homology with the gene cluster discussed here presents an opportunity to expand present studies into several other bacterial pathogens. Such comparative studies may contribute significantly to the ultimate elucidation of the disease mechanisms in bacteria-plant interactions.

ACKNOWLEDGEMENT

The assistance of R. C. Peet in the construction of the genomic library and his helpful discussion throughout this work is acknowledged. We thank Dr Brian Staskawicz for his helpful discussions and for comments during preparation of the manuscript.

REFERENCES

1. Fahy, P.C. and Lloyd, A.B., 1983, Pseudomonas: The fluorescent pseudomonads, in: "Plant Bacterial Diseases: A Diagnostic Guide", P.C. Fahy and G.J. Presley, eds., pp. 141-188, Academic Press, Sydney.
2. Schroth, M.N., Hildebrand, D.C. and Starr, M.P., 1981, Phytopathogenic Members of the Genus Pseudomonas, in: "The Prokaryotes", M.P. Starr, H. Stolp, H.G. Truper, A. Balows and H.G. Schlegel, eds., pp. 701-718, Springer-Verlag, Berlin.
3. Klement, Z., 1982, Hypersensitivity, in: "Phytopathogenic Prokaryotes", Vol. 2, M.S. Mount and G.H. Lacy, eds., pp. 150-177, Academic Press, New York.
4. Hille, J., Hoekema, A., Hooykaas, P. and Schilperoort, R., 1984, Gene organization of the Ti plasmid, in: "Plant Pathogens and Defense Mechanisms. Plant Gene Research Vol. 1. Genes Involved in Microbe-Plant Interactions", T. Hohn and D.P.S. Verma, eds., pp. 286-309, Springer-Verlag, Vienna.
5. Durbin, R.D., ed., 1981, "Toxins in Plant Disease", Academic Press, New York and London.
6. Mitchell, R.E., 1984, The relevance of non-host specific toxins in the expression of virulence by pathogens, Annu. Rev. Phytopathology, 22: 215-245.
7. Smidt, M. and Kosuge, T., 1979, The role of indole-3-acetic acid accumulation by alpha-methyl tryptophan resistant mutants of Pseudomonas savastanoi in gall formation on oleanders, Physiol. Plant Pathol., 13: 203-14.

8. Orser, C.S., Staskawicz, B.J., Panopoulos, N.J., Dahlbeck, D. and Lindow, S.E., 1985, Bacterial ice nucleation: Cloning and expression of genes in Escherichia coli, J. Bacteriol: in press.
9. Ellingboe, A.H., 1982, Genetic aspects of active defense, in: "Active Defense Mechanisms in Plants", R.K.S. Wood, ed., pp. 179-192, Plenum Press, New York.
10. Heath, M.C., 1981, A generalized concept of host-parasite specificity, Phytopathology, 71: 1121-1123.
11. Yoder, O.C., 1980, Toxins in pathogenesis, Annu. Rev. Phytopathology, 18: 411-434.
12. Panopoulos, N.J. and Staskawicz, B.J., 1981, Genetics of production, in: "Toxins in Plant Disease", R.D. Durbin, ed., pp. 79-107, Academic Press, New York and London.
13. Panopoulos, N.J., Walton, J.D. and Willis, D.K., 1984, Genetic and biochemical basis of virulence of plant pathogens, in: "Plant Pathogens and Defense Mechanisms. Plant Gene Research, Vol. 1. Genes Involved in Microbe-Plant Interactions", T. Hohn and D.P.S. Verma, eds., pp. 339-374, Springer-Verlag, Vienna.
14. Staskawicz, B.J., Dahlbeck, D. and Keen, N.T., 1984, Cloned avirulence gene of Pseudomonas syringae pv. glycinea determines race specific incompatibility of Glycine max (L.) Merr., Proc. Natl. Acad. Sci. USA, 81: 6024-28.
15. Gabriel, D.W., 1985, Molecular cloning of specific avirulence genes from Xanthomonas malvacearum, in: "Molecular Genetics of Bacteria-Plant Interactions", A. Szalay and P. Legoki, eds., Media Services, Cornell Univ. Publ., New York, in press.
16. Kobayashi, D.Y. and Keen, N.T., 1985, Cloning of a factor from Pseudomonas syringae pv. tomato responsible for a hypersensitive response on soybean, Phytopathology, 75: in press.
17. Kiraly, Z., 1980, Defenses triggered by the invader: Hypersensitivity, in: "Plant Disease-An Advanced Treatise", Vol. V, J.G. Horsfall and E.B. Cowling, eds., pp. 201-224, Academic Press, New York.
18. Turner, J.G. and Novacky, A., 1974, The quantitative relation between plant and bacterial cells involved in the hypersensitive reaction, Phytopathology, 64: 885-890.
19. Schuster, M.L. and Coyne, D.R., 1981, Biology, epidemiology, genetics and breeding for resistance to bacterial pathogens of Phaseolus vulgaris L., Hort. Rev., 3: 28-58.
20. Weiss, A.A., Hewlett, E.L., Myers, G.A. and Falkow, S. 1983, Tn5-induced mutations affecting virulence factors of Bordatella pertusis, Inf. and Immun., 42: 33-41.
21. Lindgren, P.B., Panopoulos, N.J., Willis, D.K. and Peet, R.C., 1984, Analysis of Vir^- HR^- Tn-insertion mutants of Pseudomonas syringae pv. phaseolicola, Phytopathology, 74: 837.
22. Lindgren, P.B., Peet, R.C. and Panopoulos, N.J., 1985, Cloning and analysis of genes associated with pathogenicity and hypersensitivity in Pseudomonas syringae pv. phaseolicola, Phytopathology, 75: 1355.
23. Anderson, D.M. and Mills, D., 1985, The use of transposon mutagenesis in the isolation of nutritional and virulence mutants in two pathovars of Pseudomonas syringae, Phytopathology, 75: 104-108.
24. Stapleton, J.M., Deasey, M.C. and Mattisee, A.G., 1985, Virulence mutant in Pseudomonas syringae pv. phaseolicola similar to hypersensitive response mutants, Phytopathology, 75: 1355.
25. Niepold, F., Anderson, D. and Mills, D., 1985, Cloning determinants of pathogenesis from Pseudomonas syringae pv. syringae, Proc. Natl. Acad. Sci. USA, 82: 406-410.
26. Willis, D.K., Lindgren, P.B., Peet, R.C., Lindow, S.E. and Panopoulos, N.J., 1985, Isolation and characterization of pathogenicity-specific genes in Pseudomonas syringae pv. syringae, Phytopathology, 75: 1320.

REGULATION OF BACTERIAL GENES INVOLVED IN BACTERIUM-PLANT INTERACTIONS BY PLANT SIGNAL MOLECULES

R.J.H. Okker, H. Spaink, C. Wijffelman, A.A.N. van Brussel and B. Lugtenberg

University of Leiden
Dept. of Plant Molecular Biology
Nonnensteeg 3
2311 VJ Leiden, The Netherlands

INTRODUCTION

Agrobacterium tumefaciens is the causative agent of the plant tumour, crown gall. The molecular basis of the tumour induction by A. tumefaciens is the transfer and subsequent expression of a defined part, the T-DNA, of the Ti-plasmid, a large plasmid, from the bacterium to the plant cell [1]. The processes occurring between the first bacterium-plant contact and the expression of the T-DNA in planta are almost unknown. At least eight virulence (Vir) operons are involved, whose functions are essential for tumour formation on all plants or which influence the host range. Many of the Vir genes are located on the Ti-plasmid [2] but some Vir genes are located on the chromosome [3].

The closely related genus Rhizobium is able to form root nodules on legumes in which it fixes N_2. In fast-growing Rhizobia the nodulation genes are usually located on a large plasmid, the Symbiosis plasmid [4]. Also chromosomal genes are involved in nodulation, but these have not been extensively studied as yet [5].

The Vir and Nod genes are designated as interaction genes, i.e. genes whose products are, directly or indirectly, involved in the interaction between bacterium and plant. Many interaction genes are transcribed at only a very low level or not at all in bacteria cultivated in the usual laboratory media [6, and H. Spaink et al., in prep.] As it therefore seems reasonable to assume that their expression is regulated directly or indirectly by plant signals, we have developed methods to measure the expression of these genes as a consequence of contact of bacteria with plant material.

We preferred the use of genetic 'expression-indicator' elements to the direct measurement of mRNA because of the greater sensitivity and efficiency of the expression-indicator elements. Expression-indicator elements are tailored genetic elements. They contain an indicator gene, i.e. a structural gene devoid of its own promoter, which codes for an easily detectable enzyme. The structural gene for β-galactosidase is widely used for this purpose. As a consequence of the lack of its own promoter, the structural gene can only be transcribed from a newly obtained promoter. Two types of expression-indicator elements are known, namely transposons (Fig. 1) or plasmids (Fig. 2) containing indicator genes. We will describe our experiences with both types of these elements.

NATO ASI Series, Vol. H1
Biology and Molecular Biology of Plant-Pathogen Interactions
Edited by J. Bailey

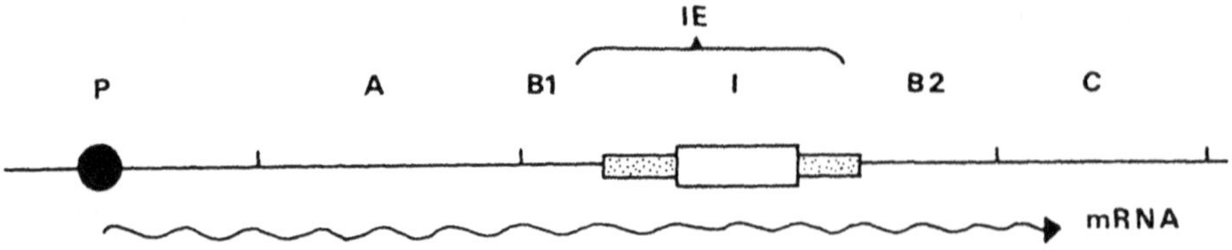

Figure 1. Transposon-type expression-indicator element. The genes A, B and C are transcribed from the promoter P. The mRNA is represented by the wavy line. The transposon element IE is inserted in gene B. Gene B is split by the insertion and the product of gene B is no longer functional. Transcription of A, B and C can be monitored now by determining the product of gene I.

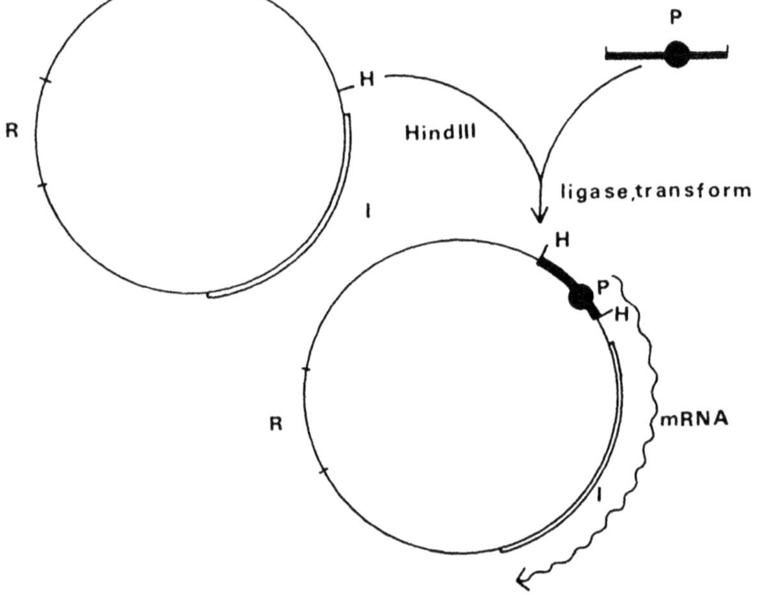

Figure 2. Plasmid-type expression-indicator element. The plasmid contains an indicator gene I, which is not expressed, a resistance marker R and an unique cloning site (e.g. *Hind*III). A promoter P is cloned into the unique site and can be tested for its ability to initiate transcription. Transcription is measured as the product of gene I.

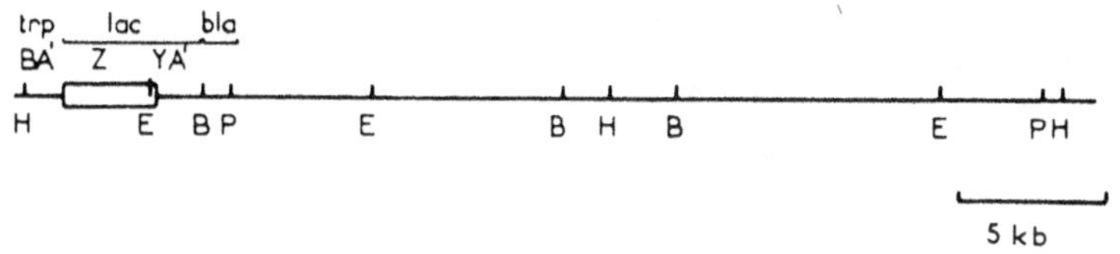

Figure 3. Structure of bacteriophage MudI (Ap^R, lac) [7]. Both ends of bacteriophage Mu are essential for transposition and therefore have been conserved. About 19.7 kb of the left part of Mu has been replaced by part of the trp operon, the structural genes of the lac operon (including the ribosome binding site but without the promoter-operator region of this operon), and the bla promotor and structural gene (Ap^R).

Restriction sites: H, HindIII; E, EcoRI; B, BamHI; P, PstI.

TRANSPOSON-TYPE EXPRESSION-INDICATOR ELEMENTS

Bacteriophage Mu dI (Ap^R, lac) [7] was studied as an example of the transposon-type element (Fig. 3). Three approaches with Mulac were followed in order to transpose Mulac to the Ti plasmid.

1. Transposition of Mulac from a suicide plasmid in an Agrobacterium background.

The small plasmid pACYC 184 does not replicate in A. tumefaciens. pMP 29::Mulac, a derivate of this plasmid, was mobilized to A. tumefaciens and the transposition in A. tumefaciens was checked. Transposition-like events occurred with a frequency of 10^{-5}, but all resulting A. tumefaciens strains showed the same level of β-galactosidase expression. This is not expected when at random transposition occurs and it therefore is probably an artefact. In conclusion, this approach with Mulac is not reliable.

2. Site-directed mutagenesis (Fig. 4).

Transposition of Mulac to the vector pMP29:: vir (a derivative of pACYC 184) occurred with a frequency of about 10^{-7}. Three isolates were analyzed for the restriction pattern. No deletions or rearrangements were detected. However, the procedure was too complicated for routine use. The main problem is the low transposition of Mulac from the chromosome to the cloning vector (Fig. 4, step A).

The recently constructed smaller transposons like mini-Mulac [8] and Tn3-Hoho 1 [9] are probably more suited for this approach since the sizes of the latter transposons are in better agreement with the sizes of the vectors used.

3. Direct mutagenesis of Agrobacterium and Rhizobium plasmid DNA intergrated in broad host range plasmids.

The Ti and Sym plasmids are not stably maintained in E. coli. However, cointegrates with broad host range R plasmids can easily be isolated and are reasonably stable in E. coli recA or E. coli rec^+ strains [10, 11].

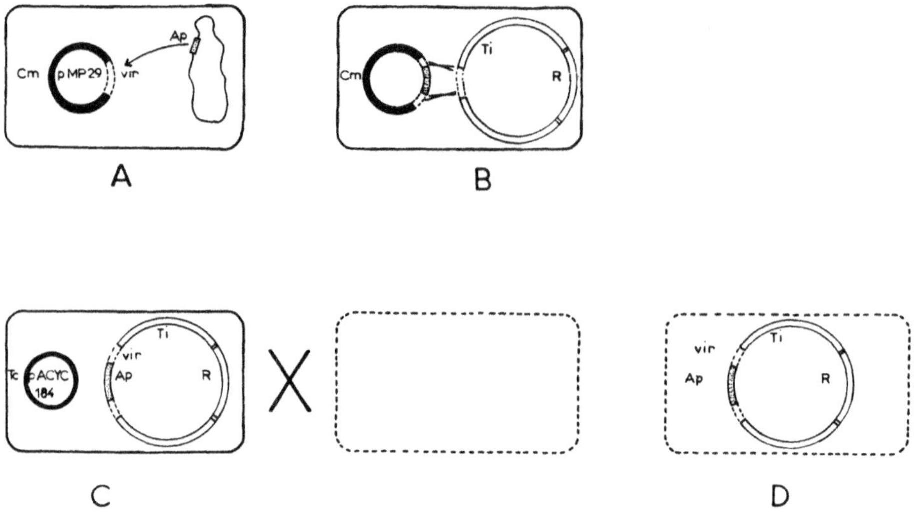

Figure 4. Outline of the site specific mutagenesis with Mulac. Bacteria with a drawn surface represent E. coli, those with a dotted surface A. tumefaciens.
A. The cloning vector pMP29, a derivate of pACYC 184 with a bom site, contains a fragement of the Vir region. Mulac (Ap^R) is located on the chromosome. Mulac was partially induced at 37 °C. pMP29 was mobilized with R702 to E. coliR05: del(lac-pro) recA trp::MuBam Rif^R. Ap^R strains of E. coliR05 were selected. The pMP29::Mulac plasmids were isolated and tested by restriction enzyme analysis for insertion of Mulac in the Vir clone.
B. Suitable plasmids were transformed into a stain of E. coli containing the stable cointegrate Ti-R772 [10].
C. The Ti-R cointegrate was crossed to E. coliR05 containing pACYC 184 (Cm^STc^R). Selection of inserted Mulac was for Ap^R. The Ap^R strains were grown for several generations in the presence of Tc to select against free pMP 29::Mulac plasmids. The resulting strains were checked for Ap^R and Cm^S.
D. The Ti::Mulac - R cointegrate was crossed to A. tumefaciens LBA 288.

The isolation of R' plasmids with parts of the chromosome of Rhizobium has also been described [12]. As a model system Mulac was transposed from the chromosome of E. coli to the Km^R marker of the broad host range plasmid R702. The R702::Mulac was crossed to A. tumefaciens, R. leguminosarum and R. trifolii with high frequency (5-6 x 10^2). From genetic and physical evidence we concluded that R702::Mulac remained intact and was stably maintained in both A. tumefaciens and Rhizobium. This is a point worth investigating since RP_4::Mu has been reported to be a highly unstable in Rhizobium [13]. In conclusion, cointegrates of Agrobacterium or Rhizobium DNA with broad host range plasmids are most likely to be suitable for mutagenesis with Mulac. The drawback of this approach is that several thousand isolates of A. tumefaciens or Rhizobium, each with a different R'::Mulac, must be screened in plants before sufficient numbers of isolates with a Mulac insert in the Vir or Nod genes have been isolated.

PLASMID-TYPE EXPRESSION-INDICATOR ELEMENTS

In this type of expression-indicator element the indicator gene is preceded by one or more unique cloning sites in which the promoter to be tested for expression can be cloned. This approach has the following advantages.

1. Promoters can, contrary to the situation in the transposon-type expression-indicator element, be tested without disruption of gene functions which are essential for the bacterium-plant interaction.

2. The copy number can be varied by choosing different plasmids.

3. The constructs are easier to make. A drawback is that the region under study must be genetically and physically reasonably known.

The β-galactosidase of the lacZ indicator gene is easily determined with a sensitive colour reaction, using O-nitrophenyl-β-D-galactopyranoside (ONPG) or 3-bromo-4-chloro-5-indolyl-β-D-galactopyranoside (X-gal) as the substrate. The usefulness of lacZ as an indicator gene in studies on bacterium-plant interactions is hampered by the presence of β-galactosidases in A. tumefaciens, Rhizobium and plant cells. However, we found that these hampering β-galactosidases are less resistent to heat-treatment than the β-galactosidase of E. coli lacZ. Treatment of A. tumefaciens, Rhizobium or plant material during 15 min. at 56 °C inactivated the endogenous β-galactosidase activities for over 99%, whereas the E. coli lacZ β-galactosidase activity was only reduced by 40%.

INDUCTION OF A. TUMEFACIENS VIRC GENE BY A PLANT FACTOR

Cells of E. coli and A. tumefaciens, containing pMP30 and the cloned virC promoter (plasmid pMP31), were tested for induction of the virC promoter [14] according to the scheme shown in Figure 5. The essential results are:

1. Cultivation with seedlings or cultivated roots induced to a higher level than exudates of these materials. The inducing factor therefore seems to be labile or to be metabolized by the bacteria.

2. Induction was found in E. coli as well as in A. tumefaciens. The presence of a Ti plasmid with a complete set of vir genes in A. tumefaciens did not alter the degree of induction. Induction of virC promoter therefore is independent of other vir genes.

3. The inducing activity of exudates was retained by dialysis membranes (cut-off M_r 7000) and was nearly destroyed by heating at 55 °C for 15 min or by treatment with pronase or trypsin. This suggests that the inducer is a protein-like substance.

4. The plant factor is produced by dicotyledons as well as by some monocotyledons.

The results show that a proteinous plant factor induces at least one of the virulence genes of A. tumefaciens and that wounding, a prerequisite for tumour formation, is not required. An obvious problem is how such a large protein-like substance is able to activate the promoter as the outer membrane of E. coli permits the entry of aqueous solutes up to a M_r of only 700 [15]. Moreover, as cell surfaces of E. coli and A. tumefaciens have little in common, it has to be explained how induction of virC was observed in both backgrounds. In theory we can envisage three possibilities (Fig. 6): (i) self-promoted permeation [16]; (ii) enzymatic cleavage of cell surface components, thereby generating the inducer, and (iii) degradation of the plant factor by bacterial cell surface or secretion products, thereby generating the inducer.

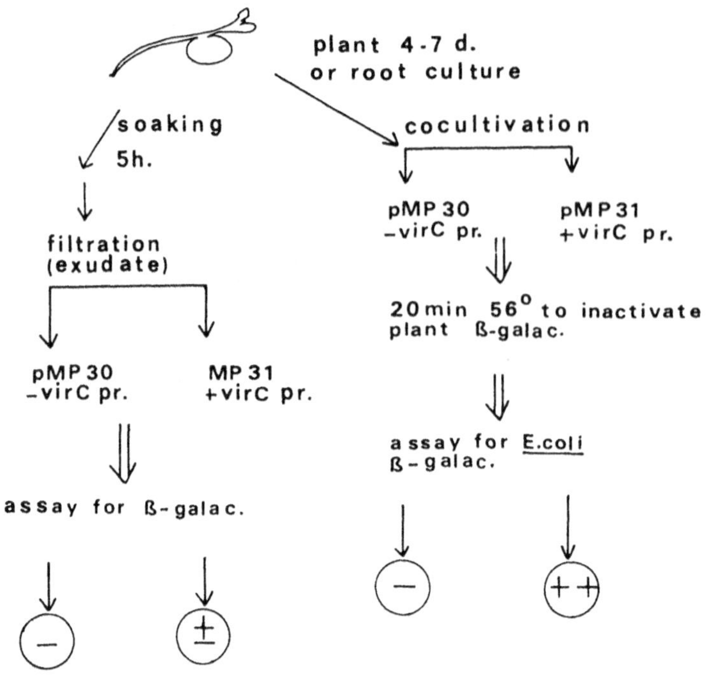

Figure 5. Experimental protocol for testing of the induction of promoter virC by plants or plant-derived material. Xgal develops a blue colour after cleavage by β-galactosidase.
-, no blue colour detectable after 16 h, 29 °C
±, blue colour detectable after 16 h, 29 °C
++, blue colour after 4 h, 29 °C

Although the results show that wounding is not required for induction of virC, we recently discovered that, when the plant is wounded 24 h prior to the addition of bacteria, the virC inducing activity of the plant is many times greater. Future experiments will assess whether the same factor is involved.

DISCUSSION

The experiments show that a plant factor(s) induces the virC gene of A. tumefaciens. It has also been shown that the common nod genes of R. leguminosarum are induced by a component of plant exudate [17]. Moreover, many or even all Sym plasmid-located nod genes are induced by a plant factor [18]. With respect to its nature, tentative results indicate that the latter factor has a low molecular weight and is found in many plants [19]. In conclusion, one of the initial interactions of Agrobacterium and Rhizobium with their host plants is initiated by plant factors which seem to be distinct in the two aforementioned cases.

It is likely that chemotaxis to root exudate and adhesion to the plant surface are steps that usually precede induction of the interaction genes and that, once the bacteria are present at the plant surface, the concentration of inducer is sufficiently high to induce the interaction genes.

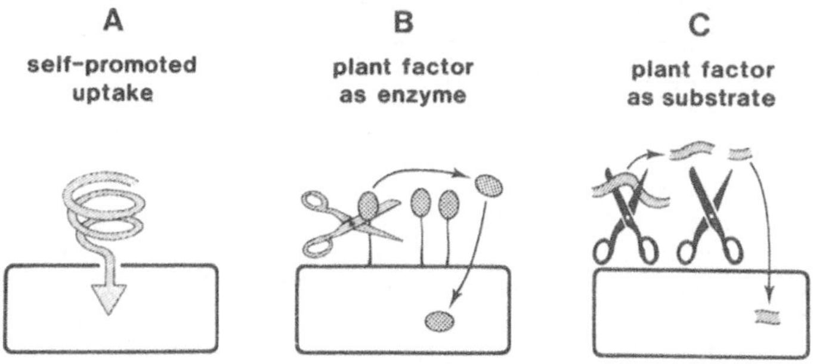

Figure 6. Models for uptake of virC inducing plant factor. For further explanation, see text.

REFERENCES

1. Hooykaas, P.J.J. and Schilperoort, R.A., 1984, The molecular genetics of crown gall tumorigenesis, Adv. Gen., 22: 209-283.
2. Hille, J., Hoekema, A., Hooykaas, P. and Schilperoort, R., 1984, Gene organization of the Ti-plasmid in: "Genes Involved in Microbe-Plant Interactions", D.P.S. Verma and T.H. Hohn, eds., pp. 287-309, Springer-Verlag, Wenen and New York.
3. Douglas, C.J., Staneloni, R.J., Rubin, R.A. and Nester, E.W., 1985, Identification and genetic analysis of an Agrobacterium tumefaciens chromosomal virulence region, J. Bacteriol., 161: 850-860.
4. Rolfe, B.G. and Shine, J., 1984, Rhizobium-leguminosae symbiosis: the bacterial point of view, in: "Genes Involved in Microbe-Plant Interactions", D.P.S. Verma and T.H. Hohn, eds., pp. 95-128, Springer-Verlag, Wenen and New York.
5. Noel, K.D., Sanchez, A., Fernardez, L., Leemans, J. and Cevallos, M.A., 1984, Rhizobium phaseoli symbiotic mutants with transposon Tn5 insertions, J. Bacteriol., 158: 148-155.
6. Gelvin, S.B., Gordon, M.P., Nester, E.W. and Aronson, A.I., 1981, Transcription of the Agrobacterium Ti plasmid in the bacterium and in crown gall tumors, Plasmid, 6: 17-29.
7. Casadaban, M.J. and Cohen, S.N., 1979, Lactose genes fused to exogenous promoters in one step using a Mu-Lac bacteriophage: in vivo probe for transcriptional control sequences, Proc. Natl. Acad. Sci. USA, 76: 4530-4533.
8. Castilho, B.A., Olfson, P. and Casadaban, M.J., 1984, Plasmid insertion mutagenesis and lac gene fusion with mini-Mu bacteriophage transposons, J. Bacteriol., 158: 488-495.
9. Stachel, S.E., An, G., Flores, C. and Nester, E.W., 1985, A Tn3 lacZ transposon for the random generation of β-galactosidase gene fusions: application to the analysis of gene expression in Agrobacterium, EMBO J., 4: 891-898.
10. Hille, J., Van Kan, J., Klasen, I. and Schilperoort, R., 1983, Site-directed mutagenesis in Escherichia coli of a stable R772::Ti cointegrate plasmid from Agrobacterium tumefaciens, J. Bacteriol., 154: 693-701.
11. Scott, D.B. and Ronson, C.W., 1982, Identification and mobilization by cointegrate formation of a nodulation plasmid in Rhizobium trifolii, J. Bacteriol., 151: 36-43.

12. Beringer, J.E., Hoggan, S.A. and Johnston, A.W.B., 1978, Linkage mapping in Rhizobium leguminosarum by means of a R plasmid-mediated recombination, J. Gen. Microbiol., 104: 201-207.
13. Beringer, J.E., Beynon, J.L., Buchanon-Wollaston, A.V. and Johnston, A.W.B., 1978, Transfer of the drug resistance transposon Tn5 to Rhizobium, Nature, 276: 633-634.
14. Okker, R.J.H., Spaink, H., Hille, J., Van Brussel, A.A.N., Lugtenberg, B. and Schilperoort, R.A., 1984, Plant-inducible virulence promoter of the Agrobacterium tumefaciens Ti plasmid, Nature, 312: 564-566.
15. Lugtenberg, B. and Van Alphen, L., 1983, Molecular architecture and functioning of the outer membrane of Escherichia coli and other gram negative bacteria, Biochim. Biophys. Acta, 737: 51-115.
16. Hancock, R.E.W., 1984, Alterations in outer membrane permeability, Annu. Rev. Microbiol., 38: 237-264.
17. Van Brussel, A.A.N., Zaat, S.A.J., Canter-Cremers, H.C.J., Wijffelman, C.A., Pees, E., Tak, T. and Lugtenberg, B.J.J., 1985, Role of plant root exudate and Sym plasmid-localized nodulation genes in the synthesis by Rhizobium leguminosarum of Tsr factor, which causes thick and short roots on common vetch, J. Bacteriol., 165: 517-522.
18. Innes R.W., Kuempel, P.L., Plazinski, J., Canter-Cremers, H., Rolfe, B.G. and Djordjevic, M.A., 1985, Plant factors induce expression of nodulation and host-range genes in Rhizobium trifolii, Mol. Gen. Genet., 201: 426-432.
19. Mulligan, J.T. and Long, S.R., 1985, Induction of Rhizobium meliloti nodC expression by plant exudate requires nodD, Proc. Natl. Acad. Sci. USA, 82: 6609-6613.

Note added in proof

Recently it has been reported that the low molecular weight substance acetosyringon and structurally related compounds are inducers of Vir operons (Stachel et al., Nature, 318: 624-629). Therefore we retested quantitatively our earlier results, which are mentioned in the present paper, with the improved expression vector pMP190. The virC promoter was induced 2-3.5-fold in E. coli by co-cultivation with Vicia sativa and 100 μM acetosyringon, and it was induced 11-fold in A. tumefaciens Ti^+ by 100 μM acetosyringon. Induction in E. coli by exudates and A. tumefaciens Ti^+ and Ti^- by co-cultivation or exudates was not observed in these recent experiments. The recently used vector pMP190 produces about 50-fold higher β-galactosidase activity under lacP control than pMP30 did. Contamination of plant β-galactosidase in the exudates used may be responsible for the earlier results which indicated a proteinaceous inducer. The low but significant induction during co-cultivation of virC in E. coli is probably due to acetosyringon or a related substance, which is masked in A. tumefaciens by other compounds exuded during co-cultivation.

INDOLEACETIC ACID AND CYTOKININS IN THE OLIVE KNOT DISEASE. AN OVERVIEW OF THEIR ROLE AND THEIR GENETIC DETERMINANTS

G. Surico

Istituto Tossine e Micotossine da Parassiti Vegetali
Consiglio Nazionale delle Ricerche
Bari
Italy

INTRODUCTION

Pseudomonas syringae pv. *savastanoi* (Smith) Young, Dye *et* Wilkie induces hyperplasia on several plant species. The pathogen is commonly associated with olive (*Olea europaea* L.) and oleander (*Nerium oleander* L.) plants but it is also pathogenic on ash (*Fraxinus excelsior* L.), privet (*Ligustrum japonicum* Thunb.), jasmine (*Jasminus* spp.) and forsythia (*Forsythia intermedia* Zab.).

Although excrescences on olive trees have been known since the time of the Greek philosopher Theophrastus (300 B.C.), it was not until the first years of this century that the bacterial origin of the disease was proved by Savastano in Italy [1]. Smith [2] described the bacterium and named it *Bacterium savastanoi*. Later [3] Stevens included this species in the genus *Pseudomonas* and Young et al. [4] finally classified all isolates of *P. savastanoi* as members of *Pseudomonas syringae* pv. *savastanoi*. A new classification of pv. *savastanoi* has been recently proposed by Janse [5] but to avoid confusion with preceding papers we shall refer to the nomenclature of Young et al. [4].

Relatively little has been published on pv. *savastanoi* and until recently almost nothing was known about the mechanism by which the bacterium induces disease. In the last decade, characterized by extensive use of genetic techniques in the study of the molecular nature of pathogenesis, decisive, but not conclusive, advances have been made in the genetics of pv. *savastanoi* and the factors involved in the expression of disease symptoms. This paper summarizes current knowledge on the role of indoleacetic acid and cytokinins in the formation of knots and on their genetic determinants. Most of the information can be found in the references cited.

DESCRIPTION OF THE DISEASE

Symptomatology

The most obvious external evidence of the disease in established olive trees is the formation of knots on the young stems and branches and, occasionally, on the leaves and fruits.

NATO ASI Series, Vol. H1
Biology and Molecular Biology of Plant-Pathogen Interactions
Edited by J. Bailey

The knots first appear as small, green-coloured protuberances which gradually grow into partly organized masses of tissue. Olive knots histologically resemble those induced by other pathogens. They are mainly composed of peridermal cells, irregularly shaped parenchimal cells and meristematic cells which give rise to vascular tissues partly joined to those of the host origin [6].

With ageing the knots crack and become partly necrotized as a result of an insufficient nutrient supply due to an inadequate tissue vascularization of the parenchimatous tissues. Thus, knot tissue may die within 6-8 months after their formation but often they last for a long time and increase in volume year after year.

On oleander shoots, leaves and fruits the symptoms of the disease are similar to those induced by the bacterium on olive trees with the difference that knots formed on leaves are often surrounded by a chlorotic halo. In addition, vertically elongated canker-like lesions can be observed on young shoots.

Aetiology and Pathogenesis

Proliferation of pv. savastanoi and formation of knots are found to occur particularly in spring and autumn. Some reports have attributed olive knot disease epidemics to inoculum formed by bacteria present as epiphytes on host plants [7].

Infection has been observed to occur mainly through wounds, particularly through leaf scars and injuries caused by freezing [8]. The bacterium invades the intercellular spaces of the nearby host cells and colonization generally proceeds no further than the infection site, with the exception of oleander plants where the pathogen moves through the laticifers and induces secondary knots [9].

Evidence from artificial inoculations on olive stems suggests that the bacterium must invade the cambium for infection to occur [6]. The first reaction of the inoculated tissues consisted in either a quick renewal, in summer and winter, or an acceleration, in spring, of the cambial activity. As a result of the continued activity of pre-formed and new differentiated cambia, mounds of callus tissue were produced at the inoculated sites. Irregular cavities containing many bacteria were formed by enzymatic action, within the developing knots. Later fissures were found to originate from these cavities and reach the surface of intact knots. The bacteria thus reached the surface of the knots and became available for other infections.

Extensive studies on oleander plants [10] have shown that knot development involves: i) colonization of the infection site by the bacterium and local formation of lysogenous cavities; ii) abnormal enlargement and proliferation of the host cells around the bacterial cavities and, iii) differentiation of xylem and, probably, phloem elements in the proliferating area. The developmental pattern of hyperplasia on both olive and oleander plants suggest that at least two different growth factors could be involved in pathogenesis.

THE ROLE OF PHYTOHORMONES

The development of a neoplastic condition generally involves factors that release the cells of the host from normal morphogenetic influences. Such factors include growth promoting substances produced by pathogens at the infection site and diffusing away from that site. This abnormal supply of phytohormones may enable the normal biosynthetic systems controlling the status or the proliferation of plant cells to resume their activity, resulting in growth responses of plants to pathogens. Earlier studies showed that many fungal and bacterial diseases often produce phenomena in plants that indicate an increase in auxin and cytokinins content. Both of these groups of growth promoting substances have been found to be produced in culture by pv. savastanoi [11,12,13].

The results of extensive studies by several authors [6,10,11,14] suggest that symptoms of the knot disease on olive and oleander plants are associated with the accumulation of hyperauxinic concentrations of indoleacetic acid (IAA) in infected tissues. The earliest work on formation of knots as a response of the plant to bacterial production of IAA was done by Smidt and Kosuge [15]. They observed that IAA deficient mutants failed to induce knots on oleander but grew in the leaves like their parent strains. However, it is not clear whether or not the amount of IAA produced by the bacterium also limits the size of mature knots. Later observations confirmed the primary role of IAA in knot induction on oleander as well on olive plants [16], and also revealed that the amount of IAA produced by the pathogen influences the period of incubation of the disease rather than the size of knots, whereas the amount of cytokinins produced in culture is related to the knot enlargement [17]. These observations were carried out with six wild pathogenic strains of pv. savastanoi: three from olive (olive strains) and three from oleander (oleander strains). IAA deficient mutants were obtained from wild type strains by selection for resistance to the tryptophan analogue α-methyltryptophan using the procedure of Smidt and Kosuge [15]. Selection for α-methyltryptophan resistance yields mutants deficient in IAA production in most strains of pv. savastanoi, probably because IAA deficient mutants have higher intracellular tryptophan. Each strain along with its respective IAA deficient mutant was investigated for phytohormone production, presence of plasmids and pathogenicity. Moreover, two oleander strains, ITM519 and NCPPB640, were further studied for IAA metabolism and cytokinin production [18,19]. The results of these investigations are discussed in the next sections.

Indoleacetic acid

The results of pathogenicity tests and phytohormone accumulation are summarized in Tables 1, 2 and 3. The analysis of the data reported show that typical knots only formed on olive and oleander plants following inoculation with pv. savastanoi strains that produced cytokinins in addition to IAA.

With respect to the role of IAA all wild type strains of pv. savastanoi reproduced the same symptoms of the disease on olive plants after a 4-6 week period. Symptoms were first seen 7-12 days after inoculation with strains NCPPB640, ITM519, PBa219 and ITM317, and 21 days after inoculation with the other two strains. Comparison of these results with IAA production indicated that the period of incubation of the disease on olive plants was related to the amount of IAA released into the culture media by pv. savastanoi. The most virulent strains (shortest period of incubation) were those which also produced most IAA.

The pathogenic behaviour of the IAA deficient mutants on olive (and oleander) plants confirmed the primary role of IAA in knot induction. In fact, olive plants inoculated with IAA deficient mutants of all olive strains did not show any symptoms of the disease. However, peculiar expressions of the disease were observed on both olive and oleander plants following inoculation with some IAA deficient mutants. Olive plants inoculated with IAA deficient mutants of oleander strains NCPPB640 and ITM519 showed atypical tissue proliferation which appeared smaller, greener, softer and more corrugated than knots produced by parental strains, while IAA deficient

Table 1. Relationship between indoleacetic acid (IAA) production in culture media and virulence of six strains of *Pseudomonas syringae* pv. *savastanoi* on olive plants [Data from Surico et al., 17].

Strain	Host of Origin	IAA accumulation in minimal medium (mg/l)	Incubation period on olive plants (days)
NCPPB640	Oleander	8.44	7
ITM317	Olive	3.18	12
PBa219	Oleander	2.90	12
ITM519	Oleander	2.49	12
NCPPB639	Olive	0.90	21
PBa225	Olive	0.0	21

mutants of the oleander strains inoculated on oleander plants induced necrotic symptoms and, occasionally, some outgrowths at the site of inoculation. Only wild type oleander strains induced typical disease symptoms on oleander plants.

The abnormal overgrowth induced on olive plants by mutants from strains NCPPB640 and ITM519 might be due to their ability to utilize plant tryptophan to synthesize IAA in the inoculated tissues and thus induce symptoms. This hypothesis is supported by three considerations: i) tryptophan-requiring mutants induce symptoms on olive plants [Surico, data not published]; ii) IAA deficient mutants are able to produce small amounts of IAA in the presence of added tryptophan and, iii) strain PBa225, which produces detectable amounts of IAA only in the presence of added tryptophan, induces small but typical knots. On the contrary, we have not yet found any explanation for the fact that, in addition to producing cytokinins, mutants PBa219-A, PBa225-A and ITM317-A also synthesize IAA in the presence of tryptophan, although they do not induce any typical or atypical overgrowths. Of course we cannot exclude that mutants NCPPB640-A and ITM519-A induce abnormal growth through the production of still unknown growth promoting substances.

The pathogenic behaviour of IAA deficient mutants of strains from oleander on oleander plants could also be due to the activity of the IAA produced in the host tissues in the presence of plant tryptophan. Actually, applications of IAA on oleander leaves induce chlorosis or necrosis, but not overgrowth, of the infiltrated tissues, depending on the concentration used [18]. The former effect of IAA at concentrations of about 10^{-5} M could explain the chlorotic halos that are often observed around knots developing on oleander leaves. The mechanism by which the growth substance induces that alteration has not yet been investigated but it is likely that the appearance of chlorotic symptoms could be mediated by ethylene, another phytohormone, to which many disease symptoms including chlorosis, necrosis, epinasty etc. have been attributed [20]. According to several reports, plants treated with IAA are stimulated to produce ethylene. Thus, the production of ethylene in oleander plants may result from the increase in IAA that occurs in infected tissues. Considerations of IAA synthesis by IAA deficient mutants are reported in a following paragraph.

IAA in Relation to the Epiphytic Survival of pv. savastanoi

The conversion of tryptophan into IAA represents a secondary pathway of the metabolism in pv. savastanoi [21]. The association of pathogenicity with this secondary metabolite allows avirulent mutants of the bacterium to grow in culture as well as in host tissues [15]. Nonetheless, it has been suggested that synthesis of IAA may confer a selective advantage to the bacterium in nature. In fact, studies [22] on the epiphytic survival on olive leaves of IAA deficient mutants from two olive strains indicated that IAA may contribute to parasitic fitness of pv. savastanoi. It was observed that the number of bacteria sprayed on olive leaves declined within 3-5 days after contamination to only a few colony forming units/cm^2 of leaf. Then, the number of wild type strains increased, reaching about 10^2-10^3 colony forming units/cm^2 30 days after contamination. IAA deficient mutants, on the contrary, failed to multiply and were either scarcely or not at all detectable. Thus, it seems that the production of IAA is associated with better epiphytic survival of pv. savastanoi, although we cannot rule out the possibility that there are other significant differences between IAA deficient mutants and wild type strains.

Cytokinins

Table 2 compares the average fresh weight of knots harvested 60 days after inoculation with the total cytokinin activity of pv. savastanoi culture media. The data suggest that the size of the knots is related to the amount of cytokinins produced in culture. The culture media of strains NCPPB640, ITM317, ITM519, and PBa225 which induced the largest knots contained the highest cytokinin activity, whereas strains NCPPB639 and PBa219, which induced small-sized knots, produced only 0.0 to 0.01 mg zeatin equivalent per litre of culture medium.

Table 2. Relationship between cytokinins production in culture media and virulence of six strains of Pseudomonas syringae pv. savastanoi on olive plants [Data from Surico et al., 17].

Strain	Host of Origin	Cytokinin activity	Fresh weight of knots (mg) 60 days after inoculation
NCPPB640	Oleander	5.3	307
ITM317	Olive	4.4	224
ITM519	Oleander	3.0	204
PBa225	Olive	2.96	180
PBa219	Oleander	0.01	142
NCPPB639	Olive	0.0	108

Cytokinin activity was determined in the cucumber cotyledons bioassay and is expressed as mg zeatin equivalents l^{-1} culture medium.

Table 3. Relationship between indoleacetic acid (IAA) and cytokinin production in culture media and virulence of IAA deficient mutants of five wild type Pseudomonas syringae pv. savastanoi strains [Data from Surico et al., 17].

IAA deficient mutants	Host of origin	IAA accumulation in minimal medium -Trp	+Trp	Cytokinin activity	Plant response to inoculation Olive	Oleander
NCPPB640-A	Oleander	1.39	1.90	5.0	AK (85)	NS
PBa219-A	Oleander	0.0	1.66	trace	-	NS
ITM519-A	Oleander	0.0	1.51	3.0	AK (71)	NS
PBa225-A	Olive	0.0	2.02	3.0	-	--
ITM317-A	Olive	0.93	2.13	4.1	-	--

AK, atypical knots; NS, necrotic swellings; -, no symptoms observed. Numbers in parentheses indicate the average fresh weight in milligrams of 24 abnormal knots.

Strain PBa219, which induced knots of very small size, accumulated an amount of IAA comparable to that accumulated by strains ITM317 and ITM519 which induced larger knots. This indicates that the effect of IAA and cytokinins occurs during different stages of disease development. Certainly cytokinins are not directly implicated in knot stimulation since IAA deficient mutants, which produce the same cytokinin activity as their parent strains, substantially fail to induce knot symptoms. Consequently, the effect of cytokinins must occur mainly after knot initiation. Previous investigations [23] on the multiplication of pv. savastanoi in olive tissues demonstrated that the size of the knots is not correlated to the number of viable bacteria present at each inoculation site and revealed that the size of knots increases slowly for the first 27-34 days after inoculation. Thereafter, knots grow rapidly while the multiplication of bacteria decreases. It was also observed that the rapid increase of the volume of knots was concurrent with the proliferating activity of meristematic regions which appeared inside the parenchimatous tissue 4-5 weeks after inoculation. The delayed differentiation of meristematic tissues and the consequent production of new tissues followed by knot enlargement could be related to the achievement of optimal concentrations of cytokinins, also in relation to the concentration of other hormones such as IAA, in host tissues. It remains to be established whether these compounds are produced by the bacterium or whether they are of host origin. At the present time we only know that, i) pv. savastanoi produces cytokinins in culture, ii) strains which produce higher amounts of cytokinins induce larger knots, and iii) the total cytokinin activity of ethyl acetate extract of diseased tissues is much greater than that of healthy tissues [24]. The isolation of cytokinin deficient mutants and the determination of their pathogenic properties will conclusively establish the role of cytokinin production in knot formation.

METABOLISM OF IAA IN PV. SAVASTANOI

Pathovar savastanoi synthesizes IAA from L-tryptophan via indoleacetamide [25]. Two enzymes are involved in IAA synthesis: tryptophan

Table 4. Distribution of indole compounds in six strains of Pseudomonas syringae pv. savastanoi and in their IAA deficient mutants.

Strain	Host of origin	IAA	IAA-Lys	Ac-IAA-Lys	Unidentified indoles
NCPPB640	Oleander	+	-	+	+
NCPPB640-A[a]	Oleander	trace	-	-	-
PBa219	Oleander	+	-	+	-
PBa219-A[a]	Oleander	-	-	-	-
ITM519	Oleander	+	-	+	-
ITM519-A[a]	Oleander	-	-	-	-
PBa225	Olive	+	-	-	-
PBa225-A	Olive	-	-	-	-
ITM317	Olive	+	-	-	+
ITM317-A[a]	Olive	trace	-	-	-
NCPPB639	Olive	trace	-	-	-
ITM317[b]	Olive	-	-	-	+

Pathovar savastanoi strains were cultured for 5 days in Woolley's medium and then sampled for indole accumulation. Strain ITM317 was also cultured for 10 days (ITM 317[b]).
[a] IAA deficient mutants; IAA, indoleacetic acid; IAA-Lys, indole-3-acetyl-ε-L-Lysine; Ac-IAA-Lys, α-N-acetyl-indole-3-acetyl-ε-L-Lysine.

2-monooxygenase and indoleacetamide hydrolase. The first catalyzes the conversion of L-tryptophan into indoleacetamide; the second, catalyzes the hydrolysis of indoleacetamide into IAA and ammonia. In addition, the pathogen possesses several systems which control the accumulation of IAA in culture. During the growth of pv. savastanoi in culture, IAA was first found to accumulate but later its concentration decreased with time until it disappeared altogether. Both IAA and indoleacetamide inhibit tryptophan 2-monooxygenase, thus regulating the rate of IAA synthesis, whereas IAA is metabolized to other compounds [21]. Two of these compounds have been identified as indoleacetyl-ε-L-lysine (IAA-Lys) [26] and α-N-acetyl-indoleacetyl-ε-L-lysine (Ac-IAA-Lys) [18] (Fig. 1). The former product of IAA bioconversion has been detected in the culture filtrate of a Californian oleander strain; the latter in the culture filtrates of Italian oleander strains. Both compounds showed less growth stimulating activity than IAA, indicating that pv. savastanoi has a means of converting IAA to less active forms although this capacity is not widespread among strains of the pathogen. Table 4 reports the results of a study on the presence of indole compounds in the culture filtrates of different strains of pv. savastanoi. The culture medium was sampled after 5 days (5 and 10 days for strain ITM317) of incubation in Woolley's medium [27] in shake culture. All wild type oleander strains accumulated in culture Ac-IAA-Lys but not IAA-Lys which was not

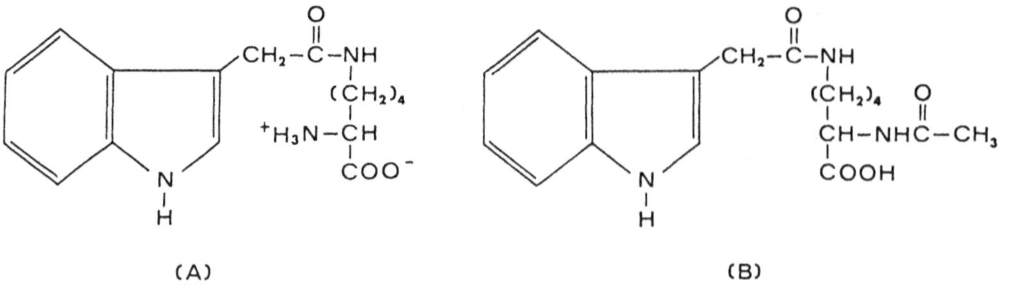

Figure 1. Structure of indole-3-acetyl-ε-L-lysine (A) and α-N-acetyl-3-acetyl-ε-L-lysine (B).

detected in any of the culture filtrates sampled. Wild type olive strains and IAA deficient mutants from olive and oleander strains did not accumulate any detectatable amount of Ac-IAA-Lys. This may be considered as an expected result for those mutants which produce little if any IAA but not for the olive strains. Thus, the behaviour of olive strains indicates that the enzymatic system which accounts for IAA degradation is not possessed by all pv. savastanoi strains or it may be of a different nature. In fact, from the culture filtrate of strain ITM317 grown for 5 or 10 days, an as yet unidentified indole compound was isolated, which induced overgrowth on potato discs and chlorosis on oleander leaves. Finally, mutant ITM519-A incubated on Woolley's medium supplemented with 0.3% of IAA did not accumulate Ac-IAA-Lys or other related compounds. This suggests, that, together with the production of IAA, mutant ITM519-A lost its capacity to transform it into other metabolites.

IAA deficient mutants analyzed for their enzyme content showed an absence of tryptophan 2-monooxygenase and indoleacetamide hydrolase. Since these mutants synthesize IAA in the presence of added tryptophan, with the only exception of mutants NCPPB640-A and ITM317-A which do the same even in the absence of added tryptophan, IAA synthesis by the IAA deficient mutants must proceed along a pathway different from the one involving the two enzymes reported before. This pathway, whose contribution for IAA synthesis appears to be less than that of the other, probably involves indolepyruvic acid as an intermediate [28].

CYTOKININ IDENTIFICATION

Strains of pv. savastanoi have been shown to release cytokinin compounds into the culture medium, although individual cytokinin contents may vary greatly between strains. Of the four cytokinins isolated from NCPPB640 culture medium (Fig. 2), two were identified as 6-(4-hydroxy-3-methylbut-trans-2-enylamino)purine (Zeatin, Z) and 6-(4-hydroxy-3-methylbut-trans-2-enylamino)-9-β-D-ribofuranosylpurine (Zeatin riboside, ZR), by comparison of their biological activities, chromatographic properties, ultraviolet- and mass-spectra with those of authentic, synthetic compounds [24]. The other two are novel compounds. They have been identified as 6-(4-hydroxy-1,3-dimethylbut-trans-2-enylamino)purine (1'-methylzeatin, 1'-MeZ) and 6-(4-hydroxy-1,3-dimethylbut-trans-2-enylamino)-9-β-D-ribofuranosylpurine (1"-methylzeatin riboside, 1"-MeZR) on the basis of their biological activities, chromatographic properties, ^{1}H and ^{13}C nuclear magnetic resonance and mass spectrometry [19,29].

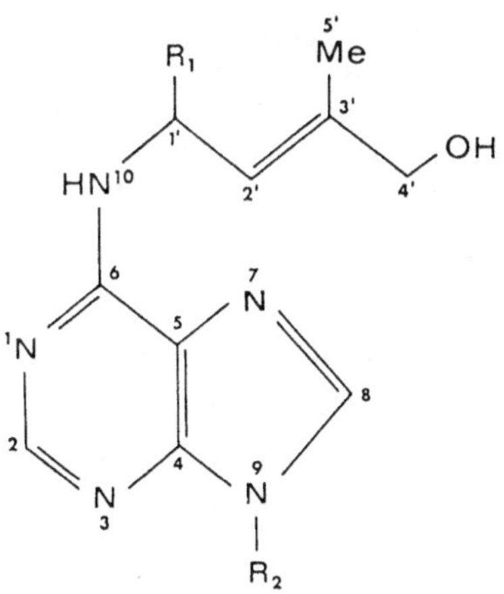

	R_1	R_2
6-(4-hydroxy-3-methylbut-trans-2-enylamino)purine	H	H
6-(4-hydroxy-1,3-dimethylbut-trans-2-enylamino)purine	CH_3	H
6-(4-hydroxy-3-methylbut-trans-2-enylamino)-9-β-D-ribofuranosylpurine	H	β-D-ribosyl
6-(4-hydroxy-1,3-dimethylbut-trans-2-enylamino)-9-β-D-ribofuranosylpurine	CH_3	β-D-ribosyl

Figure 2. List of cytokinins isolated from Pseudomonas syringae pv. savastanoi strain NCPPB640 culture medium.

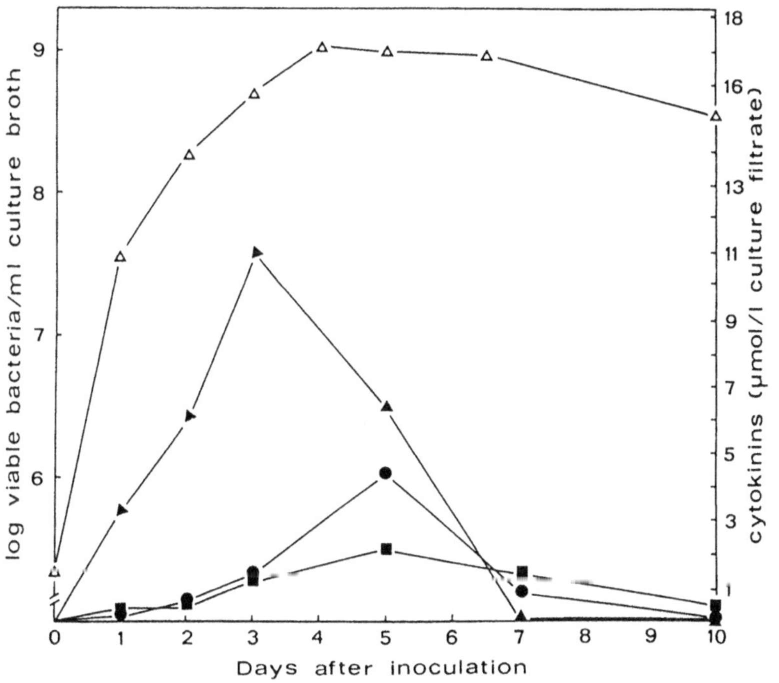

Figure 3. Relationship between culture growth and accumulation of cytokinins. Growth of Pseudomonas syringae pv. savastanoi strain NCPPB640 (—△—△—); accumulation of trans-zeatin (—▲—▲—), trans-zeatin riboside (—●—●—), and 1'-methyl-trans-zeatin plus 1"-methyl-trans-zeatin riboside (—■—■—). [Data from Surico et al. 22].

Z was found in the culture medium of all strains examined thus far [30], while ZR was detected in the culture medium of oleander strains NCPPB640, PBa204 and ITM519 but not in that of olive strains PBa222, PBa225 and ITM317. The content of the methylated derivatives of Z and ZR may also vary among strains; the culture medium of strain ITM519, for example, does not show accumulation of either compound.

Time course experiments indicate that the total cytokinin activity of the NCPPB640 culture medium obtained after 1, 2, 3, 5, 7 and 10 days of incubation increased with the growth of bacteria (Fig. 3). The relative quantities of the four cytokinins varied as well. In particular, the amount of Z present in the 3-day-old culture filtrates when bacteria are still in the growth phase, was calculated to reach approximately 11 μmol/l. Thereafter, its level decreased while the amount of the three other cytokinins increased. In fact, the level of ZR and that of 1'-MeZ plus 1"-MeZR increased progressively with bacterial growth and reached the maximum five days after inoculation. After 10 days of incubation, the total amount of the four cytokinins was only 0.63 μmol/l of culture filtrate and the predominant compounds were the methylated derivatives of Z and ZR. The variation in cytokinin content during growth in culture may to some extent explain the absence of individual cytokinins in the culture filtrates of some strains. No information is presently available on the biosynthesis of the four cytokinins. We can only infer from the time course experiments that two types of metabolic pathways may occur: interconversion pathways and degradation pathways.

PLASMIDS

Loss of IAA production in oleander strains is associated with loss or rearrangement of extrachromosomal DNA. Comai and Kosuge [31,32] reported that, in the oleander strain EW2009, the genetic loci (iaaM and iaaH) for the two enzymes which catalyze the conversion of tryptophan into indoleacetamide and then to IAA (tryptophan $\xrightarrow{\text{oaaM}}$ indoleacetamide $\xrightarrow{\text{iaaH}}$ indoleacetic acid) are borne on a plasmid called pIAA1. IAA deficient mutants of this strain selected for resistance to α-methyltryptophan or obtained after acridine orange curing, lack both iaaM and iaaH activities. To demonstrate that plasmid pIAA1 is associated with IAA synthesis, an IAA deficient mutant of strain EW2009 was transformed with pIAA1. This procedure restored iaaM and iaaH activities and the mutant became pathogenic again. Moreover, pLUC1, a recombinant plasmid, was constructed from RSF1010, a vector plasmid, and from fragments generated by restriction endonuclease digestion of pIAA1 [32]. The introduction of pLUC1 into pIAA1-less strains conferred indoleacetamide production and restored pathogenicity on oleander but not at the level restored by pIAA1. It has been suggested [21] that host enzymes converted indoleacetamide into IAA and that the latter compound caused knot formation. An alternative explanation is that the indoleacetamide of bacterial origin was not further converted into IAA in the plant and that the smaller knots were induced by the IAA synthesized through the subsidiary IAA pathway, as seen in pv. savastanoi.

A considerably different situation has been observed in the localization of IAA genes among strains of pv. savastanoi. All the oleander strains so far examined harbour a plasmid associated with IAA production but its estimated molecular weight ranges from 52 to 73 kilobase-pair (kb). Restriction endonuclease mapping and hybridization studies show that the IAA plasmid (pIAA2, 73 kb) of two Italian oleander strains (PBa205 and PBa213) do not resemble pIAA1 in its EcoRI endonuclease fragments, except in the region bearing the IAA genes [33]. The other IAA plasmids have not yet been examined in detail.

Table 5. Distribution of plasmids in some Pseudomonas syringae pv. savastanoi strains.

Strain	Host of origin	Number of plasmids	Range of molecular weights (kb)	Molecular weight of IAA plasmid (kb)
EW1017	Olive (C)	3	50-97	70
EW2009	Oleander (C)	4	22-52	52
NCPPB640	Oleander (Y)	3	48-83	73
PBa219	Oleander (I)	3	36-68	68
ITM519	Oleander (I)	6	31-75	75
PBa205	Oleander (I)	5	40-80	73
PBa213	Oleander (I)	5	40-80	53
NCPPB639	Olive (Y)	6	7-88	-
PBa225	Olive (I)	6	27-91	-
PBa202	Olive (I)	6	35-97	-
ITM317	Olive (I)	7	10-88	-
PBa230	Olive (I)	6	25-100	-
PBa207	Olive (I)	7	10-88	-
PBa209	Privet (I)	6	25-80	-
PBa215	Privet (I)	6	25-85	-
PBa217	Privet (I)	6	25-85	-
PBa218	Privet (I)	6	25-88	-

C, California; I, Italy; Y, Yugoslavia.

In the olive and privet strains tested so far, the loss of IAA production is not accompanied by the loss of any plasmid nor is there any homology of the plasmids present in these strains to the iaaM genes. Thus, genes for IAA in these strains are either located on the chromosom or on a very large plasmid which cannot be isolated by the modified Meyers et al. [34] procedure normally used [31].

Plasmid profiles show remarkable differences between oleander strains and olive and privet strains. Generally, oleander strains have 3-4 plasmids, while olive and privet strains contain 6-7 plasmids, with the exception of strain EW1017 which contains only 3 plasmids (Table 5). Plasmid analysis of this atypical olive strain which induces knot symptoms on oleander, unlike other strains from olive or privet, showed that its genes for IAA production occur on a 70 kb plasmid [16]. In fact, mutants cured of the plasmid were non-pathogenic on oleander and olive plants and did not produce IAA. On the basis of all these physiological and pathological characteristics, it was concluded that EW1017 is an oleander strain which spreads under natural conditions from oleander to olive plants. Since strains of pv. savastanoi from oleander are pathogenic on olive, it is conceivable that bacteria with characteristics from oleander may be isolated from olive knots.

CONCLUSIONS

From the foregoing discussion it may be concluded that there is strong evidence to suggest that plant growth regulators of bacterial origin are important in host-pv. savastanoi interactions. In particular, IAA has a decisive effect on the ability of pv. savastanoi to be a pathogen (i.e., to form knots) probably through the promotion of cambial activity. In this instance, IAA acts as a primary determinant or pathogenicity factor. Moreover, IAA may act as a virulence factor. In fact, more virulent strains also produce more IAA. Ability to produce IAA is also advantageous to the survival of the pathogen.

Differently from IAA, cytokinins appear to act only as a secondary determinant. Strains producing elevated levels of cytokinins induce larger knots but IAA deficient mutants, though producers of cytokinins, failed to induce disease symptoms. Higher levels of virulence were obtained when a high production of IAA and cytokinins were associated in the same strain. However, the size of knots and their anatomy is probably controlled by a balance between the amounts of the two growth regulators rather than by the absolute amount of cytokinins. It should also be remembered that no genetic analysis of the relationship between cytokinins and virulence has yet been carried out. Until this is done the possibility remains that cytokinins are necessary, but not sufficient, for knot production.

Pathovar savastanoi seems to possess two pathways for IAA synthesis. One, which is more effective in the production of IAA, involves indoleacetamide as an intermediate metabolite, and the other probably involves indolepyruvic acid. Moreover, the pathogen possesses several systems which control the rate of IAA synthesis and its successive degradation to fewer active growth-promoting substances. Cytokinins are also transformed and degraded during culture growth.

Plasmids ranging from 52 to 73 kb have been shown to carry determinants for IAA production in oleander strains. Genetic analysis has shown them to be related only in the region bearing IAA genes. Thus, IAA plasmids are quite diverse. The diversity of these plasmids, and the likely chromosomal location of the IAA genes in certain strains, raise the possibility that the IAA operon may be part of a transposable element.

Not as much is known about the location of the genetic determinants for cytokinin production. Certainly they are not located on the plasmid encoding for IAA synthesis since IAA deficient mutants lacking the IAA plasmid are still able to produce cytokinins at a level corresponding to that of their respective parents. However, recent observations indicated that genetic determinants responsible for production of trans-zeatin and its riboside are located on the larger plasmid (90 kb) of the olive strain EW1006 [Akiyoshi and Regier, unpublished data].

All these traits are similar to those of Agrobacterium tumefaciens, another plant pathogenic bacterium which induces a neoplastic disease on a number of dicotyledonous plants. Virulent strains of A. tumefaciens contain a large plasmid, the Ti plasmid (Tumour-inducing plasmid), always necessary for tumour induction. In fact, the hypertrophies that result from autonomous cell proliferation are associated with the stable integration of part of the Ti plasmid into plant nuclear DNA. In addition to genes for tumour induction and other important traits (tumour morphology, host range, utilization of opines, etc.), the Ti plasmid, which varies in size from 150 to 230 kb, carries genes for the synthesis of IAA from tryptophan and for the production of zeatin ribonucleotides. Although the exact mechanism of cellular transformation has not yet been elucidated, IAA and cytokinins, like in pv. savastanoi, are necessary in tumour formation [35].

Production of IAA by crown-gall tumours is encoded by two genes on the T-DNA, the portion of Ti plasmid that is incorporated in the plant genome. These two genes are functionally and structurally related to the IAA genes of pv. savastanoi. They encode a monooxygenase and a hydrolase and their nucleotide sequence shows an obvious relation to iaaM and iaaH of pv. savastanoi [Yamada, Palm, Brook and Kosuge, unpublished data]. This observation, together with the presence of a cytokinin production determinant on the T-DNA [36], draws a fascinating parallel between these two overgrowth-inducing plant pathogens. While pv. savastanoi produces hormones by itself in the plant, A. tumefaciens transforms the plant with genes for hormone production. The close sequence homology between the IAA genes of pv. savastanoi and those of A. tumefaciens is strong evidence that the two IAA production determinants have a common ancestor.

Considerable progress has been made towards the understanding of the biology of pv. savastanoi and of its interaction with olive and oleander. Although each advance has provided us with considerable insight into this system, new and complex questions have been raised. More work will be needed to elucidate the complexities of this fascinating knot-forming plant pathogen.

REFERENCES

1. Savastano, L., 1908, Sulla transmissibilità del bacillo della tubercolosi dell'Olivo nell'Oleandro, Bol. Arb. Ital., 4: 86-87.
2. Smith, E.F., 1911, "Bacteria in Relation to Plant Diseases", Vol. 2 The Carnegie Institution of Washington, D.C.
3. Stevens, F.L., 1913, "The Fungi Which Cause Plant Disease", Macmillan, New York.
4. Young, J.M., Dye, D.W., Bradbury, J.F., Panagopoulos, C.G. and Robbs, C.F., 1978, A proposed nomenclature and classification for plant pathogenic bacteria, N.Z.J. Agric. Res., 21: 153-177.
5. Janse, J.D., 1982, Pseudomonas syringae subsp. savastanoi (ex Smith) subsp. nov., nom. rev., the bacterium causing excrescences on Oleaceae and Nerium oleander L., Int. J. Systematic Bacteriol., 32: 166-169.
6. Surico, G., 1977, Osservazioni istologiche sui tubercoli della "rogna" dell'Olivo, Phytopathol. Medit., 16: 109-125.
7. Ercolani, G.L., 1971, Presenza epifitica di Pseudomonas savastanoi (E.F. Smith) Stevens sull'Olivo, in Puglia, Phytopathol. Medit., 10: 130-132.
8. Ciccarone, A., 1950, Alterazioni da freddo e da rogna sugli ulivi, esemplificate dai danni osservati in alcune zone pugliesi negli anni 1949-1950, Boll. Staz. Patol. Veg. Roma, 6, (1948): 141-174.
9. Wilson, E.E. and Magie, A.R., 1964, Systemic invasion of the host plant by the tumor-inducing bacterium Pseudomonas savastanoi, Phytopathology, 54: 576-579.
10. Wilson, E.E., 1965, Pathological histogenesis in oleander tumors induced by Pseudomonas savastanoi, Phytopathology, 55: 1244-1249.
11. Beltrà, R., 1961, Efecto morfogenetico observado en los extractos hormonales de los tumores del olivo, Microbiol. Espan., 14: 177-187.
12. Surico, G., Sparapano, L., Lerario, P., Durbin, R.D. and Iacobellis, N.S., 1975, Cytokinin-like activity in extracts from culture filtrates of Pseudomonas savastanoi, Experientia, 31: 929-930.
13. Surico, G., Sparapano, L., Lerario, P., Durbin, R.D. and Iacobellis, N.S., 1976, Studies on growth-promoting substances produced by Pseudomonas savastanoi (E.F. Smith) Stevens, in: "Proceedings of The 4th Congress of the Mediterranean Phytopathological Union", pp. 449-458.
14. Beltrà, R., 1958, Relacion entre la concentration de triptofano en los tallos do olivo y la localizacion de los tumores bacterianos, Microbiol. Espan., 11: 401-410.

15. Smidt, M. and Kosuge, T., 1978, The role of indole-3-acetic acid accumulation by alpha methyl tryptophan-resistant mutants of Pseudomonas savastanoi in gall formation on oleanders, Physiol. Plant. Pathol., 13: 203-214.
16. Surico, G., Comai, L. and Kosuge, T., 1984, Pathogenicity of strains of Pseudomonas syringae pv. savastanoi and their indoleacetic acid-deficient mutants on olive and oleander, Phytopathology, 74: 490-493.
17. Surico, G., Iacobellis, N.S. and Sisto, A., 1985, Studies on the role of indole-3-acetic acid and cytokinins in the formation of knots on olive and oleander plants by Pseudomonas syringae pv. savastanoi, Physiol. Plant Pathol., 26: 309-320.
18. Evidente, A., Surico, G., Iacobellis, N.S. and Randazzo, G., 1986, α-N-acetyl-indole-3-acetyl- α -L-lysine: a metabolite of indole-3-acetic acid from Pseudomonas syringae pv. savastanoi, Phytochemistry, 25: 125-128.
19. Surico, G., Evidente, A., Iacobellis, N.S. and Randazzo, G., 1985, A cytokinin from the culture filtrate of Pseudomonas syringae pv. savastanoi, Phytochemistry, 24: 1499-1502.
20. Pegg, G.F., 1976, The involvement of ethylene in plant pathogenesis, in: "Physiological Plant Pathology", R. Heitefuss and P.H. Williams, eds., pp. 582-591, Springer-Verlag, Berlin/Heidelberg/New York.
21. Kosuge, T. and Comai, L., 1982, Metabolic regulation in plant pathogen interactions from the perspective of the pathogen, in: "Plant Infection: The Physiological and Biochemical Basis", Y. Asada et al., eds., pp. 175-186, Japan Sci. Soc. Press, Tokyo/Springer-Verlag, Berlin.
22. Varvaro, L. and Surico, G., 1984, Epiphytic survival of wild types of Pseudomonas syringae pv. savastanoi and their Iaa⁻ mutants on olive leaves, in: "Proceedings of the 2nd Working Group on Pseudomonas syringae pathovars", pp. 20-22.
23. Varvaro, L. and Surico, G., 1978, Moltiplicazione di Pseudomonas savastanoi (E.F. Smith) Stevens nei tessuti dell'Olivo (Olea europaea L.), Phytopathol. Medit., 17: 179-186.
24. Iacobellis, N.S. and Surico, G., 1984, Occurrence of cytokinins in knot tissues induced by Pseudomonas syringae pv. savastanoi, in: "Proc. of the 2nd Working Group on Pseudomonas syringae pathovars", pp. 32-34.
25. Kosuge, T. Heskett, M.G. and Wilson, E.E., 1966, Microbial synthesis and degradation of indole-3-acetic acid. I. The conversion of L-tryptophan to indole-3-acetamide by an enzyme system from Pseudomonas savastanoi, J. Biol. Chem., 241: 3738-3744.
26. Hutzinger, O. and Kosuge, T., 1968, 3-indole-acetyl-L-lysine, a new conjugate of 3-indoleacetic acid produced by Pseudomonas savastanoi, Biochemistry, 7: 183-194.
27. Woolley, W.D., Pringle, R.B. and Braun, A.C., 1955, Isolation of the phytopathogenic toxin of Pseudomonas tabaci, an antagonist of methionine, J. Biol. Chem., 197: 409-417.
28. Beltrà, R., 1964, Tryptophan metabolism of Pseudomonas savastanoi, Microbiol. Espan., 17: 123-131.
29. Evidente, A., Surico, G., Iacobellis, N.S. and Randazzo, G., 1985, Isolation and identification of 1'-methyl-trans-zeatin: a new cytokinin from Pseudomonas syringae pv. savastanoi, Phytochemistry, 24: in press.
30. Surico, G., Evidente, A., Iacobellis, N.S. and Randazzo, G., 1985, On the presence and level of different cytokinins in culture filtrate of Pseudomonas syringae pv. savastanoi, in: "Proceedings of the 6th International Conference on Plant Pathogenic Bacteria".
31. Comai, L. and Kosuge, T., 1980, Involvement of plasmid deoxyribonucleic acid in indoleacetic acid synthesis in Pseudomonas savastanoi, J. Bacteriol., 143: 950-957.
32. Comai, L. and Kosuge, T., 1982, Cloning and characterization of iaaM, a virulence determinant of Pseudomonas savastanoi, J. Bacteriol., 149: 40-46.

33. Comai, L., Surico, G. and Kosuge, T., 1982, Relation of plasmid DNA to indoleacetic acid production in different strains of Pseudomonas syringae pv. savastanoi, J. Gen. Microbiol., 128: 2157-2163.
34. Meyers, J.A., Sander, D., Elwell, L.P. and Falkow, S., 1976, Simple agarose gel electrophoretic method for the identification and characterization of plasmid deoxyribonucleic acid, J. Bacteriol., 127: 1529-1537.
35. Nester, E.W., Gordon, M.P., Amasino, R.M. and Yanofsky, M.F., 1984, Crown gall: a molecular and physiological analysis, Annu. Rev. Plant Physiol., 35: 387-413.
36. Akiyoshi, D.E., Regier, D.A., Jen, G. and Gordon, M.P., 1985, Cloning and nucleotide sequence of the tzs gene from Agrobacterium tumefaciens strain T37, Nucl. Acids Res., 13: 2773-2788.

GENE EXPRESSION DURING INFECTION STRUCTURE DEVELOPMENT BY GERMLINGS OF THE RUST FUNGI

R.C. Staples[a], O.C. Yoder[b], H.C. Hoch[c],
L. Epstein[a] and S. Bhairi[b]

[a]Boyce Thompson Institute and [b]Department of Plant Pathology,
Cornell University, Ithaca, NY 14853, USA

[c]Department of Plant Pathology
New York State Agricultural Experiment Station
Cornell University, Geneva, NY14456, USA

INTRODUCTION

Stomatal penetrating fungi, typified by the bean rust fungus, Uromyces appendiculatus (Pers.) Unger., and the wheat stem rust fungus Puccinia graminis Pers. ssp. tritici, develop infection structures when a germ tube encounters a stomate on the epidermis of a host plant [1]. Infection structures develop in response to this and other specific cues from the host's surfaces which serve to ensure accurate placement of the appressorium. Here we review the current state of our knowledge regarding the differentiation response, signal transmission, and the nature of germling response to signal reception.

GERMLING RESPONSE TO SURFACE SIGNALS

When the uredospore germinates and the germ tube contacts a stomatal guard cell, infection structures form, beginning with the appressorium. The haustorial apparatus usually does not develop at this time but develops later when an infection hyphae contacts a mesophyll cell wall. Thus recognition of the host by the germ tube tip results in development of an early set of infection structures. A late set develops when the fungus recognizes certain inner tissues of the host. A set of infection structures developed by the P. graminis ssp. tritici germling after heat shock is shown in Figure 1.

While differentiation begins when a germ tube contacts a stomatal guard cell, a single scratch of sufficient dimension on most surfaces, e.g. polystyrene, glass, wax, cellophane, will also induce differentiation. Dickinson [2] suggested that the minimal unit of stimulus for germlings of P. recondita was about 120 x 1.2 nm.

The haustorial apparatus appears to arise in response to a host-specific recognition signal because few haustoria are produced on non-host tissue [3]. Recognition is confined to the inner tissues, and it involves an interaction between the walls of the cells in the inner spaces of the leaf and the infection hyphae. Using the oat rust fungus (Puccinia coronata), Mendgen [4]

NATO ASI Series, Vol. H1
Biology and Molecular Biology of Plant-Pathogen Interactions
Edited by J. Bailey

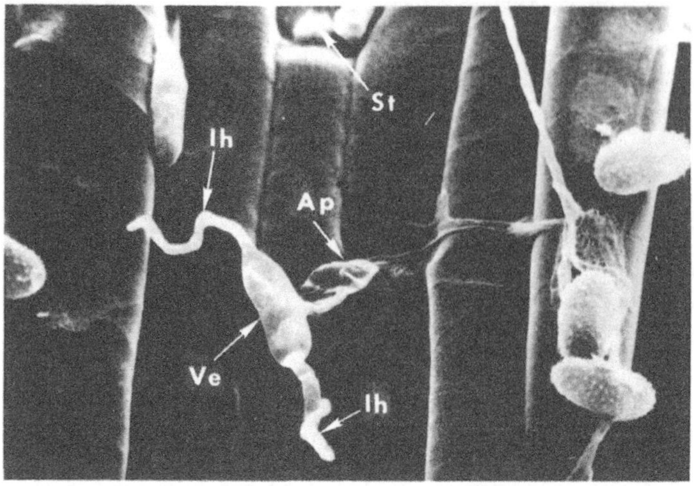

Figure 1. Scanning electron micrograph showing infection structures developed by a uredospore germling of the wheat stem rust fungus, P. graminis ssp. tritici, after induction by heat shock. Note that the appressorium is not located over the stomate. Ap, appressorium; Ih, infection hypha; St, stomate; Ve, vesicle, (x610).

observed that formation of haustoria and haustorial mother cells would occur only on the inner surface of the epidermis of excised Avena sativa coleoptiles. On the cuticular (outer) surface, haustoria mother cells formed rarely (0.5%), and haustoria were not seen.

DIFFERENTIATION-RELATED PROTEINS IN GERMLINGS

U. appendiculatus

Protein synthesis. Does gene expression change when germlings are induced to differentiate? Previous studies have suggested that development of infection structures is accompanied by an alteration in gene expression. Huang and Staples [5] showed that uredospore germlings of the bean rust fungus synthesize at least three differentiation-specific proteins during development of the infection structures. Little general change in protein synthesis appeared to occur, however, and the shifts in protein synthesis during differentiation seemed relatively limited in view of the large changes in morphology that occur.

As a consequence, the synthesis of proteins by U. appendiculatus germlings during differentiation is being re-examined using the 2-D PAGE system of O'Farrell [6], and some results are now available. We find, for example, that the synthesis of a broad spectrum of proteins is affected when the appressorium develops (3.5 to 5 h), as shown in Fig. 2A, B. Some of these appear to be synthesized de novo (cf. 2, 24, 26, 35, 36, 46, 47), while others decline or disappear (cf. 1, 4, 21, 23). Despite these alterations, the synthesis of many proteins was not affected by differentiation (cf. all proteins labelled "x"), nor did the gross content of proteins change as indicated by the relatively constant pattern of Coomassie blue staining. The studies confirm that changes in protein synthesis do occur with germling differentiation, but that the changes are much more complex than originally observed.

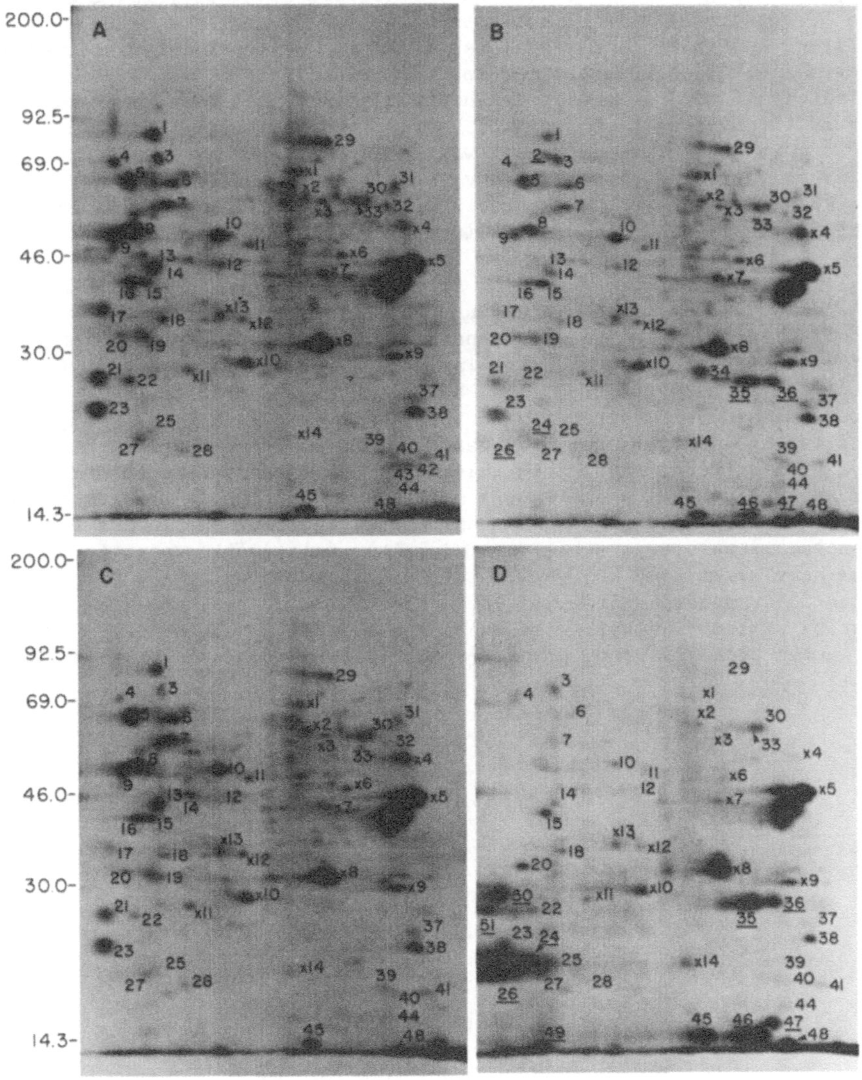

Figure 2. Autoradiograms of proteins separated by electrophoresis on two dimensional slab gels of 12.5% polyacrylamide. An ampholine-generated pH gradient was employed in the first dimension, while a 12.5% SDS gel was employed in the second dimension. The position of the molecular weight markers (kDa) is shown along the left-hand margin. Uredospore germlings on collodion membranes were misted with [^{35}S]-methionine for 1.5 h before harvest and extraction of the proteins. (A) Nondifferentiated uredospores germinated 4.5 h. (B) Differentiated uredospores germinated 4.5 h with appressoria but not vesicles. (C) Nondifferentiated uredospores germinated 6 h. (D) Differentiated uredospores germinated 6 h, with appressoria and vesicles. Paraffin oil was incorporated into the collodion membrane to induce differentiation. The synthesis of proteins labelled "x" was not affected by differentiation, whereas underscored numbers indicate proteins synthesized specifically during differentiation. Not all of the proteins have been numbered.

Further changes occur later when the vesicle develops (Fig. 2 C,D). Protein 50 appears at this time, proteins 35 and 36 continue to be synthesized, and there is a large upshift in the synthesis of proteins 24 and 26. Several small differentiation-specific proteins (46, 47, 49; each approximately 14 kDa) were also synthesized. There also was a general decline in the synthesis of proteins heavier than 69 kDa. This phenomenon, seen before in the earlier studies, was not accompanied by a loss of Coomassie blue stained proteins. Development of the vesicle, then, appears to involve an upshift in the synthesis of several smaller proteins in the range of 14-28 kDa, and a downshift in the synthesis of proteins larger than 69 kDa.

mRNA translation. The synthesis of at least some of the differentiation-specific proteins appears to result from a change in gene transcription when the appressorium is induced. Translation of poly(A)$^+$RNA in a cell-free system prepared from wheat germ showed that two differentiation-specific proteins in the 25 kDa range (proteins a,b), and two proteins in the 14 kDa range (c,d) were synthesized from template RNA prepared from differentiated germlings (Fig. 3). As reported before [5], neither of the 18.5 kDa proteins (24 and 26) was seen among the translation products, and the wheat germ system may not recognize these latter templates. Other explanations are possible. In addition, poly(A)$^+$RNA specific for proteins heavier than 46 kDa were not translated. It will be our future objective to translate spore mRNA using the cell-free reticulocyte translation system. We will also want to assay the levels of gene transcription products using protein-specific cDNA probes as these become available.

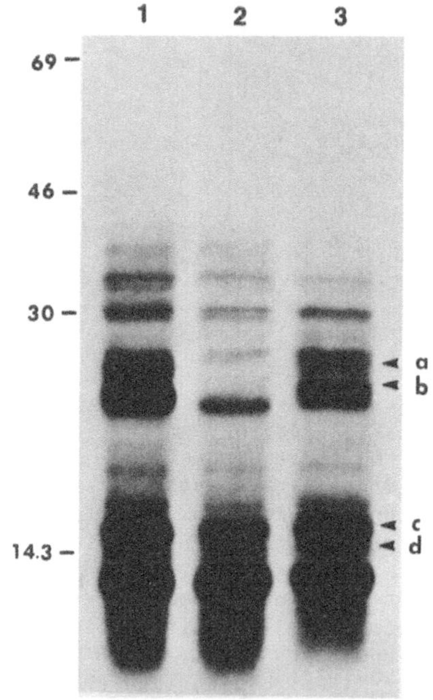

Figure 3. Proteins synthesized from [^{35}S]-methionine using a cell-free wheat germ translation system. Lane 1, poly(A)$^+$ RNA template extracted from *U. appendiculatus* germlings differentiated for 3 h (appressoria only); lane 2, poly(A)$^+$ RNA extracted from nondifferentiated uredospores germinated for 4 h; lane 3, poly(A)$^+$ RNA extracted from uredospore germlings differentiated 6 h (vesicles present). a - d, differentiation-specific proteins.

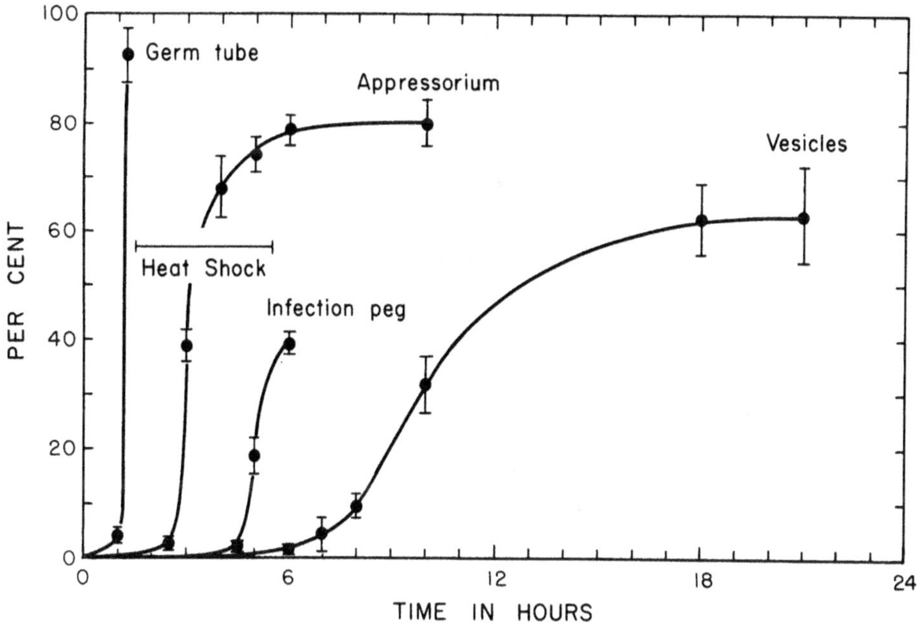

Figure 4. Time-course of development of infection structures after heat shock by uredospores of P. graminis ssp. tritici germinated on collodion membranes overlaid on 2% agar. The time of the heat shock (1.5 to 5.5 h) is shown by the horizontal bar. Vertical bars depict the standard deviation of the data. (adapted from Wanner et al., [9]).

P. graminis ssp. tritici

Uredospore germlings of the wheat stem rust fungus can be efficiently induced to differentiate with a heat shock in the absence of contact stimuli [7]. Heat shock thus represents another type of stimulus, and studies with it would allow an examination of the differentiation process from an alternative perspective [8,9]. The infection structures which developed in response to heat shock are shown in Figure 1. The time-course of differentiation is shown in Figure 4.

When germlings of P. graminis ssp. tritici are heat shocked, changes in the pattern of proteins synthesized occur beginning with appearance of the substomatal vesicles as shown using 1-D PAGE gels (Fig. 5). At this time, two differentiation-specific proteins are synthesized (34.7 and 21.9 kDa). Of great interest is the fact that the synthesis of many other proteins is shut off or very nearly ceases [8,9]. No doubt a wider range of proteins will be shown to be involved when radiolabelled proteins are employed together with 2-D PAGE gels as they have been for the bean rust fungus. Nevertheless, the earlier studies by Kim et al. [8] which used 2-D analysis also showed that two proteins somewhat similar in molecular weight to those reported in the study by Wanner et al. [9] were synthesized during differentiation even though isotopes were not used. In P. graminis ssp. tritici, detectable synthesis of differentiation-specific proteins begins with the vesicle rather than the appressorium. Further studies will be required to demonstrate that these are due to transcriptional changes of gene expression and to confirm the timing.

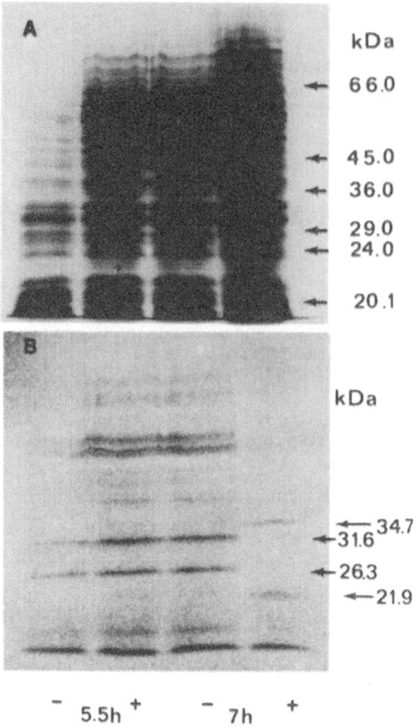

Figure 5. Autoradiograms of proteins extracted from *P. graminis* ssp. *tritici* germlings and separated by electrophoresis on a one-dimensional polyacrylamide slab gel. Left two lanes are proteins extracted from uredospores germinated for 5 h, while the right two lanes are from uredospores germinated for 7 h. (+) denotes spores heat shocked as in Figure 4, while (-) denotes nondifferentiated germlings. Labelled protein (5,000 cpm) was used in each lane. (Adapted from Wanner et al., [9]).

SIGNAL RECEPTORS

Extracellular Matrix Proteins

How is contact with a surface groove converted into gene expression by the uredospore germling? Recently, Epstein et al. [10] demonstrated that extracellular proteins are involved in thigmodifferentiation. Nuclear division of germlings incubated on normally inductive scratched surfaces were significantly reduced by treatment with either Pronase E or trypsin, but not by heat denatured enzymes (Table 1). Trypsin mixed with trypsin inhibitor or α- or β-glucosidase, α-mannosidase, or lipase likewise were not effective. Pronase E reduced adhesion of the germlings but did not decrease germination or germ tube elongation. Thus, these proteins may function in stomatal recognition by binding the germ tube to an inductive surface.

Extracellular matrix proteins in *U. appendiculatus* germlings have not been extensively characterized. Day and his associates have shown that narrow fibrillar protein polymers 7 mm wide called fimbriae are present on germ tubes of *U. appendiculatus* germlings [11,12], but their function has not been determined. Some of these proteins may possibly be glycoproteins, since germ tubes and infection structures bind lectins [13].

Table 1. Effect of selected emzymes (500 μg/ml) on nuclear division of bean rust germlings incubated on scratched surfaces[a].

Assay conditions[b]			% Nuclear division % of control[d]	
Enzyme	Buffer	pH	Active enzyme	Denatured enzyme
α-Glucosidase	Mes	7.0	117y	ND[e]
β-Glucosidase	Mes	5.0	94y	101y
α-Mannosidase	Mes	5.0	105y	ND
Pronase E	Mes	7.0	11z	96y
Pronase E	Hepes	7.5	31z	88y
Trypsin	Hepes	7.5	35z	106y
Trypsin + Trypsin inhibitor (500 μg/ml)[c]	Hepes	7.5	106y	ND
Lipase	Hepes	7.5	118y	120y

[a] Bean rust uredospores were incubated at 20 °C for 10 h on polystyrene petri dishes that were scratched with a diamond pencil; this surface normally induces nuclear division and formation of appressoria.

[b] The uredospores were misted with a 10 mM germination buffer at the pH indicated with or without active or heat denatured enzyme (20 min at 90 °C).

[c] Trypsin inhibitor (50 μg/ml) alone did not affect nuclear division of the germlings (112% of the control).

[d] Percentage nuclear division of mithramycin stained material was determined via epifluorescent microscopy by assessing 100 germlings on 3 replicate plates. Values represent percentages of a buffer control. Means followed by the same letter were indistinguishable by the Student-Newman Keuls multiple range test (P = 0.05). s.d.= 11.5.

[e] ND, not determined.

Adapted from Epstein et al., [10], reprinted with permission.

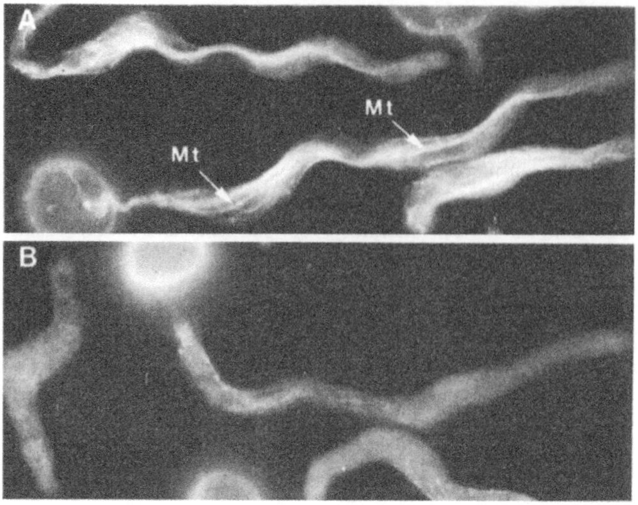

Figure 6. Uredospore germlings of U. appendiculatus fixed with 3% formaldehyde and stained for microtubules using FITC-labelled antitubulin antibodies as described by Hoch and Staples, [14]. (A) Spores germinated for 3.5 h on polyethylene sheets. (B) Uredospores germinated as in (A), but exposed to 0 °C for 60 min prior to fixation. Most of the abundant microtubules (Mt) shown in (A) have been depolymerized by the cold treatment. All x1650.

The Cytoskeleton

Uredospore germlings have a well-defined cytoskeleton [14]. For example, prominent microtubules are oriented parallel with the long axis of the germ tube (Fig. 6A). We have suggested that the cytoskeleton (microfilaments + microtubules) facilitates reception and transmission of contact stimuli to the nucleus [15]. Proof that the sensing mechanism in rust germlings involves the cytoskeleton is lacking. However, heat shock, ultrasonic vibrations, and certain drugs which interfere with cytoskeletal organization, i.e. colcemid and cytochalasin E, induce nuclear division in U. appendiculatus germlings. From this, one might predict that any disruption of the cytoskeleton would induce nuclear division as suggested originally by Crossin and Carney [16] for fibroblasts.

To test this hypothesis, Hoch and Staples [14] found that while cold treatments depolymerize the microtubules (Fig. 6B), they do not induce nuclear division. In contrast, cytochalasin E both disrupts the microfilaments and induces mitosis [17]. The data suggest that disrupting the microtubules by cold treatment is not a sufficient condition to bring about nuclear division, but disrupting the microfilaments may be.

FUTURE RESEARCH

Rust Fungi

Calmodulin. What are the differentiation-specific proteins? What is their function and where might they be located in the germling? How is their synthesis regulated by the touch-responsive sensing mechanisms in the germling? To provide answers to these questions, we are attempting to prepare monoclonal antibodies specific for each of the differentiation-specific proteins.

For example, we suspect that protein 26 contains calmodulin based on its position in O'Farrell gels. Among other functions, calmodulin synthesis is associated with the start of nuclear division in eukaryotic cells [18,19], an important event in germling differentiation. Germling extracts contain calmodulin, based on both a calmodulin-specific phosphodiesterase assay [20] and a radioimmune assay [21]. Unfortunately, we lack a calmodulin antibody that is sufficiently specific for us to identify protein 26 positively using Western blots because calmodulin-specific antibodies from such diverse sources as spinach, _Chlamydomonas_, and rat testis were either inactive or weakly active with bean rust protein.

Calmodulin gene. Because the synthesis of calmodulin is induced during differentiation, we wish to investigate the mechanism of induction of calmodulin gene expression. For this purpose, we will compare the promoter sequence of the calmodulin gene from _U. appendiculatus_ germlings with those from other tissues that have been published. To accomplish this, we are now isolating the calmodulin gene from uredospores by screening our uredospore genomic library with a heterologous cDNA calmodulin gene probe that was obtained from Dr. Igor Dawid. We have obtained several putative gene clones by _in situ_ plaque hybridization of the fungal genomic library, and we are now in the process of confirming by sequence analysis that these clones carry the calmodulin gene. A quantitative analysis of changes in the levels of calmodulin-specific mRNA during the course of infection structure development by dot-blot analysis should also be possible soon.

Anthracnose Fungi

How broadly applicable to other fungi are the data obtained with the rusts likely to be? How can one apply genetic tools for the study of infection structure development?

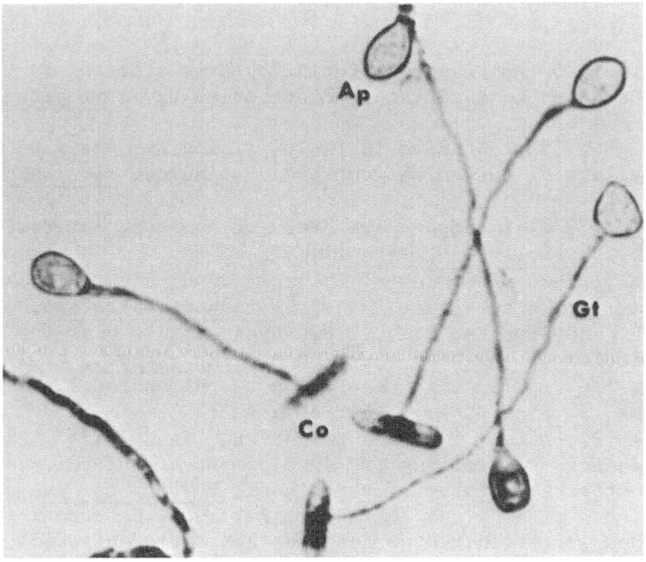

Figure 7. _Colletotrichum lindemuthianum_ conidia germinated for 12 h in a film of Mathur broth on a glass slide. Appressoria developed when germ tubes contacted the slide. Ap, appressoria; Co, conidium; Gt, germ tube. x400.

Conidial germlings of anthracnose fungi (Colletotrichum spp.) differentiate thick-walled appressoria in response to contact with a hard surface [22]. Typical appressoria are shown in Figure 7. Appressoria often develop in the grooves between cells [23], and there appears to be a sensing response to the host surface analogous to that of rust germlings.

DNA synthesis is also required for development of anthracnose appressoria [24]. In contrast with the rusts, however, replication begins with spore germination, and DNA replication may have no special relationship to the differentiation response except for the provision of new cells.

Actinomycin D reversibly inhibited the development of appressoria and the synthesis of new mRNA without affecting nuclear division [24], and it was suggested that anthracnose conidia produce appressoria in response to germ tube contact by altering the messenger programme of its germ tube nucleus. More recently, Suzuki et al. [25] showed that C. lagenarium germlings synthesize a 95 kDa polypeptide when the appressoria form. Synthesis of the polypeptide is inhibited when cyclohexamide is added after 1 h of germination, and the appressoria which form are functionless and cannot penetrate artificial membranes. Use of more specific procedures including analyses by Northern blot will be required to confirm that a change in gene expression has occurred.

Although there are important differences between rust and anthracnose fungi, e.g. modes of penetration, germ tube nucleation and DNA replication, the similarities are striking. Since some of the anthracnoses are heterothallic and easily cultured (perfect stage = Glomerella cingulata), studies on appressorium formation will be amenable to genetic analysis.

ACKNOWLEDGEMENTS

Research by the authors is supported in part by grants from the Whitehall Foundation and the US National Science Foundation.

REFERENCES

1. Wynn, W.K., 1976, Appressorium formation over stomates by the bean rust fungus: Response to a surface contact stimulus, Phytopathology, 66: 136-146.
2. Dickinson, S., 1970, Studies in the physiology of obligate parasitism, VII. The effect of curved thigmotrophic stimulus, Phytopath. Z., 69: 115-124.
3. Heath, M.C., 1977, A comparative study of non-host interactions with rust fungi, Physiol. Plant Pathol., 10: 73-88.
4. Mendgen, K., 1982, Differential recognition of the outer and inner walls of epidermal cells by a rust fungus, Naturwissenschaften, 69: 502-503.
5. Huang, B.F. and Staples, R.C., 1982, Synthesis of proteins during differentiation of the bean rust fungus, Exp. Mycol., 6: 7-14.
6. O'Farrell, P.H., 1975, High resolution two-dimensional electrophoresis of proteins, J. Biol. Chem., 250: 4007-4021.
7. Maheshwari, R., Allen, P.J. and Hildebrandt, A.C., 1967, Physical and chemical factors controlling the development of infection structures from uredospore germ tubes of rust fungi, Phytopathology, 57: 855-862.
8. Kim, W.K., Howes, N.K. and Rohringer, R., 1982, Detergent-soluble polypeptides in germinated uredospores and differentiated uredosporelings of wheat stem rust, Can. Plant Pathol., 4: 328-333.
9. Wanner, R., Förster, H., Mendgen, K. and Staples, R.C., 1985, Synthesis of differentiation-specific proteins in germlings of the wheat stem rust fungus after heat shock, Exp. Mycol., 9: 279-283.

10. Epstein, L., Laccetti, L., Staples, R.C., Hoch, H.C. and Hoose, W.A., 1985, Extracellular proteins associated with induction of differentiation in bean rust uredospore germlings, Phytopathology, 75: 1073-1076.
11. Day, A.W., Svircev, A.M., Smith, R. and Gardiner, R.B., 1984, Fungal fimbriae and host infection, Phytopathology, 74: 832 (Abstr.).
12. Kaminskyj, S. and Day, A.W., 1984, Effects of antifimbrial antisera on development of infection structures in the bean rust fungus, Phytopathology, 74: 833 (Abstr.).
13. Mendgen, K., Lange, M. and Bretschneider, L., 1985, Quantitative estimation of the surface carbohydrates on the infection structures of rust fungi with enzymes and lectins, Archiv. Microbiol., 140: 307-311.
14. Hoch, H.C. and Staples, R.C., 1985, The microtubule cytoskeleton in hyphae of Uromyces phaseoli germlings: Its relationship to the region of nucleation and to the F-actin cytoskeleton, Protoplasma, 124: 112-122.
15. Staples, R.C. and Hoch, H.C., 1982, A possible role for microtubules in the induction of nuclear division in bean rust uredospore germlings, Exp. Mycol., 6: 293-302.
16. Crossin, K.L. and Carney, D.H., 1981, Evidence that microtubule depolymerization early in the cell cycle is sufficient to initiate DNA synthesis, Cell, 23: 61-71.
17. Hoch, H.C. and Staples, R.C., 1983, Visualization of actin in situ by rhodamine-conjugated phalloin in the fungus Uromyces phaseoli, Eur. J. Cell Biol., 32: 52-58.
18. Chafouleas, J.G., Bolton, W.E., Hidaka, H., Boyd, A.E. and Means, A.R., 1982, Calmodulin and the cell cycle: involvement in regulation of cell-cycle progression, Cell, 28: 41-50.
19. Muthukumar, G., Kulkarni, R.K. and Nickerson, K.W., 1985, Calmodulin levels in the yeast and mycelial phases of Ceratocystis ulmi, J. Bacteriol., 162: 47-49.
20. Wallace, R.W., Tallant, E.A. and Cheung, W.Y., 1983, Assay of calmodulin by Ca^{2+}-dependent phosphodiesterase, Methods in Enzymology 102: 39-47.
21. Chafouleas, J.G., Dedman, J.R., Munjaal, R.P. and Means, A.R., 1979, Calmodulin: development and application of a sensitive radioimmunoassay, J. Biol. Chem., 254: 10262-10267.
22. Dey, P.K., 1919, Studies in the physiology of parasitism. V. Infection by Colletotrichum lindemuthianum, Ann. Bot., 33: 305-312.
23. Preece, T.F., Barnes, G. and Bayley, J.M., 1967, Junction between epidermal cells as sites of appressorium formation by plant pathogenic fungi, Plant Pathol., 16: 117-118.
24. Staples, R.C., Laccetti, L. and Yaniv, Z., 1976, Appressorium formation and nuclear division in Colletotrichum truncatum, Arch. Microbiol., 109: 75-84.
25. Suzuki, K., Furusawa, I., Ishida, N. and Yamamoto, M., 1981, Protein synthesis during germination and appressorium formation of Colletotrichum lagenarium spores, J. Gen. Microbiol., 124: 61-69.

ISOLATION OF STAGE- AND CELL-SPECIFIC GENES FROM FUNGI

W.E. Timberlake

Department of Plant Pathology
University of California
Davis
CA, 95616
USA

INTRODUCTION

In recent years there has been a dramatic increase in the ability to manipulate the genomes of Ascomycetous fungi such as *Saccharomyces cerevisiae* (yeast), *Neurospora crassa* and *Aspergillus nidulans*, because of the development of DNA-mediated transformation systems for these organisms [1,2,3,4,5,6]. It is now possible to clone specific fungal genes by direct complementation of mutations in yeast or *Aspergillus*. Fungal genes that have been cloned by complementation or by other procedures can be mutated *in vitro* by using recombinant DNA technology and returned to their normal genomic locations (reviewed by Botstein and Davis [7] and Struhl [8]). Thus, it is possible to determine the biological and biochemical consequences of introducing specific, pre-selected alterations into fungal genomes. For example, portions of genes or neighbouring DNA sequences can be removed or replaced to determine if they are required for regulated gene transcription. Genes can be deleted by site-directed mutation to assess the effects of their loss. Similarly, genes can be manipulated to cause over-expression or expression in an inappropriate cell type to ascertain how such deregulation affects cellular functions. Many other strategies can be adopted in attempts to understand better gene structure, regulation and function.

Recently, this technology has begun to be extended to plant pathogenic fungal species [see Yoder, this volume]. Several plant pathogenic Ascomycetes have been successfully transformed during the past year, opening up new possibilities for the study of the fungal determinants of pathogenicity, virulence and host range. One can be optimistic that the molecular genetic techniques that have been applied with great success to problems in bacterial plant pathogenesis [see Collmer and Daniels, this volume] will be available for use with a variety of fungi in the very near future.

Application of the molecular genetic approach to problems in plant pathology depends on the ability to isolate specific genes of interest. The fungi lend themselves to a variety of gene isolation strategies because of their small genomes, which are typically only a few times larger than the

NATO ASI Series, Vol. H1
Biology and Molecular Biology of Plant-Pathogen Interactions
Edited by J. Bailey

genome of Escherichia coli, and their low content of repetitive DNA sequences. In addition, fungal protein coding genes generally appear to be simple in structure, containing no or only a few introns. When present, introns are small, usually less than 0.2 kilobases in length. These properties and the availability of bacteriophage and plasmid vectors that allow convenient construction of large numbers of recombinant DNA clones have facilitated isolation of many genes from fungi. The ability to select genes from recombinant DNA libraries by complementation of mutations in heterologous hosts, such as E. coli and yeast, has permitted identification of genes encoding enzymes having known functions in cellular metabolism. However, isolation of genes that determine the pathogenic properties of fungi is problematical for at least three reasons. First, in many instances, the biochemical activities specified by pathogenicity genes are unknown, making it impossible to design rational selection schemes based on existing technology. Second, even in those instances where the biochemical functions of the genes are known, corresponding mutations in alternative hosts such as E. coli or yeast are not available, prohibiting selection of the desired clones by genetic complementation. Finally, the transcription signals of genes from many fungi appear not to be recognized by favoured cloning hosts. Thus, even if a selection strategy can be designed, efforts must be made to provide sequences that direct transcription initiation, polyadenylation and splicing in the cloning host used. In the absence of any information about the structure of the gene of interest, provision of such sequences can be quite difficult or impossible.

Two cloning strategies have been developed that may be of particular value for the isolation of genes involved in expression of fungal pathogenicity genes because they circumvent many of these problems. The first is designed to permit isolation of genes whose transcription is selectively induced or repressed at a certain stage of growth or development or in a particular cell type [9]. It involves the preparation of radiolabelled hybridization probes by depletion or "cascade" hybridization and does not depend on prior knowledge of gene function. The second cloning strategy is designed to permit isolation of genes having known functions by expression of the genes in A. nidulans [10]. Currently available data indicate that A. nidulans, unlike yeast, is capable of recognizing the transcription signals of protein coding genes from many Euascomycetes. The availability of numerous mutations in this species facilitates the design of gene selection strategies. Finally, a cosmid vector has been developed for A. nidulans that has a cloning capacity of about 40 kilobases [10]. Thus, only about two thousand recombinants are needed to represent an entire fungal genome. This means that it is possible to screen A. nidulans transformants for production of a particular gene product in those instances where direct selection for gene expression is impossible.

In this chapter, I will attempt to describe the procedures used in both cloning strategies in sufficient detail to allow investigators having basic biochemical laboratory skills to carry them out. The reader is referred to other chapters in this volume for ideas about how these techniques might be applied to problems in plant pathology.

PREPARATION OF cDNA BY DEPLETION HYBRIDIZATION

This procedure is designed to produce radiolabelled complementary DNAs (cDNAs) corresponding to mRNAs that accumulate preferentially (or are lost) at a certain stage of growth or development or in a particular cell type. The original cascade hybridization procedure was used to isolate sporulation-specific genes from A. nidulans [9,11]. Since that time, we have simplified the procedure to make it more useful to other investigators [12]. Specific cDNA probes suitable for identifying recombinant DNA clones by colony or plaque hybridization can be prepared in one to a few weeks without using any highly specialized equipment or techniques.

The theoretical basis of depletion hybridization is straightforward. Consider two populations of cells, designated A and B, having different transcriptional patterns, and a situation where one would like to clone genes that are transcribed preferentially in cell type A. This can be accomplished as follows. Poly(A)$^+$RNA is first isolated from both types of cells and radioactively labelled cDNA is synthesized by using RNA_A. This cDNA population will contain sequences corresponding to mRNAs that are common to both cell types (C-sequences) and corresponding to mRNAs that are present in only cell type A (S-sequences). It is possible to enrich for S-sequences by hybridizing the cDNA with an excess of poly(A)$^+RNA_B$ and removing cDNA/RNA hybrids by chromatography on hydroxyapatite (HAP). The S-cDNA can then be used as a probe in colony or plaque hybridization. Although this approach is conceptually simple, difficulties arise because of incomplete cDNA hybridization. This led to the development of a cascade hybridization procedure in which $cDNA_A$ was challenged several times with increasing amounts of RNA_B [9]. The many steps involved in this procedure were technically demanding and resulted in poor yields of cDNA. We have found that, for purposes of clone identification, many of these steps can be omitted. The most rapid procedure we have used to prepare S-cDNA probes is described below.

Isolation of Poly(A)$^+$RNA

A variety of RNA isolation procedures can be used successfully with fungi, but we find the following procedure to be rapid, simple and reliable, even for species producing high amounts of RNase. No special precautions are taken to make glassware, plasticware and so forth RNase free. Cells are grown and harvested by appropriate means, for example by filtration through Miracloth (Calbiochem) or cheesecloth for hyphae, or by centrifugation for spores or yeast cells. A dependable method for disrupting fungal cells involves use of an instrument called a "Bead Beater" (Biospec Products, P.O. Box 722, Bartlesville, Oklahoma 74003, USA). The stainless steel blending vessel is filled to the top with 75-150 μM glass beads (Sigma) that have been washed with 0.5N HCl and doubly distilled or distilled deionized H_2O (ddH_2O) and dried at 60-70 °C. The beads should be chilled to 4 °C before use. The cells are suspended in a minimal volume of lysis buffer, made up as follows:

Solution 1: 1.0M Tris-HCl, pH 8.5
1.25M NaCl
0.25M EGTA
(store at -20 °C.)

Solution 2: 12% (w/v) sodium para-aminosalicylate (Sigma)
(make fresh).

Solution 3: 2% (w/v) tri-iso-propylnapthalenesulphonate (TNS; Eastman).
(make fresh).

With constant stirring, 80 ml of solution 2 is mixed with 80 ml of solution 3. Solution 1 (40 ml) is added and the pH is adjusted to 8.9. The lysis buffer should be prepared on the day on which it is to be used and stored on ice.

The cell suspension is transferred to the blending vessel, which is then filled to near the top with lysis buffer. The impeller is attached, and the unit is placed on the blender base. Ice is packed around the blender vessel and the cells are homogenized by several 1 min periods of blending followed by 1 min cooling periods. The effectiveness of homogenization should be checked occasionally by microscopic examination. In rare instances, it may be necessary to include 0.5 vol of water-saturated, redistilled phenol during homogenization to prevent degradation of the RNA; the Bead Beater may be damaged by organic solvents, so a stainless steel, Waring-type blender is substituted for the Bead Beater. The blender cup is cooled by submerging it in ice at 1 min intervals. For smaller amounts of cells, disruption can be accomplished by placing the cell suspension in a Corex centrifuge tube with glass beads and vortexing them at high speed [13].

The lysate is decanted into a flask and the beads are rinsed with additional lysis buffer. The volume of lysate is determined (= 1 vol) and 0.5 vol of redistilled, water-saturated, neutralized phenol containing 0.1% (w/v) 8-hydroxyquinoline is added. The flask is sealed with a silicon rubber stopper and shaken vigorously for about 2 min. One-half vol of chloroform is added and the flask is again shaken for 2 min. The lysate is transferred to 250 ml polypropylene centrifuge bottles and centrifuged at 5,000-10,000 x g for 10 min at 4 °C. The aqueous phase is aspirated off and transferred to a flask on ice. The organic phase and interface are transferred to a flask, a small amount of lysis buffer is added (0.1-0.2 vol) and the flask is shaken and heated to 68 °C for 5 min. The mixture is again shaken for about 2 min and centrifuged as above. The aqueous phase is removed and combined with the first aqueous phase. The 68 °C extraction is repeated again and the aqueous phase is combined with the first two. The organic phase and protein interface are discarded. The combined aqueous phases are re-extracted twice with 1 vol of phenol:chloroform (1:1) and RNA is precipitated by addition of 8M or 10M LiCl to a final concentration of 2M and incubation overnight at 0 °C. The LiCl solution is made up as follows. LiCl is weighed out and dissolved in ddH_2O so that the volume is somewhat less than the desired final volume. The solution is filtered through a Whatman GF/C glass filter and autoclaved. After cooling to room temperature, the volume is adjusted with sterile ddH_2O. The LiCl solution is stored at room temperature.

The RNA is collected in four 30 ml Corex centrifuge tubes by centrifugation in a swinging bucket rotor (15 min, 15,000 x g, 4 °C). The supernatants are decanted and the tubes are refilled until the entire sample has been processed. The pellets are washed three times with 70% ethanol containing 0.1M sodium acetate (pH 6.0) by using about 15 ml per pellet, and dried under vacuum. They are then dissolved in about 5 ml of 10mM Tris-HCl, pH 7.6, 0.1mM EDTA, 0.2% SDS (BDH, specially purified for biochemical use) (TES). A 1:100 dilution of the RNA solution is made and a UV absorption spectrum is taken. The concentration of RNA is calculated as $[(A_{260} - A_{320}) \times 40 \times 100]$ μg/ml. A 5 mg/ml solution of proteinase-K (EM Biochemicals) in 10mM Tris-HCl, pH 7.6, is prepared, incubated at 37 °C for about 10 min and added to the RNA solution to a final concentration of 100 μg/ml. The RNA can be stored indefinitely at -70 °C at this stage.

Poly$(A)^+$RNA is purified from total RNA by multiple passages over oligo(dT)-cellulose. The following equipment items and buffers are needed.

Equipment:	Water jacketed chromatography column, 2.0-2.5cm I.D. Circulating water bath Fraction collector
Buffers and Solutions:	TS; 10mm Tris-HCl, pH 7.6, 0.2% SDS NaOH; 0.1M NaOH NETS; 10mm Tris-HCl, pH 7.6, 500mM NaCl, 1mM EDTA, 0.2% SDS

TN; 10mM Tris-HCl, pH 7.6, 500mM NaCl; treat with 2 μl/ml diethyl pyrocarbonate (DEP), autoclave, cool to room temperature, adjust volume

ddH_2O; treat ddH_2O with 2 μl DEP and autoclave

5M NaCl; weigh out NaCl, dissolve in ddH_2O, treat with DEP, autoclave, cool to room temperature, adjust volume.

Oligo(dT)-cellulose (0.5-1.0g of PL Biochemicals, Type 7) is suspended in ddH_2O, any fines are decanted away and the slurry is loaded into the column. The temperature of the column is maintained at 25 °C by water circulation. The column is drained and the cellulose is washed with 10-20 ml of NaOH, 50 ml ddH_2O and then NETS until the pH of the effluent is 7.6. The column is once again drained. The RNA solution (containing up to 100 mg RNA) is heated to 68 °C for 5 min, poured into the column and 5M NaCl is added to bring the final concentration to 0.5M. The RNA solution and cellulose are mixed by gentle stirring with a glass rod. The vessel that contained the RNA solution is washed with a small amount of NETS buffer which is transferred to the column. The column bed is stirred occasionally over a 10 min period to keep the cellulose suspended. The column is then drained slowly and washed with NETS buffer (with occasional stirring) until the A_{260}-A_{320} of the effluent is less than 0.05. Residual NETS buffer is drained away, 2 ml of TS buffer is added to the top of the column bed and the column is heated to 55 °C. The column output is connected to a fraction collector and 10, 1-2 ml fractions are collected, adding TS to the column as required. The A_{260}-A_{320} for each column fraction is determined, fractions having values greater than 0.05 are pooled and the total volume of the fractions is determined.

The column is cooled to 25 °C and washed with NaOH, ddH_2O and NETS as described above. The RNA solution is heated to 68 °C, added to the column and adjusted to 0.5M NaCl. The column is then washed and eluted as before. This entire process through the NETS washing step is repeated once again. The column bed is then washed with TN until the SDS has been completely removed. If fractions are collected by drop counting, removal of the SDS is evidenced by a large increase in the fraction size. Fractions may also be tested for the presence of SDS by addition of a drop of 8M potassium acetate, which causes the formation of a potassium SDS precipitate. The column is finally heated to 55 °C and eluted with ddH_2O as described above for the TS elution steps. Fractions having an A_{260}-A_{320} of greater than 0.05 are pooled, the volume of the solution is determined and it is divided and transferred to two polyallomer SW27 ultracentrifuge tubes (or equivalent). RNA is precipitated by addition of 0.1 vol of 3M sodium acetate, pH 6.0, and 2 vols of absolute ethanol, the tubes are filled with 70% ethanol, 0.1M sodium acetate solution, sealed with plastic laboratory film and incubated at -20 °C overnight. The RNA is collected by centrifugation in an SW27 rotor (or equivalent) at 25,000 rpm for 1 h at 4 °C. The supernatants are decanted, the pellets are dried under vacuum and dissolved in ddH_2O at a concentration of about 1 mg/ml. Poly$(A)^+$RNA prepared in this way can be stored at -70 °C for at least several years. The integrity of the poly$(A)^+$ RNA should be checked by fractionating about 3 μg in a denaturing agarose gel [15] followed by staining with ethidium bromide and photography.

Synthesis of cDNA

Radioactively labelled cDNA is synthesized by using poly$(A)^+$RNA from cell population A. The reaction conditions are determined by the amount and specific radioactivity of cDNA desired. We generally use ^{32}P-labelled cDNA, having a specific radioactivity of 2×10^8 DPM/ μg and try to start with at least 0.25 μg of cDNA. The specific activity of the cDNA is determined by the specific radioactivity of the deoxynucleoside triphosphate (dNTP) substrates used, which in turn is determined by the amounts of the substrates and the amounts of radioactivity added. The concentrations of the dNTPs should be greater than or equal to 5 μM. Thus, if 50 Ci of [α-^{32}P] dCTP is used, the dCTP concentration is 5 μM, the reaction volume is 100 μl and the concentration of the other, unlabelled dNTPs is high (500 μM), the specific radioactivity would be calculated as follows:

Total DPM = 50 μCi x 2.22×10^6 DPM/μCi = 1.11×10^8 DPM

Total dCTP = 5nmole/ml x 0.1ml = 0.5nmole

Total relevant dNTP = 0.5nmole x 4 = 2nmole

μg dNMP = 2nmole x 330 ng/nmole* = 660 ng = 0.66 μg

Specific radioactivity of cDNA = 1.11×10^8 DPM/0.66 μg =

1.7×10^8 DPM/μg

(* approximate molecular weight of a dNMP)

In practice, there is considerable variation in the efficiency of cDNA synthesis based on the quality of the template and reverse transcriptase used. The reaction conditions described below represent what we have found appropriate for *Aspergillus* poly(A)$^+$RNA, isolated by using the procedure given in the previous section, and reverse transcriptase from Life Sciences, Inc. In many cases, a trial reaction may have to be done to determine the precise reaction conditions.

The reaction conditions we use are as follows:

Tris-HCl, pH 8.3 at 43 °C	50mM
$MgCl_2$	10mM
2-mercaptoethanol	30mM
KCl	140mM
oligo$(dT)_{12-18}$	100 μg/ml
dCTP	20μM, 250 μCi
dATP	0.5mM
dGTP	0.5mM
TTP	0.5mM
Poly(A)$^+$RNA	200 μg/ml
reverse transcriptase	10 units/μg RNA
total volume	100 μl

Total DPM = 250 μCi x 2.22 x 10^6 DPM/μCi = 5.55 x 10^8 DPM

Total dCTP = 20nmole/ml x 0.1 ml = 2nmole

Total relevant dNTP = 2nmole x 4 = 8nmole

μg dNMP = 8nmole x 330 ng/nmole = 2640 ng = 2.64 μg

Specific radioactivity = 5.55 x 10^8 DPM/2.64 μg = 2.1 x 10^8

The cDNA synthesis reaction is set up and processed as follows. The following stock solutions are prepared in advance:

05.M Tris-HCl, pH 8.3 at 43 °C, 0.1M $MgCl_2$

1.0M KCl

1.0M 2-mercaptoethanol (freshly prepared)

10mM (each) dATP; dGTP; TTP

1mM dCTP

1 mg/ml oligo$(dT)_{12-18}$

Twenty μg of poly$(A)^+$RNA is placed into a 1.5 ml microfuge tube, the solution is heated to 68 °C for 5 min and chilled in an ethanol-ice bath. The other reaction components are added to this tube:

Tris-HCl, $MgCl_2$	10 μl
2-mercaptoethanol	3 μl
KCl	14 μl
oligo$(dT)_{12-18}$	10 μl
dATP, dGTP, TTP	5 μl
[α -32P]dCTP, 2000 Ci/mmole, 10 Ci/ 1	25 μl
ddH_2O	to 100 μl less volume of enzyme solution

The solution is mixed, AMV reverse transcriptase (200 units) is added and the reaction is incubated at 43 °C for 30 min. EDTA is added to a final concentration of 20mM to terminate the reaction. The RNA is hydrolyzed by adding 2 μl of 10M NaOH and incubating the sample at 68 °C for 30 min. The solution is neutralized by addition of 2 μl of 10M HCl, 20 μg of yeast tRNA is added, and the mixture is diluted into 0.5 ml of TES buffer. The cDNA is then separated from unincorporated label by chromatography through a Sephadex G-100 column developed in TES buffer. The column we use is 1 x 30 cm and is filled with 3 mm, siliconized glass beads to within 3-5 cm of the top. The column is filled with pre-swelled Sephadex G-100-120 so that the bed is 1-2 cm above the glass beads. A thin layer of Bio-Gel P2 is added to the top of the bed. One ml fractions are collected and the fractions containing cDNA (excluded) are identified by using a Geiger counter. The fractions (usually 5 or 6) are then combined into a siliconized, polyallomer SW27.1 centrifuge tube (or equivalent) and 0.1 vol of 3M sodium acetate, pH 6.0, and 1.1 vols of 2-propanol are added. The tube is then incubated at -20 °C overnight.

The cDNA is pelleted by centrifugation at 25,000 rpm at 4 °C, the supernatant is decanted and the pellet is dried under vacuum. TES buffer (0.5 ml) is added to the pellet and the cDNA is dissolved by agitating the tube on a rotating shaker for more than 2 h. The solution is transferred to a 1.5 ml microfuge tube and the ultracentrifuge tube is rinsed with 0.1 ml of TES. The cDNA is reprecipitated by addition of 0.1 vol of sodium acetate, pH 6.0, and 1.1 vol of 2-propanol and incubation at -70 °C for 0.5-1.0 h and pelleted by centrifugation in a microfuge for 15 min at 4 °C. The supernatant is decanted and the pellet is dried under vacuum. At this stage, the cDNA is ready for hybridization with an excess of poly(A)$^+$RNA from cell type B.

Poly(A)$^+$RNA-excess/cDNA Hybridization

In the original cascade hybridization procedure, cDNA was sequentially hybridized with a 20-fold, then a 50-fold, and finally a 100-fold mass excess of poly(A)$^+$RNA in order to remove virtually all sequences that were common to both cell populations. In some instances, for example clone selection by colony or plaque filter hybridization, this extensive enrichment appears to be unnecessary. Instead, the cDNA need only be challenged once with a 100- to 200-fold excess of RNA.

Several stock solutions should be prepared before starting the hybridization reactions. Phosphate buffer (2M NaPB) is made up as follows. One mole of Na_2HPO_4 and one mole of NaH_2PO_4 are dissolved in less than one litre of ddH_2O. About 5 g of chelating resin (e.g. Chelex, Biorad) are added and the mixture is stirred for 1-3 h. The solution is then filtered through a Whatman GF/C filter, autoclaved and adjusted to 1 litre with sterile ddH_2O. The buffers needed for HAP chromatography are 0.12M and 0.50M NaPB containing 0.1% SDS and are made up from 2M NaPB and 20% SDS stocks. RNase A (10mg/ml) is made by dissolving or diluting RNase A in 0.05M NaPB and heating it to 90 °C for 10 min. The RNase A stock solution is stored at -20 °C. Hydroxyapatite can be purchased from a number of suppliers, but we have found that the fast flowing type from CalBiochem is quite satisfactory. A 1 x 10 cm water jacketed chromatography column capable of withstanding temperatures of up to 100 °C is also needed. Such columns can be made by a glass blower by using a coarse scintered glass filter disc (Kontes) and a microbore stopcock.

Several parameters need to be considered in setting up the hybridization reaction, the most important of which are the amount of poly(A)$^+$RNA relative to the amount of cDNA and the RoT value that needs to be generated. As discussed above, a 100-fold mass excess of poly(A)$^+$RNA is sufficient for many purposes, but a greater excess may be used if more of the poly(A)$^+$RNA is easily obtained. The complexities of mRNA populations from fungi generally fall in the range of 4,000-10,000 diverse sequences, and therefore a RoT value of 1000 (M RNA nucleotides x sec) is generally sufficient to drive homologous poly(A)$^+$RNA-excess/cDNA hybridization reactions to completion. For purposes of enrichment of stage- and cell- specific cDNA probes, we generally incubate reactions to RoT 1500-3000. RoT values can be determined directly by calculating the molar concentration of RNA nucleotides and multiplying by incubation time in seconds, taking into account effects of cation concentration on hybridization rates [14]. A rate correction of 1.0 is used for incubations carried out in 0.12M PB and a rate correction of 5.0 is used for incubations carried out in 0.41M PB. A convenient formula for calculating approximate RoT values is

$$\text{RoT} = \frac{\mu\text{g RNA/ml}}{90} \times \text{salt correction factor} \times \text{h.}$$

The hybridization reaction is carried out as follows. First, the amount of RNA to be used (a 100-fold excess) is calculated from the amount of cDNA obtained. The RNA (poly(A)$^+$RNA from cell type B) is added to the cDNA pellet in the microfuge tube and is precipitated by addition of 0.1 vol of 3M sodium acetate, pH 6.0, and 2.5 vol of 95% ethanol and incubation for 1 h at -70 °C. The RNA is pelleted by centrifugation in a microfuge for 15 min at 4 °C and dried under vacuum. Appropriate amounts of ddH_2O and 20% SDS are added (final SDS concentration is 0.2%) and the RNA and cDNA are dissolved by vortexing. Then 2M NaPB is added to give the desired final NaPB concentration, usually 0.41M. The mixture is vortexed, about 25 µl of mineral oil added, and the tube is spun briefly in the microfuge to bring the solution to the bottom of the tube. The tube is then sealed tightly with TFE (Teflon) tape and incubated for 5 min at 102-104 °C in a boiling salt water bath. It is immediately transferred to a 68 °C water bath and incubated for the appropriate time.

As an example, assume that there are 1 x 10^8 DPM of cDNA having a specific radioactivity of 2 x 10^8 DPM/µg (0.5 µg cDNA). RNA (50 µg) would be added and precipitated as described above. If a RoT of 2000 was desired, one might wish to carry out the reaction in a total volume of 50 µl in 0.41M NaPB so that the RNA concentration would be 1000 µg/ml:

$$\text{RoT/h} = \frac{1000\ \mu\text{g/ml}}{90} \times 5 = 55.5\ \text{RoT/h}.$$

In this example, the incubation time would be 40 h, but could be increased or decreased by changing the RNA or NaPB concentrations. The reaction would be set up by adding 0.5 µl of 20% SDS, 39.25 µl of ddH_2O and 10.25 µl of 2M NaPB. When the reaction is complete, the tube is quickly chilled in a dry ice-ethanol bath and stored at -20 °C.

HAP Chromatography

Once the hybridization reaction is complete, it is ready to be fractionated by HAP chromatography. The water jacketed column is set up, attached to a circulating water bath and equilibrated to 68 °C. The column outlet is attached to a fraction collector with thin (-1.5 mm I.D) TFE tubing. A suspension of 100-150 µM glass beads is made in 0.5N HCl and enough beads are added to the column to form a 2-3 mm-deep layer. The column is then washed with about 40 ml of HCl and 40 ml of ddH_2O. While washing the column, the 0.12M NaPB, 0.1% SDS and 0.50M NaPB, 0.1% SDS solutions should be placed in a water bath and equilibrated to 68 °C. HAP (0.4g) is suspended in about 5 ml of 0.41M NaPB, 0.1% SDS, and any fines are decanted away. The HAP suspension is added to the column and the buffer is drained off. It is then washed twice with about 1 ml of 0.50M NaPB, 0.1% SDS, once with about 15 ml of the same buffer, twice with about 1 ml of 0.12M NaPB, 0.1% SDS and once with about 20 ml of the same buffer. During the final wash, the fraction collector should be calibrated to collect 1 ml fractions. It should not be necessary to apply pressure to the column when using commercially available, fast flowing lots of HAP. Following preparation, the column can be allowed to stand for several hours before use.

The hybridization reaction is diluted into 1 ml of NaPB buffer so that the final NaPB concentration is 0.24M. In the example given above, the reaction would contribute 0.02mmole of NaPB, so it would be diluted as follows:

2M PB	0.11 ml	(0.22mmole NaPB)
ddH_2O	0.84 ml	
hybridization reaction	0.05 ml	(0.02mmole NaPB)
	1.00 ml	(0.24mmole NaPB).

One µl of 10 mg/ml RNase A solution is added and the tube is incubated for 1 h at 25 °C to digest single stranded RNA. The solution is then adjusted to a total volume of 2 ml and to 0.1% SDS. Residual buffer is drained from the HAP column and the sample is carefully applied to the top of the HAP. The tube is washed with 0.12M NaPB, 0.1% SDS, the wash is added to the column and the sample is allowed to equilibrate to 68 °C (about 2 min). The column is run at 68 °C to denature poly(rA)-poly(dT) duplexes. The sample is then allowed to pass through the HAP, collecting 1 ml fractions in 13 x 100 borosilicate tubes. The column is washed by adding small amounts (about 0.5 ml) of 0.12M NaPB, 0.1% SDS, until 10-15 fractions have been collected. Bound cDNA is finally eluted by washing the column with small amounts of 0.50M NaPB, 0.1% SDS, until 7-10 fractions have been collected. Fractions containing unbound and bound cDNA are identified by using a Geiger counter and are pooled separately. Radioactivity in the samples is then determined by counting 1 µl of each in a liquid scintillation counter. The overall yield and amount of cDNA hybridized to RNA is calculated from these data.

Concentration and Desalting of Unbound Fractions

The sample of HAP-bound cDNA can be saved at -20 °C for use as a control probe for colony or filter hybridization. The sample of unbound cDNA, however, must be processed further, because it contains cDNA sequences complementary to poly(A)$^+$RNAs that are specific to or are present at higher concentrations in cell type A than in cell type B and "non-reactive" cDNA sequences. The cDNA sequences that do not hybridize under the conditions used for HAP fractionation might, however, hybridize to DNA (or RNA) in filter hybridizations, and must therefore be removed. This is accomplished by hybridizing the cDNA to poly(A)$^+$RNA from cell type A. To do this, the cDNA must be desalted and concentrated. This is accomplished as follows. Yeast tRNA (20 µg) is added to the unbound cDNA. It is then extracted once with phenol:chloroform (1:1) and concentrated to about 1 ml by extractions with dry, secondary butanol (SB). The sample is transferred to a 15 ml Corex centrifuge tube and an equal volume of SB is added. The tube is sealed with a silicon rubber stopper, shaken for a few minutes, and centrifuged for a few minutes in a table-top clinical centrifuge. The SB (upper phase) is removed and the process is repeated until the desired sample volume is obtained. The concentrated sample is then applied to the Sephadex G-100 column used for fractioning the cDNA synthesis reaction and processed as described above, ending with a dry cDNA pellet in a siliconized, 1.5 ml microfuge tube.

Poly(A)$^+$RNA$_A$-excess/cDNA Hybridization

The unbound cDNA is reacted with a 50-100-fold excess of poly(A)$^+$RNA from cell type A to a RoT of 500-1000 as described above for the poly(A)$^+$RNA$_B$-excess hybridization. The amount of RNA needed will, of course, be substantially less than the amount needed for the first hybridization reaction.

Final Preparation of the Probe and Filter Hybridization

The final hybridization reaction is prepared for HAP fractionation as before, applied to a fresh HAP column (HAP is not reused) and the column is then washed with 0.12M NaPB, 0.1% SDS, collecting 1 ml fractions. The temperature of the column is then raised to 98 °C and the bound cDNA is eluted by addition of small amounts of 0.12M NaPB, 0.1% SDS, collecting 1 ml fractions. A glass rod may be used to stir the HAP if gas bubbles form. The bound and unbound fractions are pooled separately, samples are counted and overall yield and fraction hybridized are calculated. The unbound sample may be discarded and the bound sample (usually 3-5 ml) is stored at -20 °C.

Filter hybridizations are done by using standard procedures, e.g. [15]. We usually hybridize duplicate filters with i) the "A-specific" cDNA probe and ii) the "A-non-specific" cDNA probe (cDNA bound during first HAP fractionation). The A-non-specific probe is prepared by diluting an amount equal to the A-specific probe into an equivalent volume of 0.12M NaPB, 0.1% SDS. Carrier nucleic acids [e.g. salmon DNA, poly(rA)] are added to both samples. Then 50-100 μg of poly(A)$^+$RNA from cell type B is added to the A-specific probe as a competitor for contaminating A-non-specific sequences. The samples are denatured by heating for 5 min to 102-104 °C in a boiling salt water bath and quickly cooled in an ethanol-ice bath. The probes are then added directly to the colony or plaque filters. We frequently include RNA gel blots [15] containing samples of the two poly(A)$^+$RNAs to estimate the degree of enrichment afforded by the procedure. Following hybridization, the filters are washed and autoradiographed. Positive clones are then picked and purified by using standard microbiological techniques. Selected clones should be re-tested by using them as hybridization probes for RNA gel blots containing samples of poly(A)$^+$RNA from each cell type.

SELECTION OF GENES BY EXPRESSION IN *ASPERGILLUS*

A very useful method for cloning genes that have known functions involves introducing complete gene libraries into a genetically marked host and selecting for complementation of the genetic defect. Such an approach, utilizing *E. coli* mutants as recipients, has been used to clone specific genes from a variety of fungi, including yeast and *Aspergillus*. Unfortunately, many, perhaps most, unmodified fungal genes are not expressed in *E. coli*, and failures have been more common than successes with this method. Similarly, simple complementation of yeast mutants with genomic libraries of DNA from Euascomycetes has been unsuccessful. For this and other reasons, we have constructed a cloning vector that can be used to isolate genes by complementation of mutations in *A. nidulans*. Interestingly, this vector has proved to be useful for cloning genes from other species of filamentous Ascomycetes; *A. nidulans* is apparently capable of recognizing the transcriptional signals of genes from these organisms and expressing the genes at levels sufficient to allow their detection. This property should be of value for investigators wishing to clone genes having known functions from pathogenic Euascomycetes. Many of the procedures used for selection of genes are fairly standard, and are presented in detail in a laboratory manual [15]. These are only summarized here. A procedure for isolating high molecular weight DNA from fungi is less standard and is given in greater detail.

The pKBY2 Vector

The vector in current use for cloning genes by complementation of *Aspergillus* mutants, designated pKBY2 [10], is a cosmid: that is, it contains the bacteriophage lambda *cos* site to permit *in vitro* assembly of defective lambda phage particles. As it contains an origin of replication for *E. coli* and genes for antibiotic resistance, it can be maintained and amplified readily in a bacterial host. It also contains the *trpC* gene of *A. nidulans*, to permit selection of transformants in that host, and a unique *BamHI* site, to permit introduction of DNA fragments made by digestion with *MboI*, *Sau3A*, *BglII* or *BamHI*. Although recombinants constructed in this vector and then selected by transformation of *Aspergillus* become integrated into the *Aspergillus* chromosomes by homologous recombination, we have found that they can be recovered by treatment of total *Aspergillus* DNA with a bacteriophage lambda *in vitro* packaging lysate and transduction of *E. coli* cells to antibiotic resistance. It is likely that this is possible because tandem duplications formed by integration are resolved in some nuclei by intrachromosome recombination, to produce free cosmid molecules.

Construction of Libraries in pKBY2

Most of the procedures used to construct recombinant DNA libraries in pKBY2 are standard and have been described in detail elsewhere [10,15]. The first step is to isolate high molecular weight chromosomal DNA. It is important to prepare very large DNA, because 35-40 kilobase fragments having ligatable sites on both ends are needed. We prepare such DNA from purified nuclei. The following buffers and solutions are neded for isolation of nuclei [16,17]:

10X SSE Salts:	Spermidine HCl	40mM
	Spermine HCl	10mM
	KCl	1.0M
	EDTA	100mM
	Tris base	100mM
	adjust to pH 7.0 with HCl	

phenylmethylsulphonylfluoride (PMSF) 100mM in 95% ethanol

2-mercaptoethanol

Nonidet P-40 (NP40)

20% sodium-N-lauroyl sarcosinate

SSE 0.5 buffer (1 litre) is made by combing 100 ml of 10X SSE salts, 1 ml of 2-mercaptoethanol, 10 ml of PMSF and 171.2 g of sucrose. Miracloth filters are prepared by boiling squares three times in 10mM Na_4EDTA, four times in ddH_2O and then rinsing the squares several times with ddH_2O.

Nuclei may be isolated from spores, hyphae or yeast cells by using a Bead Beater to homogenize the cells. The Bead Beater blending cup is partially filled with glass beads as described above for RNA isolation, and a thick slurry of cells suspended in SSE 0.5 buffer is transferred to the cup. The cells are disrupted by several 1 min homogenizations, the lysate is transferred to a flask on ice, the beads are rinsed with SSE 0.5 buffer and the rinse buffer is combined with the lysate. The lysate is filtered through four layers of Miracloth and centrifuged at 6000 x g for 20 min at 4 °C. The supernatant is decanted and the pellet is suspended in 40 ml of SSE 0.5 buffer by using a Dounce homogenizer with a loose fitting pestle. The nuclei are pelleted as before, resuspended in SSE 0.5 buffer containing 0.2% NP40 detergent and repelleted. Washing with SSE 0.5 plus 0.2% NP40 is continued until a clear supernatant fraction is obtained. The nuclei are again pelleted and suspended in about 5 ml of SSE 0.5. The suspension is adjusted to 1% sodium N-lauroyl sarcosinate and 100 μg/ml proteinase-K to lyse the nuclei. Following lysis, great care should be taken to avoid shearing of the DNA.

The lysate is adjusted to 49% (w/w) CsCl and 100 μg/ml ethidium bromide and a final volume of 10.6 ml. It is then gently loaded into Beckman VTi65 tubes (or equivalent), the tubes are sealed and centrifuged in a VTi65 rotor (or equivalent) at 47,000 rpm for 12-16 h at 25 °C. The tubes are removed from the rotor, a hole is punched in the tops of the tubes and the DNA band is removed by side puncture with a syringe and an 18 gauge hypodermic needle, illuminating with long wave UV light if necessary. Exposure to light should be minimized. If the DNA band is not well separated from other materials, the DNA may be rebanded under the same conditions. The DNA is placed in a 15 ml Corex centrifuge tube, diluted to about 5 ml with 10mM Tris-HCl, pH 7.6, 10mM NaCl, 1mM EDTA (TNE) and then extracted several times with iso-amylalcohol by gentle mixing, until the ethidium bromide is removed. The

DNA solution is transferred to a dialysis tube by using a large bore pipette and dialyzed against 3-4 changes of TNE buffer (500-1000 ml of buffer per change). The dialysis tubing should be prepared by boiling several times in 10mM Na_4EDTA and then ddH_2O and is stored in 70% ethanol at 4 °C. The solution is removed, a 1:50 dilution is made in TNE buffer and a UV absorption spectrum is taken (50 μg DNA = 1.0 OD_{260}). If the DNA is too dilute for convenient use, it can be precipitated by addition of 0.1 vol of 3M sodium acetate, pH 6.0, and 1.1 vol of 2-propanol, collected by centrifugation and redissolved in TNE. Alternatively, it can be returned to a dialysis tube and concentrated by ultrafiltration, by using Sephadex (Pharmacia) or Aquacide (CalBiochem). The DNA is stored at 4 °C.

The size of the DNA can be estimated by electrophoresis in dilute (0.2-0.4%) agarose gels with large DNA markers (e.g. undigested bacteriophage lambda or T4 DNA). For preparation of 35-30 kilobase Sau3A or MboI partial digestion products, the DNA is treated and analyzed as described [14].

Preparation of Vector DNA

The cosmid vector pKBY2 can be grown in the presence of 50 μg/ml ampicillin and/or chloramphenicol and DNA is prepared by using standard procedures [14]. The cosmid can be amplified by addition of spectinomycin to 100-200 μg/ml. Following purification, 50-100 μg of DNA are digested to completion with BamHI in the presence of 20mM Tris-HCl, pH 7.6, 7mM $MgCl_2$, 100mM KCl, 2mM 2-mercaptoethanol and 100 μg/ml autoclaved gelatin. The reaction mixture is extracted with redistilled phenol that has been saturated with 10mM Tris-HCl, pH 7.6, 1mM EDTA (TE) and is precipitated by addition of 0.1 vol of 3M sodium acetate, pH 6.0, and 2 vol 95% ethanol and incubation at -70° C for 30-60 min. It is collected by centrifugation in the microfuge at 4° C for 15 min and then washed twice with 95% ethanol by centrifugation. The DNA is redissolved in 500 μl of 10mM Tris-HCl, pH 8.0, and treated with one unit of calf intestinal alkaline phosphatase (Boehringer-Manheim, molecular biology grade, NH_4^+-free) for 30-60 min at 37 °C. EDTA is added to 5mM and the solution is phenol extracted, precipitated and washed as before. It is then dissolved in TNE buffer at a concentration of about 1 mg/ml and stored at 4 °C. Before use, the vector should be tested for ability to transform *E. coli* cells before and after ligation with and without prior phosphorylation (see [15], for reaction conditions and transformation protocols). The transformation efficiency with phosphorylated, ligated vector should be at least 100 times greater than with dephosphorylated, ligated vector and at least 50% that obtained with BamHI-digested (but otherwise untreated) religated vector.

Formation of Libraries

Partially digested chromosomal DNA is ligated with dephosphorylated DNA precisely as described by Yelton [10]. The ligation mixture is treated with a commercially available lambda packaging extract by using the procedures recommended by the supplier. Control packaging efficiencies with lambda CI857 DNA should exceed 1 x 10^8 plaque forming units/μg DNA. The packaging mixtures are titred by making serial dilutions and plating on medium containing 50-100 μg/ml ampicillin. The library is then amplified by plating transductants at a density of about 500 colonies per 10 cm Petri dish. The number of colonies needed for a complete library depends on the genome size of the organism, but for most fungi 10,000 transductants is more than adequate. The "quality" of the library can be estimated by isolating small amounts of cosmid DNA from about 30 transductants ("mini-prep"), digesting with DNA with, for example, BamHI and fractionating the digests by electrophoresis in agarose gels. All of the clones examined should be different from one another, and their sizes should range from 40-50 kilobases, including the vector-containing fragments.

The libraries formed by using these procedures are pooled by suspending the cells, after 12-24 h of growth, in liquid medium by using a glass spreader. The suspension is adjusted to 30% with sterile glycerol or 7% with dimethylsulphoxide and 1 ml samples are stored at -70 °C. Isolation of DNA from cosmid pools and recovery of cosmids from Aspergillus transformants has been described in detail [10]. Transformation of Aspergillus or related organisms is carried out precisely as described by Yelton [6]. Aspergillus transformants can be selected by complementation of mutations, but it should be kept in mind that complementation may be due to the activity of the target gene or may result from suppression. The identity of the isolated gene must be confirmed by independent methods. As only one thousand to a few thousand transformants are needed to represent a fungal genome, individual Aspergillus transformants can also be screened for a particular phenotype, for example production of an enzyme or metabolite. Once a cosmid has been obtained, small DNA fragments containing the complementing activity can be identified by using a rapid co-transformation technique [18]. Evidence is accumulating that the Aspergillus cosmid complementation approach can be used effectively with recombinant libraries formed with DNA from Eusascomycetes other than Aspergillus nidulans.

FUTURE PROSPECTS

There is now good evidence that the procedures and experimental strategies described in this chapter can have wide application to problems in plant pathology. The results from experiments with fungi such as A. nidulans and N. crassa may form paradigms for other Euascomycetes. In the near future, one can expect to see the development of "broad host-range" vectors for use with fungi for which sophisticated genetic systems do not exist. The development of transformation vectors containing dominant drug resistance genes would be of particular value, because appropriate auxotrophic mutations are not available in many plant pathogenic species. The work with A. nidulans should provide the rationale for the design of such vectors.

ACKNOWLEDGEMENTS

I thank Drs. Forrest Chumley, Barbara Valent, Olen Yoder and Hans Van Etten for communicating results before their publication, and Drs. Carlene Raper and Richard Staples for their stimulating visits to my laboratory. I am especially grateful to the graduate students and postdoctoral associates in my laboratory for their many contributions to the development of the concepts and techniques described in this chapter. Our work has been supported by grants from the National Institutes of Health, National Science Foundation and United States Department of Agriculture.

REFERENCES

1. Hinnen, A.H., Hicks, J.B. and Fink, G.R., 1978, Transformation of yeast, Proc. Natl. Acad. Sci. USA, 75: 1929-1933.
2. Beggs, J., 1978, Transformation of yeast by a replicating hybrid plasmid, Nature, 275: 104-109.
3. Case, M.E., Schweizer, M., Kushner, S.R. and Giles, N.H., 1979, Efficient transformation of Neurospora crassa by utilizing hybrid plasmid DNA, Proc. Natl. Acad. Sci. USA, 76: 5259-5263.

4. Ballance, D.J., Buxton, F.P. and Turner, G., 1983, Transformation of Aspergillus nidulans by the orotidine-5'-phosphate decarboxylase gene of Neurospora crassa, Biochem. Biophys. Res. Commun., 112: 284-289.
5. Tilburn, J., Scazzocchio, C., Taylor, G.T., Zabicky-Zissman, J.H., Lockington, R.A. and Davies, R.W., 1983, Transformation by integration in Aspergillus nidulans, Gene, 26: 205-221.
6. Yelton, M.M., Hamer, V.E. and Timberlake, W.E., 1984, Transformation of Aspergillus nidulans by using a trpC plasmid, Proc. Natl. Acad. Sci. USA, 81: 1470-1474.
7. Botstein, D. and Davis, R.W., 1981, Principles and practice of recombinant DNA research, in: "The Molecular Biology of the Yeast Saccharomyces II. Metabolism and Gene Expression", J.N. Strathern, E.W. Jones and J.R. Broach, eds., Cold Spring Harbor Laboratory, New York.
8. Struhl, K., 1983, The new yeast genetics, Nature (London), 305: 391-397.
9. Timberlake, W.E., 1980, Developmental gene regulation in Aspergillus nidulans, Dev. Biol., 78: 497-510.
10. Yelton, M.M., Timberlake, W.E. and van den Hondel, C.A.M.J.J., 1985, A cosmid for selecting genes by complementation in Aspergillus nidulans: Selection of the developmentally regulated yA locus, Proc. Natl. Acad. Sci. USA, 82: 834-838.
11. Zimmermann, C.R., Orr, W.C., Leclerc, R.F., Barnard, E.C. and Timberlake, W.E., 1980, Molecular cloning and selection of genes regulated in Aspergillus development, Cell, 21: 709-715.
12. Raper, C.A. and Timberlake, W.E., 1985, Exp. Mycol., in press.
13. Van Etten, J.L. and Freer, S.N., 1978, Simple procedure for disruption of fungal spores, Appl. Environ. Microbiol., 35: 622-623.
14. Britten, R.J., Graham, D.E. and Neufeld, B.R., 1974, Analysis of repeating DNA sequences by reassociation, in: "Methods in Enzymology", Vol. 29. S.P. Colowick and N.O. Kaplan, eds., pp. 363-418, Academic Press, New York.
15. Maniatis, T., Fritsch, E.F. and Sambrook, J., 1982, Molecular Cloning: A Laboratory Manual", Cold Spring Harbor Laboratory, New York.
16. Gealt, M.A., Sheir-Neiss, G. and Morris, N.R., 1976, The isolation of nuclei from the filamentous fungus Aspergillus nidulans, J. Gen. Microbiol., 94: 204-210.
17. Timberlake, W.E., 1978, Low repetitive DNA content in Aspergillus nidulans, Science, 202: 773-775.
18. Timberlake, W.E., Boylan, M.T., Cooley, M.B., Mirabito, P.M., O'Hara, E.B. and Willett, C.E., 1985, Exp. Mycol., in press.

AN ANTIBIOTIC RESISTANCE MARKER FOR FUNGAL TRANSFORMATION

G. Turner and M. Ward

Department of Microbiology
The Medical School
University of Bristol
Bristol, UK

INTRODUCTION

Transformation of *Aspergillus nidulans* has been reported by a number of groups using a variety of cloned, prototrophic genes to transform equivalent nutritional mutants [1-5]. The disadvantage of this approach is that in order to transform other desired strains, the nutritional mutation must always be transferred, usually by a sexual cross, to the new strain. Antibiotic resistance transformation markers would obviate this necessity, and enable any strain of *Aspergillus nidulans* to be transformed directly.

Oligomycin, an inhibitor of the mitochondrial ATP synthase enzyme, is an effective growth inhibitor of *Aspergillus nidulans* at 3 µg/ml, and spontaneous resistant mutants can be easily isolated [6]. Although a wide range of phenotypes are seen, all map at 2 genetic loci, one in the mitochondrial genome and one on linkage group VII [7]. Mutations at both loci confer *in vitro* resistance on the mitochondrial ATPase activity. Analogy with data from *Neurospora crassa* [8] suggested that the nuclear locus was probably the gene for subunit 9 of the ATP synthetase complex. We have used *Neurospora* subunit 9 cDNA [9] as a hybridization probe to identify and isolate the equivalent *Aspergillus* sequence from a lambda bank constructed from total DNA of a nuclear oligomycin-resistant mutant of *A. nidulans*.

RESULTS AND DISCUSSION

Identification and Isolation of the Gene

Total DNA isolated from an oligomycin resistant strain carrying the *oliC31* mutation was used to construct a gene bank in lambda EMBL 4. A lambda clone carrying DNA homologous (approximately 80%) to the *Neurospora crassa* probe was purified and mapped. The region responsible for homology was subcloned into an *Aspergillus* transformation vector pDJB1 [5] and named pMW2 (Fig. 1).

Transformation of A. nidulans

A. nidulans can be transformed by pMW2 using either of the selectable markers which it carries, *pyr4* or *oliC31*. Selection for *pyr4*, which is a

NATO ASI Series, Vol. H1
Biology and Molecular Biology of Plant-Pathogen Interactions
Edited by J. Bailey

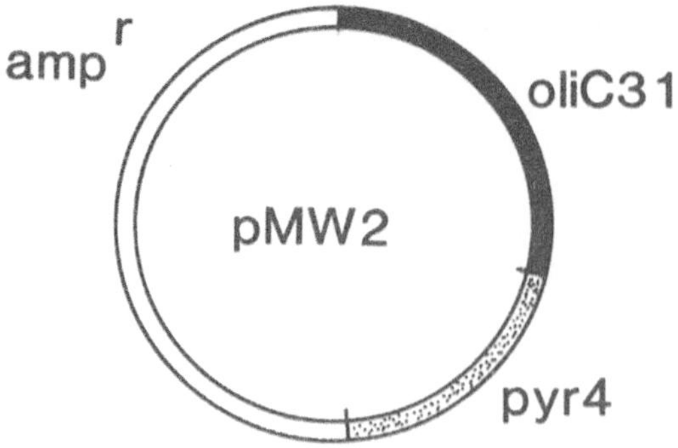

Figure 1. Transformation vector pMW2.

Neurospora crassa gene coding for orotidine-5'-phosphate decarboxylase, requires that the recipient Aspergillus carries a mutation in the equivalent pyrG gene, making it a uridine-requiring auxotroph [1,5]. However, any strain of A. nidulans can be transformed with pMW2 by selection of transformants on oligomycin. Addition of the drug immediately to transformed protoplasts prevented their regeneration, so protoplasts are initially plated on drug-free regeneration medium, and an overlay of agar containing oligomycin added after 20 h. Only the oligomycin-resistant transformants are able to grow through the overlay.

Phenotypes of Oligomycin-Resistant Transformants

Subculture of pMW2 transformants and their subsequent analysis revealed 3 major phenotypic classes, depending on the nature of the initial integration event:

1. Fully-resistant colonies (Oli^r)

2. Semi-resistant colonies (Oli^{sr}) which frequently gave rise to Oli^r sectors when subcultured on drug-containing medium ("unstable" Oli^{sr}).

3. Semi-resistant colonies which remained as such even when subcultured on drug-containing medium ("stable" Oli^{sr}).

Molecular analysis of these classes was essential to an understanding of the integration events and post-transformation behaviour.

Analysis of Transformants

DNA was prepared from transformants representative of the three major classes, and analyzed in two ways:

i) Partial digestion of DNA with restriction endonuclease PstI was followed by religation and transformation of E. coli to ampicillin resistance. This approach, known as marker rescue, rescues the bacterial plasmid beta-lactamase sequence from the transformant DNA together with flanking sequences from the Aspergillus chromosome. The resulting plasmids can be reisolated from E. coli and mapped, giving a picture of the original integration event.

ii) Agarose gel electrophoresis of restriction endonuclease digested DNA, Southern blot transfer to nitrocellulose, and probing with radiolabelled pMW2 DNA.

In addition, the location of the integrated oliC31 and/or pyr4 could be ascertained by genetic analysis, using either the parasexual cycle, to allocate the markers to a particular linkage group, or the sexual cross to demonstrate whether or not the markers were linked to the wild-type oliC allele of the recipient strain.

The data obtained from these approaches was consistent with the events outlined below. A complete account of this can be found in [10].

Fate of Transforming DNA

Stable transformants of A. nidulans result from integration of transforming DNA into the nuclear genome by one or more recombination events. This recombination can be homologous or non-homologous [1-3,5] with some preference for homologous recombination. pMW2 can therefore recombine at oliC on linkage group VII, the only possible homologous event, or elsewhere in the genome by non-homologous recombination. Note that the pyr4 sequence, derived from Neurospora crassa, has no significant homology with any A. nidulans sequence [5].

Integration at sites other than oliC leads to class 3 transformants (stably semi-resistant). The semi-resistance results from the presence of both wild-type and resistant alleles in the same strain. It had been already shown, by the use of diploids, that nuclear mutations to oligomycin resistance are semi-dominant [6]. Although such a transformant carries a duplication, the distance between the duplicated sequences is such that the frequency of recombination between them is negligible, hence the stability on oligomycin. The recombination event resulting in non-homologous integration can be via any of the plasmid sequences, though this will often inactivate one of the selectable markers. For instance, a proportion of the class 3 transformants selected on oligomycin in the presence of uridine carry an inactivated pyr4 as a result of recombination via that sequence and its resulting disruption.

Instability of type 2 transformants when plated on oligomycin results from recombination events leading to loss of the wild-type, $oliC^+$ (sensitive) allele. There are two possible recombination events (Fig. 2). Homologous recombination (a) across the duplicated oliC sequences results in reversal of the integration process, but with loss of the wild-type allele $oliC^+$ along with the rest of the pMW2 sequence, and leaving a single oliC31 gene in the strain. Gene conversion (b) results in the presence of two copies of oliC31, and retention of the intervening pMW2 sequences. Retention or loss of these sequences can be rapidly tested by scoring fully-resistant sectors for ability to grown on uridine-free medium. This showed that homologous recombination:gene conversion occur in a ratio of about 2:3, similar to that observed in yeast [11].

Analysis of class 2 transformants propagated on drug-free medium shows that these recombination events are quite rare during vegetative growth (less than 1 in 10^3 spores), but these rare events are easily detected on oligomycin medium as a result of the selection pressure favouring the Oli^r mycelium over Oli^{sr}.

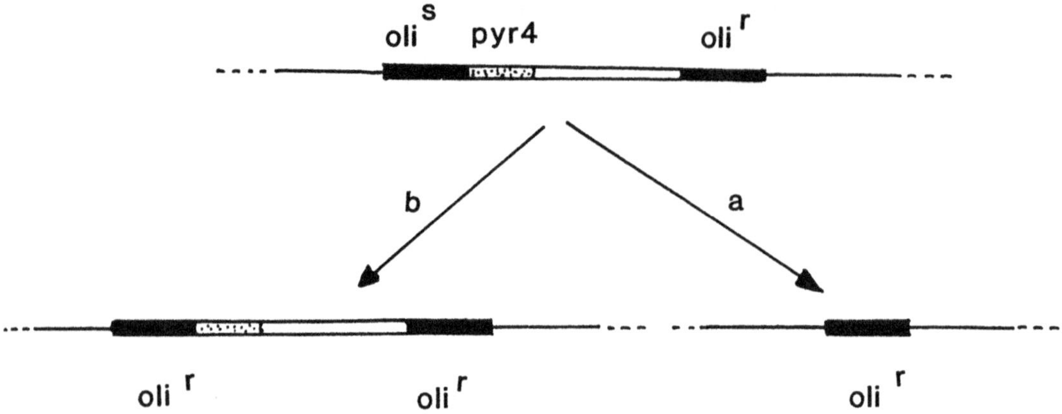

Figure 2. Two possible recombination events; (a) homologous recombination, (b) gene conversion.

A further feature of the $oliC^r$ allele is that it confers hypersensitivity to another ATP synthase inhibitor, triethyltin [7]. It is therefore possible to select for Oli^s derivatives, resulting from loss of oliC31 by recombination, by plating class 2 transformants on medium containing triethyltin.

POTENTIAL APPLICATIONS

While it is unlikely that the *A. nidulans* $oliC^r$ gene would function as a transformation marker in another fungal species, the high degree of conservation of this subunit means that it is relatively easy to find and isolate the equivalent gene from other fungi by using the *Aspergillus* or *Neurospora* genes as hybridization probes. Furthermore, oligomycin is effective against a wide range of filamentous fungi at low concentrations, and it is not difficult to isolate resistant mutants. Since mutations in the mitochondrial DNA can also confer oligomycin resistance, a routine precaution is to carry out a heterokaryon test [6] to eliminate these.

The ability to select for or against the resistant *A. nidulans* gene is a feature which we hope to exploit in devising gene replacement techniques similar to those used in *Saccharomyces cerevisiae* [12].

SUMMARY

We have isolated a gene conferring oligomycin resistance from a mutant of *Aspergillus nidulans*. It can be used as a selectable marker for transformation of wild-type (sensitive) strains of *A. nidulans*. The resistance mutation resides in the gene for subunit 9 of the mitochondrial ATPase complex. Following transformation and chromosomal integration, the location of the gene can be inferred from the phenotype of the transformants, and subsequent recombinational behaviour has been analyzed.

Subunit 9 is highly conserved, thus the *Aspergillus* gene can be used as a hybridization probe to identify and isolate the equivalent gene from resistant mutants of other fungal species.

REFERENCES

1. Ballance, D.J., Buxton, F.P. and Turner, G., 1983, Transformation of Aspergillus nidulans by the orotidine-5'-phosphate decarboxylase gene of Neurospora crassa, Biochem. Biophys. Res. Comm., 112: 284-289.
2. Tilburn, J., Scazzocchio, C., Taylor, G.G., Zabricky-Zissman, J.H., Lockintgon, R.A. and Davies, R.W., 1983, Transformation by integration in Aspergillus nidulans, Gene, 26: 205-221.
3. Yelton, M.M., Hamer, J.E. and Timberlake, W.E., 1984, Transformation of Aspergillus nidulans by using a trpC plasmid, Proc. Natl. Acad. Sci. USA, 81: 1470-1474.
4. Johnstone, I.L., Hughes, S.G. and Clutterbuck, A.J., 1985, Cloning of an Aspergillus developmental gene by transformation, EMBO J., 4: 1307-1311.
5. Ballance, D.J. and Turner, G., 1985, Development of a high-frequency transforming vector for Aspergillus nidulans, Gene, 36: 321-331.
6. Rowlands, R.T. and Turner, G., 1973, Nuclear and extranuclear inheritance of oligomycin resistance in Aspergillus nidulans, Molec. Gen. Genet., 126: 201-216.
7. Rowlands, R.T. and Turner, G., 1977, Nuclear-extranuclear interactions affecting oligomycin resistance in Aspergillus nidulans, Molec. Gen. Genet., 154: 311-318.
8. Sebald, W. and Hoppe, J., 1981, On the structure and genetics of the proteolipid subunit of the ATP synthase complex, Curr. Top. Bioenerg., 12: 1-64.
9. Viebrock, A., Perz, A. and Sebald, W., 1982, The imported pre-protein of the proteolipid subunit of the mitochondrial ATP synthase from Neurospora crassa, Molecular cloning and sequencing of the mRNA, EMBO J., 1: 565-571.
10. Ward, M., Wilkinson, B. and Turner, G., 1986, Transformation of Aspergillus nidulans with a cloned, oligomycin-resistant ATP synthase subunit 9 gene, Molec. Gen. Genet., 202: 265-270.
11. Jackson, J.A. and Fink, G.R., 1981, Gene conversion between duplicated genetic elements in yeast, Nature, 292: 306-311.
12. Struhl, K., 1983, Direct selection for gene replacement events in yeast, Gene, 26: 231-242.

ANALYSIS OF TRANSCRIPTION-CONTROL SIGNALS IN *ASPERGILLUS*

C.A.M.J.J. van den Hondel, P.J. Punt, B.L.M. Jacobs-Meijsing, W. van Hartingsveldt, R.F.M. van Gorcom and P.H. Pouwels.

Medical Biological Laboratory TNO
Rijswijk
The Netherlands

INTRODUCTION

Genetic studies in *Aspergillus* have shown that expression of a number of genes is controlled by trans-acting elements [1,2,3]. These elements may interact with controlling regions located adjacent to structural genes [4, 5]. However, little is known about the molecular mechanisms of gene expression and gene regulation. The recent cloning of genes whose expression is regulated such as the *amdS* gene [6], *trpC* gene [7], *yA* gene [8] and *brlA* gene [9] opens the possibility to study gene expression and gene regulation at the molecular level. A powerful and convenient method for such a study is fusion of the gene of interest to the *lacZ* gene of *E. coli* [10]. Studies in *E. coli* [10], *Saccharomyces cerevisiae* [10] and animal cells [11] have shown that the expression and regulation of genes fused to the *lacZ* gene can be assayed qualitatively and quantitatively.

Recently we have made a fusion of the *E. coli lacZ* gene and the *trpC* gene of *A. nidulans*. This fusion gene is expressed in *A. nidulans* and its expression is controlled by the *A. nidulans trpC* transcription/translation control sequences [12]. This result indicates that the *lacZ*-fusion system can also be used in *A. nidulans*.

In this paper we describe a set of promoter-probe vectors which enable the isolation and analysis of transcription-control sequences. With the aid of these vectors we have analyzed in *A. nidulans* and *A. niger* the transcription control sequences of the *A. nidulans gap* gene, encoding glyceraldehyde-3-phosphate dehydrogenase.

VECTORS FOR ISOLATION AND ANALYSIS OF TRANSCRIPTION-CONTROL SIGNALS

We have constructed a fusion of the *A. nidulans trpC* gene and the *E. coli lacZ* gene [12]. Based on this fusion gene (see Fig. 1, pAN 923-21A) a set of promoter probe vectors was constructed. These vectors contain a unique *BamHI* site in three different reading frames in front of either the truncated *trpC-lacZ* gene (Fig. 1, pAN923-31, 32 and 33) or the truncated *lacZ* gene (pAN923-41, 42 and 43). In addition the vectors carry the *A. nidulans argB* gene as a selection marker. Cloning of part of a gene, including its transcription/translation control signals, in the *BamHI* site of these vectors will result in an in-phase *lacZ* fusion which can be analysed in *A. nidulans*.

NATO ASI Series, Vol. H1
Biology and Molecular Biology of Plant-Pathogen Interactions
Edited by J. Bailey

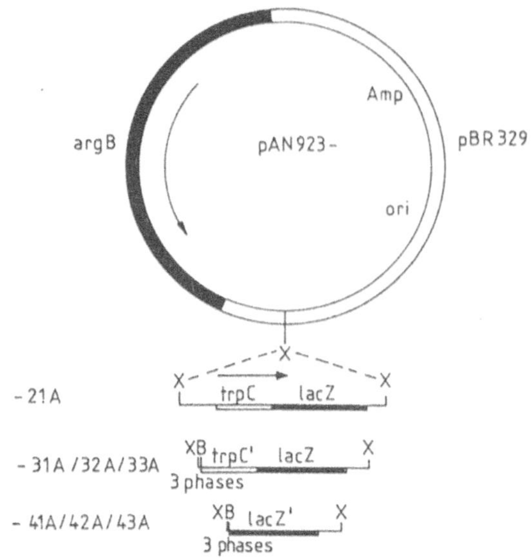

Figure 1. Schematic representation of vectors for isolation of transcription and translation signals. argB denotes the A. nidulans arginine B gene, lacZ the E. coli β-galactosidase gene and trpC the A. nidulans tryptophan C gene. The truncated genes are indicated with a mark('). The construction of the trpC-lacZ fusion gene of 21A was described previously [12]. The arrow indicates the direction of transcription. In the A-version of the plasmids, the transcription direction of the argB gene and the truncated lacZ-fusion gene is counter clockwise. In the B-version, the transcription directions have opposite polarity. The unique BamHI (B) site is located in three different reading frames in front of the truncated lacZ fusion gene. In the pAN923-vectors the original BamHI site of pBR329, located in the tet gene, was replaced by an XhoI (X) site.

Since autonomously replicating plasmids are not available for A. nidulans, intact and stable integration of the fusion gene into the chromosome is a prerequisite for analysis of transcription/translation control signals or comparison of promoter strength. To exclude possible interference of different surroundings of the fusion gene on its expression it also might be necessary to integrate the fusion gene at a specific site in the chromosome, preferably as a single copy. To test whether the constructed vectors meet these requirements, plasmids pAN923-21A and 21B were transformed into an A. nidulans argB mutant and the Arg+ transformants were analyzed. The results of these experiments revealed that in 50% of the Arg+ transformants a single copy of the plasmids was integrated at the argB locus, indicating that these vectors fulfil the prerequisites to study gene expression and gene regulation.

ISOLATION OF THE ASPERGILLUS GPD GENE

To analyze the transcription/translation signals of a gene that is strongly expressed in A. nidulans we have isolated the A. nidulans gpd gene. A Southern blot analysis of EcoRI-digested A. nidulans DNA hybridized with a [32P]-labelled DNA fragment containing the S. cerevisiae GAPDH gene, showed strong hybridization with a 6.9 kb and 2.5 kb DNA fragment (Fig. 2).

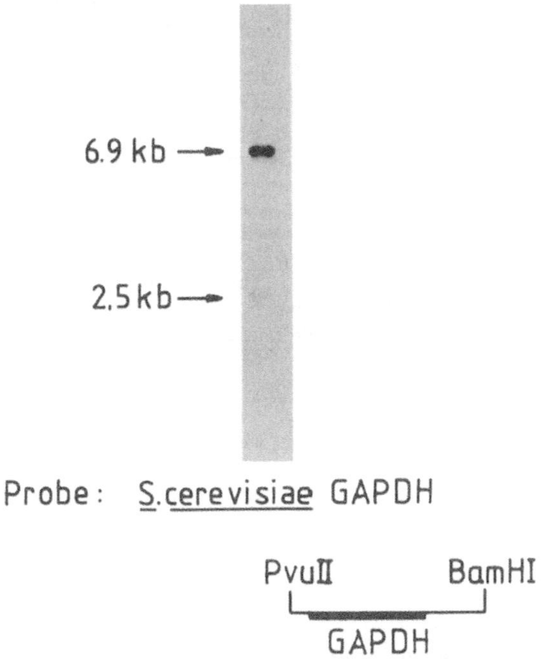

Figure 2. Southern blot hybridization of EcoRI-digested A. nidulans chromosomal DNA probed with a [^{32}P]-labelled PvuII-BamHI fragment of pGAP63 containing the GAPDH gene of S. cerevisiae [13]. The hybridization was carried out in 6.6 x SSC at 56 °C followed by washing at the same temperature and at a stringency of 3 x SSC.

For the isolation of the A. nidulans gpd gene a genomic library of A. nidulans DNA cloned in the vector λ charon 4A [14] was screened with the same labelled S. cerevisiae GAPDH probe. Three independently isolated clones were recovered containing overlapping DNA fragments of a single region of the Aspergillus genome. The S. cerevisiae GAPDH fragment hybridized to a single EcoRI-HindIII fragment of these clones, suggesting that the putative A. nidulans gpd gene in this region the nucleotide sequence of about 600 nucleotides around the central EcoRI- and StuI-sites was determined (Fig.3). Comparison of the nucleotide and the amino acid sequence of the putative A.nidulans gpd gene with those of the S. cerevisiae GAPDH gene revealed a homology of about 50% for both nucleotide and amino acid sequence, indicating that we had isolated the A. nidulans gpd gene. This comparison also suggested the presence of several introns in the region that encode for the first 42 amino acids.

ANALYSIS OF TRANSCRIPTION-CONTROL SIGNALS OF THE A. NIDULANS GPD GENE IN A. NIDULANS AND A. NIGER.

Partial sequence analysis of the StuI fragment (Fig.3) revealed that this fragment contains the N-terminal part of the gpd gene, including several introns, and upstream transcription-control sequences. To analyze these sequences the StuI fragment was cloned in the vectors pAN923-31A, 31B, 32A, 32B and in pAN923-41A, 41B, 42A, 42B. From the nucleotide sequence of the

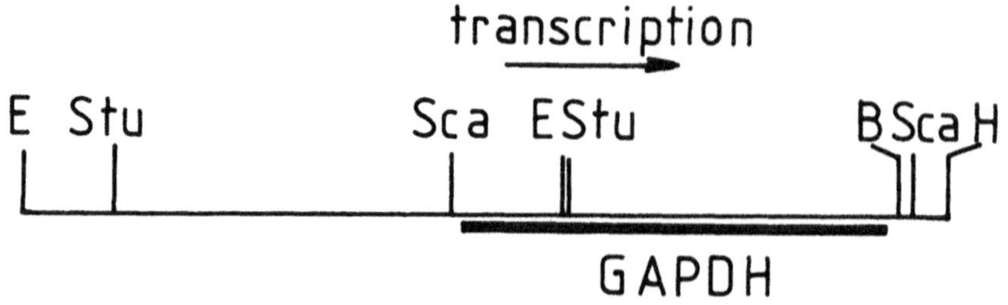

Figure 3. Restriction map of the A. nidulans genomic DNA fragments which contains the A. nidulans gpd gene. The solid line indicates the position of the gpd gene on the map. The symbols of the restriction enzymes are E=EcoRI, Stu=StuI, Sca=ScaI, B=BamHI and H=HindIII.

StuI fragment we could predict that insertion in the 31- and 41-vectors should be in-phase. Transformation with these constructions of an A. nidulans argB mutant and analysis of the Arg+ transformants on X-gal plates revealed that more than 50% of the transformants, containing the StuI fragment in the 31- and 42 vectors gave colourless colonies only. Moreover the intensity of the colour of the colonies was much stronger than of colonies containing the A. nidulans trpC-lacZ gene fusion (pAN923-21A, -21B). This result confirms that 31- and 41-constructions are in-phase. From these results we also conclude that the efficiency of expression directed by the transcription/ translation signals of the gpd gene is much higher than that of the trpC gene. Experiments are in progress to localize more precisely the transcription-control signals of the gpd gene by deletion and linker-screening studies.

To test whether the promoter-probe vectors can be used in A. niger as well, pAN923-41BSA (containing the StuI fragment in-phase) was introduced in A. niger by co-transformation with the selection plasmid p3SR2 according to the procedure of Kelly and Hynes [15]. Analysis of AmdS$^+$ transformants on X-gal plates revealed that more than 50% of the colonies were very blue, whereas transformants containing p3SR2 only gave colourless colonies. This result indicates that the promoter-probe vectors can also be used in A. niger. Furthermore we can conclude that the transcription/translation signals of the A. nidulans gpd gene are functional in A. niger and that the A. nidulans introns are properly excised. Similar results have been found for the A. nidulans amdS gene [15].

SUMMARY

We have constructed a set of promoter-probe vectors. With these vectors the transcription-control signals of the Aspergillus nidulans glyceraldehyde-3-phosphate dehydrogenase (gpd) gene were analyzed in A. nidulans and Aspergillus niger. Preliminary results indicate that (1) a relatively strong promoter is present in front of the gpd gene, (2) the A.nidulans gap promoter is functional also in A. niger and (3) A. nidulans introns are properly excised in A. niger.

REFERENCES

1. Pateman, J.A. and Kinghorn, J.R., 1977, in: "Genetics and Physiology of Aspergillus", J.E. Smith and J.A. Pateman, eds., pp. 203-241, Academic Press, London.
2. Cove, D.J., 1979, Genetic studies of nitrate assimilation in Aspergillus nidulans, Biol. Rev., 54: 291-327.

3. Arst, H.N., Jr., 1981, in: "Society for General Microbiology Symposium 31", S.W. Glover and D.A. Hopwood, eds., pp. 131-160, Cambridge University Press.

4. Hynes, M.J., 1979, Fine-structure mapping of the acetamidase structural gene and its controlling gene in Aspergillus nidulans, Genetics, 91: 381-392.

5. Sharma, K.K. and Arst, H.N., Jr., 1985, The product of the regulatory gene of the proline catabolism gene cluster of Aspergillus nidulans is a positive-acting protein, Curr. Genet., 9: 299-304.

6. Hynes, M.J., Corrick, C.M. and King, J.A., 1983, Isolation of genomic clones containing the amdS gene of Aspergillus nidulans and their use in the analysis of structural and regulatory mutations, Mol. Cell Biol., 3: 1430-1439.

7. Yelton, M.M., Hamer, J.E., De Souza, E.R., Mullaney, E.J. and Timberlake, W.E., 1983, Developmental regulation of the Aspergillus nidulans trpC gene, Proc. Natl. Acad. Sci. USA., 80: 7576-7580.

8. Yelton, M.M., Timberlake, W.E. and van den Hondel, C.A.M.J.J., 1985, A cosmid for selecting genes by complementation in Aspergillus nidulans : Selection of the developmentally regulated yA locus, Proc. Natl. Acad. Sci. USA., 82: 834-838.

9. Johnstone, I.L., Hughes, S.G. and Clutterbuck, A.J., 1985, Cloning an Aspergillus nidulans developmental gene by transformation, EMBO J., 4: 1307-1311.

10. Casadaban, M.J., Martinez-Arias, A., Shapira, S.K. and Chou, J., 1983, in: "Methods in Enzymology", Vol. 100, Recombinant DNA, Part B, R. Wu, L. Grossman and K. Moldave, eds., pp. 293-308, Academic Press, New York.

11. Hall C.V., Jacob, P.E., Ringold, G.M. and Lee, F., 1983, Expression and regulation of Escherichia coli lacZ gene fusions in mammalian cells, J. Mol. Appl. Genet., 2: 101-109.

12. van Gorcom, R.F.M., Pouwels, P.H., Goosen, T., Visser, J., van den Broek, H.W.J., Hamer, J.E., Timberlake, W.E. and van den Hondel, C.A.M.J.J., 1985, Gene, 40: 99-106.

13. Holland, J.P. and Holland, M.J., 1980, Structural comparison of two nontandemly repeated yeast glyceraldehyde-3-phosphate dehydrogenase genes, J. Biol. Chem., 255: 2596-2605.

14. Zimmerman, C.R., Orr, W.C., Leclarc, R.F., Banard, E.C. and Timberlake, W.E., 1980, Molecular cloning and selection of genes regulated in Aspergillus development, Cell, 23: 709-715.

15. Kelly, J.M. and Hynes, J.H., 1985, Transformation of Aspergillus niger by the amdS gene of Aspergillus nidulans, EMBO J., 4: 475-479.

TECHNOLOGY FOR MOLECULAR CLONING OF FUNGAL VIRULENCE GENES

O.C. Yoder, K. Weltring, B.G. Turgeon, R.C. Garber and H.D. VanEtten

Dept. Plant Pathology
Cornell University
Ithaca, NY 14853, USA

INTRODUCTION

One approach to understanding the genetic and molecular basis of fungal pathogenicity is to clone the genes involved so that their physical structures, metabolic functions, and mechanisms of regulation can be analyzed both *in vivo* and *in vitro*. Before this goal can be achieved, molecular technology specific for filamentous fungi must be developed. An essential component of such technology is a transformation system based on a gene that is selectable in fungal pathogens. We have identified two genes that are attractive as selectable markers for genetically undeveloped fungi because they can be used to transform wild type cells, thereby eliminating the need for particular mutants and their complementary genes. In addition, we have shown that a fungal virulence gene can be cloned and expressed in a heterologous fungal host.

This chapter is divided into three sections, The first and second sections describe transformation of wild type fungal cells using two different genes as selectable markers. The third section reviews the evidence that a particular group of fungal genes are causally involved in fungal virulence and outlines the steps we have taken to clone one of these genes.

The development of transformation vectors utilized the maize pathogen *Cochliobolus heterostrophus* (anamorph: *Bipolaris maydis* = *Helminthosporium maydis* = *Dreschlera maydis*) as a model fungus. We have used this fungus for a number of molecular investigations [1,2,3,4] because it is easy to culture [5] and can be genetically manipulated both sexually [5,6] and asexually [7,8,9].

COCHLIOBOLUS TRANSFORMATION WITH THE *E. COLI HYGB* GENE

The antibiotic hygromycin B (Fig. 1) binds to both 70S and 80S ribosomes, inhibiting growth of both prokaryotes and eukaryotes. It is produced by the prokaryote *Streptomyces hygroscopicus*, which protects itself by synthesizing the enzyme hygromycin B phosphotransferase. This enzyme phosphorylates the drug making it inactive in the host [10]. Genes encoding hygromycin B phosphotransferase have been cloned from both *S. hygroscopicus*

NATO ASI Series, Vol. H1
Biology and Molecular Biology of Plant-Pathogen Interactions
Edited by J. Bailey

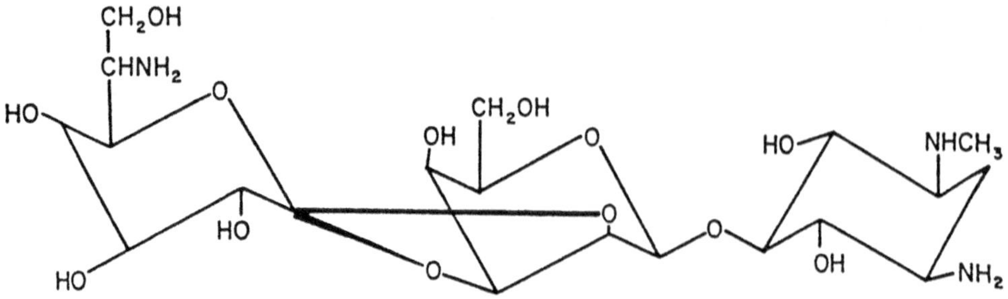

Figure 1. Structure of hygromycin B.

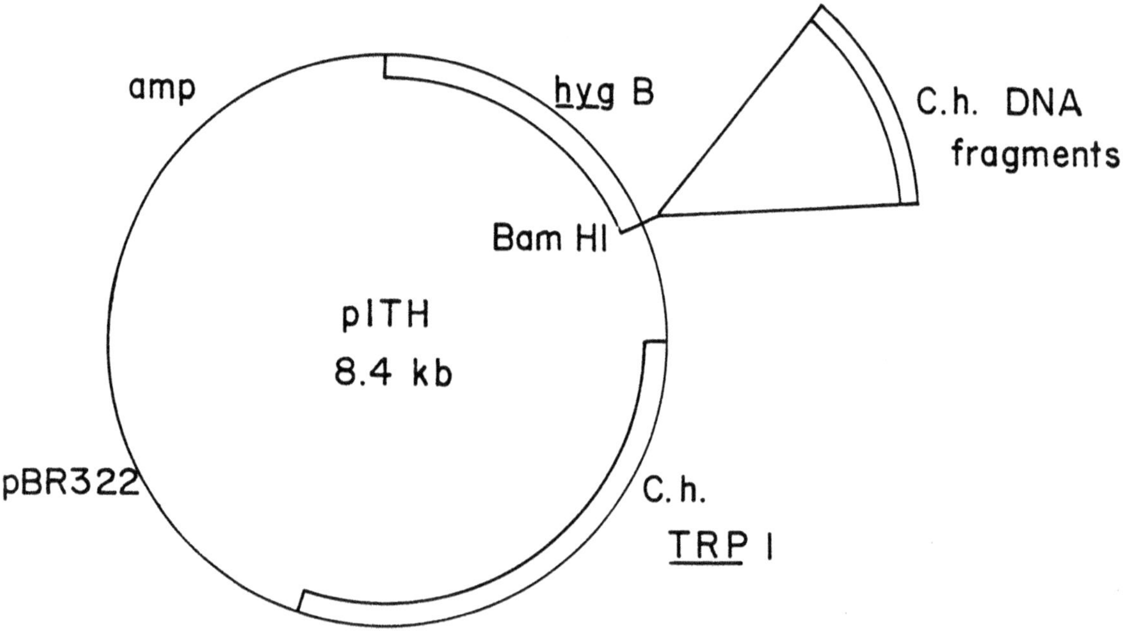

Figure 2. The C. heterostrophus promoter library in pITH. To construct pITH, yeast sequences and the Km^R gene were deleted from pIT213 [13] by HindIII digestion and a SalI fragment carrying part of the C. heterostrophus TRP1 gene [19] was added to provide homology with the C. heterostrophus nuclear genome. The remainder of the plasmid is pBR322 carrying the E. coli ampicillin-resistance gene and origin of replication. To construct the library, C. heterostrophus genomic DNA was partially digested with MboI and fractionated on a sucrose gradient to yield fragments of 0.5-1.5 kb, which were then ligated into the BamHI site located at the 5' end of the coding region of the promoterless hygB gene.

[11] and E. coli [12,13]. The gene (hygB) from E. coli confers hygromycin B-resistance to E. coli when under control of a prokaryotic promoter and to yeast when fused to a yeast promoter [14,15]. Moreover, hygB has been coupled to appropriate eukaryotic promoters and shown to function as a selectable marker for transformation of mammalian cells [16,17] and plant cells [18]. These observations suggested to us that hygB might function in cells of filamentous fungi, such as C. heterostrophus, if placed under control of a fungal promoter.

C. heterostrophus is completely inhibited by hygromycin B at 100 μg/ml. Since there was no readily available strong Cochliobolus promoter to fuse to hygB, we constructed a "promoter library" of Cochliobolus genomic DNA to serve as a source of promoters. Random fragments from a MboI partial digest of genomic DNA were fractionated on a sucrose density gradient and 0.5-1.5kb fragments ligated into a unique BamHI site located at the 5' end of a hygB gene that lacked its own promoter (Fig.2), on a vector modified from one constructed by Kaster et al. [15]. Protoplasts of Cochliobolus were transformed with the library and plated in nonselective regeneration agar that was overlaid several hours later with agar containing hygromycin B. After 6-8 days at 32 °C colonies appeared on plates containing protoplasts exposed to library DNA but not on plates containing control protoplasts. When these colonies were transferred to fresh hygromycin B-containing medium, they continued to grow at near-normal rates whereas wild type cells did not grow at all.

DNA was isolated [1] from transformed and untransformed colonies, digested and probed with pBR322. There was no hybridization of the probe to DNA of untransformed cells but distinct hybridization to DNA from transformed cells. In all cases the transforming DNA appeared to be integrated into a chromosome, either as a single copy or as multiple copies, perhaps in a tandem array.

Genetic analysis of transformants was performed by crossing transformants with wild type, using standard procedures [5]. Random ascospores were isolated, and colonies derived from them tested for sensitivity to hygromycin B. The progeny of hygTX1 segregated 28 resistant:29 sensitive and the progeny of hygTX2 segregated 16 resistant: 19 sensitive. Neither ratio is significantly different from 1:1, indicating that in each transformant the transforming DNA integrated at a single chromosomal locus.

Our predominant interest in transformants was to identify the Cochliobolus promoters isolated by this procedure and compare their abilities to drive transcription of the hygromycin B-phosphotransferase gene in C. heterostrophus. We recovered the transforming DNA by preparing lambda libraries of genomic DNA from several transformants. Plaques were probed with pBR322 to identify clones that contained transforming sequences. DNA was prepared from lambda clones that hybridized with the probe, and mapped with restriction enzymes. A fragment carrying the hygB gene fused to the Cochliobolus promoter insert, the E. coli amp gene and the E. coli origin of replication was eluted from a gel, circularized by ligation, cloned in E. coli, and used to retransform Cochliobolus protoplasts. Thousands of colonies appeared within 3-4 days on plates containing protoplasts treated with the cloned SalI fragment whereas no colonies were found on plates containing protoplasts treated with the vector without the Cochliobus promoter fragment. This evidence indicates that we now have a cloned Cochliobolus promoter which will drive the hygB gene in Cochliobolus so that the recombinant plasmid can be used as the basis of an efficient gene isolation vector.

COCHLIOBOLUS TRANSFORMATION WITH THE ASPERGILLUS NIDULANS AMDS GENE

Wild type strains of A. nidulans carry the $amdS^+$ gene, which permits them to grow on medium containing acetamide as a sole source of nitrogen. The amdS gene encodes acetamidase, an enzyme that converts acetamide to acetate and ammonia. The gene has been cloned into pBR322 by Hynes et al. [20] as a SalI-EcoRI fragment of A. nidulans DNA to give the recombinant plasmid p3SR2 (Fig. 3). The amdS gene in p3SR2 has been used as a selectable marker to transform $amdS^-$ mutants of A. nidulans [21] and wild type strains of A. niger [22], a species that lacks acetamidase activity.

We have found that Cochliobolus grows poorly on acetamide medium and that its genomic DNA shows no significant homology with the cloned amdS gene. Thus, it is a good candidate for transformation using amdS as a selectable marker. Cochliobolus protoplasts were transformed with p3SR2 and plated in a minimal medium containing only acetamide as a nitrogen source. Colonies appeared on plates with DNA-treated protoplasts approximately a week later, whereas plates with control protoplasts had no colonies [4]. When transferred to complete medium, the colonies grew at about the same rate as wild type colonies whereas on acetamide medium growth of transformants was much faster than wild type (Fig. 4). To test for mitotic stability, transformant colonies were grown to maturity on complete (nonselective) medium, then returned to acetamide medium where they grew at the normal transformant rate, indicating that they were stable under nonselective conditions.

The organization of the transforming DNA was determined in five transformants grown in either acetamide or complete medium; similar results were obtained with both media. DNAs were prepared [1], digested with either BamHI (one site in p3SR2) or XhoI (no sites in p3SR2), separated on an

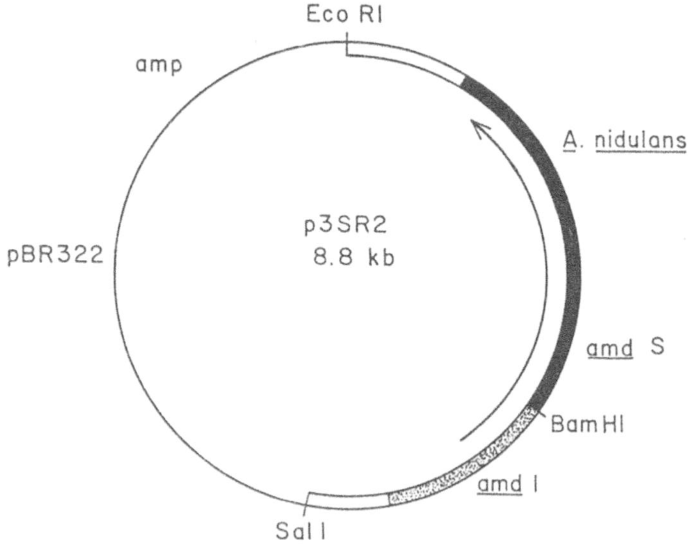

Figure 3. Structure of p3SR2. Single line = pBR322 carrying the E. coli ampicillin-resistance gene and origin of replication; double line = A. nidulans sequence; solid bar = amdS (the coding region); stippled bar = amdI (the 5' regulatory region). Note that there is a single BamHI site, no XhoI site, and that the size of p3SR2 is 8.8 kb. The arrow indicates the discretion of transcription. For a description of the original A. nidulans clone see Hynes et al. [20].

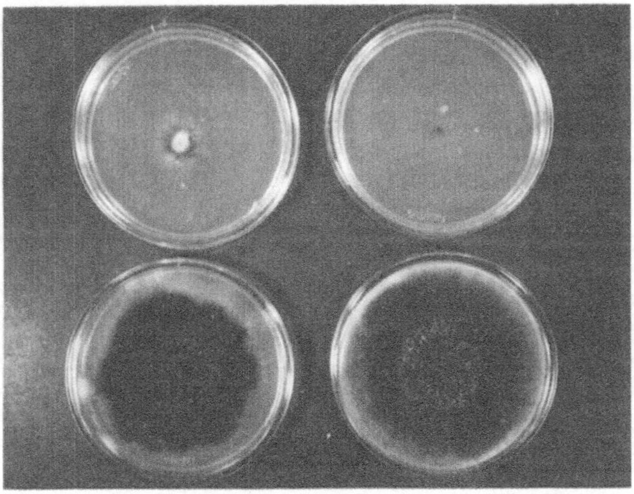

Figure 4. Growth of C. heterostrophus wild type and amdS transformant on complete (nonselective) and acetamide (selective) medium. Clockwise from lower right: wild type strain C3 on complete medium, amdTX1 on complete medium, amdTX1 on acetamide medium, wild type strain C3 on acetamide medium.

agarose gel, blotted to nitrocellulose paper, and probed with ^{32}P-labelled plasmid. Identical results were obtained using either pBR322 or p3SR2 as the probe.

Under conditions of moderate stringency, there was no detectable homology between either probe and wild type Cochliobolus genomic DNA. If genomic DNA from transformed mycelium was undigested, only the high molecular weight DNA hybridized to the probe; there was no evidence that p3SR2 existed as a free plasmid (Fig. 5). There was clear hybridization between either probe and genomic DNA from each transformant. Two patterns of hybridization were observed. In the first case (not shown), if genomic DNA was digested with BamHI, two hybridizing bands were seen, neither of which was 8.8 kb (the size of p3SR2). If digested with XhoI, one band was found which was larger than 8.8 kb but smaller than 23 kb. Our interpretation of these two observations is that in this class of transformant a single copy of p3SR2 integrated at one chromosomal locus. In the second class, BamHI digestion of genomic DNA yielded a very strongly hybridizing band of 8.8 kb and two weaker bands of various sizes. Digestion with XhoI resulted in one band of high molecular weight (Fig. 5). These results indicate that this second class of transformant has multiple copies of p3SR2, arranged in a head-to-tail orientation, integrated at a single locus in a chromosome. So far we have not found multiple locus integrations of p3SR2 in Cochliobolus. Such events have been observed in A. nidulans [23] and A. niger [22], along with the single copy integrations and tandem repeats we see in Cochliobolus.

To determine whether or not Cochliobolus transformants were heterokaryons, single conidia, which resolve heterokaryons in Cochliobolus [7], were isolated from transformant and wild type colonies and plated on acetamide medium. Transformant condidia were 70% amdS$^+$, 20% amdS$^-$, and 10% nonviable. Conidia from wild type were 100% amdS$^-$. Additional single

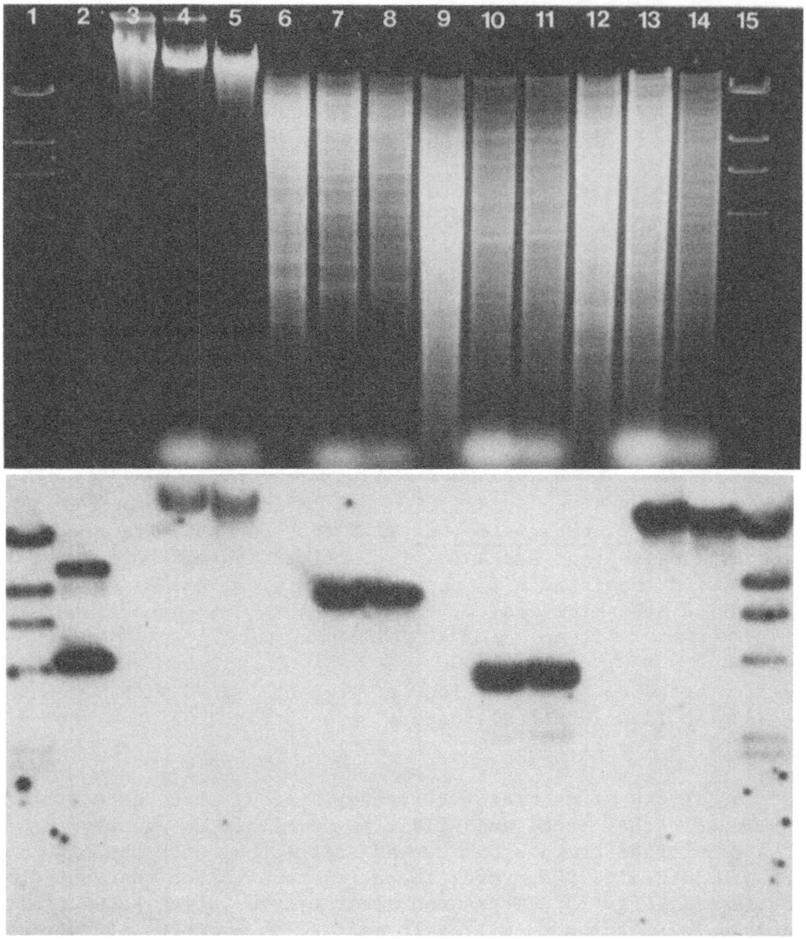

Figure 5. Tandemly repeated transforming DNA in a *C. heterostrophus* *amdS* transformant. An agarose gel containing wild type or transformant DNA (top) was blotted to nitrocellulose paper and probed with ^{32}P-labelled p3SR2 (bottom). Lanes 1 and 15: lambda DNA digested with *HindIII*; Lane 2: undigested p3SR2. DNA from untransformed *C. heterostrophus* is in lanes 3 (undigested), 6 (*BamHI*-digested), 9 (*PstI*-digested), and 12 (*XhoI*-digested). DNA from transformed *C. heterostrophus* is in lanes 4 and 5 (undigested), 7 and 8 (*BamHI*-digested), 10 and 11 (*PstI*-digested), and 13 and 14 (*XhoI*-digested). Prolonged exposure of the paper to film revealed two fainter bands in the *BamHI*-digested lanes 7 and 8, but no additional bands in any of the other lanes except *PstI*-digested lanes 10 and 11. Note: *XhoI* does not cut p3SR2, *BamHI* cuts once, and *PstI* cuts several times.

conidia were isolated from the purified transformant and wild type colonies and plated on acetamide medium. Of those from the transformant, 100% were now *amdS*$^+$, whereas those from wild type remained 100% *amdS*$^-$. Thus, it appears that only a portion of the nuclei in each cell of a transformant actually carry foreign DNA. The remainder may be maintained by activity of the transformed nuclei. It is easy, however, to purify transformants by isolation of single conidia. A. nidulans transformants can also be heterokaryotic, with the frequency of transformed nuclei ranging from 0.1% to 100% [23].

Meiotic segregation of the amdS gene was analyzed by crossing Cochliobolus $amdS^+$ transformants carrying either single or multiple copies of p3SR2 with wild type. Ascospores were isolated randomly or as complete tetrads (4 sets of twins). Progeny of a cross between a single copy transformant and wild type segregated 35 $amdS^+$:43 $amdS^-$ and progeny of a cross between a multiple copy transformant and wild type segregated 31 $amdS^+$:29 $amdS^-$. Both of these ratios approximate 1:1, indicating that amdS segregates as a single gene in genomes carrying either a single copy or multiple copies of p3SR2. All complete tetrads that were scored segregated 4:4 ($amdS^+$:$amdS^-$), as expected when a single gene is involved.

Two independent $amdS^+$ transformants were crossed with each other to determine if integration of p3SR2 was at the same chromosomal locus in each transformant. Ten tetrads were dissected and scored for amdS and for Alb1, a gene controlling pigment production that was heterozygous in the cross. When ratios of $amdS^+$:$amdS^-$ among the ten tetrads were scored, three were found to be 8:0, four were 6:2, and three were 4:4. These frequencies indicate that two genes were segregating and permit the conclusion that p3SR2 integrated at different, unlinked loci in these two transformants. Segregation at Alb1 was 4:4 for all ten of the tetrads.

CONSTRUCTION OF A COCHLIOBOLUS GENE-ISOLATION VECTOR

Now that we have two genes, amdS and hygB, that can be used as selectable markers for Cochliobolus transformation, we will construct a vector that transforms Cochliobolus at high frequency and will be suitable for construction of Cochliobolus genomic libraries from which virulence genes can be isolated by complementation of Cochliobolus itself. A possible construction for this purpose is shown in Figure 6. In the cosmid shown, the selectable marker is hygB fused to a Cochliobolus promoter. There is a unique BamHI site that would accept DNA fragments after partial digestion with Sau3A or MboI. The E. coli amp gene and origin of DNA replication would permit cloning in E. coli, and the lambda cos site [24] would modify the vector to accept only large (35-45 kb) DNA fragments so that the number of transformants needed to represent the entire Cochliobolus genome would be relatively small (approximately 2000). A small library is important when transformants must be screened for the presence of a particular virulence gene since virulence assays are often inefficient, especially if they involve inoculation of plants.

To isolate a virulence gene using such a vector (Fig. 6), two strains of the pathogen would be chosen - one which had a dominant allele of the virulence gene and one which had a recessive allele. It would be desirable to have the two strains near-isogenic with each other, a condition that could be achieved by backcrossing. A library of DNA fragments from a partial digest prepared from genomic DNA of the strain with the dominant allele would be inserted into the BamHI site. The library would be used to transform protoplasts of the strain with the recessive allele and transformants would be selected by their resistance to hygromycin conferred by hygB. Approximately 2000-3000 transformants would then be screened for a change in phenotype that reflected expression of the virulence gene. When such a transformant is found, the transforming cosmid would be recovered from the genome by lambda packaging and transduction of E. coli as described for Aspergillus [24]. The fragment bearing the virulence gene itself would be subcloned by the co-transformation technique [25].

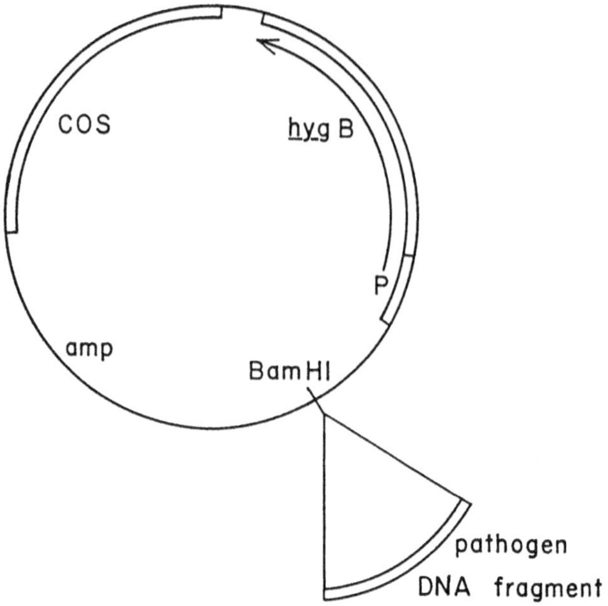

Figure 6. Hypothetical gene isolation vector. Essential elements include a marker selectable in filamentous fungi, eg. the hygB gene fused to a suitable promoter, a marker selectable in E. coli, eg. amp, a cos site to permit insertion of large (35-40 kb) DNA fragments, and a unique cloning site, eg. BamHI, that is convenient for insertion of genomic DNA fragments.

CLONING A FUNGAL VIRULENCE GENE BY EXPRESSING IT IN ASPERGILLUS

An alternative to isolation of fungal virulence genes by complementation in the homologous host is to express the gene in a heterologous system. This approach depends on the ability of the gene to function normally in the heterologous host. We have used this approach to isolate and clone a virulence gene from Nectria haematoccocca Berk. and Br. mating population VI, a fungal pathogen of pea. The product of the gene we have cloned is a cytochrome P-450 monooxygenase responsible for detoxification of the phytoalexin pisatin by O-demethylatin (Fig. 7) [26,27]. Phytoalexins are antimicrobial compounds produced by plants in response to infection and are considered to be responsible for an active mechanism of disease resistance in plants [28,29,30,31,32,33,34]. The ability of a plant pathogen to detoxify its host's phytoalexin is thought to be one factor required for pathogenicity in some host-parasite interactions [35,36,37,38].

A survey of field isolates of N. haematococca revealed several different phenotypes with regard to pisatin demethylating ability (pda) [39, 40, 41]. The phenotypes are referred to as Pda^- (no ability to demethylate), Pda^i (inducible Pda), and Pda^n (noninducible Pda), distinguished by their rates of demethylation after preinduction with pisatin of 0, 260, and 15 pmol/min/mg fresh wt., respectively. These Pda phenotypes are correlated with virulence of N. haematococca on peas in that Pda^i isolates are highly to moderately virulent while almost all Pda^n isolates are of low virulence (Fig. 8). Pda^- isolates are virtually nonpathogenic. Genetic analysis has verified this correlation and led to the identification of several unlinked genes that independently confer a specific Pda phenotype in N. haematococca [42,43].

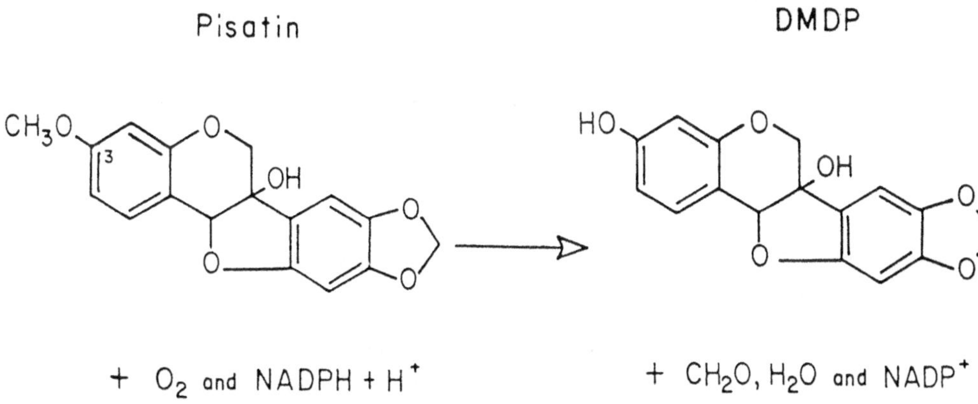

Figure 7. Detoxification of the phytoalexin pisatin by pisatin demethylase via O-demethylation to 3, 6a-dihydroxy-8,9-methylenedioxy pterocarpan.

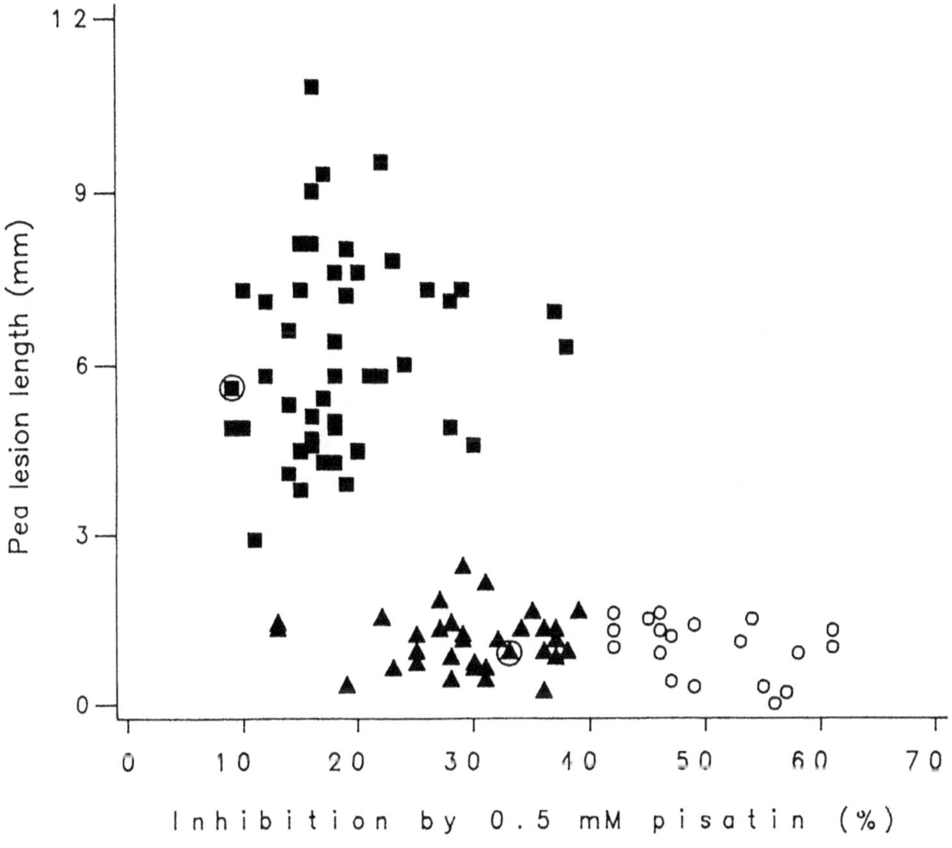

Figure 8. Virulence and pisatin tolerance of random ascospore progeny from a cross between a Pda^i parent (■) and a Pda^n parent (▲); Pda^i progeny (■); Pda^n progeny (▲); Pda^- progeny (○). Pisatin was added at 161 $\mu g\ ml^{-1}$, and tolerance was measured in a radial growth assay, expressed as a percentage inhibition of the control growth rate (% I growth rate).

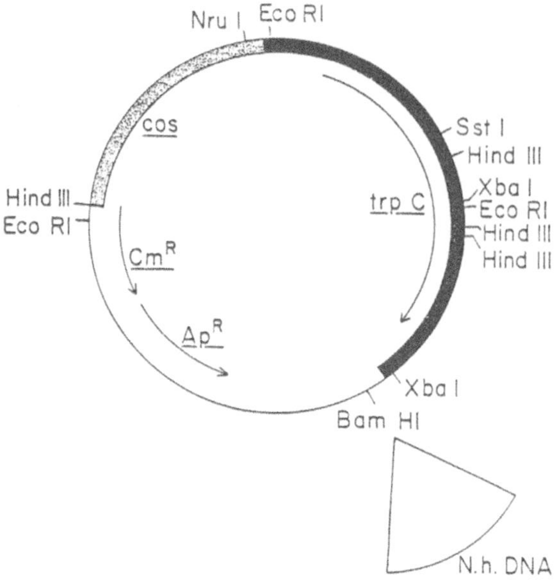

Figure 9. Construction of pNhT9. The recombinant molecule consists of the 9.9 kb plasmid pKBY2 with 30-40 kb of N. haematococca DNA inserted at the BamHI site.

We have begun to isolate the Pda genes to examine whether the different Pda loci code for different structural genes or one structural gene and several regulatory genes. A genomic library of N. haematococca isolate T-9 (which is Pda^{i}) was constructed by ligating 30-40 kb genomic DNA fragments into the cosmid vector pKBY2 (Fig. 9). This cosmid contains the trpC gene of Aspergillus nidulans and was constructed for transformation of $trpC^{-}$ mutants of A. nidulans to allow direct isolation of fungal genes [24]. The availability of this vector and the fact that A. nidulans does not demethylate pisatin made it possible to look for expression of a PDA gene in Aspergillus. Moreover, we knew from reconstruction experiments that A. nidulans has a nonspecific NADPH cytochrome P-450 reductase (required for pda) that will complement the N. haematococca cytochrome P-450 in vitro such that pisatin demethylating activity is produced.

Transformation of A. nidulans strain UCDI ($trpC^{-}$) with the Nectria library, yielded transformants at a rate of about 3 per µg DNA. Individual transformants were transferred to medium containing [3-O-methyl ^{14}C]-pisatin and incubated for one week. One transformant (III-202) among 1250 transformants tested, was able to demethylate pisatin, as indicated by a decrease in radioactivity (Fig. 10). The rate of demethylation after preinduction of the A. nidulans transformant was less than that of Nectria strain T-9, the source of the clone gene. Transformant III-202 was less sensitive than A. nidulans Pda^{-} transformants to growth inhibition by pisatin (Fig. 11).

The cosmid containing the PDA gene, designated pNhT9 (47kb), was recovered by lambda-packaging of total DNA from transformant III-202 as described by Yelton et al. [24]. After retransformation of A. nidulans strain UCD1 with pNhT9, 99% of the $trpC^{+}$ transformants had the Pda^{+}

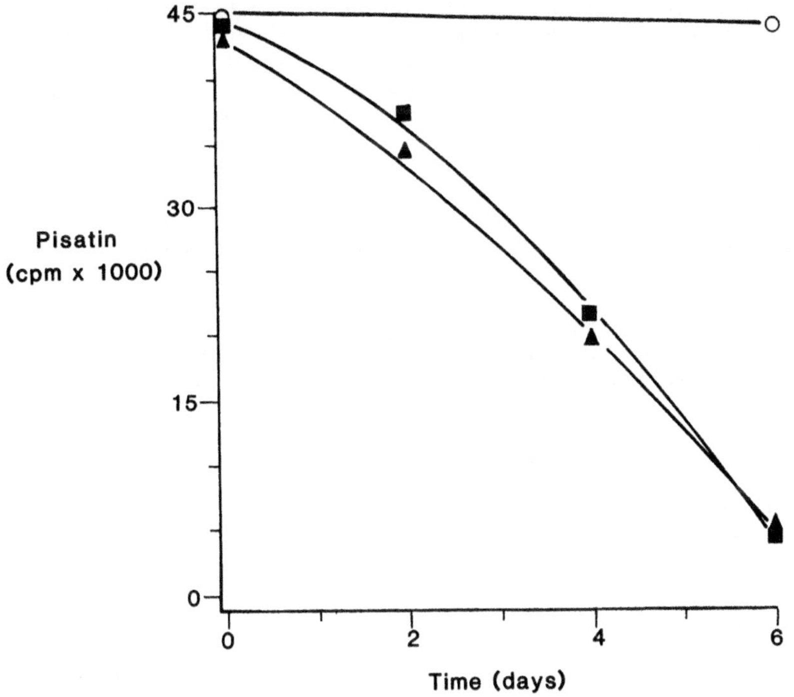

Figure 10. Time course of pisatin O-demethylation by *A. nidulans* transformant III-202 on agar medium containing [3-methyl-^{14}C]-pisatin ($160\ \mu g\ ml^{-1}$)
(□) original colony of transformant III-202
(▲) colony of transformant III-202 after one subculture
(○) arbitrarily chosen *trpC*+ Pda^{-} transformant

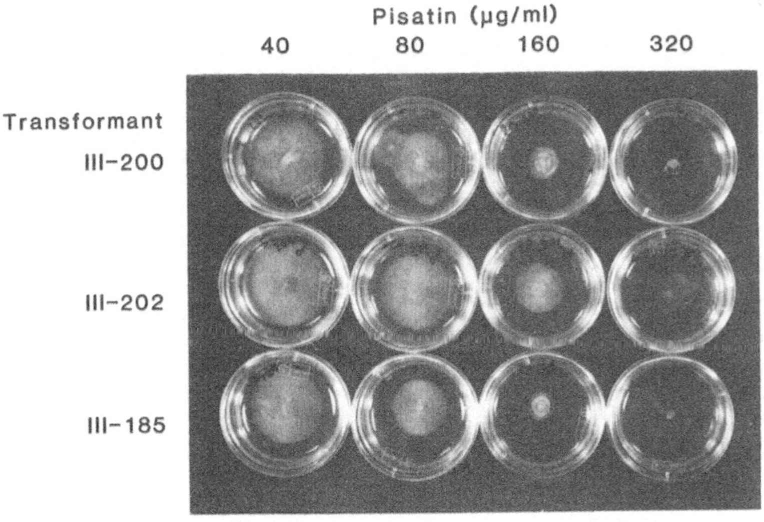

Figure 11. Effect of pisatin on the growth of *A. nidulans* Pda^{+} transformant III-202 and two *A. nidulans* Pda^{-} transformants, III-200 and III-185.

phenotype, indicating that the insert carried a gene that determined pda. The presence of vector DNA in the transformant was verified by Southern hybridization of a restriction enzyme digest of genomic DNA from transformant III-202, with pKBY2 as a probe. Furthermore, in Southern hybridization experiments only pNhT9, but not pKYB2, hybridized to Nectria strain T-9 genomic DNA.

After subcloning of the 37 kb insert of pNhT9 to obtain the smallest fragment containing the functional PDA gene it will be possible to isolate and compare other PDA genes from strains with different Pda phenotypes. Furthermore the isolation of a PDA gene is the first step towards the most stringent test for evaluating the importance of pisatin demethylation in virulence: the introduction of isolated PDA genes into N. haematococca isolates with different Pda phenotypes, i.e. construction of pairs of isolates that are isogenic except for a particular PDA gene.

In addition the molecular study of the Pda genes will contribute to the general understanding of cytochrome P-450 systems and the genes that control their function. Knowledge of the structure and regulation of fungal cytochrome P-450s should also facilitate comparative studies on the evolution of these genes in prokaryotic and eukaryotic organisms.

ACKNOWLEDGEMENTS

The experimental investigations described in the article were supported by grants to O.C. Yoder from the US National Science Foundation, US Dept. of Agricultural Competitive Research Grants Office, and Pioneer Hibred International and to H. D. VanEtten from the US Dept. of Energy and the US Dept. of Agriculture Competitive Research Grants Office. K. Weltring was supported by German Academic Exchange Service (DAAD). We wish to thank Thomas Ingolia for providing pIT213 with a promoter-less hygB gene, Michael Hynes for p3SR2 with amdS, and William Timberlake for the cosmid pKBY2 and Aspergillus nidulans strain UCDI.

REFERENCES

1. Garber, R.C. and Yoder O.C., 1983, Isolation of DNA from filamentous fungi and separation into nuclear, mitochondrial, ribosomal, and plasmid components, Anal. Biochem., 135: 416-422.
2. Garber, R.C. and Yoder O.C., 1984, Mitochondrial DNA of the filamentous ascomycete Cochliobolus heterostrophus. Characterization of the mitochondrial chromosome and population genetics of a restriction enzyme polymorphism, Curr. Genet., 8: 621-628.
3. Garber, R.C., Turgeon, B.G. and Yoder O.C., 1984, A mitochondrial plasmid from the plant pathogenic fungus Cochliobolus heterostrophus, Mol. Gen. Genet., 196: 301-310.
4. Turgeon, B.G., Garber R.C. and Yoder, O.C., 1985, Transformation of the fungal maize pathogen Cochliobolus heterostrophus using the Aspergillus nidulans amdS gene, Molec. Gen. Genet., 201: 450-453.
5. Leach, J., Lang, B.R. and Yoder, O.C., 1982, Methods for selection of mutants and in vitro culture of Cochliobolus heterostrophus, J. Gen. Microbiol., 128: 1719-1729.
6. Yoder, O.C. and Gracen, V.E., 1975, Segregation of pathogenicity types and host-specific toxin production in progenies of crosses between races T and O of Helminthosporium maydis (Cochliobolus heterostrophus), Phytopathology, 65: 273-276.
7. Leach, J. and Yoder, O.C., 1982, Heterokaryosis in Cochliobolus heterostrophus, Exp. Mycol., 6: 364-374.

8. Leach, J. and Yoder, O.C., 1983, Heterokaryon incompatibility in the plant pathogenic fungus, Cochliobolus heterostrophus, J. Hered., 74: 149-152.
9. Leach, J., Tegtmeier, K.J., Daly, J.M. and Yoder, O.C., 1982, Dominance at the Tox1 locus controlling T-toxin production by Cochliobolus heterostrophus, Physiol. Plant Pathol., 21: 327-333.
10. Pardo, J.M., Malpartida, F., Rico, M. and Jimenez, A., 1985, Biochemical basis of resistance to hygromycin B in Streptomyces hygroscopicus - the producing organism, J. Gen. Microbiol., 131: 1289-1298.
11. Malpartida, F., Zalacain, M., Jimenez, A. and Davies J., 1983, Molecular cloning and expression in Streptomyces lividans of a hygromycin B phosphotransferase gene from Streptomyces hygroscopicus, Biochem. Biophys. Res. Commn., 117: 6-11.
12. Rao, R.N., Allen, N.E., Hobbs, J.N., Jr., Alborn, W.E., Jr., Kirst, H.A. and Paschal, J.W., 1983, Genetic and enzymatic basis of hygromycin B resistance in Escherichia coli, Antimicrob. Agents Chemother., 24: 689-695.
13. Kaster, K.R., Burgett, S.G., Rao, R.N. and Ingolia, T.D., 1983, Analysis of a bacterial hygromycin B resistance gene by transcriptional and translational fusions and by DNA sequencing, Nucleic Acids Res.,11: 6895-6911.
14. Gritz, L. and Davies, J., 1983, Plasmid-encoded hygromycin B resistance: the sequence of hygromycin B phosphotransferase gene and its expression in Escherichia coli and Saccharomyces cerevisiae, Gene, 25: 179-188.
15. Kaster, K.R., Burgett, S.G., and Ingolia, T.D., 1984, Hygromycin B resistance as dominant selectable marker in yeast, Curr. Genet., 8: 353-358.
16. Blochlinger, K. and Diggelmann, H., 1984, Hygromycin B phosphotransferase as a selectable marker for DNA transfer experiments with higher eukaryotic cells, Mol. Cell. Biol., 4: 2929-2931.
17. Santerre, R.F., Allen, N.E., Hobbs, J.N., Jr., Rao, R.N. and Schmidt, R.J., 1984, Expression of prokaryotic genes for hygromycin B and G418 resistance as dominant-selection markers in mouse L cells, Gene, 30: 147-156.
18. Waldron, C., Murphy, E.B., Roberts, J.L., Gustafson, G.D., Armour, S.L. and Malcolm, S.K., 1985, Resistance to hygromycin B: a new marker for plant transformation studies, Plant Mol. Biol, 5: 103-108.
19. Turgeon, B.G., MacRae, W.D., Garber, R.C., Fink, G.R. and Yoder, O.C., 1986, A cloned tryptophan synthesis gene from the Ascomycete Cochliobolus heterostrophus functions in Escherichia coli, yeast, and Aspergillus nidulans, Gene, 42: 79-87.
20. Hynes, M.J., Corrick, C.M. and King, J.A., 1983, Isolation of genomic clones containing the amdS gene of Aspergillus nidulans and their use in the analysis of structural and regulatory mutations, Mol. Cell. Biol., 3: 1430-1439.
21. Tilburn, J., Scazzocchio, C., Taylor, G.G., Zabicky-Zissman, J.H., Lockington, R.A. and Davies, R.W., 1983, Transformation by integration in Aspergillus nidulans, Gene 26: 205-221.
22. Kelly, J.M. and Hynes, M.J., 1985, Transformation of Aspergillus niger by the amdS gene of Aspergillus nidulans, EMBO J., 4:475-479.
23. Wernars, K., Goosen, T., Wennekes, L.M.J., Visser, J., Bos, C.J., van den Broek, H.W.J., van Gorcom, R.F.M., van den Hondel, C.A.M.J.J. and Pouwels, P.H., 1985, Gene amplification in Aspergillus nidulans by transformation with vectors containing the amdS gene, Curr.Genet., 9: 361-368.
24. Yelton, D.E., Timberlake, W.E. and van den Hondel, C.A.M.J.J., 1985, A cosmid for selecting genes by complementation in Aspergillus nidulans: Selection of the developmentally regulated yA locus, Proc. Natl. Acad. Sci. USA, 82: 834-838.

25. Timberlake, W.E., Boylan, M.T., Cooley, M.B., Mirabito, P.M., O'Hara, E.B. and Willett, C.E., 1985, Rapid identification of mutation-complementing restriction fragments from Aspergillus nidulans cosmids, Exp. Mycol., 9: 351-355.
26. VanEtten, H.D., Pueppke, S.G. and Kelsey, T.C., 1975, 3,6a-Dihydroxy-8,9-methylenedioxypterocarpan as a metabolite of pisatin produced by Fusarium solani f. sp. pisi, Phytochemistry, 14: 1103-1105.
27. Matthews, D.E. and VanEtten, H.D., 1983, Detoxification of the phytoalexin pisatin by a fungal cytochrome P-450, Arch. Biochem. Biophys., 224: 494-505.
28. Albersheim, P. and Valent, B.S., 1978, Host-pathogen interactions in plants. Plants, when exposed to oligosaccharides of fungal origin, defend themselves by accumulating antibiotics, J. Cell Biol., 78: 627-643.
29. Bailey, J.A. and Mansfield, J.W., eds., 1982, "Phytoalexins", Blackie and Son, Glasgow and London, 334 pp.
30. Darvill, A.G. and Albersheim, P., 1984, Phytoalexins and their elicitors - A defense against microbial infection in plants, Annu. Rev. Plant Physiol., 35: 243-275.
31. Dixon, R.A., Dey, P.M. and Lamb, C.J., 1983, Phytoalexins: enzymology and molecular biology, Adv. Enzymol., 55: 1-136.
32. Kuć, J. and Rush, J.S., 1985, Phytoalexins, Arch. Biochem. Biophys., 236: 455-472.
33. Paxton, J., 1980, A new working definition of the term "Phytoalexin", Plant Disease, 64: 734.
34. Smith, D.A. and Ingham, J.L., 1981, Phytoalexins and plant disease resistance, Biologist, 28: 69-74.
35. Smith, D.A., Harrer, J.M. and Cleveland, T.E., 1982, Relation between production of extracellular kievitone hydratase by isolates of Fusarium and their pathogenicity on Phaseolus vulgaris, Phytopathology, 72: 1319-1323.
36. Smith, D.A., Wheeler, H.E., Banks, S.W. and Cleveland, T.E., 1984, Association between lowered kievitone hydratase activity and reduced virulence to bean in variants of Fusarium solani f. sp. phaseoli, Physiol. Plant Pathol., 25: 135-147.
37. VanEtten, H.D., 1982, Phytoalexin detoxification by monooxygenases and its importance for pathogenicity, in: "The Physiological and Biochemical Basis of Plant Infection", Y. Asada, W.R. Bushnell, S. Ouchi and C.P. Vance eds., pp. 315-327 Japan Scientific Societies Press and Springer-Verlag.
38. VanEtten, H.D., Matthews, D.E. and Smith, D.A., 1982, Metabolism of phytoalexins, in: "Phytoalexins", J.A. Bailey and J.W. Mansfield, eds., pp, 181-217 Blackie and Son Ltd., Glasgow and London.
39. Tegtmeier, K.J. and VanEtten, H.D., 1983, The role of pisatin tolerance and degradation in the virulence of Nectria haematococca on peas: a genetic analysis, Phytopathology, 72: 608-612.
40. VanEtten, H.D. and Matthews, P.S., 1984, Naturally-occurring variation in the induction of pisatin demethylating ability in Nectria haematococca mating population VI, Physiol. Plant Pathol., 25: 149-160.
41. VanEtten, H.D., Matthews, P.S., Tegtmeier, K.J., Dietert, M.F. and Stein, J.I., 1980, The association of pisatin tolerance and demethylation with virulence on pea in Nectria haematococca, Physiol. Plant Pathol., 19: 257-271.
42. Kistler, H.C. and VanEtten, H.D., 1984, Regulation of pisatin demethylase in Nectria haematococca and its influence on pisatin tolerance and virulence, J. Gen. Microbiol. 130: 2605-2613.
43. Kistler, H.C. and VanEtten, H.D., 1984, Three nonallelic genes for pisatin demethylation in the fungus Nectria haematococca, J. Gen. Microbiol., 130: 2595-2603.

MOLECULAR EVOLUTION OF FUNGAL POLYGALACTURONASE

F. Cervone[a], G. De Lorenzo[a], G. Salvi[a] and L. Camardella[b]

[a]Dipartimento di Biologia Vegetale
Università "La Sapienza"
Piazzale A. Moro 00100 Roma, Italy

[b]Istituto Internazionale di Genetica e Biofisica
Via Marconi 10
Napoli, Italy

INTRODUCTION

In plant-fungi interactions the establishment of basic compatibility occurs when a potential pathogen acquires functions that allow colonization of host plant species and/or suppression or neutralization of host resistance responses. Cultivar and race-specific resistance (specific compatibility) is generally considered as superimposed in host-parasite systems which have already acquired basic compatibility. Specific compatibility is achieved with the matching of two molecules which are products of one gene in the host and one gene in the parasite [1,2]. However a plausible and economical evolutionary strategy could utilize, for specific compatibility, the same molecules involved in basic compatibility, if their structure and function are flexible enough.

This paper deals with the structural and functional variability of fungal polygalacturonases (PG). The early interaction between PG and plant cell walls will be discussed in relation to both basic and specific compatibility. Finally, the approach used in our laboratory for cloning a gene coding for fungal PG will be described.

SOME CHARACTERISTICS OF FUNGAL POLYGALACTURONASE

The importance of PG in pathogenesis is now well established for certain plant diseases characterized by a rapid and extensive degradation of cell wall, increase in permeability and death of protoplasts [3]. The role of PG, although less important, is also acknowledged for diseases, mainly caused by biotrophs, where only a minimal breakdown of cell wall occurs during penetration and colonization by an infecting fungus [4]. Of the many polysaccharide degrading enzymes, PG must act before other enzymes can attack other substrates [5] and it is the first polysaccharide-degrading enzyme secreted by certain pathogens cultured on isolated cell walls [6].

NATO ASI Series, Vol. H1
Biology and Molecular Biology of Plant-Pathogen Interactions
Edited by J. Bailey

Table 1. Amino acid composition of fungal polygalacturonases.

Amino acid	*A. niger* PG determined	*A. niger* PG nearest integer	*F. oxysporum* PG[12]	*R. fragariae* PG[10]	*V. albo-atrum* PG[13]
Lysyne	17.8	18	22	18	20
Hystidine	5.3	5	8	8	8
Arginine	3.3	3	2	3	4
Aspartic acid	38.6	39	62	38	41
Threonine	31.6	32	36	38	24
Serine	32.2	32	40	40	26
Glutamic acid	20.9	21	13	17	13
Proline	9.6	10	13	8	7
Glycine	40.9	41	47	49	36
Alanine	19.2	19	17	25	20
Half-Cystine	8.9	9	7	5	7
Valine	22.0	22	29	30	21
Methionine	1.7	2	0	0	0
Isoleucine	23.2	23	27	17	22
Leucine	16.6	17	20	16	16
Tyrosine	8.4	8	10	8	2
Phenylalanine	9.0	9	12	13	5
Tryptophan	6.2	6	4	4	5

Although it is known from pioneering work [3] that PG has a central role in pathogenesis, the amount of information on its structure and function is minimal. This is due to the technical difficulty of preparing suitable quantities of homogeneous polygalacturonase. Most of the physiological studies involving PG, in fact, have been performed with preparations containing contaminating proteins or enzymes.

The amino acid composition of homogeneous polygalacturonases so far obtained is given in Table 1. The compositions of the various enzymes appear to be quite similar, suggesting that the basic protein structure is maintained among different fungi. The maintenance of a structure common to different fungi is also suggested by the double immunodiffusion tests performed in our laboratory by using various PGs and specific rabbit antisera prepared against homogeneous PGs from *Aspergillus niger* and *Fusarium moniliforme* [unpublished results]. Also the molecular weight, ranging from 32 to 37 kilodaltons, seems to be maintained among polygalacturonases from different fungi (Table 2). Higher values of molecular weight occur when the enzyme forms complexes with pectinesterase [7], exopolygalacturonase [8] or itself [9]. PG may also form complexes with plant cell wall glycoproteins and this will be discussed in the next section. The capability of forming complexes with other proteins is likely to depend on the glycoprotein nature of fungal polygalacturonases [10].

Polygalacturonase exhibits a great variability of isoenzymatic forms, easily distinguishable by their isoelectric points (Table 2). In one case, isoelectric polymorphism of PG has been ascribed to the glycosidic moiety of the protein, but no explanation is as yet available on the genetic basis of this polymorphism [10]. The variability of PG as well as that of lyases [see Collmer, this volume] and its significance in pathogenesis and specificity of microorganisms deserves more experimental studies. The possibility that different isoenzymes from the same microorganism have different functional characteristics remains to be investigated. In one case (*Erwinia atroseptica*), three isoenzymes of polygalacturonate *trans*-eliminase showed several differences in their functional parameters. Moreover, one of them

Table 2. Structural characteristics of fungal polygalacturonases.

Organism	Enzyme	Molecular weight (kilodaltons)	pI	References
A. niger	endo PG	33.5	5.1	this work
F. moniliforme	endo PG	39.5	6.7	this work
G. candidum	endo PG	38.0	7.8	[14]
R. fragariae	endo PG-1	36.0	6.76	[15]
	endo PG-2	36.0	7.08	
T. koningii	endo PG-1	32.0	6.41	[16]
	endo PG-2	32.0	6.57	
R. solani	endo PG-1		5.55	[17]
	endo PG-2		5.98	
	endo PG-3		6.66	
	endo PG-4		6.78	
	endo PG-5		7.05	
	endo PG-6		7.32	
	endo PG-7		7.66	
C. lindemuthianum	endo PG	70[a] 37[b]	9.75[c]	a[8]b[18]c[19]
V. albo-atrum	endo PG	34.5	9.7	[20]
F. oxysporum	endo PG-1	37		[12]
	endo PG-2	37		
B. cinerea	endo PG	69	8.3	[7]
		24.5-32	8.8	[21]
S. sclerotiorum	endo PG	29-33*	4.8-8.3*	[22]

* depending on the culture conditions.

was extremely sensitive to inactivation by chlorogenic acid; another was extremely toxic towards the plant cell although it exhibited reduced macerating activity [11]. The biological significance of PG isoenzymes would be greater if they were not only functionally different but also if their syntheses were differentially regulated.

By comparing the specific activities (i.e. catalytic efficiencies) of homogeneous PGs (Table 3), it can be noted that those from *Aspergillus niger* and *Geotrichum candidum*, two post-harvest pathogens, are the most efficient enzymes. The other fungi in the table are field-pathogens and some of them are biotrophs at a certain stage of their life cycle. The low efficiency of their PG may be the result of an evolutionary mechanism. Pathogens such as obligate parasites, in order to guarantee their own survival, cannot become too aggressive. The evolution of PG towards less efficient enzymatic forms should be advantageous for this kind of pathogen. It would be interesting, also, to know the efficiency of PG from a symbiotic fungus.

INTERACTION OF POLYGALACTURONASE WITH THE PLANT CELL WALL IN RELATION TO SPECIFICITY

The structural complexity and the functional wealth of plant cell walls are now being clarified. Polysaccharides capable of regulating gene expression [24], morphogenesis [25] and defence responses against invading pathogens [26] are localized in the plant cell wall. Enzymes and proteins presumably involved in host defence mechanisms, are also localized in the cell wall [27, see Mazau, this volume]. Cell wall and its components are the primary target of polygalacturonase.

A few years ago we tested the hypothesis that PG activity follows the absorption of PG on to the plant cell wall when a successful infection

Table 3. Functional characteristics of fungal polygalacturonases.

Organism	Enzyme	Sp.activity of homogeneous enzyme at 30°C (RVU/mg)*	Sp.activity of homogeneous enzyme at 30°C (RGU/mg)*	References
A. niger	endo PG	4,370	2,300	this work
F. moniliforme	endo PG	1,170	-	this work
G. candidum	endo PG	-	3,026	[14]
R. fragariae	endo PG-1	250	-	[15]
	endo PG-2	255	-	[15]
C. lindemuthianum	endo PG	-	234	[23]
V. albo-atrum	endo PG	-	2,151**	[13]
F. oxysporum	endo PG-1	-	194	[12]
	endo PG-2	-	148	[12]

* RVU = Relative viscosimetric units; RGU = reducing groups units [15]

** The enzyme assay was carried out at 35 °C. A lower value should be expected at 30 °C.

occurs. Lack of PG absorption would explain lack of pathogenicity of phytopathogenic fungi towards non-host plants and of saprophytic fungi towards all plants. Thus, PG from Rhizoctonia fragariae, the causal agent of black root rot of strawberry plants, bound to strawberry but not to potato, carrot and beet tissues [10]. Similarly, PG from Colletotrichum lindemuthianum, the bean anthracnose agent, was absorbed by the host bean tissue but not by tissues of non-host plants. Bean tissue, on the other hand, absorbed PG from C. lindemuthianum but not PGs from several fungi non-pathogenic to French bean [19]. PG from Trichoderma koningii, a saprophytic fungus, was not absorbed by any plant tissue tested [16]. The nature of the interaction between PG and the plant tissue is not clear. It could be explained either by trivial charge attraction or by a more complex interaction between PG and cell wall polysaccharides or glycoproteins. Cell wall PG-inhibiting proteins may be involved (see later in this section). In this regard, a possible interpretation is that, since PG elicits plant defence responses [28,29], its inhibition would reduce elicitor production and thus prevent the defence response. The interaction between PG and cell walls should be seen in this case as one of those functions which suppress the host resistance response when a basic compatibility between a plant and a fungus is established. It is difficult, however, to accept the fact that a molecule meant to destroy the cell wall barrier is so easily inactivated when a successful infection occurs. Regardless of the molecular explanation, the above results suggest that the interaction between PG and plant cell wall is one of the positive functions that allow plant colonization.

We have recently purified by affinity chromatography two proteins from Phaseolus vulgaris cell walls which form a complex with polygalacturonase. The two proteins exhibit a slight inhibiting activity against polygalacturonases from A. niger, F. moniliforme and C. lindemuthianum, suggesting that they are proteins having the same kind of function as the PG-inhibiting glycoprotein identified by Albersheim and Anderson [23] and recently characterized by Laffitte et al. [30]. Nevertheless the molecular weights of the two proteins (55 and 65 kd) are higher than that (46 kd) reported by Laffitte et al. We are confident that the proteins we have purified represent a primary binding site of PG in the bean cell wall and that they may modulate the PG activity in planta. A partially inhibited PG activity may regulate the array of compounds which are released from the cell wall upon the action of the PG itself. Saturating quantities of the proteins do not exert 100% inhibition on homogeneous PGs from A. niger and F. moniliforme. Similar results have been obtained by Hoffman and Turner [31]. They reported that saturating amounts of PG inhibitor from pea leaflets only partially inhibit polygalacturonases from the pea pathogens Mycosphaerella pinodes, Phoma medicaginis and Aphanomices eutiches. They also reported that polygalacturonases from different fungi show different sensitivity to pea inhibitor indicating the existence of differences in the relative protein-protein interaction.

PGs not only degrade plant cell walls but may also kill protoplasts [3,4]. The toxic effect of PG may be due to one or more of these factors: 1) direct action on the plasma membrane [32]; 2) the release of a cell wall enzyme which alters the membrane functions [33]; 3) the release of cell wall oligosaccharides with toxic activity [34]. A degraded cell wall, as pointed out at the beginning of this section, may also release proteins presumed to be involved in defence, or oligosaccharides which elicit active defence mechanisms, such as synthesis of phytoalexins [26,28,35], proteinase inhibitors [29,36,37], hydroxyproline-rich glycoproteins [38] and lignin [see Robertsen, this volume]. A fine balance between the side-effects of cell wall degradation could regulate the rate or the timing of the plant reaction. It is worth mentioning in this context that resistant and susceptible reactions are, in some cases, similar phenomena [39]. The resistant response, in these cases, is characterized by a rapid hypersensitive necrosis, restricting the microbial growth soon after the penetration. The susceptible response is, on the other hand, characterized by a delayed host cell death.

Polygalacturonases from α and β races of C. lindemuthianum have indistinguishable enzymological properties. Nevertheless, they show a differential rate of absorption to tissue from resistant and susceptible bean cultivars. The rate of enzyme absorption is faster in the resistant combinations, and slower in the susceptible ones [unpublished data]. The rapid PG absorption is consistant with the rapid cell death and necrosis of the tissue surrounding the infection site in bean cultivars resistant to C. lindemuthianum. Further work will clarify whether a rapid PG absorption also causes a rapid phytoalexin synthesis.

In conclusion, we would like to emphasize that evidence is presently being gathered on the following points: the fate of a plant-fungus interaction may depend on the side-effects caused by the plant cell wall degradation; the rate or the timing of these side-effects may be regulated by differential activity of PG; PG activity may be modulated by PG-binding proteins localized in the cell wall.

MOLECULAR CLONING OF A FUNGAL GENE CODING FOR POLYGALACTURONASE

Progress in biochemistry of fungal polygalacturonase has been restrained by technical difficulties and by the conviction that the enzyme, just one of the many cell wall degrading enzymes, would not have any particularly interesting characteristics. Some characteristics of the enzyme which seem to be very important in relation to pathogenesis and specificity have been discussed in this paper. Technical difficulties which prevent further biochemical characterization of the enzyme may somehow be circumvented now by using the advanced procedures of molecular genetics. The cloning of the gene coding for PG is the first step for studying the molecular evolution of the enzyme by studying the distribution, organization, structure and regulation of the gene for PG in various fungi. The nucleotide sequence of the gene will also allow the primary structure of the enzyme to be determined. Comparative primary structures may direct comparative enzymological studies.

High levels of extracellular PG are induced in the fungus *Fusarium moniliforme* by the presence of pectin in the culture media. Glucose-grown fungal cultures do not synthesize the enzyme. Poly(A)$^+$ RNA was purified from total RNA extracted from 50 g of fungus grown for 2 days in Czapek-dox liquid in which glucose had been replaced by 1% pectin. The total RNA was passed several times through oligo (dT) cellulose column to generate approx. 400 μg of poly(A)$^+$ mRNA. Double stranded DNA complementary to the poly(A)$^+$mRNA (cDNA) was prepared according to standard procedures [40] with slight modifications. ^{32}P first strand cDNA was synthesized from 10 μg of *F. moniliforme* poly(A)$^+$ mRNA by reverse transcriptase. The efficiency of the first strand synthesis in our experiments was about 17-30%. ^{32}P-labelled second strand cDNA was generated by *E. coli* DNA polymerase (Klenow fragment). This procedure produced ds cDNA ranging from 0.5 to 2.5 kb. The ds DNA was flush ended and ligated with *Eco*RI linkers. The ds cDNA with linkers was restricted with *Eco*RI and the excess linker was removed by several cold ethanol precipitations and washing with 70% ethanol. The ds cDNAs were ligated with the vector DNA. We utilized the phage λ gt 10 as a vector [41]. After the ligation, the recombinant DNA was *in vitro* packaged. The unamplified library was plated on C600 for 8-9 hours at a plaque density of about 500-700 plaques for 100 x 15 mm petri dish.

The specific inducibility of poly (A)$^+$mRNA for PG in pectin medium has been exploited for selecting possible PG cDNA clones in the total cDNA library. Replica nitrocellulose filters containing plaques of the fungal cDNA library were hybridized with labelled cDNA probes transcribed from poly(A)$^+$mRNA isolated from either cultures induced to produce PG or uninduced cultures. A total of 1.0×10^4 plaques were differentially screened. We are now going to examine 85 clones exhibiting differential hybridization by hybrid selected translation.

This work was supported by the "V.V. Landi" Foundation and by I.P.R.A. (publ. no. 676).

REFERENCES

1. Callow, J.A., 1984, Cellular and molecular recognition between higher plants and fungal pathogens, *in*: "Encyclopedia of Plant Physiology-Cellular Interactions" Vol. 17, H.F. Linkens and J. Heslop-Harrison, eds., pp. 212-237, Springer-Verlag, Berlin.
2. Keen, N.T., 1982, Specific recognition in gene-for-gene host-parasite systems, *in*: "Advances in Plant Pathology" Vol. 1, D.S. Ingram and P.H. Williams, eds., pp. 35-82, Academic Press, New York.
3. Bateman, D.F. and Basham, H.G., 1976, Degradation of the plant cell walls and membranes by microbial enzymes, *in*: "Encyclopedia of Plant Physiology - Physiological Plant Pathology", Vol. 4, R. Heitefuss and P.H. Williams, pp. 316-355, Springer-Verlag, Berlin.
4. Cooper, R.M., 1984, The role of cell wall-degrading enzymes in infection and damage, *in*: "Plant Diseases: Infection, Damage and Loss", R.K.S. Wood and G.J. Jellis, eds., pp. 13-27, Blackwell Scientific Publications, Oxford.
5. Karr, A.L. and Albersheim, P., 1970, Polysaccharide-degrading enzymes are unable to attack plant cell walls without prior action by a "wall-modifying enzyme", *Plant Physiol.*, 46: 69-80.
6. Jones, T.M., Anderson, A.J. and Albersheim, P., 1972, Host-pathogen interactions. IV. Studies on the polysaccharide-degrading enzymes secreted by *Fusarium oxysporum* f. sp. *lycopersici*, *Physiol. Plant Pathol.*, 2: 153-166.
7. Urbanek, H. and Zalewska-Sobczak, J., 1975, Polygalacturonase of *Botrytis cinerea* E-200 Pers., *Biochim. Biophys. Acta*, 377: 402-409.

8. English, P.D., Maglothin, A., Keegstra, K. and Albersheim, P., 1972, A cell wall degrading endopolygalacturonase secreted by Colletotrictum lindemuthianum, Plant Physiol., 49: 293-297.
9. Fielding, A.H. and Byrde, R.J.W., 1969, The partial purification and properties of endopolygalacturonase and α-L-arabinofuranosidase secreted by Sclerotinia fructigena, J. Gen. Microbiol., 58: 73-84.
10. Cervone, F., Scala, A. and Scala, F., 1978, Polygalacturonase from Rhizoctonia fragariae: further characterization of two isoenzymes and their action towards strawberry tissue, Physiol. Plant Pathol., 12: 19-26.
11. Quantick, P., Cervone F. and Wood, R.K.S., 1983, Isoenzymes of a polygalacturonate trans-eliminase produced by Erwinia atroseptica in potato tissue and in liquid culture, Physiol. Plant Pathol., 22: 77-86.
12. Strand, L.L., Corden, M.F. and MacDonald, D.L., 1976, Characterization of two endopolygalacturonase isozymes produced by Fusarium oxysporum f. sp. lycopersici, Biochim. Biophys. Acta, 429: 870-883.
13. Wang, M.C. and Keen, N.T., 1970, Purification and characterization of endopolygalacturonase from Verticillium albo-atrum, Arch. Biochem. Biophys., 141: 749-757.
14. Barash, I., Zilberman, E. and Marcus, L., 1984, Purification of Geotrichum candidum endopolygalacturonase from culture and from host tissue by affinity chromatography on cross-linked polypectate, Physiol. Plant Pathol., 25: 161-169.
15. Cervone, F., Scala, A., Foresti, M., Cacace, M.G. and Novello, C., 1977, Endopolygalacturonase from Rhizoctonia fragariae: purification and characterization of two isoenzymes, Biochim. Biophys. Acta, 432: 379-385.
16. Fanelli, C., Cacace, M.G. and Cervone, F., 1978, Purification and properties of two polygalacturonases from Trichoderma koningii, J. Gen. Microbiol., 104: 305-309.
17. Scala, A., Camardella, L., Scala, F. and Cervone, F., 1980, Multiple forms of polygalacturonase in two strains of Rhizoctonia solani, J. Gen. Microbiol. 116: 207-211.
18. Barthé, J.P., Cantenys, D. and Touzé, A., 1981, Purification and characterization of two polygalacturonases secreted by Colletotrichum lindemuthianum, Phytopath. Z., 100: 161-171.
19. Cervone, F., Andebrhan, T., Coutts, R.H.A. and Wood, R.K.S., 1981, Effect of French bean tissue and leaf protoplasts on Colletotrichum lindemuthianum polygalacturonase, Phytopath. Z., 102: 238-246.
20. Mussel, H.W. and Strouse, B., 1972, Characterization of two polygalacturonases produced by Verticillium albo-atrum, Can. J. Biochem., 50: 625-632.
21. Di Lenna, P. and Fielding, A.H., 1983, Multiple forms of polygalacturonase in apple and carrot tissue infected by isolates of Botrytis cinerea, J. Gen. Microbiol., 129: 3015-3018.
22. Marciano, P., Di Lenna, P. and Magro, P., 1982, Polygalacturonase isoenzymes produced by Sclerotinia scelerotiorum in vivo and in vitro, Physiol. Plant Pathol., 20: 201-212.
23. Albersheim, P. and Anderson, A.J., 1971, Proteins from plant cell walls inhibit polygalacturonase secreted by plant pathogens, Proc. Nat. Acad. Sci. USA., 68: 1815-1819.
24. McNeil, M., Darvill, A.G., Fry, S.C. and Albersheim, P., 1984, Structure and function of the primary cell walls of plants, Annu. Rev. Biochem. 53: 625-663.
25. Tran Thanh Van, K., Toubart, P., Cousson, A., Darvill, A.G., Gollin, D. J., Chelf, P. and Albersheim, P., 1985, Manipulation of the morphogenetic pathways of tobacco explants by oligosaccharins, Nature, 314: 615-617.
26. West, C.A., Moesta, P., Jin, D.F., Lois, A.F. and Wickham, K.A., 1985, The role of pectic fragments of the plant cell wall in the response to

biological stresses, in: "Cellular and Molecular Biology of Plant Stress", J.L. Key and T. Kosuge, eds., pp. 335-349, Alan R. Liss Inc., New York.
27. Federico, R. and Angelini, R., 1986, Occurrence of diamine oxidase in the apoplast of pea epicotyls, Planta, 167: 300-302.
28. Bruce, R.J. and West, C.A., 1982, Elicitation of casbene synthetase activity in Castor bean. The role of pectic fragments of the plant cell wall in elicitation by fungal endopolygalacturonase, Plant Physiol., 69: 1181-1188.
29. Walker-Simmons, M., Hadwiger, L. and Ryan, C.A., 1983, Chitosans and pectic polysaccharides both induce the accumulation of the antifungal phytoalexin pisatin in pea pods and antinutrient proteinase inhibitors in tomato leaves, Biochem. Biophys. Res. Commun., 110: 194-199.
30. Laffitte, C., Barthé, J.P., Montillet, J.L. and Touzé, A., 1984, Glycoprotein inhibitors of Colletotrichum lindemuthianum endopolygalacturonase in near isogenic lines of Phaseolus vulgaris resistant and susceptible to anthracnose, Physiol. Plant Pathol., 25: 39-53.
31. Hoffman, R.M. and Turner, J.G., 1984, Occurrence and specificity of an endopolygalacturonase inhibitor in Pisum sativum, Physiol. Plant Pathol., 24: 49-59.
32. Cervone, F. and De Lorenzo, G., 1985, Pectic enzymes as phytotoxins: Absorption of polygalacturonase from Colletotrichum lindemuthianum to French bean protoplasts, Phytopathologia Mediterranea, in press.
33. Hislop, E.C., Keon, J.P.R. and Fielding, A.H., 1979, Effect of pectin lyase from Monilia fructigena on viability, ultrastructure and localization of acid phosphatase on cultured apple cells, Physiol. Plant Pathol., 14: 371-381.
34. Yamazaki, N., Fry, S.C., Darvill, A.G. and Albersheim, P., 1983, Host-pathogen interactions. XXIV. Fragments isolated from suspension-cultured sycamore cell walls inhibit the ability of cells to incorporate ^{14}C- leucine into proteins, Plant Physiol., 72: 864-869.
35. West, C.A., Bruce, R.J. and Jin, D.F., 1984, Pectic fragments in plant cell walls as mediators of stress response, in: "Structure, Function and Biosynthesis of Plant Cell Wall", W.M. Dugger and S. Bartnicki-Garcia, eds., pp. 359-379, American Society of Plant Physiologists.
36. Bishop, P.D., Makas, D.J., Pearce, G. and Ryan, C.A., 1981, Proteinase inhibitor-inducing factor activity in tomato leaves resides in oligosaccharides enzymically released from cell walls, Proc. Nat. Acad. Sci. USA., 78: 3536-3540.
37. Brown, W.E. and Graham, J.S., 1985, Pectic fragments regulate the expression of proteinase inhibitor genes in plants, in: "Cellular and Molecular Biology of Plant Stress", J.L. Key and T. Kosuge, eds., pp. 319-334, Alan, R. Liss Inc., New York.
38. Esquerré-Tugayé, M.T., Mazan, D., Pélissier, B., Roby D., Rumeau, D. and Toppan, A., 1985, Induction by elicitors and ethylene of proteins associated to the defence of plants, ibidem pp. 459-473, New York.
39. Kuć, J., 1975, Phytoalexins and the specificity of plant-parasite interaction, in: "Specificity in Plant Disease", R.K.S. Wood and A. Graniti, eds., pp. 253-268, Plenum Press, New York.
40. Maniatis, T., Fritsch, E.F. and Sambrook, J., 1982, in: "Molecular Cloning - A Laboratory Manual", Cold Spring Harbor Laboratory, New York.
41. Huynh, T.V., Young, R.A. and Davis, R.W., 1984, in: "DNA Cloning Techniques: A Practical Approach", D. Glover, ed., IRL Press, Oxford.

THE D^2-FACTOR IN OPHIOSTOMA ULMI : EXPRESSION AND LATENCY

H.J. Rogers[a], K.W. Buck[a] and C.M. Brasier[b]

[a]Department of Pure and Applied Biology
Imperial College of Science and Technology
London, SW7 2BB
UK

[b]Forest Research Station
Alice Holt Lodge
Farnham
Surrey, GU10 4LH
UK

INTRODUCTION

D-factors are genetic elements that have the potential to exert a deleterious effect on the growth and reproductive fitness of *Ophiostoma ulmi*, the causative agent of Dutch elm disease [1,2]. They show cytoplasmic inheritance and are readily transmitted by hyphal anastomosis to other isolates in the same vegetative-compatibility (v-c) group; less efficient transmission is also possible between isolates in different v-c groups [3]. When a d-infected donor isolate and a "healthy" recipient isolate are paired on elm sapwood agar (ESA) medium, altered sectors of growth (d-reactions) are observed in the recipient isolate on either side of the area where the two colonies initially merged. In its most severe form a d-reaction can result in almost complete cessation of growth of the recipient culture.

D-factors are common components of populations of the North American (NAN) and Eurasian (EAN) races of the aggressive strain and of the non-aggressive strain of *O. ulmi*. There is evidence that d-factors in these three sub-groups of the fungus, and possibly also some of those within a sub-group, differ qualitatively [2]. The first reported d-factor, found in a French NAN isolate [1], is now named the d^1-factor and other d-factors have been named d^2, d^3, d^4 etc. as they have been discovered.

D-factors can affect the pathogenicity of *O. ulmi* in the xylem phase and the reproduction and survival of the fungus in the bark phase of the pathogenic cycle [2]. However progress in manipulating d-factors as possible biological control agents for Dutch elm disease will depend on progress in establishing the molecular nature of d-factors [4], and in understanding how their expression and stability are affected by host sexual and asexual reproduction, by transmission to different host genetic backgrounds and by changes in environmental conditions. In this paper current knowledge of the nature, stability and expression of the d^2-factor is reviewed.

NATO ASI Series, Vol. H1
Biology and Molecular Biology of Plant-Pathogen Interactions
Edited by J. Bailey

Table 1. Fungal phenotypes associated with cytoplasmic inheritance.

Fungus	Phenotype	Genetic determinant
Agaricus bisporus	Diseased	Viral ds and/or ssRNA
Aspergillus amstelodami	Ragged	MtDNA
Ophiostoma ulmi	D-infected	dsRNA?
Cochliobolus heterostrophus	Senescent	MtDNA
Endothia parasitica	Hypovirulent	dsRNA?
Helminthosporium victoriae	Diseased	Viral dsRNA?
Kluyveromyces lactis	Killer	DNA plasmid
Neurospora crassa	Poky	MtDNA
	Stopper	MtDNA
Podospora anserina	Senescent	MtDNA
Podospora curvicolla	Senescent	MtDNA
Rhizoctonia solani	Diseased	dsRNA?
	Pathogenic	Viral dsRNA?
Saccharomyces cerevisiae	Cold sensitivity	Satellite dsRNA
	Drug resistance e.g. Chloramphenical, erythromycin	MtDNA
	Killer	Satellite dsRNA
	Petite	MtDNA
Ustilago maydis	Killer	Satellite dsRNA

ASSOCIATION OF THE D^2-FACTOR WITH SPECIFIC SEGMENTS OF DOUBLE-STRANDED RNA

Cytoplasmically inherited traits in fungi (Table 1) may be caused by (i) alterations in mitochondrial DNA (mtDNA), (ii) DNA plasmids, (iii) double-stranded RNA (dsRNA) or single-stranded RNA viruses (ssRNA), (iv) satellite dsRNAs, which are dependent on helper viruses for their replication and encapsidation and (v) double-stranded RNAs, not associated with traditional viruses [5,6-9,10]. Several lines of evidence have shown that the d^2-factor in O. ulmi is closely associated with one or more specific segments of dsRNA.

(a) The d^2-factor was discovered as the result of an experiment in which as elm tree was inoculated with a O. ulmi isolate, W2 tol 1, which carried a single nuclear gene for resistance to MBC fungicide [2]. Following the death of the tree the bark became colonized by beetles carrying wild-type (MBC-sensitive) O. ulmi genotypes and a proportion of MBC-tolerant isolates subsequently recovered was found to be d-infected, presumably as a result of transfer from a wild-type O. ulmi individual. The d^2-infected isolates were shown to contain ten dsRNA segments, with molecular weights ranging from 2.40×10^6 to 0.30×10^6, while W2 tol 1 and other recovered non-d^2-infected MBC-tolerant isolates contained only two of these, namely segments 1 and 8 (Fig. 1). Acquisition of the d^2-factor by W2 tol 1 therefore was accompanied by acquisition of eight segments of dsRNA.

(b) Transmission of the d^2-phenotype from one d^2-infected isolate to eight other healthy (non-d^2-infected) recipient isolates in the same v-c group (the NAN super-group [4]) was always accompanied by faithful transmission of all ten dsRNA segments.

(c) Six wild, non-d^2-infected, O. ulmi isolates in the NAN supergroup were found to have either no dsRNA or a few (up to four) dsRNA segments. Collectively dsRNA segments with electrophoretic mobilities of segments 1, 2, 4, 6, 8 and 9 were observed in these isolates, although further tests are

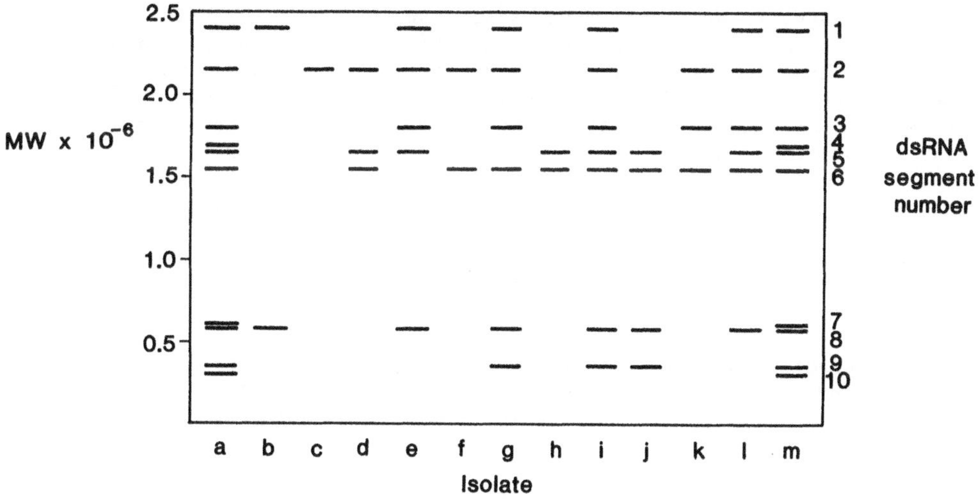

Figure 1. Double-stranded RNA segments in d^2-infected and healthy isolates of O. ulmi.
a and m, d^2-infected isolates; b, W2 tol 1 ; c, healthy single ascospore isolate derived from a cross between a d^2-infected isolate as the female parent and a healthy isolate; e, healthy isolate recovered from a tree inoculated with a d^2-infected isolate; f to l, healthy single conidial isolates derived from a d^2-infected isolate.

needed to confirm the identity of these segments with those in d^2-infected isolates. However the results suggest that, if the d^2-phenotype is caused by dsRNA, segments 3, 5, 7 and 10 are the most likely candidates.

(d) Single ascospore progeny derived from a genetic cross between a d^2-infected isolate as the female parent and a non-d^2-infected isolate were all found to be non-d^2-infected and to contain either no dsRNA or, rarely, only dsRNA segment 2 (Fig. 1). Hence loss of dsRNA correlates with loss of the d^2 phenotype, but because of the large number of segments lost, this experiment does not help in attempting to correlate the d^2-factor with one or more specific dsRNA segments.

Traits associated with altered mtDNA generally show maternal inheritance [5]. The absence of maternal inheritance of the d^2-factor suggests that it is probably not associated with mitochondrial DNA. However, formation and maturation of perithecia can be impaired as a result of d-infection [2] and the possibility must still be considered that d^2-infected isolates could contain both normal and altered mtDNA and that protoperithecia may only be formed from cells containing a predominance of normal mtDNA.

(e) When nine elm trees were infected by inoculation with four different d^2-infected O. ulmi isolates (one isolate per tree) and re-isolations were made from the lowest infected branches of each tree 12 weeks later, all of the re-isolates were found to be non-d^2-infected and to contain only two to six of the original ten dsRNA segments of the d^2-infected isolates. It is noteworthy that all of the isolates had lost dsRNA segments 4, 7 and 10.

(f) Following conidiogenesis, a d^2-infected isolate gave rise to a proportion of non-d^2-infected single conidial progeny (see below). These isolates retained only two to seven of the original ten dsRNA segments and again segments 4, 7 and 10 were the only ones consistently lost in all of the non-d^2-infected isolates (Fig. 1).

In summary, acquisition of the d^2-factor by isolate W2 tol 1 was accompanied by acquisition of dsRNA segments 2, 3, 4, 5, 6, 7, 9 and 10 and loss of the d^2-factor in a variety of situations was also accompanied by loss of three or more dsRNA segments which always included segments 4, 7 and 10. Additionally dsRNA segments 7 and 10 have not been found in any naturally occurring non-d^2-infected isolates in the NAN super v-c group and the d^2-factor could consist of one or both of these segments. However segment 4 could also be involved because its identity with a segment of the same mobility in a naturally occurring non-d^2-infected isolate (see section c above) has not yet been confirmed. Unequivocal proof of the identity of the d^2-factor with specific dsRNA segments will require transformation experiments with isolated dsRNA segments.

The association of the d^2-factor with specific segments of dsRNA appears to be similar to the situation in killer strains of *Saccharomyces cerevisiae* and *Ustilago maydis* in which specific segments of dsRNA encode protein toxins [6,9]. Segment 7 in d^2-infected isolates is similar in size to the dsRNA segment encoding protein toxin in *U. maydis*. Possibly one or more dsRNA segments in d^2-infected isolates encode a protein which is toxic to *O. ulmi*. Most killer strains of *S. cerevisiae* and *U. maydis* are resistant to the protein toxins which they produce, as a result of resistance genes encoded chromosomally or on a dsRNA segment. However poorly growing "suicide" strains have been described which are sensitive to their own toxins [6].

The dsRNAs which encode killer toxins in *S. cerevisiae* and *U. maydis* are satellites in that they are dependent on helper dsRNA viruses for their replication and encapsidation. Hence, although non-killer strains of these two fungi containing helper virus dsRNA but lacking satellite dsRNA are common, strains containing only satellite dsRNA have never been found. dsRNA segments 4, 7 and 10 in d^2-infected isolates could also be satellites in that they have not been found alone in any isolate. The most stable dRNAs in d^2-infected isolates, i.e. those which are the least frequently lost, appear to be segments 2 and 6 and these are possible candidates for "helper" dsRNAs. Whether the dsRNAs in d^2-infected isolates are encapsidated in isometric virus particles, like those in killer strains of *S. cerevisiae* and *U. maydis*, is not yet known.

LATENTLY D^2-INFECTED ISOLATES

Single conidial progeny derived from a slow growing MBC-tolerant d^2-infected *O. ulmi* isolate were found to be either slow growing like their parent isolate or fast growing like typical non-d^2-infected, healthy isolates. Analysis of the slow growing conidial isolates showed that they retained all ten dsRNA segments and gave strong d-reactions when paired against non-d^2-infected isolates, resulting in transmission of the d^2-factor to the recipient. It is clear that during conidiogenesis the d^2-factor had been transmitted to the slow growing conidial isolates in a form in which it was overtly expressed.

The faster growing conidial isolates were of two types, non-d^2-infected and latently d^2-infected. The former failed to give d-reactions when paired against non-d^2-infected isolates, but were able to act as recipients in d-reactions when paired against d^2-infected isolates. Furthermore they could not be induced on storage or growth on different media with different recipients to give d-reactions, did not transmit the d^2-phenotype to other isolates by hyphal anastomosis, gave rise only to faster growing single conidial progeny and, as discussed above, retained only two to seven of the original ten dsRNA segments. It is evident that such isolates had lost the d^2-factor during conidiogenesis.

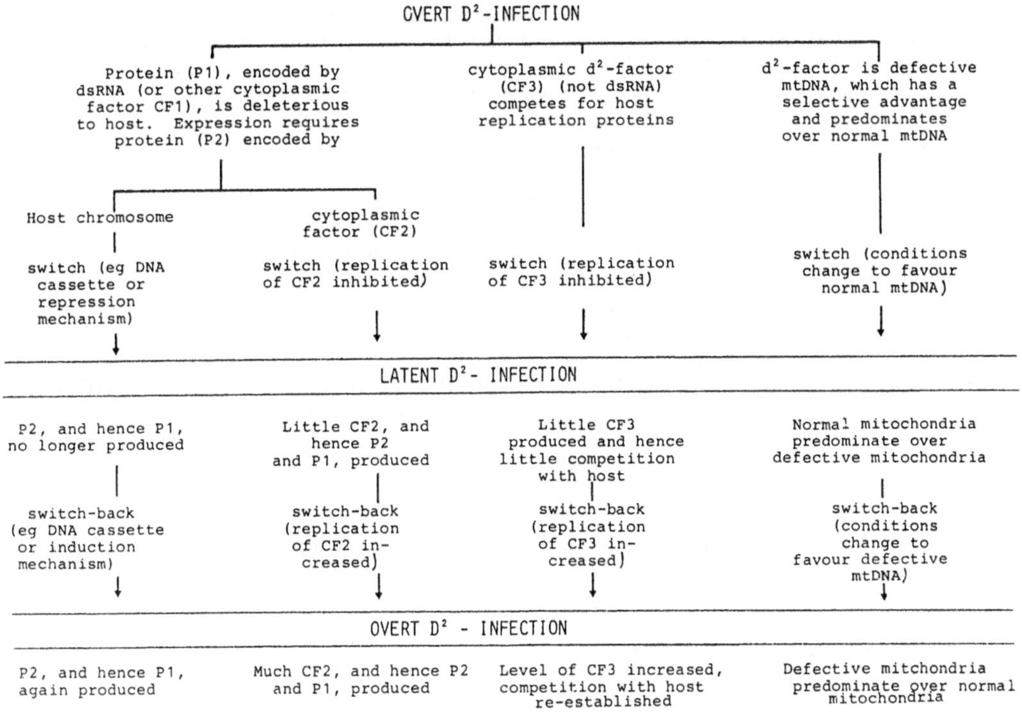

Figure 2. Possible mechanisms for the switch from overt to latent d-infection.

The latently d^2-infected isolates were similar to the non-d^2-infected isolates in their growth rates, failure (initially) to give d-reactions when paired against non-d^2-infected isolates and ability to act as recipients in d-reactions when paired against (overtly) d^2-infected isolates. However they differed from the non-d^2-infected single conidial isolates in that d^2-infection could be demonstrated in one or more of the following ways: (i) ability to give a d-reaction (eventually) either after merely storing at 4 °C for several weeks, by growing on various media prior to testing or by using different recipients; (ii) by pairing against MBC-sensitive recipient isolates and making isolations from putative d-reaction zones; often such MBC-sensitive recipient isolates had acquired the ability to give a d-reaction; (iii) by a further cycle of conidiogenesis, when a proportion of the conidial progeny were slow growing and overtly d^2-infected.

Twenty-seven latently d^2-infected conidial isolates contained all the ten dsRNA segments of the parent isolate. However one isolate with only seven dsRNA segments (segments 4, 7 and 10 lost) which could not be induced to give a d-reaction by methods (i) and (ii) gave rise, following a second cycle of conidiogenesis, to three slow growing progeny (out of 56 tested), one of which appeared to be overtly d^2-infected with the ability to transfer this phenotype to a recipient isolate. Four other (non-d^2-infected) conidial isolates with seven dsRNA segments gave rise only to fast growing conidial progeny. Out of 280 single conidial progeny from the five isolates with seven dsRNA segments only one appeared to be overtly d^2-infected. This isolate would appear to be a rare exception to the observation that loss of dsRNA segments 4, 7 and 10 is accompanied by loss of the d^2-factor. However sensitive hybridization tests, capable of detecting one dsRNA molecule in 500 hyphal compartments [4] will be required to confirm the complete absence of these three segments. Another possibility is that a mutation could have occurred in one of the remaining seven dsRNAs converting it _de novo_ into a new d-factor.

Possible explanations of the switch, on conidiogenesis, from overt d^2-infection and _vice versa_, are summarized in Figure 2. Although available evidence favours a dsRNA d^2-factor, in the absence of unequivocal proof for this, explanations are based on the possibilities that the d^2-factor might be dsRNA, a DNA plasmid or defective mitochondrial DNA. Any explanation has to take into account that latently d^2-infected isolates can act as recipients in d-reactions with overtly d^2-infected isolates i.e. latent d^2-infection appears to be a recessive characteristic. Hence we are looking for a product which is produced in overt d^2-infection, but in much smaller amounts, or not at all, in latent d^2-infection. This would exclude hypotheses that in the latent state the host produces a protein to make it resistant to the d^2-factor or that latency might be due to integration of the d^2-factor (or a DNA copy) into the host chromosome and production of a host repressor of d^2-factor expression.

The explanation for latency based on a dsRNA d^2-factor assumes that the product of a host gene is required for expression of dsRNA. This is feasible because it is well established that the products of host _kex_ genes are required for killer expression in _S. cerevisiae_ [6]. The explanation based on defective mtDNA assumes that d^2-infected mycelium will contain a mixture of normal and defective mtDNA, but that in vegetative growth defective mtDNA has a selective advantage. This is true for senescence in _Podospora anserina_ and probably also for abnormal phenotypes caused by defective mtDNA in other fungi (Table 1) [5].

POTENTIAL OF THE D^2-FACTOR FOR THE BIOLOGICAL CONTROL OF DUTCH ELM DISEASE

A cytoplasmically transmitted factor causing hypovirulence would seem to be the ideal biological control agent for a disease since it would have the capacity to spread to virulent genotypes of the pathogen converting them to hypovirulence. A good example of this is cytoplasmically transmitted hypovirulence in _Endothia parasitica_, which is closely associated with dsRNA. In Italy hypovirulent strains have become predominant, leading to a host-parasite balance and re-establishment of the chestnut [10].

However the likely effect of d-factors on Dutch elm disease is far more complex because the fungus has (i) a long saprophytic phase in elm bark which is the equivalent of the _Endothia_ canker phase (ii) a beetle vector transmission phase and (iii) a vascular wilt or pathogenic phase in elm xylem. The potential of d-factors for the biological control of Dutch elm disease has been considered in detail by Brasier [2]. Their greatest impact may come from their reducing the effectiveness of spore inoculation on the vectors, so reducing the number of resulting infections and breaking the cycle of disease. Here, some of the observations which need to be considered in assessing the potential with regard to the d^2-factor are summarized.

(a) The bark beetles which transmit the disease carry conidia and ascospores. These propagules arise in the beetle breeding galleries during the saprophytic or bark phase of the disease [12].

(b) The extensive mycelial growth which occurs on the bark will favour the spread of d-factors by hyphal anastomosis. Such spread will be limited to various degrees between isolates in different v-c groups but will be facilitated by the occurrence of large v-c groups such as the NAN v-c super-group, in O. ulmi populations [2].

(c) The viability of conidia from mycelia of d^2-infected isolates is greatly reduced [2]. A majority of the remaining viable germlings are d^2-infected. However, many of these may be latently infected. The frequency of latency or the frequency of switching from latent to overt d^2-infection and vice versa in natural populations is unknown. The ecological impact of latency may be lessened by the tendency for O. ulmi conidia to be dispersed in clumps, such that a proportion of overtly infected germlngs are always likely to develop and fuse with their neighbours. Nevertheless, latent d^2-infections are important because they act as reservoirs of the d^2-factor yet they appear phenotypically similar to non-d^2-infected isolates [12].

(d) D^2-factors are transmitted infrequently, if at all, into ascospore progeny which appears to be a means whereby the fungus can free itself from d-infection and most of its dsRNA. Ascospore formation is, however, probably reduced in those d^2-infected isolates which show retarded growth [2,12].

(e) D^2-infected isolates appear to show a greater spread of pathogenic ability than healthy isolates in the xylem phase. However the d^2-factor and some dsRNA bands tend to be lost during yeast-stage sporogenesis on the xylem [2,12].

D-factors occur commonly in natural populations of O. ulmi. Given the widespread devastation of the elm during the current Dutch elm disease epidemics, it would be difficult to argue that d-factors have so far had any noticeable effect in reducing the disease. However, their potential impact may be greater in the forthcoming post-epidemic period [12]. An ideal d-factor distribution system is available in the form of the beetle vector which can be bred and released artificially. If we can increase our understanding of the molecular biology of the d^2-factor, including the molecular basis of latency and the mechanism of switching to and from overt d-infection, it may be possible to manipulate the d^2 and other d-factors artificially as a means of disease control.

REFERENCES

1. Brasier, C.M., 1983, A cytoplasmically transmitted disease of Ceratocystis ulmi, Nature, 305: 220-223.
2. Brasier, C.M., 1986, The d-factor in Ceratocystis ulmi: its biological characteristics and implications for Dutch elm disease, in: "Fungal Virology", K.W. Buck, ed., CRC Press, Boca Raton, Florida, USA, in press.
3. Brasier, C.M., 1984, Inter-mycelia recognition systems in Ceratocystis ulmi: their physiological properties and ecological importance, in: "The Ecology and Physiology of the Fungal Mycelium", D.H. Jennings and A.D.M. Rayner, eds., pp. 451-497, Cambridge University Press, Cambridge.
4. McFadden, J.J.P., Buck, K.W. and Rawlinson, C.J., 1983, Infrequent transmission of double-stranded RNA virus particles but absence of DNA proviruses in single ascospore cultures of Gaeumannomyces graminis, J. Gen. Vir., 64: 927-937.
5. Bockelmann, B., Osiewacz, H.D., Schmidt, F.R. and Schulte, E., 1986, Extrachromosomal DNA in fungi : organisation and function, in: "Fungal Virology", K.W. Buck, ed., CRC Press, Boca Raton, Florida, USA, in press.

6. Bruenn, J., 1986, The killer systems of *Saccharomyces cervisiae*, *in*: "Fungal Virology", K.W. Buck, ed., CRC Press, Boca Raton, Florida, USA, in press.
7. Buck, K.W., 1986, Fungal virology: an overview, *in*: "Fungal Virology", K.W. Buck, ed., CRC Press, Boca Raton, Florida, USA, in press.
8. Ghabrial, S.A., 1986, A transmissible disease of *Helminthosporium victoriae*: evidence for a viral etiology, *in*: "Fungal Virology", K.W. Buck, ed., CRC Press, Boca Raton, Florida, USA, in press.
9. Koltin, Y., 1986, The killer systems of *Ustilago maydis*, *in*: "Fungal Virology", K.W. Buck, ed., CRC Press, Boca Raton, Florida, USA, in press.
10. Van Alfen, N.K., 1985, Hypovirulence of *Endothia* (*Cryphonectria*) *parasitica* and *Rhizoctonia solani*, *in*: "Fungal Virology", K.W. Buck, ed., CRC Press, Boca Raton, Florida, USA, in press.
11. Webber, J.F. and Brasier, C.M., 1984, The transmission of Dutch elm disease : a study of the processes involved, *in*: "Invertebrate - Microbial Interactions", J.M. Anderson, A.D.M. Rayner and D.W.H. Walton, eds., pp. 271-306, Cambridge University Press, Cambridge, UK.
12. Rogers H.J., Buck K.W. and Brasier, C.M., 1985, The molecular nature of the d-factor in *Ceratocystis ulmi*, *in*: "Fungal Virology", K.W. Buck, ed., CRC Press, Boca Raton, Florida, USA, in press.

THE EFFECTS OF PASSAGING ON THE HOST SPECIFICITY OF SEPTORIA NODORUM: A CAUTIONARY TALE

A.E. Osbourn[a], C.E. Caten[a] and P.R. Scott[b]

[a]Department of Genetics
University of Birmingham
Birmingham, UK

[b]Plant Breeding Institute
Trumpington
Cambridge, UK

INTRODUCTION

This paper is subtitled "A Cautionary Tale" because it illustrates the risks of contamination where successive cycles of selective propagation are involved. Selective techniques are central to microbial and molecular genetics and are likely to be important in the application of the methodology of these disciplines to the analysis of plant pathogenesis. Where selective procedures are being carried out on plants, control over contamination becomes a major practical consideration.

Septoria nodorum (teleomorph Leptosphaeria nodorum) is the causal agent of a leaf spot and glume blotch disease of wheat and also infects barley and various wild grasses [1,2,3,4]. Cross infection tests indicate that isolates are generally adapted to individual host species; in particular, isolates from wheat are strongly pathogenic to wheat but only weakly pathogenic to barley, while some, but not all, isolates from barley show the reverse specificity [1,4,5]. In contrast to this host-species adaptation there is no clear pathogen-race/host-cultivar specificity within the S. nodorum/wheat pathosystem [6], although occasional isolate by cultivar interactions have been reported [3].

Because of the potential significance of this wide host range as a source of inoculum, various workers have examined the effects of serial reinfection (passage) of an isolate originally adapted to one host species (own host) on a host to which it was not adapted (alien host). At least four research groups have reported that adaptation to wheat or to barley can be changed by passaging through the alien host [3,7,8,9]. In many cases the initial isolates were of single pycnidiospore origin and there is good evidence that such cultures should be genetically homogeneous [10,11]. The observed changes during passaging therefore suggested that adaptation to wheat or to barley is an inducible property of a single genotype of S. nodorum. If this were so, conversion from one form to the other might involve induction and repression of host-specific genes or a change in the state of a plasmid or transposable element. Whatever the mechanism, the

NATO ASI Series, Vol. H1
Biology and Molecular Biology of Plant-Pathogen Interactions
Edited by J. Bailey

Table 1. Pathogenicity of six isolates of *Septoria nodorum* to wheat and to barley[a] in detached leaf and seedling tests.

Isolate No.	Origin	Detached leaves[b]		Seedlings[c]		Host adaptation
		Wheat	Barley	Wheat	Barley	
4	Barley	1.8 ± 0.8	12.5 ± 0.8	0.0 ± 0.0	61.1 ± 6.4	Barley
11	Barley	0.2 ± 0.2	8.1 ± 0.6	NT[d]	NT	Barley
154	Barley	2.6 ± 0.8	7.3 ± 0.8	1.6 ± 0.4	69.1 ± 6.9	Barley
171	Barley	8.8 ± 0.8	0.7 ± 0.8	30.0 ± 4.7	3.2 ± 0.8	Wheat
189	Wheat	10.3 ± 0.6	2.2 ± 0.6	39.2 ± 4.9	4.5 ± 0.8	Wheat
241	Wheat	7.2 ± 0.4	3.0 ± 0.4	NT	NT	Wheat

[a] Wheat = cv. Maris Ranger; Barley = cv. Maris Trojan. [b] Lesion length (mm). [c] % leaf area infected. [d] Not tested.

apparent inter-convertibility of adaptation suggested that its genetic control was simple and potentially amenable to experimental attack.

We set out to determine the genetic basis of the adaptation of S. nodorum to wheat or to barley and the mechanism by which it is changed by host passage. Our take-off point was to establish a collection of isolates showing clear adaptation to wheat or to barley and then to change these specificities by passaging.

ADAPTATION OF S. NODORUM TO WHEAT AND TO BARLEY

Thirty-six cultures of S. nodorum isolated from wheat, barley or triticale in the UK, Ireland, Scandinavia and the USA, were obtained. Immediately on receipt each isolate was propagated through a single pycnidiospore to ensure genetic homogeneity. The adaptation of all 36 isolates was tested by inoculation on detached leaves of wheat and barley following the method of Benedikz et al. [12]. Some isolates were also tested by spraying spore suspensions on to seedlings at the four-leaf stage. Results for a sample of six isolates are shown in Table 1. Three of these six isolates were clearly adapted to wheat in that they produced more severe symptoms on this host than on barley, while the remaining three were adapted to barley. Although this host specificity is quantitative rather than qualitative it is nevertheless a very clear effect (Table 1). Of the 36 isolates, 23 were wheat-adapted, eight were barley-adapted and five showed no clear preference for wheat or barley (the last isolates appeared generally weakly pathogenic, producing mild symptoms on both hosts). All barley-adapted isolates came from barley, whereas wheat-adapted isolates came from wheat, barley or triticale (171 in Table 1 is an example of a wheat-adapted isolate originating from barley). These results clearly substantiated the existence of wheat-adapted (W) and barley-adapted (B) forms within S. nodorum.

Previous workers had noted that the W and B forms differed in cultural morphology, with the former producing grey-green colonies and the latter pink colonies [5,8]. Although there was considerable variation in colony morphology within each form, the same tendency could be discerned in our sample of isolates. Cunfer and Youmans [5] noted that W isolates fluoresce under UV light while B isolates do not. This association also was apparent in our isolates, although one W isolate which did not fluoresce and one B isolate which did were found.

PASSAGING ON SEEDLINGS

Passage Experiments

On the basis of the above results, several clear W and several clear B isolates were selected for passaging in an attempt to induce changes in host adaptation. Some of these isolates were marked with a nitrate reductase deficient mutation [10] to enable identification after passaging and to facilitate genetic analysis of any changed reisolates. Passaging was initially carried out on seedlings of wheat (cv. Maris Ranger) and barley (cv. Maris Trojan) grown on an isolation plant propagator. The seedlings were inoculated by spraying with a spore suspension in a laminar flow cabinet. After symptoms developed, infected leaves were removed and incubated under conditions favouring sporulation of the fungus. A new spore suspension was then produced and used to repeat the process. Appropriate precautions to minimize the risk of cross contamination were taken throughout [13].

Table 2. Pathogenicity to wheat and to barley in detached leaf and seedling tests of parent isolates and of reisolates after five passages through their own and alien host.

Isolate/ reisolate	Detached leaves		Seedlings		Host adaptation
	Wheat	Barley	Wheat	Barley	
154	1.6±0.4	7.4±0.5	1.6±0.4	69.1±6.9	Barley
154/W/I[a]	4.9±0.6	2.7±0.2	54.5±5.0	3.4±0.5	Wheat[b]
154/W/II	1.0±0.4	9.4±1.1	0.2±0.1	20.9±6.2	Barley
154/B/I[a]	1.3±0.4	8.4±0.8	0.5±0.1	9.0±2.7	Barley
154/B/II	1.6±0.5	6.7±0.6	0.5±0.1	24.2±6.0	Barley
189	2.6±0.3	1.4±0.3	15.8±1.8	0.9±0.4	Wheat
189/W/I	3.3±0.4	1.4±0.1	24.7±3.6	1.4±0.2	Wheat
189/W/II	3.6±0.3	1.7±0.1	39.2±4.9	4.5±0.8	Wheat
189/B/I	0.4±0.3	8.0±0.9	1.2±0.3	71.8±6.0	Barley[b]
189/B/II	2.9±0.4	4.2±0.2	47.8±3.3	14.8±2.9	Wheat & Barley[b]

[a] Duplicate line I of isolate 154 passaged five times through wheat (W) or barley (B)

[b] Change in host adaptation. 189/B/II was pathogenic to both species and showed no host adaptation.

Five W and five B isolates were each passaged through wheat and through barley for five transfers before being reisolated from each host and assayed for host adaptation. Independent duplicate lines (denoted I and II) were used for two W and two B isolates so that altogether 14 lines were passaged. At the end of each experiment three forms of each isolate were available for testing: (i) maintained on agar throughout the passaging (c 10 weeks), (ii) passaged through own host and (iii) passaged through alien host. In all cases the cultures maintained on agar and those passaged through the own host retained the original adaptation. However, six of the 14 alien-passaged lines changed, four switching their adaptation to the alien host and two becoming pathogenic to both hosts [13]. A typical set of results is shown in Table 2. These results resemble the published reports of changes in host-species adaptation during passaging [3,7,8,9].

Analysis of Reisolates

In addition to pathogenicity, the reisolates were examined for colony morphology and fluorescence. All reisolates unchanged in pathogenicity retained the other characteristics of their parent isolate, while those reisolates changed in pathogenicity were also changed in colony morphology and, in many cases, fluorescence too [13]. These associated changes during passaging had been observed by previous workers and interpreted as pleiotropic responses to host adaptation [7,8]. In our experiments however, we also noted that where genetically marked isolates were involved there were changes in the markers, with isolates both losing and gaining markers. Furthermore, each parent isolate had a unique morphology and the changed reisolates now resembled one of the other parent isolates used in the experiment [13]. Together all these observations suggested that, despite the precautions taken, cross contamination had occurred and was responsible for the observed changes in host-species adaptation.

We attempted to confirm this contamination hypothesis by testing the heterokaryon compatibility of the changed reisolates. In many filamentous fungi only isolates carrying identical alleles at a number of somatic compatibility loci will form heterokaryons, consequently heterokaryon compatibility provides a sensitive test for genetic relatedness [14,15]. In S. nodorum heterokaryon compatibility/incompatibility is indicated by the presence/absence of complementation between physiologically-distinct auxotrophic mutations [10]. All changed reisolates were heterokaryon incompatible with their own parent isolate, indicating a lack of parent-reisolate relatedness, but compatible with one of the other parent isolates used in the experiment. In each case the compatible parent was the one which the reisolate resembled in morphology [13]. These results therefore not only confirmed the contamination hypothesis but also allowed identification of the contaminants.

PASSAGING ON DETACHED LEAVES

We were considerably disturbed by the high incidence of contamination and to improve control we moved to a system of passaging on detached leaves. The experimental strategy was similar to that used in the seedling experiments but all manipulations were carried out in a laboratory and the different treatments were maintained in separate boxes in a controlled environment room. Inoculation was either by placing a 5 μl drop of spore suspension (2×10^6 spores per ml) on the leaf, or by dipping it into the spore suspension. Five experiments were carried out and seven W and six B isolates were each passaged through wheat and barley; independent replicate lines were used with several of these isolates so that altogether 27 lines were passaged. In one experiment only W isolates were used and in another only B isolates. Reisolates were made after five transfers and examined in the same manner as those from the seedling experiments.

No changes in adaptation were detected following maintenance in culture or passage through the own host. However, two changes (both B to W) were observed following passage through the alien host; both these occurred in experiments where W and B isolates were being passaged simultaneously and changes were never observed where isolates of only one adaptation type were being handled. As in the seedling experiments, the two changed isolates resembled in morphology, fluorescence, marker genotype and heterokaryon compatibility, not their parent but another isolate of appropriate adaptation which was in use in the laboratory at the same time. Thus, the only instances of changes in host adaptation during passage on detached leaves also appeared to have arisen through contamination.

CONCLUSIONS

The studies confirmed the existence within S. nodorum of distinct groups of isolates adapted to wheat or to barley. These two groups differ for a number of characters in addition to host adaptation (Table 3) and their genetic divergence is further indicated by recent studies of restriction-fragment-length polymorphism in the species. Total DNA digested with XhoI was probed with a random clone from a λ EMBL3 DNA library of S. nodorum. Six W isolates gave identical bands and three B isolates gave identical bands, but the two sets of bands were clearly distinct from each other (A.C. Newton, personal communication). These observations indicate that the W and B forms of S. nodorum are genetically distinct populations, which warrant taxonomic recognition below the species level.

Table 3. Characteristics of wheat- and barley-adapted isolates of *Septoria nodorum*.

Host adaptation	Colony colour	Fluorescence[a]	Growth[a,b] at 30 °C	Isozymes[a,b,c]	
				HK	AP
Wheat	Grey to green	+	High	Fast	Fast
Barley	Pink	-	Low	Slow	Slow

[a] exceptional isolates occur, [b] A.C. Newton, personal communication,

[c] HK = hexokinase, AP = alkaline phosphatase.

Given the extensive phenotypic and genotypic differences between the W and B types it would be very surprising if they could be interconverted by passaging. All changes of adaptation in our experiments arose through contamination and, since our precautions against cross contamination were at least as stringent as those generally adopted in experimental work with *S. nodorum*, these results imply that the similar changes in initially pure cultures observed by other groups [7,8,9] originated in the same way. The plausibility of contamination as an explanation is supported by titration experiments on detached leaves which indicate that, on the own host, *S. nodorum* has a high infection efficiency and that a single spore can probably initiate a lesion [16]. Furthermore, an adapted isolate initially present at a low frequency (1%) in an artificial mixture with an unadapted isolate can increase rapidly in frequency and may apparently replace the original majority component after five transfers [16]. Thus the inherent multiplication rates and selection pressures of the system are consistent with the contamination hypothesis. The two reisolates which had become pathogenic to both hosts (e.g. 189/B/II in Table 2) are believed to be mixtures of the original parent with an alien-adapted contaminant. Where a complete switch in host adaptation was observed, the contaminant is presumed to have replaced the parent.

Our results provide no evidence that adaptation to wheat or to barley is an inducible property of single genotypes of *S. nodorum* and consequently it is not necessary to speculate about the involvement of specific gene regulation, plasmids or transposons. Rather there are extensive differences between the two forms and adaptation to these two hosts may be under complex genetic control.

Our experience provides a clear demonstration and warning of the power of selection. In retrospect, the experimental precautions were clearly not sufficient to entirely eliminate contamination, and selection can magnify the consequences of very rare events. Without the subsequent checks on the changed reisolates provided by the genetic markers and the heterokaryon-compatibility tests, contamination would have been considered an unlikely explanation with organisms like *Septoria*. Complete control over contamination is very difficult when pathogens are being handled on plants; in these circumstances the use of natural or induced markers to confirm the identity of selected genotypes is strongly recommended.

ACKNOWLEDGEMENTS

The award of an RCCA studentship by the Science and Engineering Research Council to AEO is gratefully acknowledged. This research was supported in part by a grant (AG 6/97) from the Agricultural and Food Research Council and was carried out under MAFF licence numbers PHF 78/128, PHF 78/177, PHF 78/213 and PHF 78/215.

REFERENCES

1. Holmes, S.J.I. and Colhoun, J., 1970, Septoria nodorum as a pathogen of barley, Trans. Brit. Mycol. Soc., 55: 321-325.
2. King, J.E., Cook, R.J. and Melville, S.C., 1983, A review of the Septoria diseases of wheat and barley, Ann. Appl. Biol., 103: 345-373.
3. Rufty, R.C., Hebert, T.T. and Murphy, C.F., 1981, Variation in virulence in isolates of Septoria nodorum, Phytopathology, 71: 593-596.
4. Smedegård-Petersen, V., 1974, Leptosphaeria nodorum (Septoria nodorum), a new pathogen on barley in Denmark and its physiological specialization on barley and wheat, Friesia, 10: 251-264.
5. Cunfer, B.M. and Youmans, J., 1983, Septoria nodorum on barley and relationships among isolates from several hosts, Phytopathology, 73: 911-914.
6. Allingham, E.A. and Jackson, L.F., 1981, Variation in pathogenicity, virulence and aggressiveness of Septoria nodorum in Florida, Phytopathology, 71: 1080-1085.
7. Cunfer, B.M., 1984, Change of virulence of Septoria nodorum during passage through barley and wheat, Ann. Appl. Biol., 104: 61-68.
8. Fitzgerald, W. and Cooke, B.M., 1982, Response of wheat and barley isolates of Septoria nodorum to passage through barley and wheat cultivars, Plant Pathol., 31: 315-324.
9. Sharma, H.S.S., Brown, A.E. and Swinburne, T.R., 1982, Relationship of spore germination and appressoria formation in isolates of Septoria nodorum to pathogenicity on wheat and barley leaves, Trans. Brit. Mycol. Soc., 79: 519-526.
10. Newton, A.C. and Caten, C.E., 1985, Heterokaryosis and heterokaryon incompatibility in Septoria nodorum, in: "Septoria of Cereals", A.L. Scharen, ed., pp. 13-15, United States Department of Agriculture, ARS-12.
11. Shaw, D.E., 1953, Cytology of Septoria and Selenophoma spores, Proceedings of the Linnean Society of New South Wales, 78: 122-130.
12. Benedikz, P.W., Mappledoram, C.J. and Scott, P.R., 1981, A laboratory technique for screening cereals for resistance to Septoria nodorum using detached seedling leaves, Trans. Brit. Mycol. Soc., 77: 667-669.
13. Osbourn, A.E., Scott, P.R. and Caten, C.E., 1986, The effects of host passaging on the adaptation of Septoria nodorum to wheat or barley, Plant Pathol., 35: 135-145.
14. Croft, J.H. and Jinks, J.L., 1977, Aspects of the population genetics of Aspergillus nidulans, in: "Genetics and Physiology of Aspergillus", J.E. Smith and J.A. Pateman, eds., pp. 339-360, Academic Press, London.
15. Jinks, J.L., Caten, C.E., Simchen, G. and Croft, J.H., 1966, Heterokaryon incompatibility and variation in wild populations of Aspergillus nidulans, Heredity, 21: 227-239.
16. Osbourn, A.E., 1985, Host adaptation and variation in Septoria nodorum, Ph.D. Thesis, University of Birmingham, UK.

PARTICIPANTS

Bailey, J.A.
University of Bristol, Long Ashton Research Station, Long Ashton, Bristol, BS18 9AF, UK.

Brewin, N.
Dept. of Genetics, John Innes Institute, Colney Lane, Norwich, NR4 7UH, UK.

Buck, K.
Dept. of Pure & Applied Biology, Imperial College of Science & Technology, Prince Consort Road, London, SW7 2BB, UK.

Callebaut, A.
Institut voor Scheikundig Onderzoek, B 1980 Tervuren, Museumlaan 5, Belgium.

Caten, C.
School of Biological Sciences, The University of Birmingham, P.O. Box 363, Birmingham, B15 2TT, UK.

Cervone, F.
Dipartimento di Biologia Vegetale, Universita Degli Studi di Roma "La Sapienza", Citta Universitaria, 00100 Rome, Italy.

Clarke, D.D.
Dept. of Botany, The University, Glasgow, G12 8QQ, UK.

Collmer, A.
University of Maryland, Division of Agricultural and Life Sciences, College Park, Maryland 20742, USA.

Daniels, M.
John Innes Institute, Colney Lane, Norwich, NR4 7UH, UK.

Davidse, L.C.
Agricultural University Wageningen, Dept. of Phytopathology, 9 Binnenhaven, 6709 PD Wageningen, The Netherlands.

Dixon, R.A.
Dept. of Biochemistry, Royal Holloway College (University of London), Egham Hill, Egham, Surrey, TW20 0EX, UK.

Donaldson, I.
Dept. of Chemistry, Carlsberg Laboratory, Gamle Carlsbergveg 10, DK-2500 Copenhagen Valby, Denmark.

Esquerré-Tugayé, M.-T.
Centre de Physiologie Végétale, Université Paul Sabatier, Laboratoire Associe au C.N.R.S. No. 241, 118 route de Narbonne, 31062 Toulouse Cedex, France.

Friend, J.
Dept. of Plant Biology, University of Hull, Hull, HJ6 7RX, UK.

Gianinazzi-Pearson, V.
Station D'Amélioration de Plantes, INRA, BV 1540, 21034 Dijon Cedex, France.

Gurr, S.
Dept. of Biochemistry & Microbiology, University of St. Andrews, Irvine Building, North Street, St. Andrews, Fife, KY16 9AL, UK.

Hadwiger, L.A.
Molecular Biology of Disease Resistance Laboratory, Dept. of Plant Pathology, Washington State University, Pullman, Washington, 99164-6430, USA.

Hargreaves, J.A.
University of Bristol, Long Ashton Research Station, Long Ashton, Bristol, BS18 9AF, UK.

Heath, M.C.
Dept. of Botany, University of Toronto, Toronto, Ontario, M5S 1A1, Canada.

Higgins, V.J.
Dept. of Botany, University of Toronto, Toronto, Ontario, M5S 1A1, Canada.

van den Hondel, A.M.J.J.
Recombinant DNA Research Group, Medical Biology Laboratory - TNO, P.O. Box 45, 2280 AA Rijswijk, The Netherlands.

Howes, N.
Research Station, 195 Dafoe Road, Winnipeg, Manitoba, R3T 2MG, Canada.

Ingram, D.S.
Botany School, Downing Street, Cambridge, CB2 3EA, UK.

Keen, N.T.
Dept. of Plant Pathology, University of California, Riverside, California, 92521, USA.

Kolattukudy, P.E.
Institute of Biological Chemistry, Washington State University, Pullman, Washington 99164-6340, USA.

Kombrink, E.
Max-Planck-Institut für Zuchtungsforschung, Abteilung Biochemie, 5000 Köln 30, Federal Republic of Germany.

Lindgren, P.
Dept. of Plant Pathology, University of California, Berkeley, California, 94720, USA.

Lugtenberg, E.J.J.
University of Leiden, Dept. of Plant Molecular Biology, Nonnensteeg 3, 2311 VJ Leiden, The Netherlands.

Manocha, M.S.
Dept. of Biological Sciences, Brock University, St. Catherines, Ontario, L2S 3A1, Canada.

Mansfield, J.
Wye College (University of London), Wye, Ashford, Kent, TN25 5AH, UK.

O'Connell, R.J.
University of Bristol, Long Ashton Research Station, Long Ashton, Bristol, BS18 9AF, UK.

Oliver, R.
School of Biological Sciences, University of East Anglia, Norwich, NR4 7TJ, UK.

Petitprez, M.
Institut National Polytechnique de Toulouse, Ecole Nationale Supérieure Agronomique, 145 Av. de Muret, 31076 Toulouse Cedex, France.

Reisener, H.J.
Lehrstuhl und Institut für Biologie III, RWTH Aachen, Worringer Weg, 5100 Aachen, Federal Republic of Germany.

Ricci, P.
I.N.R.A., Centre de Recherches D'Antibes, Station de Botanique et de Pathologie Végétale, Villa Thuret-62, Bd.de Cap (Ch.Raymond), B.P.78-F 06606 Antibes Cedex, France.

Robertsen, R.
University of Tromsø, Institute of Biology and Geology, P.O. Box 3085, Guleng, N-9001 Tromsø, Norway.

Ryder, T.
Plant Biology, The Salk Institute, P.O. Box 85800, San Diego, California, 92138-9216, USA.

Showalter, A.M.
Dept. of Biology, Washington University, Campus Box 1137, St. Louis, Mo. 63130, USA.

Staples, R.C.
Boyce Thompson Institute for Plant Research, Cornell University, Tower Road, Ithaca, New York, 14853, USA.

Surico, G.
Consiglio Nazionale delle Richerche, Istituto Tossine & Micotossine da Parasiti Vegetali, 70126 Bari, Via G. Amendola, 197/F, Italy.

Timberlake, W.
University of California, College of Agricultural and Environmental Sciences, Dept. of Plant Pathology, Davis, California 95616, USA.

Tjamos, E.C.
Benaki Phytopathological Institute, Kifissia, Athens, Greece.

Turner, G.
Dept. of Microbiology, The University, Bristol, UK.

Vian, B.
Université Pierre et Marie Curie, Ecole Normale Supérieure,
24, Rue Lhomond, 75231 Paris Cedex 05, France.

Ward, E.W.B.
Research Centre, University Sub Post Office, London,
Ontario, N6A 5B7, Canada.

de Wit, P.J.G.M.
Agricultural University Wageningen, Dept. of Phytopathology,
9 Binnenhaven, 67009 PD Wageningen, The Netherlands.

Wood, R.K.S.
Dept. of Pure and Applied Biology, Imperial College,
London, SW7 2BB, UK.

Yoder, O.C.
Dept. of Plant Pathology, Cornell University, 334 Plant Science
Building, Ithaca, New York, 14853, USA.

INDEX

Abscisic acid, 198-201
Adhesion
  mycoparasites, 61-63
  plant-bacterial, 80-83, 312
  plant-mycorrhizal, 29-31
Acetamidase, 374
Agglutination, 87
Agrobacterium tumefaciens, 71, 78, 297, 307-312
Aminooxyacetic acid, 142
Appressoria, 15, 39, 331
Anthracnose, 39-44, 207-212, 339
Arachidonic acid, 195, 197-201
Arbuscules, 33
Aspergillus nidulans, 343-353, 359-362, 365-368

Bacteroids, 99-106
Biology of plant-pathogen interactions, 15-20
Biological control, 398
Biotrophy, 1, 10, 18, 34, 40, 74, 100, 107
Brassica campestris, 291

Calmodulin, 339
Carbohydrates
  histochemistry, 40-42, 49-55, 63-66
  binding, 65-68
Carnation, 191-195
Cell cultures, 117, 207, 221, 254
Cell wall modifications, 133-136, 245-249
  isolation of, 136-138
Chalcone isomerase, 208, 221
Chalcone synthase, 208-212, 221-231
Chitinase, 212, 245, 256, 265
Chitosan, 182, 263-6
Cinnamic acid-4-hydroxylase, 208
Cladosporium cucumerinum, 177
Cladosporium fulvum, 4, 164, 171-176
Cochliobolus heterostrophus, 371-379
Colletotrichum lagenarium, 245-249
C. lindemuthianum, 4, 39-44, 116-119, 207-212, 221-230, 235, 238-241, 339

Compatibility
  basic, 9, 120
  plant-pathogen, 35, 120-122, 149, 210-212, 241
Coumarate CoA ligase, 143, 208, 254-258
Cucumber, 177-182
Cultivar specificity, 118, 120-121 209, 238, 266, 271, 304
Cytokinins, 319-322
Cytoskeleton, 338

Differentiation of fungal germlings 331-338
DNA
  complementary, 209, 236, 347
  complexes, 265
  preparation by depletion hybridization, 344-353
  recombinant technology, 11, 21-23, 284-285, 293, 371
Dutch elm disease, 393

Eicosapentaenoic acid, 197
Elicitors, 2, 21, 91, 118, 150, 178, 208, 221, 236, 249, 253
  constitutive, 152, 257
  endogenous, 152, 257
  microbial, 110, 145, 151, 177-180, 185-188, 197-201, 205
  oligogalacturonides, 176-180
  polygalacturonase, 180
Enzyme cytochemistry, 50
Enzymes of phenylpropanoid biosynthesis, 222
Erwinia amylovora, 74, 89
Erwinia carotovora, 75, 278
Erwinia chrysanthemi, 277
Ethylene, 109
Extracellular polysaccharides, 73

Fulvia fulva, 4, 164, 171-176
Furanocoumarins, 257
Fusarium solani, 263-268

Gene cloning, 22, 343, 361
glyceraldehyde-3-phosphate dehydrogenase, 365-368
pathogenicity genes, 292, 302, 367
pectic enzymes, 281-283
pisatin demethylase, 378-382
polygalacturonase, 389
resistance genes, 209, 236
vectors, 353-356
Gene expression in fungi, 353
Gene-for-gene systems, 1, 3-4, 45, 120-121, 149, 156-161, 185
Gene polymorphism, 213-216, 228
Gene transcription, 209, 258
Gene transformation, see gene cloning
Genetic engineering, 24
Gibberellic acid, 198-201
Glucans, 120, 151, 237
Glycoproteins, 66, 101, 138, 145

Haustoria, 15, 142, 331
Histochemistry, 49-52, 79, 85, 134
Hydroxyproline-rich glycoproteins, 111, 212, 235, 245-8
DNA sequence, 236
Hygromycin resistance, 371
Hypersensitivity, 44, 79, 87-90, 120, 141-145, 150, 209, 241, 261, 298

Immunocytochemistry, 43, 52, 101
gold labelling, 52, 101-103
fluorescence, 101
Incompatibility, 210-212, 241, 293
Indoleacetic acid, 201, 317, 320
Induced resistance, 1, 89, 133, 149, 175, 207, 212
Induced susceptibility, 6, 172
Infection vesicles, 42
Isoflavonoid biosynthesis, 222-226

Kievitone, 224

Lectins
role in resistance, 154-5
use in histochemistry, 50, 60, 63, 275
Lesion formation, 43
Lignin, 113-114, 141-143, 152, 177-180
Lipopolysaccharide, 102-105
Lubimin, 187, 197-201

Matrices, 30, 43
Membrane, see plasmalemma
Models of plant-pathogen interactions, 1, 90, 156-162
Monoclonal antibodies, 100-103
Mortierella spp., 59-68

Mutagenesis,
transposon, 292, 300, 309
marker exchange, 283
site directed, 309, 343
Mutants, 23, 75, 77, 99, 172, 283, 291
Mycorrhizae,
ericoid, 29-34
vesicular arbuscular, 29-35

Necrosis, 21, 44, 88, 107, 119, 141, 205
Nectria haematococca, 22, 378-382
Nodulation, 307
Non steroidal anti-inflammatory drugs, 187-189

Oligomycin resistance, 359-361
Olive knot disease, 315
Ophiostoma ulmi, 393-399

Papillae, 44, 133-136, 152
Passaging, 403-405
Pathogenesis, 20
mechanisms of, 73, 277-283, 298
Pectate lyase, 75-76, 278-281
Pectic enzymes, 277-285, 385
regulation of synthesis, 279-281
Peroxidase, 112, 145
Petroselenium hortense, 209, 253
Phaseollin, 44, 207, 224
Phaseolus vulgaris, 39-44, 207-212, 221
Phenylalanine ammonia-lyase, 143, 208-216, 221-231, 254-258
Phenylpropanoid pathway, 113, 222
Phosphatase
ATPase, 33
Phytoalexins
role in resistance, 5, 21, 46, 89, 114-119, 143
biosynthesis, 40, 113, 187, 207-211, 221-226
effects of hormones, 197-201
Phytophthora capsici, 191-3
Phytophthora cryptogea, 192-3
Phytophthora infestans, 108, 151, 185-188, 197-201, 205
Phytophthora megasperma, 4, 116, 119, 151, 225, 237, 253
Piptocephalis virginiana, 59-68
Pisatin, 378-379
Plasmalemma
depolarization, 107, 121
histochemical changes, 245
specialization, 33, 41, 43, 99-106, 225-227
Plasmid transfer, 74
Prolyl hydroxylase, 225-228

Polygalacturonase, 180, 277
Polygalacturonic acid, 178
Polypeptides
  specificity, 271
Pseudomonas spp., 71, 80-83, 89-91
  P. fluorescens, 82
  P. syringae pv. coronafaciens, 80-84
  P. syringae pv. phaseolicola, 71-73, 79-81, 292, 297-303
  P. syringae pv. pisi, 263-266
  P. syringae pv. savastanoi, 315-327
  P. syringae pv. tabaci, 80-84
Puccinia graminis, 141-147, 271-275, 331

Recognition, 29, 35, 55, 90, 293
  models, 90, 156-162
Resistance
  effect of lectins, 145
  host, 20, 401
  mechanisms of, 17-21, 44, 79, 87-90, 107, 143
  non-host, 9, 20, 253, 263
  race non-specific, 8-9
  race specific, 1, 120, 209, 266, 304
  ultrastructural aspects, 34, 44-46, 79-86
Rhizobium, 99-105, 307-310
Rishitin, 187, 197-201
RNA
  changes during infection, 210, 229, 248, 257, 265
  changes during germling differentiation, 334
  in vitro translation, 208, 248
  isolation from fungi, 345
Rust fungi, 141-147, 271-275, 331-335

Saprophytes, 10
Septoria nodorum, 401-407
Sesquiterpenoids, 115, 197
Staining procedures, 41, 49-53
Stress metabolites, 114, 210
Suppressors, 153
Susceptibility,
  mechanisms of, 17, 120-121, 143, 297
  ultrastructural aspects, 31-34, 40-43, 63-65, 79-86, 99-104

Ti plasmids, 76-77, 307
Tobacco, 191-194
Toxins,
  host-specific, 7
  bacterial, 297
Transformation, see gene cloning

Uromyces appendiculatus, 331-332

Vectors, see gene cloning
Virulence genes, 311

Wound responses, 108

Xanthomonas campestris, 88, 291, 298

NATO ASI Series H

Vol. 1: **Biology and Molecular Biology of Plant-Pathogen Interactions.**
Edited by J. A. Bailey. 415 pages. 1986.

GPSR Compliance

The European Union's (EU) General Product Safety Regulation (GPSR) is a set of rules that requires consumer products to be safe and our obligations to ensure this.

If you have any concerns about our products, you can contact us on ProductSafety@springernature.com

In case Publisher is established outside the EU, the EU authorized representative is:

Springer Nature Customer Service Center GmbH
Europaplatz 3
69115 Heidelberg, Germany

Batch number: 10761697

Printed by Printforce, the Netherlands